T0320644

NEW HORIZONS IN TIME-DOMAIN ASTRONOMY

IAU SYMPOSIUM No. 285

COVER ILLUSTRATION by P. Marenfeld
National Optical Astronomy Observatory (Tucson, AZ, USA)

The IAU Symposium 285 Scientific Organizing Committee dedicates this volume to
Elizabeth Griffin in honour of her 70th birthday

INTERNATIONAL ASTRONOMICAL UNION

UNION ASTRONOMIQUE INTERNATIONALE

NEW HORIZONS IN TIME-DOMAIN ASTRONOMY

PROCEEDINGS OF THE 285th SYMPOSIUM OF THE
INTERNATIONAL ASTRONOMICAL UNION
HELD IN OXFORD, UNITED KINGDOM
SEPTEMBER 19 – 23, 2011

Edited by

R. ELIZABETH GRIFFIN
Dominion Astrophysical Observatory, Canada

ROBERT J. HANISCH
Space Telescope Science Institute and Virtual Astronomical Observatory, USA

and

ROBERT L. SEAMAN
National Optical Astronomical Observatory, Tucson AZ USA

CAMBRIDGE
UNIVERSITY PRESS

Shaftesbury Road, Cambridge CB2 8EA, United Kingdom

One Liberty Plaza, 20th Floor, New York, NY 10006, USA

477 Williamstown Road, Port Melbourne, VIC 3207, Australia

314–321, 3rd Floor, Plot 3, Splendor Forum, Jasola District Centre, New Delhi – 110025, India

103 Penang Road, #05–06/07, Visioncrest Commercial, Singapore 238467

Cambridge University Press is part of Cambridge University Press & Assessment, a department of the University of Cambridge.

We share the University's mission to contribute to society through the pursuit of education, learning and research at the highest international levels of excellence.

www.cambridge.org
Information on this title: www.cambridge.org/9781107019850

First published 2012

A catalogue record for this publication is available from the British Library

ISBN   978-1-107-01985-0   Hardback

Table of Contents

Day 1: Can Our Data Meet the Challenges?

Day 2: Explosive or Irreversible Changes

Day 3: Things That Tick

Day 4: Irregular and Aperiodic Changes

Day 5: Preparing for the Future

Contents

Workshop Reports

Poster Papers

Introduction

Studies of variability constitute prolific and profitable sources of information about how objects in the cosmos form, exist, and evolve. Variability can be periodic, aperiodic, spasmodic, or secular; it can involve times-scales from a millisecond to a century and beyond, and it can embrace the whole electromagnetic spectrum or just one portion of it. Studies of radial-velocity variables require well-tried techniques and only need dedication and persistence to yield new information; stellar pulsations and exoplanets are tough to identify and demand special observing techniques, while previously unknown cases of variability require new data-mining techniques applied to large data collections. Progress and innovation thus depend on fully-supported, open, and coherent database management systems plus appropriate data extraction and analysis tools. Discoveries and observations of variability can drive theory (e.g., supernovæ light-echoes or binary-star mergers), while theory can inspire searches for phenomena that would otherwise pass undetected (e.g., in asteroseismology).

Whereas it was once believed that variable stars were exceptional, every celestial object actually varies to some degree. Our plate stores proved essential for studies of spectroscopic and astrometric variability, and the AAVSO (operational since 1911) has likewise been invaluable for studies of photometric variability; both sources of historic data are in fact growing in value as their respective time-bases increase. At the same time, the substantial developments in theory and in observational techniques have enabled studies of group similarities to graduate to the finer details suggested by their differences. That progress owes as much to rapid access to digital data, to database mining techniques, and to tools developed by the Virtual Observatory, as to the increased power of telescopes and systems, improved detector sensitivity, innovative technology, and phenomenology-focused research.

But while modern ingenuity and proficiency have opened up new fields of study such as transients, blazars, gamma-ray bursts, active galactic nuclei and quasars, the puzzles posed by more traditional longer-term variability have not been laid to rest. On the contrary, objects with substantially long variability characteristics are now returning to the scene; some even reveal stellar evolution itself. At the same time, networks of observers are nowadays "hotwired" to alerts of new events rather than relying on telegraph or mail. Studies of variability are thus burgeoning in all respects, and the principle of free, open-access data is a core factor throughout.

On first detection all processes are mysterious, all objects unknown. As time-series data accumulate and archival cross-matching proceeds, empirical inferences emerge from the mist, and the first physical models are developed and debated. Available information on the plethora of variability types has now become overwhelming, to the point where whole symposia focus on just one type and its associated research community. But while specialist studies are undeniably important for refining theory, the significance of occurrences of similar phenomena in different objects may get overlooked, so key astrophysics can be missed.

This Symposium focused on the different manifestations of variability, and sought to shed light on new scientific insights which are not apparent when one type of object is studied in isolation. It therefore crossed previously recognized boundaries because the need is precisely to erase those boundaries, to think outside the box. Astrophysics transcends disciplines. Structures such as disks and jets, or processes such as pulsations and occultations, appear in different guises at different scales in different cosmic contexts.

The phenomenology of time-varying measurements that drives the empirical charac-terization of dramatically diverse astrophysical objects recurs time and again, but the cross-boundary links are less well aired. Phenomena of SNe, for example, were inves-tigated for many years before it was realised that an apparently single class of object actually was composed of several very different celestial events. Recent digitizing of early plates of the Harvard collection has revealed objects that vary over periods of several tens of years with amplitudes of almost a magnitude. Can the cause(s) of those variations be linked to ones that manifest variations or pulsations of similar amplitudes but very much shorter period, and—if so—are there false constraints in current models? Similarly, can high-energy studies influence current concepts of stellar evolution and variability in the AGB zone? On the other hand, scientific progress in the phenomenology itself is still handicapped by very incomplete information regarding the frequency of events. How can observers make better use of tools, technologies and techniques in order to capture a greater percentage of transients, novæ, etc.?

The core question, "How can technology and collaboration be better harnessed to enhance the science requirements?" was fundamental to the Symposium's planning. The full potential of new observing opportunities and techniques, new capabilities to revisit historic data, and new interpretative tools will not be realized unless the user community acquires the necessary skills to manage relevant data in diverse forms. "Showing and telling" are a vital element of the learning process, but are insufficient if performed only generically or are deemed to be the province of the specialist. Summarising the principles of applying the tools is inadequate without real examples. "What" and "when" are vital complementary ingredients of the banquet offered by variability, but achieve little without the all-important recipes for "how". Neither database managers nor researchers can be maximally productive if working in isolation and without appropriate feedback. As well as highlighting what is actually new and what is promised, the Symposium included a strong didactic content in the form of topical workshops focusing on practical skills and knowledge.

The timing for a cross-discipline symposium in time-domain astronomy was highly favourable, and as at least one participant noted, this may be the last time that a Sym-posium of such broad scope will be feasible. Major new transient surveys are coming on-line as soon as the next year or two, and their data will drive the respective fields substantially forward at all wavelengths. On-line data from projects such as the Palomar Transient Factory, SkyMapper, Pan-STARRS, and LOFAR will revolutionize studies of (at least) supernovæ, novæ, AGNs (quasars/blazars), variable stars and pulsars. These projects (and many others) will lay the groundwork for even greater time-domain in-vestigations over the next decade, including the truly massive Large Synoptic Survey. At the same time, high-speed digitizers to scan photographic plates are revealing the fascinating pervasiveness of even "more obvious" photometric and spectrum variability by harnessing the past to the present over long temporal baselines.

The Symposium was organized into daily themes:

- Monday: Can our data meet the challenges?
- Tuesday: Explosive or irreversible changes
- Wednesday: Things that tick
- Thursday: Irregular and aperiodic changes
- Friday: Preparing for the future

On each day we examined commonalities in the science as revealed by certain types of variability, crossing frequency and time-scale boundaries in the process, and including presentations from database experts on the present and projected status of analysis tools. Talks from different sub-disciplines were intentionally interleaved in order to avoid specialist-level isolation, and speakers rose to the challenge and presented talks that were accessible to a broad audience. Some 110 poster papers were displayed in two multi-day sessions, leading to stimulating discussions over coffee and evening refreshments.

Afternoons were set aside for topical workshops, each organized by participants in the Symposium and structured as they saw fit for discussion of the challenges facing a particular subset of time-domain studies. Topics ran the gamut from Extreme Physics and Gravitational Waves to Stellar Variability, Astrotomography, Light Echoes, Historical Data, and Data Management.

An additional highlight of the Symposium was the Monday evening public lecture given by Professor Sir Martin Rees (Baron Rees of Ludlow), FRS and Astronomer Royal, entitled "From Microseconds to Æons—How Our Complex Cosmos Emerged." Held in the auditorium of the Oxford University Museum of Natural History, the talk attracted a full house and was followed by a lively question-and-answer session.

Some in the community were concerned that a Symposium of such breadth, and structured as a hybrid between presentations of new research results, visions of the future, and a practicum of research tools, would not succeed on any of these fronts. In fact, comments from participants after the meeting were unanimous in their acclaim. More than anything else, perhaps, the meeting opened up lines of communication and collaboration that had not existed before. On the first day of the meeting a common comment was "I barely know 20% of the people here," whereas by the end of the week people were saying "I've met at least three-quarters of the people, and have started new collaborations that would not have happened otherwise." The welcoming environment of St. Catherine's College, wherein nearly all participants of the conference were housed, encouraged many side discussions that often continued in the convivial pubs of Oxford.

While there may indeed not be another conference on time domain astronomy that casts such a broad net, this one certainly accomplished its goal of being integrative and enabling of cross-cutting research. The organizers thank the participants for their willingness to share their knowledge—and to appreciate the knowledge of others in different fields—as a means to understanding the many mysteries of the time domain.

Robert Hanisch
  Space Telescope Science Institute and Virtual Astronomical Observatory, USA
Elizabeth Griffin
  Dominion Astrophysical Observatory, Canada
December 2011

Foreword

The divergences from convention which IAU S285 introduced affected not only the scheduling of each day's communications but also the layout of these Proceedings. In compiling them we have aimed to reflect the contents of the week's science in a manner that is both informative and useful as a research document. To that end, we laid emphasis on capturing not only the new but also the slightly speculative, welcoming opinions and ideas as well as journal-style research papers. Speakers were given the option of not submitting a full write-up if the content of a talk had been, or would be, published in its entirety elsewhere; 28% so chose, and for those we have reproduced here just an abstract, slightly modified into a summary.

The most severe divergence from a conventional schedule was the introduction of workshops (*a.k.a.* breakout sessions or focus discussions) on the three full afternoons. For each we have included here a report, written in whatever style its author(s) selected; some are statements condensed from the discussions which constituted the Workshop, some are brief scientific papers, while for just a few—those with a predominantly pedagogical element—we have reproduced instead the "paragraphs" which told visitors to our Wiki page what were a workshop's objectives and (possibly) an outline programme.

That daily schedule did not permit as many contributed talks as would have been the case in a more conventional programme—only 1 in 9 applicants could be thus accommodated; most of the rest prepared posters instead. We therefore offered all 110 poster presenters the opportunity to submit short write-ups of their posters. One-half accepted, while for the rest we have included their abstracts, once again modified into summaries.

While endeavouring to maintain fidelity between these Proceedings and what was communicated and discussed at IAU S285, we have tried hard to make the book *readable*. However, while the laurel of that achievement could be shared by the two closing papers of the Symposium (*q.v.*), we also recommend making time to heed the thought-provoking views in the after-dinner speech, which is reproduced intact on page 455. Without question, the pulse of change that is so clearly revealed by the perceptive and thought-provoking view of the younger generation will be incorporated into the way in which we conduct scientific research in the future.

As many remarked, ours was a star-studded cast, and—alas—the promised full-length write-up by the opening speaker of the first session could not be completed because its author was subsequently called to receive the Nobel Prize. The diversity of the topics which appeared to be touched by variability astonished even the organizers, and the capacity number of participants whom they attracted ran the whole gamut from senior academics and researchers to programmers and database experts and a blind graduate from an ethnic minority.

We thank the participants of IAU Symposium 285 for their contributions to this Proceedings, and for making the conference such a success.

Acknowledgements

It is a real pleasure to thank the colleagues and organizations whose assistance, co-operation and advice contributed vitally towards the overall success of the Symposium:

• Mark Sullivan and Aris Karastergiou (co-Chairs of the LOC), who undertook the lion's share of the Website and Wiki management, handled all the housekeeping duties such as Registration and funding administration, and organized many invisible details with perfection,

• the LOC team recruited from Oxford Astrophysics: Vanessa Ferraro-Wood (Departmental Secretary) and graduate students Sarah Blake, Tom Evans, Ian Heywood, Kate Maguire, Amy McQuillan, Yen-Chen Pan and Kimon Zagkouris. Between them they coped efficiently with all the back-stage duties and manifold preparations and tasks, both before and during the event,

• Amy McQuillan, for managing the "Variability" mug,

• Pete Marenfeld, whose design of the Symposium poster captured cryptically and exquisitely the breadth and diversity of the meeting,

• Sehar Tahir, for providing invaluable technical support with Latex,

• the IAU, for generously sponsoring S285 through travel grants to 28 participants,

• the Royal Astronomical Society, for a grant to run the public lecture—and of course to Professor Rees for giving it,

• St. Catherine's College—so near the centre of this historic and vibrant University city, yet far enough away as to not be inconvenienced by its bustle—for accommodating the Symposium when we wanted it, and for being amazingly co-operative and flexible. The impressive organizational skills of their dining-room staff in serving an excellent lunch each day speedily yet graciously was a practical lesson in parallel processing,

• Oxford Astrophysics, for financial sponsorship and for lending us personnel,

• Oxford University Press and Springer Scientific Publications, for books to display,

• the Lord Mayor of Oxford (Cllr. Elise Benjamin), for opening the Symposium, and the City Council for laying on a reception, and

• most importantly of all, the participants, for donating time, energy, enthusiasm, ideas and good-will.

Elizabeth and Bob
Co-Chairs, IAU S285

CONFERENCE PHOTOGRAPH

Symposium participants, in the grounds of St. Catherine's College, Oxford

Day 1:

Can Our Data Meet the Challenges?

New Horizons in Time-Domain Astronomy
Proceedings IAU Symposium No. 285, 2011
R.E.M. Griffin, R.J. Hanisch & R. Seaman, eds.

© International Astronomical Union 2012
doi:10.1017/S1743921312000117

The Power of the Unexpected

Brian Warner[1,2]

[1] Department of Astronomy, University of Cape Town, Cape Town, South Africa
[2] School of Physics and Astronomy, Southampton University, Southampton, UK
email: `Brian.Warner@uct.ac.za`

Invited Talk

Abstract. The history of astronomy has provided variable sources of unexpected kinds—from novae recorded in oriental archives, rapid radio, optical and high-energy changes in white dwarfs, neutron star and black hole binaries, to recent discoveries with satellites. A brief overview is given, as a prelude to a conference that anticipates a tidal wave of observations and discoveries to be made at all wavelengths during the next five to ten years.

Keywords. history and philosophy of astronomy, miscellaneous, surveys

1. Introduction

This is a conference with many dimensions: one of frequencies or energies from radio to TeV, a range of transients' time-scales from microseconds to decades, recurrence times from milliseconds to tens of years, a range of amplitudes from KEPLER's 10^{-6} precision to the most energetic supernovae and gamma ray bursts, and a variety of observational cadences from nanosecond logging of individual photons, future All-Sky better-than hourly coverage, oriental archival records, and even the potential of Antarctic ice cores to record sunspot cycles and supernovae (Motizuki *et al.* 2009), currently over centuries but perhaps eventually spanning millennia. Some of these are still relatively unexplored parts of phase space but which within a few years will undoubtedly reveal unexpected phenomena.

Many major discoveries in astronomy have derived from chance observations—and many have required the development of new theories for their explanation. As Andy Fabian has emphasised, the role of serendipity in the progress of astronomy can hardly be overestimated (Fabian 2010). And as is the case throughout science, the importance of technological advances is often paramount—new, more sensitive or more accurate instrumentation enables exploration at new wavelengths, energies or time-scales, where unexpected phenomena are lurking. In astronomy at almost all wavelengths the first signs of variability—the theme of this conference—were usually incidental to the main objectives, but often initiated deliberate, more widespread, searches.

2. Early History

It was of course known, since the observation of a solar flare by Carrington and Hodgson in 1859, that minor astronomical brightness variations could occur (accompanied, puzzlingly at the time, by strong aurorae and geomagnetic storms), but the earliest examples of major changes had in fact come more than two millennia before with the oriental discoveries of novae (Fig. 1), then largely of prime interest to Emperors' astrologers who wished to keep their heads.

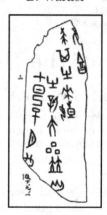

Figure 1. Chinese record of a nova in 1300 BC (Ronan 1994). An early Astrologers' Telegram. From Ronan (1994).

In the West there were very early visual discoveries of periodic variable stars, with Mira in 1596, Algol in 1667 and a total of 10 known by 1786. Only 230 years later, thanks to improvements of detectors and technique, the crop of variable stars we expect from GAIA is about 18 million (Eyer & Cuypers 2000).

Karl Jansky's announcement in 1932 that the Sun radiates radio waves is a prime example of exploration at a previously untested wavelength, and is the foundation of all radio astronomy, but it was not until 1962 that a strongly variable extrasolar source was recognised (this was the supernova remnant Cas A, first detected at radio wavelengths in 1947). Within the solar system the Sun itself was early known to have variable radio flux, but it is indicative of the then adolescent state of radio astronomy that decametric waves from Jupiter, detected using the most sensitive long wavelength apparatus available to Bernie Burke in 1955 and immediately found to be highly variable, were first reported in the Journal of Geophysical Research (Burke & Franklin 1955) and only later in the Astronomical Journal (Franklin 1959). This work helped lay foundations for later understanding of the magnetosphere of Jupiter; in particular the modulation of its radio emission by the position of satellite Io in orbit led to a uni- (or homo-) polar inductor model (Goldreich & Lynden-Bell 1969) with later relevance not only to volcanism on Io but more widely with applications to compact binaries, black holes, pulsars, the moon surrounded by the solar wind, the solar wind itself, sunspots, and planetary magnetic tails. This array of phenomena, illustrating the astrophysical occurrence of the Faraday Disc, making its appearance in a laboratory in 1831, is an example of the universality and wide applicability of basic physics.

Before the 1950s there were only a few indications that the universe contained dynamic events on short time-scales. For example, the term "short period variable star" then usually indicated RR Lyrae or δ Scuti stars with pulsation periods around 1.5 hours. During the late 1940s the first hints of much more rapid non-periodic luminosity variations appeared—surprisingly from *visual* observations of cataclysmic variable stars: in recurrent nova T CrB in 1947, in the nova-like variable AE Aqr in 1949, and in VV Pup in 1950. These stars' emergence onto the rapid variability stage included recognition that they are all short period binaries, e.g. UX UMa at 4.7 hours (Linnell 1949) and later ones as short as 1.4 hours, with the implication that at least one of the components must be a white dwarf. This was reinforced by the discovery in 1954 of a 71-second brightness modulation of high stability in the old nova DQ Herculis by Merle Walker (1956), which was of unprecedented shortness, and in retrospect was the first indication of rapid rotation

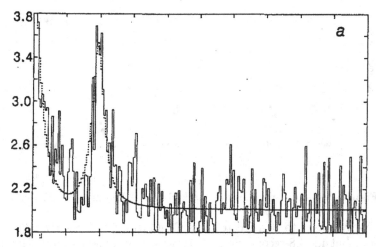

Figure 2. QPOs in X-Ray Binaries; the horizontal axis displays a frequency range from 0 to 100 Hz (van der Klis *et al.* 1985).

in some white dwarfs. Later recognition, from X-Ray observations starting in the 1970s, of binary systems containing neutron stars and black holes produced a wide range of the very short-time-scale phenomena naturally associated with these even more compact objects.

At radio frequencies, where compact extra-galactic sources were principal objects of study, observations initially trailed optical progress by one or two decades, so it was only in 1968 that a review of their variability was commissioned and written by Kellerman & Pauliny-Toth (1968). The key conclusions were that *"until recently, it had been generally felt that, apart from occasional supernova explosions or fluctuations in variable stars detected in nearby galaxies, no significant changes either in the radio or in the optical luminosity of extragalactic objects could take place on a time-scale shorter than the lifetime of a human observer"*, and that *"the unexpected discovery of remarkably intense fluctuations in the radio and optical luminosities of a number of quasi-stellar radio sources ... suggests that the previous conceptions of the processes involved in the generation of strong radio extragalactic radio sources ... were unjustified"*. Even more remarkably, just as the review was being written, it was upstaged by galactic radio sources, which generated a note added in press: *"The discovery of an unexpected and entirely new class of variable radio source, announced in February 1968, has produced an unprecedented flurry of activity and publications"*. This was of course the announcement of the serendipitous discovery of pulsars (Hewish et al. 1969), found with a new radio telescope specifically constructed to search for rapid interplanetary scintillation which thus had the ability also to explore higher pulsation or rotation frequencies. In contrast to the 71-second modulation of DQ Her, which involved a white dwarf and caused only a short-lived gasp, the pulsars provided the first evidence for the long-predicted neutron stars, and as a result have become permanent star attractions. Other unexpected discoveries in this field were millisecond binary pulsars, of which several are now known, and planets in orbit around a millisecond pulsar, which was an example of a system so unexpected as to require almost a paradigm shift of interpretation.

Historically, observations at high energies gave strikingly unexpected results, with the first variable extra-solar X-Ray source, Sco X-1, found in 1962, the first X-Ray transients found in the late 1960s, and TeV gamma-ray flares from the blazar Markarian 501 found independently in 1997 by several ground-based Cherenkov telescopes (see Fig. 3). The

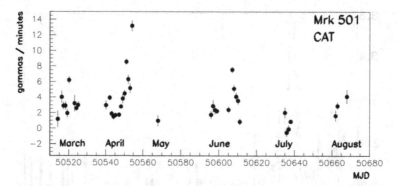

Figure 3. Gamma-ray flares in Markarian 501 (Protheroe *et al.* 1997).

first purely X-ray pulsar to be detected was Centaurus X-3, observed with a sounding rocket in 1967, but its 4.8-second periodicity was only found in 1971 with the Uhuru satellite—though by that time the Crab pulsar was already known to have pulsed X-Rays.

3. Later Developments

There is space-time here to discuss only a couple of the many unanticipated observational results that have created entire fields of research. Among degenerate stars, where the shortest time-scale variations are expected, highly stable rotation periods from seconds down to milliseconds were found among the neutron stars, down to ~ 20 seconds

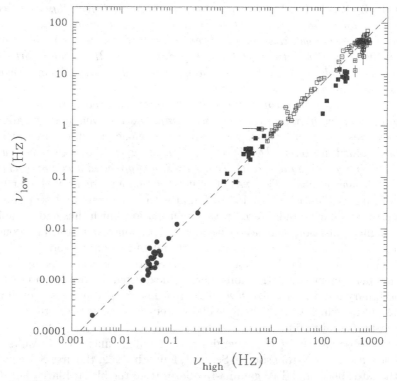

Figure 4. QPOs in Cataclysmic variables and X-Ray Binaries (Warner & Woudt 2004).

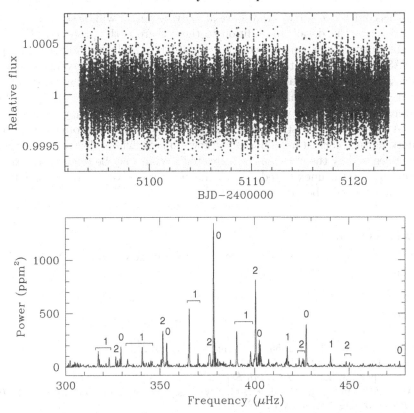

Figure 5. Solar-like oscillations in a red giant observed by KEPLER (Di Mauro *et al.* 2011).

among white dwarfs, plus non-radial pulsations typically $\sim 100 - 1000$ seconds in some of them; but by searching, several kinds of quasi-periodic oscillations (QPOs) were discovered, including among black hole binaries. They were discovered first in cataclysmic variables (Warner & Robinson 1972) and later in X-Ray binaries (van der Klis *et al.* 1985; see also Fig. 3). Currently these are probably the least well understood of the rapid phenomena that occur among the close binaries that contain accretion discs. In particular, the high-frequency and low-frequency luminosity oscillations (Fig. 2) show similar behaviour in all sources—the relevant physics to connect these disparate sources is not known (Warner 2006). Some binge thinking is required here.

During the past year results from the Kepler (2011) satellite, expected through asteroseismology to confirm and extend knowledge of non-radial acoustic pulsations in main sequence stars, and perhaps even to detect gravity-mode oscillations in red giants, have created a major new industry. Red giants turn out to rival dwarfs in the richness of their oscillations—this is now the era of ensemble asteroseismology: of 16 500 giants observed in the KEPLER field in the first month 71% had oscillation spectra that could be analysed for period spacings. From the gravity-mode periods a clear distinction is obtained between those giants that are purely hydrogen shell-burning (period spacings ~ 50 seconds) and those that are in addition helium core-burning (spacings $\sim 100 - 300$ seconds) (Bedding et al. 2011). And there is even an unexpected bonus beyond this remarkable result: excess power at low frequencies in the Fourier transforms of ~ 1000 red giants is attributable to surface granulation—this is the first direct detection of the effects of surface convection in giant stars (Mathur *et al.* 2011).

During this conference surveys that have already begun, or will be starting within a few years, are described in detail. In addition to these there is the proposed SUPERWASP Variable star survey, which should contain $\sim 800\,000$ periodic variables—and a possible SUPERWASP all-sky survey to record ~ 1 billion objects every 10 seconds over the whole sky, which will be rather data-intensive, generating ~ 100 Tb per day (Norton 2011).

These are all surveys for which we have *high expectations* of getting unexpected results; there are many smaller surveys and targeted observations which may *unexpectedly* find unexpected results. We have still to find evidence for some theoretic objects such as quark stars, quark novae, black stars, Thorne-Żytkow objects, whose discovery, however, may simply be relegated by theoreticians as 'expected'. What may be required to excite them will be the *unexpected* unexpected. So, as the subject evolves over the next few years, *expect* the unexpected—be prepared for it.

Acknowledgements

My research is supported by the University of Cape Town and the National Research Foundation.

References

Bedding, T. R., *et al.* 2011, *Nature*, 471, 608.
Burke, B. F. & Franklin, K. L. 1955, *J. Geophys. Res.*, 60, 213
Di Mauro, M. P., *et al.* 2011, *MNRAS*, 415, 3783
Eyer, L. & Cuypers, J. 2000, *ASP Conf. Ser.*, 203, 71
Fabian, A. C. 2010, in: M. de Rond & I. Morley (eds.), *Serendipity: Fortune and the Prepared Mind* (Cambridge University Press), p. 73
Franklin, K. L. 1959, *AJ*, 64, 37
Goldreich, P. & Lynden-Bell, D. 1969, *ApJ*, 156, 59
Hewish, A., *et al.* 1969, *Nature*, 224, 472
Kellerman, K. I. & Pauliny-Toth, I. I. K. 1968, *Ann. Rev. Astr. Astrophys.*, 6, 417
Kepler 2011, http://kepler.nasa.gov/
Linnell, A. P. 1949, *Sky & Tel.*, 8, 166
Mathur, S., *et al.* 2011, *arXiv:1109.1194*
Motizuki, Y., *et al.* 2009, *arXiv:0902.3446*
Norton, A. 2011, http://www.star.le.ac.uk/conf50/talks/S01_1215_Andy_Norton.pdf
Protheroe, R. J., *et al.* 1997, *arXiv:astro-ph/9710118*
Ronan, C. A. 1994, *The Shorter Science and Civilization in China*, Cambridge University Press
van der Klis, M., *et al.* 1985, *Nature*, 316, 225
Walker, M. W. 1956, *ApJ*, 123, 68
Warner, B. 2006, *PASP*, 116, 115
Warner, B., Robinson E. L. 1972, *Nature Phys. Sci.*, 239, 2
Warner, B. & Woudt, P. A. 2004, *IAU Colloq.*, 315, p. 87

New Horizons in Time-Domain Astronomy
Proceedings IAU Symposium No. 285, 2011
R.E.M. Griffin, R.J. Hanisch & R. Seaman, eds.

© International Astronomical Union 2012
doi:10.1017/S1743921312000129

Optical Transient Surveys

Brian Schmidt

Research School of Astronomy and Astrophysics, Australian National University,
Weston Creek, ACT 2611, Australia
email: brian@mso.anu.edu.au

Invited Talk

Abstract. The last five years have seen an explosion of activity to monitor the sky at optical wavelengths. The following summarizes an overview of the range of experiments currently being, or soon to be, undertaken in both cataloguing and monitoring the sky, and suggests scientific opportunities for the short- to medium- term in the arena of optical transient astronomy. In so doing, it applies the philosophy described by Warner (page 3) to the gamut of variability studies that are now burgeoning in observational astronomy.

1. History

Studies of transient objects have played a key role in the historical development of Astronomy. Henrietta Swan Leavitt's discovery of the Cepheid period–luminosity relationship, and its use to set the scale of the universe, is an early example. The first studies of extra-galactic transients, such as supernovæ, were empowered by dedicated hardware such as the 18-inch Schmidt telescope used by Zwicky to scan the night skies for supernovæ by "blinking" large photographic plates. This was the same technique used 50 years later by the Calan-Tololo SN search to define the set of Type Ia supernovæ which paved the way for discovering the accelerating universe. Over the past 20 years, a new set of optical survey telescopes, powered by increasingly large digital CCD detectors, are revolutionising the field.

2. Methodology

Transient detection is typically done in two ways. One method is through the comparison with a catalogue of objects, where variability is monitored and new objects catalogued. This is computationally less demanding, is good where very high precision is required, but results in poor detection efficiency near the detection threshold, or in crowded regions. The other method—image subtraction—where images are matched to a template and the template subtracted, is more computationally demanding and has poorer absolute precision, but leads to much better transient detection efficiency across a survey.

3. Science Drivers

Currently, there is a range of science themes that drive transient surveys. They include **transits**, where planets are identified by their transits in front of their host stars, **gravitational microlensing**, which is used to undertake novel studies of stars and planets, and to measure the stellar column densities towards the Milky Way and other nearby galaxies, **discoveries of asteroids**, to quantify the dynamics of the solar system and to look for potential Earth-colliding bodies, **variable stars**, which are used to understand the physics of stars and stellar systems, **extragalactic transients**, which are used to explore the violent physical processes involved in stellar explosions, black hole accretion and the unknown, and **cosmology**, where transient objects are used to probe the scale and content of the universe.

4. Figure of Merit

The figure of merit for a transient search, the *etendue*, depends precisely on the objectives of a survey, but can generically be described as the rate a search can observe objects to a fixed brightness limit. In the sky-limited case, it is interesting to note that it was not until the commissioning of the MACHO Camera on the Great Melbourne Telescope in 1992 that a digital system exceeded the etendue of the photographic Schmidt telescopes. The state-of-the art etendue currently resides with the CFHT+MegaCam instrument, but will soon eclipsed by the CTIO-4m+DECCAM, by the Subaru+Hyper Suprime Cam, and eventually, by LSST. These instruments will be 150–2200 times quicker at surveying the sky than was the 48-inch Schmidt Telescope. But many transient science cases are not about just going deep quickly. Planetary transits and nearby supernova searches are cases in point, where high precision and large areal coverage are the key factors. Ultimately, the success of a project depends on its execution with respect to its scientific goals—good execution can make up for a lot of etendue.

New Horizons in Time-Domain Astronomy
Proceedings IAU Symposium No. 285, 2011
R.E.M. Griffin, R.J. Hanisch & R. Seaman, eds.

© International Astronomical Union 2012
doi:10.1017/S1743921312000130

The Scientific Potential of LOFAR for Time Domain Astronomy

Rob Fender[1], on behalf of the LOFAR Transients Key Science Project

[1]School of Physics & Astronomy, University of Southampton, SO17 1BJ, UK For TKSP
member list see: www.transientskp.org/team.shtml
email: r.fender@soton.ac.uk

Invited Talk

Abstract. LOFAR is a ground-breaking low-frequency radio telescope that is currently nearing completion across northern Europe. As a software telescope with no moving parts, enormous fields of view and multi-beaming, it has fantastic potential for the exploration of the time-variable universe. In this brief paper I outline LOFAR's capabilities, our plans to use it for a range of transient searches, and some crude estimated rates of transient detections.

Keywords. accretion, stars:binaries, stars:pulsars, stars:supernovae:general, ISM: jets and out-flows, radio continuum: general

---

1. Introduction

LOFAR, the Low Frequency Array, is a large low-frequency radio telescope in northern Europe, led by ASTRON. Construction of the array, which has its core collecting area in The Netherlands with international stations in France, Germany, Sweden and the UK, is nearly complete, and astronomically interesting data are now being taken. LOFAR operates in the 30–80 and 120–240 MHz frequency ranges. The 80–120 MHz frequency gap corresponds to the FM radio bands—frequencies at which astronomical observations are impossible†. Construction of the array is almost complete; Fig. 1 indicates the distribution of operating LOFAR stations across Europe. In addition, observations are occasionally possible to frequencies as low as 15 MHz.

LOFAR has six *key science projects* (KSPs), one of which is *Transients* (principal investigators Fender, Stappers & Wijers). The remit of the TKSP covers all transient and variable astrophysics, including commensal searches of all data (ultimately in near-real-time, although that functionality is not yet implemented). The TKSP covers both time-series and image-plane searches for transients and variables, including pulsars (Stappers *et al.* 2011). The adoption of transients and variables as key science drivers for the project is a theme for most of the large SKA pathfinders and precursors, and in that respect is generally unlike older radio facilities. However, time-series and image-plane transients have been separated for both ASKAP (which has *CRAFT* and *VAST* respectively) and MeerKAT (*TRAPUM* and *ThunderKAT*). That makes some sense from the aspect of techniques, although there is some overlap in the science.

---

† Unless northern Europe could be persuaded to stop night-time FM radio broadcasts for a few weeks in the interests of finding the *Epoch of Reionisation* signal..

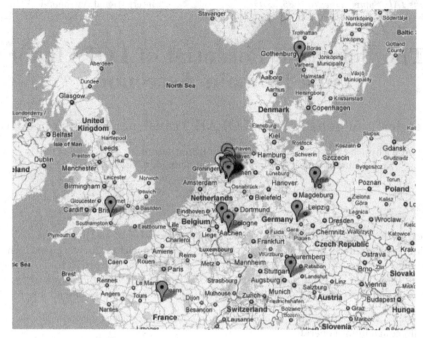

Figure 1. The distribution of complete LOFAR stations across Europe.

2. Types of Radio Transients

What are these transients that we're looking for? In planning transients searches, whether via targeted surveys or blind commensal studies, we can crudely separate events into astrophysics and technology.

Incoherent synchrotron emission arises from the acceleration of relativistic electrons in a magnetic field. It has a brightness temperature (intensity) limit of $\sim 10^{12}$K, and is associated with practically every event which injects kinetic energy into the ambient medium (e.g. jets of all types, nova and supernova explosions). Those events tend to evolve relatively slowly, especially at low frequencies (where the optical depth is higher), and so with LOFAR we do not expect to see variability on time-scales shorter than the standard imaging timescale of 1–10 sec.

Coherent emitters, such as radio pulsars or masers, can have much higher brightness temperatures. Like synchrotron events they can at times be associated with extreme astrophysical environments. The much higher brightness temperatures means that the variability time-scales can be much shorter, and even astrophysically distant objects can vary on time-scales much shorter than the standard imaging one. However, such short bursts are dispersed and scattered by the intervening interstellar medium, and need to be corrected for in studying the intrinsic properties of the burst (a well-understood problem in the field of radio pulsars). Note that there can be considerable overlap between these two sets of objects and techniques: for instance, variability in coherent sources can be detected and tracked in images.

Fig. 2 summarises this dichotomy in radio transients, and is indicative of how early attempts at automated classification pipelines might make an early branch based upon variability time-scales and polarisation characteristics.

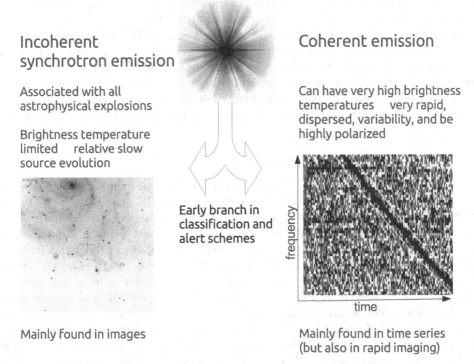

Incoherent synchrotron emission

Associated with all astrophysical explosions

Brightness temperature limited   relative slow source evolution

Early branch in classification and alert schemes

Mainly found in images

Coherent emission

Can have very high brightness temperatures   very rapid, dispersed, variability, and be highly polarized

Mainly found in time series (but also in rapid imaging)

Figure 2. Transients can be divided into incoherent synchrotron and coherent events, which also corresponds roughly to a divide in the techniques used to find them (image plane *vs.* time series).

3. Finding Transients with LOFAR

LOFAR can operate in a variety of modes, all of which can be important for the study of transients and variables. Furthermore, all of those modes can be operated at a variety of levels, from a single station to the entire pan-European array.

Interferometric mode. LOFAR is a 'software telescope' which in effect has no moving parts. Pointing of the array and/or individual stations is done by introducing delays appropriate to a certain direction on the sky (phased array). Different frequencies can be therefore set to observe in different directions by introducing different delays. LOFAR can already, as a standard imaging mode, produce 8 beams each of 6 MHz bandwidth. In the low band those beams can be placed anywhere on the sky; in the high band they are limited by the beam of the high-band tiles, an analogue beamformer. That allows for an extraordinary instantaneous field of view: $8 \times 90 = 720$ deg^2 in the low band and $8 \times 25 = 200$ deg^2 in the high band. In other words, the entire northern hemisphere could be mapped in the low band in less than 30 sets of pointings (with sparse tiling). Initial processing of wide-field surveys for transients, including the MSSS (Multifrequency Snapshot Sky Survey) due between late 2011 and early 2012, will only localise sources to a few arcmin, but later and/or responsive observations could localise interesting sources (including transients) to arcsecond precision.

Timing mode. LOFAR also has high-time-resolution ('pulsar') modes, which can achieve 10s of ns time resolution and can map either a full field of view with sensitivity $s \propto N^{-1/2}$ (incoherent sum), or the synthesised beam with sensitivity $s \propto N^{-1}$ (where smaller s is better). Recently it has been possible to record data from over 100 coherent tied-array beams simultaneously and to tile out the entire HBA field. Stappers *et al.* (2011) give more details about searches for fast transients with LOFAR.

Direct storage. The LOFAR Transients Buffer Boards (TBBs) can be used to record up to several seconds of full bandwidth antenna level data (or longer, in a trade-off with bandwidth), *before* the beam-forming stage. That means that beams can be formed retrospectively in a certain direction anywhere in the sky (LBA) or tile beam (HBA) upon receipt of an 'internal' alert (from LOFAR itself) or an 'external' one (e.g. from VOEvent). That mode is currently being developed by the *Cosmic Rays* KSP (PI Falcke).

4. LOFAR in a Global Context

As noted above, LOFAR, ASKAP and MeerKAT have all embraced the science of radio transients as part of their Key Science Programmes. To that list we hope to add APERTIF, the focal-plane array upgrade to WSRT, to which several transients-oriented proposals have already been submitted as statements of interest for its KSP programme.

In a global context, LOFAR has the widest field of view of any of the major facilities, and although its suffers in terms of raw mJy sensitivity compared to (say) EVLA and MeerKAT, when a spectral correction is made it can be shown to be a very powerful facility. That is illustrated by Fig 3, where sensitivity and field of view are compared for a range of world-class radio facilities. The solid diagonal lines correspond to constant figures of merit (FoM, defined as $FoM \propto \Omega s^{-2}$, where Ω is the field of view and s is the sensitivity; smaller s implies increased sensitivity). The four new or upgraded GHz facilities—EVLA, MeerKAT, APERTIF and ASKAP—all have a comparable FoM. For LOFAR it is not until a spectral correction is made that its survey power becomes apparent, as the Figure caption explains in more detail.

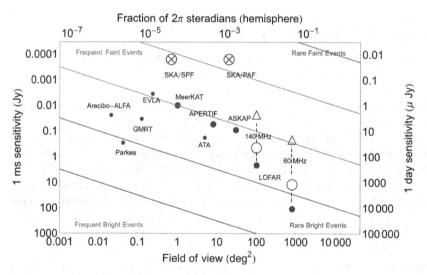

Figure 3. A comparison of sensitivity vs. field of view for a range of existing and planned radio telescopes. The solid lines represent a constant survey figure of merit (FoM, $\propto \Omega s^{-2}$ where Ω is the field of view and s the sensitivity; smaller s implies higher sensitivity). For LOFAR, the dots indicate the raw sensitivities, the open circles represent a spectral correction for a spectral index of -0.7 (where spectral index α is in the sense that $S_\nu \propto \nu^\alpha$), as is appropriate for optically thin synchrotron emission. The open triangles correspond to a correction for a spectral index of -2.0, corresponding to the steepest (most aged) synchrotron sources, as well as some coherent radio sources such as pulsars and other flavours of neutron star.

5. Proposed Approach

The LOFAR TKSP have considered several approaches for the detection of transients and variables, and propose a programme of targeted searches, wide-field blind surveys, and commensal searches for transients.

Targeted surveys. Within the TKSP there is a number of targeted observations and surveys planned (and in some cases, under way). They include observations of well-known high-energy astrophysics sources with variable radio counterparts (e.g. the binary SS 433, the pulsar PSR 0329+54, the blazar PKS 1510-089). They also include targets beyond the usual suspects, which may provide breakthroughs in some research areas (e.g. a search of nearby stars for radio bursts from 'hot Jupiters'). It is also our intention to probe transient parameter space by searching for concentrations of mass and/or exotic objects at a range of distances (and hence luminosities). They include globular clusters within our own galaxy, M31, the core of the Virgo cluster and beyond.

Wide-field blind surveys. An example of a wide-field blind survey planned for LOFAR is the zenith monitoring programme, which is a key component of the 'Radio Sky Monitor' (RSM) programme. Fig. 4 illustrates how a small number of 7-pointing tiles could cover the entire zenithal strip (Dec +54 for LOFAR), with maximum sensitivity (it peaks at the zenith for the dipoles). For example, with the LBAs at around 60 MHz, a 10-degree wide strip could be covered in 16 tiles, surveying around 2000 square degrees (10% of the northern hemisphere; note that this is for fairly dense tiling). Thus a dedicated phase of RSM observing (which may be likely to happen once or twice per year for a few weeks, in coordination with other multiwavelength facilities) with 100% of resources spent on it, would correspond to ~ 1.5 hr on each field, resulting in an expected r.m.s. of a few mJy. At the lowest frequencies it only takes two more tiles to cover the entire northern galactic plane, which may well sample a different population of transients. At present we have been monitoring the single part of the zenithal field that contains the bright pulsar PSR 0329+54 and using that to constrain and estimate the rate of low frequency transients (Bell, 2011). In fact AARTFAAC (Amsterdam-ASTRON Radio Transients Facility and Analysis Centre; PI Wijers) plans to use the LOFAR core in

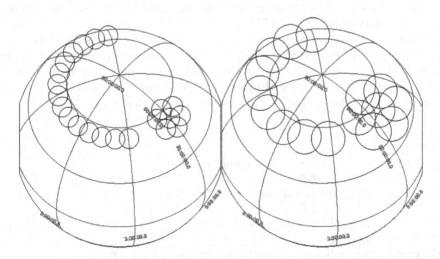

Figure 4. The proposed zenith monitoring programme for LOFAR. During intense periods of monitoring, some 2000 square degrees of the sky, centred on Dec +54, could be imaged to $\sim$mJy sensitivity every 24 hr. This Figure demonstrates the enormous field of view of LOFAR.

parallel with other observing programmes to perform a quasi-continuous, low-angular-resolution survey of the entire northern hemisphere for transients.

Commensal surveys. Since the initial design of LOFAR it has been a stated goal of the TKSP that LOFAR data be imaged on short ($\leqslant$ 10 sec) time-scales in almost real time, in order to search for bright transients and variables. That functionality has yet to be implemented, but is in the design and commissioning plan for 2012. In the meantime the TKSP aims to search all other observations performed by LOFAR in a hunt for transients. That will naturally give us a range of wide-field and deep surveys (such as those planned by the *Surveys* and *Epoch of Reionisation* KSPs) with which to probe transient phase space. It is not unreasonable to assume that the entire northern hemisphere will effectively be surveyed several times per year, besides making deeper pointings with much higher cadence.

6. Predicted Rates

Until recently, the best estimates for predicted rates were around 0.1 sources $\deg^{-1}$ for two-epoch transients and a flux-density limit of $\sim$ mJy (Bell 2011, based on Bower *et al.* 2007); however, according to Frail et al. (2011) that may be an overestimate by up to an order of magnitude. For the zenith monitoring programme outlined above, we could have 2000 $\deg^2$ surveyed to approximately the same r.m.s. per 24 hr, resulting in a transient detection rate—from this programme alone—of 20 (Frail) to 200 (Bower) events *per day*. Considering commensal searches of all data, which may on any given day be going wide, or deep, or some combination, we can conservatively expect more than ten high-significance transient events per day. One major unknown is the fraction of events that will prove to be repeaters, so the true number of distinct astrophysical variables found may be lower. It is our stated policy to report these events as widely as possible, probably via some version of VOEvent (Seaman *et al.* 2011), for example `skyalert.org`.

7. Summary

LOFAR has a wide range of diverse capabilities—multi-beaming, simultaneous timing and imaging modes, splitting the array between large numbers of individual stations, re-imaging past observations with the TBBs—all of which are ideally suited for exploring transient parameter space. Just as importantly, there is the will to support this science: a Key Science Project is dedicated precisely to that exploration. According to current estimates, it should find many thousands of interesting radio transients per year, thus furnishing a huge target list for multiwavelength follow-ups and providing new tests of our ability to detect, classify and report automatically such events efficiently.

References

Bell, M. E. 2011, *PhD thesis, University of Southampton*
Bower, G. C., *et al.* 2007, *ApJ*, 6, 346
Frail, D. A., *et al.* 2011, *ApJ*, submitted
Seaman, R., *et al.* 2011, *IVOA Recommendation*, arXiv:1110.0523
Stappers, B. W., Hessels, J. W. T., Alexov, A., *et al.* 2011, *A&A*, 530, 80
LOFAR: `www.lofar.org`
LOFAR Transients Key Science Project: `www.transientskp.org`

New Horizons in Time-Domain Astronomy
Proceedings IAU Symposium No. 285, 2011
R.E.M. Griffin, R.J. Hanisch & R. Seaman, eds.

© International Astronomical Union 2012
doi:10.1017/S1743921312000142

Kepler, CoRoT and MOST: Time-Series Photometry from Space

Hans Kjeldsen[1] and Timothy R. Bedding[2]

[1] Department of Physics and Astronomy, Aarhus University, DK-8000 Aarhus C, Denmark,
email: hans@phys.au.dk
[2] Sydney Institute for Astronomy, School of Physics, University of Sydney, Australia

Invited Talk

Abstract. During the last 10 years we have seen a revolution in the quality and quantity of data for time-series photometry. The two satellites MOST and WIRE were the precursors for dedicated time-series missions. CoRoT (launched in 2006) has now observed more than 100,000 targets for exoplanet studies and a few hundred stars for asteroseismology, while KEPLER (launched in 2009) is producing extended time-series data for years, aiming to discover Earth-size planets in or near the habitable zone. We discuss the accuracy of some of the parameters one may extract from the high-quality data from such photometric space missions, including the prospects for detecting oscillation-period changes due to real-time stellar evolution.

Keywords. techniques: photometric, stars: oscillations (including pulsations)

1. The Science Goals of Photometry from Space

The science goals for all the photometric space missions are high-precision asteroseismology and search for and characterization of exoplanet transits. Reaching those goals will require high-precision time-series photometry with high duty cycle for extended periods of time.

Launched on 2003 June 30, the Canadian space mission MOST became the first satellite dedicated to high-precision time-series photometry (Walker *et al.* 2003). The first indication of the quality of photometry that can be obtained from space had already been given by the NASA WIRE-mission, launched on 1999 March 4. Owing to the loss of cryogen needed for cooling the main detector, WIRE was unable to carry out its primary science mission, and operations were changed so as to use the on-board 52 mm star tracker for continuous time-series monitoring of bright stars for one to six weeks per target (Bruntt & Buzasi 2006). The results showed that even a small telescope in space may be able to reach a photometric quality which cannot be obtained from the ground because of atmospheric scintillation (Kjeldsen & Frandsen 1992).

Improved photometric accuracy (compared to WIRE) was the achieved by the MOST satellite, which contains a 150 mm aperture Rumak-Maksutov telescope that feeds two frame-transfer CCDs. MOST is orbiting in a low-Earth Sun-synchronous polar orbit, allowing stars to be viewed continuously for up to 60 days (Walker *et al.* 2003). While both WIRE and MOST focused on a limited number of stars per field of view, the French (CNES) CoRoT mission, launched into a low-Earth polar orbit on 2006 December 27, made it possible to monitor several thousand stars at once. CoRoT (Auverne et al. 2009) was therefore the first mission capable of searching for exoplanet transits from space. CoRoT can follow a large number of stars simultaneously and continuously for up to

150 days. The CoRoT telescope has a diameter of 27 cm, and is thus more sensitive than MOST and WIRE.

A low-Earth orbit will reduce the field-of-view, and will restrict the length of time for which a given target can be followed continuously. Searching for the transit signatures of planets in orbits around stars with periods longer than a year will therefore require significantly longer time-series data than any of the above-mentioned missions can provide.

The NASA KEPLER mission is specifically designed to survey a portion of our Milky Way galaxy to discover Earth-size planets in or near the habitable zone. The KEPLER mission therefore requires higher accuracy and longer observing periods than WIRE, MOST and CoRoT can achieve.

The KEPLER instrument (Koch *et al.* 2010) contains a 95-cm Schmidt telescope with a 105 deg^2 field of view. In order to ensure a continuous time-series for each target for several years, KEPLER is placed in an Earth-trailing heliocentric orbit. From that orbit KEPLER is capable of attaining the required photometric precision for long uninterrupted periods of time.

2. The Revolution in Data Quality and Quantity

MOST and WIRE should be regarded as the precursors of dedicated time-series missions focussed on bright targets. CoRoT has now observed more than 100,000 targets for exoplanet studies and a few hundred stars for asteroseismology studies, while KEPLER is producing extended time-series data over a period of years, on targets with visual magnitude between 5 and 17.

The literature clearly shows how photometric space missions have revolutionized the quantity and quality of observational data. WIRE produced the first series of new results (Bruntt 2007; Bruntt & Southworth 2007), and publications by the MOST team reveal the high-quality photometry that can be obtained through the use of even a small telescope in space; an example is the discovery of a super-Earth transit by Winn *et al.* (2011). CoRoT has measured *p*-mode oscillations and granulation in a number of stars hotter than our Sun (Michel *et al.* 2008), and has discovered non-radial oscillations with long life-times in RGB stars (De Ridder et al. 2009). The first exoplanet discovered by CoRoT in 2007 was a hot Jupiter (Barge *et al.* 2008).

A milestone for CoRoT in particular and for exoplanet research in general was the discovery of the hot super-Earth, CoRoT-7b (Leger et al. 2009), with a diameter smaller than 2 Earth radii, in a 0.853 day orbit. The discovery of CoRoT-9b in an orbit with a period of 95.3 days showed how these space missions are capable of discovering exoplanets with star–planet distances that imply equilibrium temperatures near or below 300 K (Deeg *et al.* 2010). The higher precision of the KEPLER photometry, combined with the longer uninterrupted time series data obtained by that mission, enables the KEPLER team to search for exoplanets with smaller sizes and larger orbital semi-major axes than can any of the other space missions. The recent discoveries by KEPLER clearly demonstrate this capability (e.g., Batalha *et al.* 2011; Borucki et al. 2012).

In asteroseismology KEPLER has improved the quality of available data by at least an order of magnitude, resulting in measurements of new properties in several types of stars. The most striking example is probably the measurement of hydrogen shell and helium core fusion in RGB stars (Beck et al. 2011; Bedding *et al.* 2011).

If we compare the measured properties of the photometry obtained by the existing space missions, we can list the following issues that should be considered for the future space missions:

- Field crowding shows a clear effect in the data quality.
- There is a huge gain in combining the science goals for the same time-series data (asteroseismology and transit studies).
- The quality of the data analysis is improved if one can analyse the data on the ground (images).
- It is better to saturate the detector than to defocus or spread light by using a prism.
- Bright stars are preferred in order to follow-up from the ground (may require a larger field-of-view).
- For hotter stars (O and B stars) one needs observations in the plane of the Milky Way (CoRoT sees more hot stars than KEPLER does).

3. Observational Asteroseismology

3.1. KEPLER *photometry on bright stars*

An example of KEPLER time-series data for a star bright enough to saturate the detector is given by the KEPLER data on θ Cygni, which has a magnitude $V = 4.86$. Although that star is too bright for the instrument by a factor of 1000, the photometry for a 100-day time-series achieves a noise level of 0.13 parts per million in amplitude at frequencies near 7 mHz. Figs. 1 and 2 both show the amplitude spectrum of θ Cygni at a frequency between 1 and 2.5 mHz, but in Fig. 2 the spectrum has been smoothed by a Gaussian smoothing (FWHM of $4\,\mu$Hz). In both figures we see the excess from p-mode oscillations. The broadness of the individual oscillation peaks indicates that the mode lifetime is only a few hours, which makes accurate frequency determination difficult.

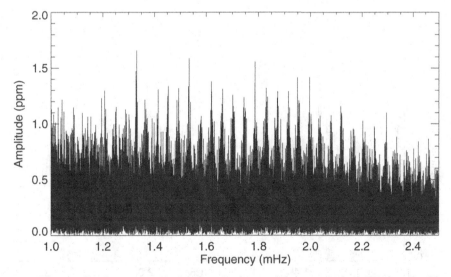

Figure 1. The amplitude spectrum of θ Cygni, observed for a period of 100 days by KEPLER. The spectrum clearly shows p-mode oscillations near 1.8 mHz despite the fact that the raw data saturated the detector by a factor of 1000.

3.2. *Asteroseismic observables*

The scientific impact which such high-precision photometric time-series data obtained from space can have on asteroseismology depends on the precision with which certain asteroseismic observables can be measured. Those observables are:

- Oscillation frequencies, frequency differences (splittings) and frequency ratios.
- Oscillation mode identification (degree, order and mode type; $g/p/f$, mixed)
- Oscillation mode properties (amplitudes, amplitude ratios, phases, phase differences, lifetimes)
- Variations (short-term and long-term) in mode parameters (frequencies, amplitudes, lifetimes)

The accuracy with which one can measure the above observables depends on the noise level per minute of observation and the length of the time series. We first consider data containing noise plus a number of coherent oscillations:

$$\mathrm{data}(t) = \mathrm{noise}(t) + \sum_{i=1}^{n} a_i \, \sin(2\pi \, f_i \, t - \phi_i). \tag{3.1}$$

We can estimate the uncertainties in the frequencies (f), phases (ϕ) and amplitudes (a), following Kjeldsen & Frandsen (1992) and Montgomery & ODonoghue (1999):

$$\sigma(a) = \sqrt{\frac{2}{\pi}} \langle A_{\mathrm{noise}} \rangle = \sqrt{\frac{\langle P_{\mathrm{noise}} \rangle}{2}} \approx 0.80 \, \langle A_{\mathrm{noise}} \rangle \tag{3.2}$$

$$\sigma(\phi) = \frac{\sigma(a)}{a} \tag{3.3}$$

$$\sigma(f) = \sqrt{\frac{3}{\pi^2}} \frac{1}{T} \sigma(\phi) = \sqrt{\frac{6}{\pi^3}} \frac{\langle A_{\mathrm{noise}} \rangle}{a \, T} \approx 0.44 \frac{\langle A_{\mathrm{noise}} \rangle}{a \, T}, \tag{3.4}$$

where

$$\langle A_{\mathrm{noise}} \rangle = \sqrt{\frac{\pi}{N}} \, \sigma_{\mathrm{noise}} \propto T^{-1/2} \tag{3.5}$$

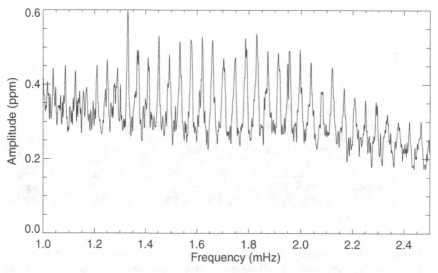

Figure 2. Same as Fig. 1, but smoothed by a Gaussian filter of FWHM of $4\,\mu$Hz.

is the mean noise level in the amplitude spectrum at frequency f, T is the length of the time series, N is the number of data points in the time series and σ_{noise} is the noise per data point. If we define σ_{noise} as the noise level per minute in the time series (at frequency f) and assume that the time series is uninterrupted, we find for coherent oscillations that:

$$\sigma(a) \propto \sigma_{\text{noise}}\, T^{-1/2} \tag{3.6}$$

$$\sigma(\phi) \propto \sigma_{\text{noise}}\, a^{-1}\, T^{-1/2} \tag{3.7}$$

$$\sigma(f) \propto \sigma_{\text{noise}}\, a^{-1}\, T^{-3/2} \tag{3.8}$$

If instead we consider a damped and re-excited oscillation (like the one found in the Sun), we then find:

$$\sigma(f) \propto \sigma_{\text{noise}}\, a^{-1}\, T^{-1/2}. \tag{3.9}$$

For a mode lifetime of τ (which is infinite for coherent oscillations), we find the frequency accuracy to be:

$$\sigma(f) \approx 0.44\, \langle A_{\text{noise}} \rangle\, a^{-1}\, \sqrt{T^{-2} + \tau^{-2}}. \tag{3.10}$$

If we consider the bright targets observed by KEPLER , we may use this last equation to estimate the accuracy in frequency obtained for solar-like oscillations. For stars brighter than magnitude $V = 8$, the noise level in the KEPLER data is 25 ppm per minute (Gilliland et al. 2010). The mean noise level in the amplitude spectrum will then be 0.13 ppm after three months, 0.04 ppm after three years and about 0.025 ppm after seven years. Table 1 shows the frequency accuracy for these values of T, for various values of amplitude and mode lifetime.

Table 1. Frequency accuracy for a $V = 8$ target observed by KEPLER

	Coherent		Damped		
			$\tau = 10$ d	$\tau = 3$ d	$\tau = 6$ hr
T	$a = 0.01$ ppm	$a = 100$ ppm	$a = 50$ ppm	$a = 5$ ppm	$a = 5$ ppm
90 d	0.7 pHz	70 pHz	0.0013 μHz	0.040 μHz	0.50 μHz
3 yr	0.02 pHz	2 pHz	0.0004 μHz	0.013 μHz	0.16 μHz
7 yr	0.005 pHz	0.5 pHz	0.0002 μHz	0.008 μHz	0.10 μHz

3.3. Evolution-related variations in the oscillation period

One of the main goals for long uninterrupted observations of stars with a large amplitude is to detect stellar evolution of main-sequence stars. That can be done by detecting slow variations in the oscillation period (Breger & Pamyatnykh 1998). The expected change in the quantity

$$\frac{1}{P}\frac{dP}{dt} \tag{3.11}$$

is larger than $10^{-8}\,\text{yr}^{-1}$ for δ Scuti Stars (Breger & Pamyatnykh 1998). According to the values in Table 1, the accuracy of the period determination for coherent oscillations increases rapidly with the length of observing time. For observing periods of 5–7 years, we should therefore be able to make a definite detection of stellar evolution in δ Scuti stars.

The accuracy with which one can determine a change in the oscillation period can be estimated from the equations above (for coherent oscillations):

$$\sigma\left(\frac{1}{P}\frac{dP}{dt}\right) \approx \frac{\sigma(f)}{f}\frac{1}{T} \propto \sigma_{\text{noise}}\, a^{-1}\, f^{-1}\, T^{-5/2}, \tag{3.12}$$

where σ_{noise} is the noise per minute of observations. As can be seen, we will increase by a factor of 100 the accuracy with which we can determine the effect of stellar evolution if observations are maintained continuously for 6 years compared to a 1-year campaign. The increase in accuracy from 3 to 7 years is a factor of 10. Such gain from long uninterrupted space observations clearly provides a strong incentive for extending the KEPLER mission beyond its nominal 3.5-year lifetime.

4. Acknowledgement

I acknowledge the huge work of the team behind KEPLER. Funding for the KEPLER Mission is provided by NASAs Science Mission Directorate.

References

Auvergne, M., et al. 2009, AA, 506, 411
Barge, P., et al. 2008, AA, 482, L17
Batalha, N. M., et al. 2011, ApJ, 729, 27
Beck, P. G., et al. 2011, Science, 332, 205
Bedding, T. R., et al. 2011, Nature, 471, 608
Borucki, W. J., et al. 2012, accepted for publication in ApJ
Breger, M. & Pamyatnykh, A. A. 1998, AA, 332, 958
Bruntt, H. 2007, Communications in Asteroseismology, 150, 326
Bruntt, H. & Buzasi, D. L. 2006, Mem. della Soc. Astron. Italiana, 77, 278
Bruntt, H. & Southworth, J. 2007, in: W. I. Hartkopf, E. F. Guinan, & P. Harmanec (eds.), Binary Stars as Critical Tools and Tests in Contemporary Astrophysics, Proc. IAUS. 240 (Cambridge, UK: Cambridge University Press), p. 624
Deeg, H. J., et al. 2010, Nature, 464, 384
De Ridder, J., et al. 2009, Nature, 459, 398
Gilliland R. L., et al. 2010, ApJ Letters, 713, L160
Kjeldsen, H. & Frandsen, S. 1992, PASP, 104, 413
Koch, D. G., Borucki, W. J., & Basri, G. et al. 2010, ApJ Letters, 713, L79
Leger, A., et al. 2009, AA, 506, 287
Michel, E., et al. 2008, Nature, 322, 558
Montgomery, M. H. & O'Donoghue, D. 1999, DSSN, 13, 28
Walker, G., et al. 2003, PASP, 115, 1023
Winn, J. N., et al. 2011, ApJ Letters, 737, L18

New Horizons in Time-Domain Astronomy
Proceedings IAU Symposium No. 285, 2011
R.E.M. Griffin, R.J. Hanisch & R. Seaman, eds.

© International Astronomical Union 2012
doi:10.1017/S1743921312000154

Long-term Monitoring with Small and Medium-sized Telescopes on the Ground and in Space

P. A. Charles[1,2,3], M. M. Kotze[2,3] and A. Rajoelimanana[2,3]

[1]School of Physics & Astronomy, University of Southampton, Southampton, UK
email: P.A.Charles@soton.ac.uk

[2]South African Astronomical Observatory, Observatory 7935, South Africa

[3]Department of Astronomy, University of Cape Town, Cape Town, South Africa

Invited Talk

Abstract. The last 20 years have seen revolutionary developments of large-scale synoptic surveys of the sky, both from the ground (e.g., the MACHO and OGLE projects, which were targetted at micro-lensing studies) and in space (e.g., the X-ray All-Sky Monitor onboard RXTE). These utilised small and medium-sized telescopes to search for transient-like events, but they have now built up a huge database of long-term light-curves, thereby enabling archival research on a wide range of objects that has not been possible hitherto. This is illustrated with examples of long time-scale optical and X-ray variability studies from the field of X-ray binary research: the high-mass BeX binaries in the SMC (using MACHO and OGLE), and the bright galactic-bulge X-ray sources (mostly LMXBs, using RXTE/ASM). As such facilities develop greater capabilities in future and at other wavelengths (developments in South Africa will be described), real-time data processing will allow much more rapid follow-up studies with the new generation of queue-scheduled large telescopes such as SALT.

Keywords. telescopes, astronomical data bases: miscellaneous, X-rays: binaries, accretion, accretion disks

1. Introduction

The last 20 years have seen remarkable developments in, and expansion of, the number of very large ground-based optical/IR telescopes now available to the global astronomy community (Charles 2011). However, the growth over the last two decades in the number of *small* (< 2-m) telescopes has been on an even larger scale, although such facilities rarely attract high-level media attention. This has occurred even as a number of major observatories have closed or retired many of their smaller telescopes as a result of financial pressures on operating costs. But the latest generation of small telescopes are very different from those older facilities, which had been designed and built as multi-purpose telescopes capable of tackling a wide range of scientific programmes, and whose time was allocated to observers in *classical* mode (i.e., in blocks of nights or even a week). Instead, the latest small telescopes (and sometimes entire telescope networks) are robotic in operation, usually have a single observing mode, and can be maintained over the Internet. What has made this possible is a consequence of "Moore's Law" in the inexorable rise of computing and electronics performance over the last 30 years, thereby providing automatic telescope control and data processing on scales that would once have been the province of supercomputers but yet are available now for not much more than the cost of a domestic washing machine. Such a setup has removed at a stroke one of the major costs of any telescope operation, namely the provision of local staffing. A parallel factor has been the availability of low-cost, panchromatic, large-format CCD detectors, resulting in instruments capable of undertaking programmes that would have been impossible just 20 years ago.

A glimpse of what is possible from dedicated useage of small telescopes (albeit older ones) was provided by the MACHO and OGLE projects beginning in the 1990s (Alcock *et al.* 1996, Udalski *et al.* 1997), in their quest for gravitational microlensing events. As a result of the extremely rich star-fields of their main target areas in the Magellanic Clouds and the Galactic Bulge, their archives now provide > 18-year light-curves of an enormous number of stars.

Among space missions there are analogies of small telescopes that provide general-purpose survey capabilities, one of the most valuable being the All-Sky Monitor (ASM) on NASA's Rossi X-ray Timing Explorer (RXTE). Since early in 1996, and lasting until the end of 2011, the RXTE/ASM has scanned the entire accessible X-ray sky daily, producing an (immediately available) on-line archive of the light-curves of the bright, mostly Galactic, X-ray sources (Levine *et al.* 1996).

These ground and space-based facilities have made it possible to study the truly long-term variability of luminous X-ray binaries for the first time. Whilst the spin and orbital periods have been known for many of them ever since their discovery, it is now becoming clear that many, if not all, display so-called "super-orbital" modulations on time-scales of months to years, and that these modulations can occur for different reasons—as summarised by Charles *et al.* (2010) in their Table 1. These will be used as examples of the kind of science that these monitoring facilities can undertake.

2. High-Mass X-ray Binaries

2.1. *BeX systems*

The Be X-ray binaries (BeX systems) are a remarkable population in that they all have neutron-star compact objects (they are pulsars), and approximately half of those known lie in the SMC (Coe 2000). That means that many of them are in the MACHO and OGLE databases, and thus led to the discovery by McGowan & Charles (2003) of an extraordinary, large-amplitude (∼0.5 mag) 421-day super-orbital modulation in the optical light-curve of the BeX system A0538-66. Since the 16.6-day orbital period of A0538-66 was well established (on the basis of strictly periodic X-ray and optical flares as the neutron star in its eccentric orbit interacted with the equatorial envelope of the rapidly rotating Be star), McGowan & Charles (2003) were led to propose that the equatorial envelope (or disk) grew and dissipated on the 421-day timescale.

The MACHO and OGLE databases then provided the opportunity to investigate how widespread such behaviour was amongst the SMC BeX systems, and Fig. 1 (from Rajoelimanana *et al.* 2011) shows a selection of SMC light-curves from the 50 or so examined. Only a handful display the same behaviour as A0538-66, but that is explained as an inclination effect, with A0538-66 being close to edge-on. Furthermore, many super-orbital (and orbital) modulations were found in this dataset, and Fig. 2 shows how they are correlated. This is to be expected, with the orbit of the compact object restricting the extent to which the equatorial disk of the Be star can expand.

2.2. *Dynamical Power Spectra*

The potential of X-ray monitoring missions to yield remarkable new insights into these long-term physical processes is very well demonstrated in the RXTE/ASM light-curve of SMC X-1 (Fig. 3). This eclipsing HMXB consists of a 0.7-sec X-ray pulsar in a 3.9-day orbit about its B0I companion, which has long demonstrated a ∼60-day super-orbital

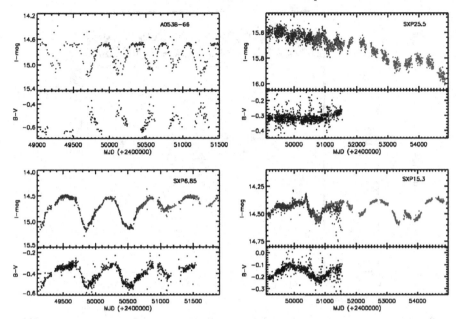

Figure 1. 18-year optical lightcurves of A0538-66 and 3 selected SMC BeX systems that display large-amplitude long-term modulations. (The different colours seen in the on-line version refer to the MACHO (blue), OGLE-I (red) and OGLE-II (green) archives.) From Rajoelimanana *et al.* (2011).

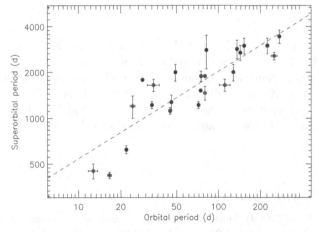

Figure 2. Correlation of super-orbital and orbital periods of SMC BeX systems obtained from the survey of Rajoelimanana *et al.* (2011), and updated with the results of Reig (2011).

modulation in its X-ray output. However, in the first years of RXTE it became apparent that the period of this modulation was shortening, an observation which led Clarkson *et al.* (2003) to exploit the long timebase provided by RXTE by applying a "dynamical power spectrum" (DPS) analysis to the SMC X-1 light-curve. The DPS takes a data window of 400 days, derives a periodogram, slides it forward by 50 days, derives a new periodogram, and so on. In this way a time-dependent power spectrum (or DPS) is obtained, which reveals how any super-orbital modulations change on long timescales. That work has been brought up to date by Kotze & Charles (2011), who utilised the entire RXTE/ASM light-curves of X-ray binaries (for which such modulations have been

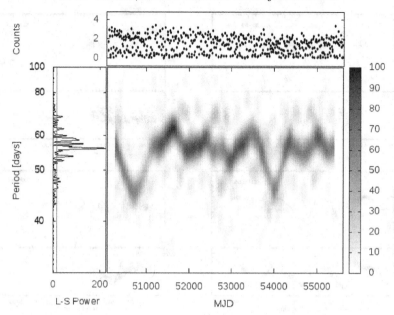

Figure 3. The top box shows the full RXTE/ASM 18-year X-ray light-curve of SMC X-1. The periodogram of the complete light-curve is shown at left in the lower box, and contains multiple significant peaks. The reason for this is demonstrated in the main plot in which the time-dependent periodogram (the Lomb-Scargle power is represented by the grey-scale) reveals how SMC X-1's super-orbital period itself varies with time. From Kotze & Charles (2011).

reported) to derive their DPS. That of SMC X-1 is shown in Fig. 3, where the time-scale for the drifting of the super-obital modulation is clearly around 5–10 years.

Such super-orbital periods are attributed to a tilted, precessing disk around the compact object, driven by X-ray irradiation. The stability of the disk is a function of the binary mass ratio and separation (Ogilvie & Dubus 2001). SMC X-1 lies in a region where a warp can be expected to form, leading to the kind of behaviour shown here.

3. Low-Mass X-ray Binaries

When Harris *et al.* (2009) found an almost sinusoidal, ∼4 year modulation in the RXTE/ASM light-curve of the 4.2-hour LMXB GX9+9, it led Kotze & Charles (2010) to examine all LMXBs for such very long-term variations. They indeed found similar variations in almost all Atoll and Z sources, but with the amplitude much greater in the Atoll, compared to the Z, sources. This is demonstrated in Fig. 4 where the correlation with observed average flux levels is very clear. While accurate distances are not known for many of these sources, they are all in the Galactic Bulge (i.e., at distances ∼5–12 kpc), indicating that such behaviour is probably related to the intrinsic luminosity level (since Z sources are close to L_{Edd}, but Atoll sources are ∼10 times weaker). Consequently, Kotze & Charles (2010) have proposed that the modulation is actually due to changes in the mass-transfer rate from the donor, as a result of a solar-cycle-like variation. Similar variations have been seen in CVs, for which this process was first suggested by Applegate & Patterson (1987).

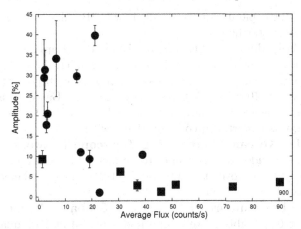

Figure 4. Relationship between the very long-term variability amplitude and average flux levels for Atoll sources (dots) and Z sources (squares), based on the results in Kotze & Charles (2010).

4. International Robotic Facilities in South Africa

The advent of a democratic South Africa has resulted in a dramatic expansion of international telescope projects at the SAAO's observing site (in Sutherland), almost all of them being robotic. They exploit South Africa's "astronomical geographic advantage" (leading to the Sutherland area now being protected by Act of Parliament), but also bring benefits for scientific development within the South African community. The following have all come to pass since 2000:

• **IRSF, the IR Survey Facility**, a 1.4-m IR telescope $(1–2.5\mu)$, that began operations in 2000 (Glass & Nagata 2000). This Japan/South Africa collaborative project conducts IR surveys of the southern skies, concentrating on the Galactic Centre and Bulge regions, as well as the Magellanic Clouds. The results have been the deepest and highest-resolution IR surveys of these regions to date (Kato *et al.* 2007). IRSF is not robotic, but is operated in classical observing mode, with a mix of Japanese and South African observers.

• **SuperWASP, the Wide Angle Search for Planets**, is a UK-led project which has been operating robotically at Sutherland since 2005 (there is also a node in La Palma), with the aim of detecting extra-solar planets via observations of transits (Smith *et al.* 2009). SuperWASP uses large-format CCDs attached to wide-angle lenses to survey the same regions of the sky continuously, reaching $V \sim 14$ out of the Galactic plane, even with $\sim$15-arcsec pixels. The resulting database of hundreds of billions of observations has revealed more than 30 transiting exoplanets, and is the only such source of bright objects $(V < 13)$ in the southern hemisphere; they can then be followed up with high-resolution spectroscopy of the parent star. The huge database also provides an archive for independent research programmes, one of which has been determining the detailed properties of contact binaries (Skelton 2010).

• **MONET, the MOnitoring NETwork**, is a pair of 1.2-m fast-slewing, remotely-operated telescopes for CCD imaging and fast photometry (Hessman 2001). They are part of the HET/SALT partnership, one being located at the HET site in Texas and the other at Sutherland. They aim to (i) monitor targets to determine their variability state, and hence suitability for related HET/SALT spectroscopy, and (ii) provide a facility for real-time education and outreach work at schools around the world, even involving schools in the real-time operations of the telescope. MONET should begin full operations in 2012. Its fast-slewing capability makes it ideal for GRB follow-up, for exoplanet transit timing and for variable star light-curves.

• **KELT, the Kilo-degree Extremely Little Telescope**, is run by Vanderbilt University and is also a copy of a US-based northern cousin (Kuhn et al. 2010). Similar to SuperWASP in scientific concept, KELT aims to monitor stars brighter than SuperWASP can (in particular $V \sim 8$–10 mag) for transiting exoplanets. Based on a medium-format 35-mm camera lens attached to a large-format CCD, KELT is low-cost (only $\sim$\$50K). The commissioning was undertaken collaboratively with UCT.

• **KMTNet, the Korean Microlensing Telescope Network**, which consists of three identical $\sim$1.8-m wide-field telescopes to be located in Australia, Chile and South Africa, is to be constructed over the interval 2011–18 and has planet-searching as its prime scientific goal (Kim et al. 2010).

• **LCOGT, the Las Cumbres Observatory Global Telescope** is a network of (ultimately, at least 6, possibly 8) telescope nodes placed at high-quality sites around the world (Shporer et al. 2011), of which SAAO will be the second main node after Chile (Cerro Tololo). Each node will consist of at least two 1-m telescopes, and at least four 0.4-m telescopes, all of which operate robotically, with the intention of providing continuous all-sky coverage. These telescopes will focus on *time-domain astrophysics*, having the potential to follow-up GRBs (Gamma-Ray Bursters), supernovæ, microlensing events, planetary transits and a host of other objects rapidly, thereby providing a uniquely powerful facility for research as well as for education.

Acknowledgements

MK and AR acknowledge bursaries from NASSP and the South African SKA Project.

References

Alcock, C., et al. 1996, *ApJ*, 461, 84
Applegate, J. H. & Patterson, J. 1987, *ApJ*, 322, L99
Charles, P., Kotze, M., & Rajoelimanana, A. 2010, *AIP Conf. Series*, 1314, 303
Charles, P. 2011, in: J.-P. Lasota (ed.), *Astronomy at the Frontiers of Science* (Dordrecht: Springer), p. 209
Clarkson, W. I., et al. 2003 *MNRAS*, 339, 447
Coe, M. J. 2000, *ASP Conf. Series*, 214, 656
Glass, I. S. & Nagata, T. 2000, *MNASSA*, 59, 110
Harris, R. J., et al. 2009 *ApJ*, 696, 1987
Hessman, F. V. 2001, *ASP Conf.Series*, 246, 13
Kato, D., et al. 2007, *PASJ*, 59, 615
Kim, S.-L., et al. 2010, *SPIE*, 7733, 107
Kotze, M. M. & Charles, P. A. 2010, *MNRAS*, 402, L16
Kotze, M. M. & Charles, P. A. 2011, *MNRAS*, online
Kuhn, R., et al. 2010, *BAAS*, 42, 288
Levine, A.M., et al. 1996, *ApJ*, 469, L33
McGowan, K. E. & Charles, P. A. 2003, *MNRAS*, 339, 748
Ogilvie, G. I. & Dubus, G. 2001, *MNRAS*, 320, 485
Rajoelimanana, A. F., Charles, P. A., & Udalski, A. 2011, *MNRAS*, 413, 1600
Reig, P. 2011, *Ap.Sp.Sci.*, 332, 1
Shporer, A., et al. 2011, in: A. Sozzetti, M. G. Lattanzi & A. P. Boss (eds.), *The Astrophysics of Planetary Systems*, IAU Symp. 276, Cambridge, UK: Cambridge University Press, p. 553
Skelton, P. 2010, in: I. Basson & A. E. Botha (eds.), *Modelling of Eclipsing Binaries*, 55th SAIP Ann. Conf., (Pretoria: CSIR)
Smith, A. M. S., et al. 2009, *MNRAS*, 398, 1827
Udalski, A., Kubiak, M., & Szymanski, M. 1997, *Acta Ast.*, 47, 319

New Horizons in Time-Domain Astronomy
Proceedings IAU Symposium No. 285, 2011
R.E.M. Griffin, R.J. Hanisch & R. Seaman, eds.

© International Astronomical Union 2012
doi:10.1017/S1743921312000166

Opening the 100-Year Window for Time-Domain Astronomy

Jonathan Grindlay, Sumin Tang, Edward Los, and Mathieu Servillat

Harvard Observatory & Center for Astrophysics, Cambridge, MA 02138, USA
email: `jgrindlay@cfa.harvard.edu`

Abstract. The large-scale surveys such as PTF, CRTS and Pan-STARRS-1 that have emerged within the past 5 years or so employ digital databases and modern analysis tools to accentuate research into Time Domain Astronomy (TDA). Preparations are underway for LSST which, in another 6 years, will usher in the second decade of modern TDA. By that time the Digital Access to a Sky Century @ Harvard (*DASCH*) project will have made available to the community the full sky Historical TDA database and digitized images for a century (1890–1990) of coverage. We describe the current *DASCH* development and some initial results, and outline plans for the "production scanning" phase and data distribution which is to begin in 2012. That will open a 100-year window into temporal astrophysics, revealing rare transients and (especially) astrophysical phenomena that vary on time-scales of a decade. It will also provide context and archival comparisons for the deeper modern surveys.

Keywords. astronomical data bases: catalogues, surveys; STARS: variables; galaxies: active

1. Introduction

As was clear from the 2010 US *Astronomy and Astrophysics Decadal Survey*, which ranked LSST as the highest-priority project for ground-based astronomy, Time-Domain Astronomy (TDA) has emerged as a key field of current astronomy and astrophysics. The temporal domain offers new routes to astrophysical understanding of extreme phases of stellar and galaxy evolution through studies of novæ, supernovæ, gamma-ray bursts and AGN, to list only a few. Wide-field and/or temporal surveys of long duration can discover and study binaries in various phases of evolution, some with exotic stellar components such as black holes, or add to the increasingly rich harvest of exoplanets. But while these extreme events or phenomena have short time-scales (hours to months), astronomical time-scales are predominantly long, often measured in millenia or even 10^{3-6} times longer still. Time-scales appreciably longer than the duration of a given survey require the assimilation of large samples of objects at different evolutionary (or binary) phases in order to piece together temporal histories. Now that we have the possibility to digitize archived plate collections of large and (nearly) continuous duration and totalling many images, it is possible to initiate and conduct studies in the emerging field of *Historical TDA* for very large samples of objects on time-scales that are at least an order of magnitude longer than was possible before. That was the prime motivation for the *DASCH* project (Digital Access to a Sky Century @ Harvard).

2. Development of *DASCH* and Current Status

The Harvard College Observatory plate collection is the world's largest, containing some 450,000 direct plates. As described elsewhere in these Proceedings—see page 243, it has approximately uniform full-sky coverage from 1890 to 1990. The *DASCH* project (http://hea-www.harvard.edu/DASCH/) was initiated to digitize all the plates and make their digital images (~400 Tb) and reduced photometric data available on line (Grindlay *et al.* 2009; Grindlay *et al.*, in preparation). *DASCH* incorporates the world's fastest and most astrometrically precise plate scanner (Simcoe *et al.* 2006) and a powerful astrometric and photometric reduction pipeline (Laycock *et al.* 2010). Over the 5-year development phase of the scanner and reduction software, both the astrometry (Servillat *et al.* 2011) and the photometry (Tang *et al.*, in preparation) were further optimized while scanning about 19,500 plates from five fields selected for having calibrated sequences and a range of stellar densities. A semi-automated plate-cleaning machine is in the final stages of development; it will clean the glass back-side of each plate more quickly than the 80 seconds that it takes for an operating sequence of loading–scanning–unloading a standard-size plate of 30 cm × 25 cm (they are scanned two at a time). Full "production scanning" and processing of some 400 standard plates per day can commence later this year when two such cleaning machines are ready. Scanning of the whole collection of direct plates, together with associated photometric reduction and population of the MySQL database, will require approximately 4 years. For each resolved object the database contains positions (J2000), magnitudes, and a list of processing flags. Full processing into the database for all 400 plates scanned each day can be accomplished overnight.

3. *DASCH* vs. Current and Planned TDA Surveys

In about 2 years the northern high latitude ($|b| > 10°$) full sky from *DASCH* will be on-line and available for access and analysis. The 100-year temporal coverage, compared with <10 years of coverage by PTF and CRTS and the several epochs of SDSS, will enable new studies of long time-scale phenomena. Several examples are shown in Fig. 1, where variable classes are plotted by their approximate ranges of absolute magnitude (M_V) against *recurrence* time-scale. Recurrence time-scales are chosen instead of variability time-scales because they signal better which objects can be discovered and measured. Recurrence (or occurrence) is unambiguous, whereas a variability time-scale can only be measured if there is nearly complete temporal coverage; in practice, that is only available for a continuous-viewing space mission like Kepler. The PTF and PS1 surveys have typical observing cadences (for a given sky region) of at least 1 day, though more frequent sampling is being achieved for limited fields with PTF. LSST will be comparable, whereas for *DASCH* the average observation cadence is about 2 weeks, though can be as short as 30 minutes on plates with multiple (offset) exposures of a given field. That sets the minimum recurrence time-scale that can be measured; the maximum is limited by the total duration of the survey. All the surveys can cover the same M_V range, but of course the corresponding distance range differs by the apparent magnitude limit of each survey. We plot on the right of Fig. 1 an approximate distance scale out to which each class of variable or recurrently variable object can be detected. For *DASCH* we adopted an approximate limiting magnitude of $m_V = 15$, which is appropriate for an assumed $(B - V) = 1$; it corresponds to the mean *DASCH* limiting photographic (B) magnitude m_B ~14 measured on the 19,500 plates scanned thus far.

The recurrence times for various classes are taken from the literature, and are either uncertain by the ranges shown (e.g., the recurrence time-scale range for black-hole "novæ" is based on just 3 objects) or are an estimated range given by systematic effects (such as the metallicity and and red-shift of SN Ia hosts). Recurrence-time ranges ending at the right side of the plot (e.g., for bright blazar flares) are simply lower limits. The recurrence times for large flares (Δm >1) from flare stars or QSOs are estimates based on variability studies against luminosity. The overall conclusion is simply that by expanding

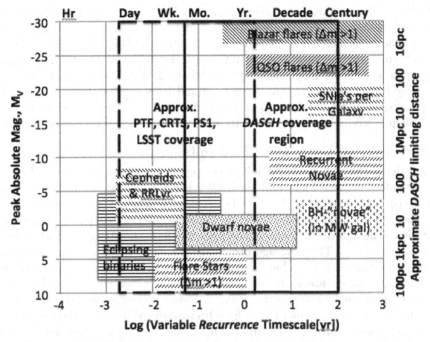

Figure 1. Representative classes of variables and transients vs. their recurrence time that can be measured for a complete sample with *DASCH* (right) vs. PTF, CRTS, Pan-STARRS-1 and LSST (dashed box, left) or jointly (overlap region). The distance scale (right axis) is for a conservative *DASCH* limiting magnitude $B = 14$, which corresponds to $V = 15$ for typical colours. For PTF, CRTS and LSST, the distance-scale axis for any object is increased by a factor of 10–30 owing to their deeper limiting magnitudes. Note that shorter variability time-scales may be measured by all those surveys, but the short time-scale limits shown for each object class or survey are approximately those for the variability to *recur*.

TDA surveys to time-scales that are 1 or 2 orders of magnitude longer than those reached by current or immediately projected modern surveys, a range of fundamental classes of objects can be studied as "individual" objects in well-defined (complete) samples. Those could include measuring the SN Ia rate in the Virgo cluster, or the outburst recurrence times for the full set of the 20 or so currently known black-hole transients in the Galaxy. For each, recurrence time-scales can be measured or limited in the "local" Universe and then tested for red-shift dependence using much more distant samples of differing objects in the modern TDA surveys.

4. Current *DASCH* Processing

4.1. *Photometry*

Photometric analyses developed by Laycock *et al.* 2010, and recently further improved by Tang *et al.* (in preparation), yield rms uncertainties of ~0.10 mag over the full range of century of data from the 9 or so series of plates that contribute to a typical light curve, despite differences in plate scale, image quality and any systematic effects. The basic approach employs SExtractor as the object detection and isophotal photometry engine for instrumental magnitude determination by using the now global Hubble Guide Star catalogue (GSC2.3) for a large sample of calibrators present on every plate, thereby allowing both global and local calibrations. Calibration curves are first derived in annular bins to account for vignetting, by fitting instrumental magnitudes against GSC2.3

magnitudes (B) for an initial photometric solution. That initial calibration is followed by local corrections to remove spatially-dependent plate effects (usually in the emulsion) or sky-related effects (atmospheric extinction and clouds). The GSC2.3 catalogue is not ideal since its photometric precision is only $\pm$ 0.2 mag and it is predominantly in a single band (photographic B). Fortunately, the all-sky APASS CCD survey (www.aavso.org/apass) described by Henden (page 95) has Johnson B and V as well as Sloan g', r' and i', and will improve significantly both the precision and, particularly, the colour corrections for *DASCH* photometry. APASS is now partly available and should be full-sky by 2013.

5. Representative Early *DASCH* Results

The five fields scanned in the *DASCH* development phase are centred on M44, 3C273, Baade's Window, Kepler Field and the LMC. For the first three, plates were simply selected if they contained the object (or the centre of Baade's Window) on the plate, and at least 1 cm from the plate edge, or approximately interior to "bin 9", the outermost annular bin or the outermost 5 mm of the plate where both astrometry and photometry can have large errors. The coverage obtained from that mode of plate selection was increasingly incomplete with radial distance from the target object. For both the Kepler Field and LMC, which are each extended regions, a wider boundary of plate-centre coordinates was applied when selecting plates to ensure more complete coverage of the full object. Only the M44 and Kepler Fields have been analyzed in detail, though exploratory results have been obtained for the other 3 fields and will be reported soon: for 3C273 (Grindlay *et al.*, in preparation) and for a classical nova discovered in Baade's Window (Tang *et al.*, in preparation). The M44 study led to the discovery of long-term dimming in a population of K giants (Tang *et al.* 2010). As a follow-up, more slowly-variable K giants were found in the Kepler Field and studied with the higher time-coverage of Kepler data (Tang *et al.*, in preparation). The variability study of the Kepler Field also led to the discovery of a dust-accretion event in the binary star KU Cyg (Tang *et al.* 2011).

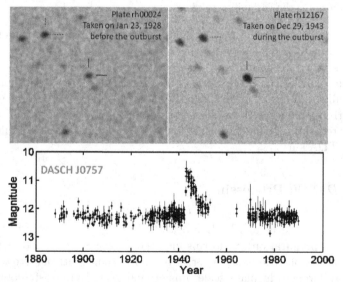

Figure 2. Symbiotic nova images and light curve discovered by Tang *et al.* 2012. Upper panel: *DASCH* images, showing a quiescent phase (left) and an outburst (right). The symbiotic nova (marked) is in the centre; a comparison star (also marked) is to the upper left. Lower panel: Light curve derived from *DASCH* scans. The outburst might be due to H shell-burning on the white-dwarf companion of the M0 III giant in the system, although an accretion-powered flare cannot be ruled out.

A third, and perhaps most dramatic, example from *DASCH* is a new type of stellar variability, also discovered in the M44 field (Tang *et al.* 2012), and is illustrated in Fig. 2. Observations with the telescopes at the Harvard-Smithsonian FLWO observatory in Arizona for spectroscopic classification of *DASCH* variables revealed that this star was an M0 III giant. Comparison with the ASAS CCD photometric survey (Pojmanski 2002) revealed it to be a semi-detached binary with a 119.2-day period and an amplitude of 0.16 mag in the *V* band. As described in detail by Tang *et al.* (2012), the cool giant's companion is a white dwarf, and we surmise that the remarkable flare in 1942 and its subsequent 10-year decline was most probably due to nuclear H-shell burning. The lack of emission lines from this symbiotic nova is new, but is consistent with the behaviour of some other old novæ and may also imply little or no ejection of the envelope mass ($\sim$3 $\times$ 10^{-5} $M_\odot$) that is required for ignition on what is probably a $\sim$0.6-$M_\odot$ white dwarf.

We include a brief summary of the *DASCH* light curve for the bright quasar/blazar 3C273, which is discussed in detail by Grindlay *et al.* (in preparation). Its 100-year light curve (Fig. 3) demonstrates the long-term variability of this luminous AGN. The inset shows a "flare" in 1941 with characteristic rise and fall times of about 3 days (and there are several other examples). Time-scales of 1–30 days for brightening are evident in the overall light curve, and are comparable to the optical and infra-red variability for 3C273 reported by Courvoisier *et al.* (1988), who found that the fastest variations were of the order of 1 day but their observations were too short to measure the longer-term variations detected with *DASCH*. The dominant power in the *DASCH* variability spectrum is over

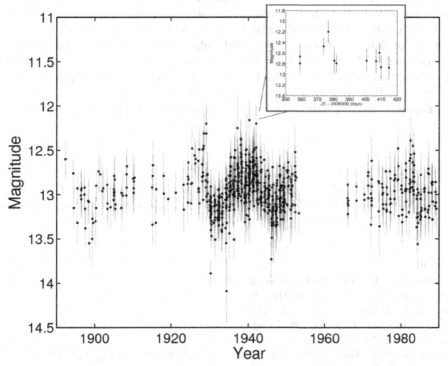

Figure 3. Light curve of 3C273 from 786 points (of 1494 total) measured by *DASCH* (Grindlay *et al.*, in preparation). The 708 points not plotted are within 0.75 mag of their respective plate limiting magnitude and may therefore have larger errors. One example of a flare event over a 3-day time-scale is shown in the inset. Note the absence of *DASCH* measurements during the "Menzel gap" period from approximately 1955–1970, as discussed by Grindlay and Griffin (page 243)

longer time-scales like 0.5–2 years (such as the abrupt decline in flux from about 1927–1930), and provides constraints on the size of the optical emission region.

6. *DASCH* Database

The *DASCH* Pipeline (Los *et al.* 2010) and database software run on a high-speed computer cluster and RAID disk system. In production mode it can process (overnight) the full Pipeline for the nominal 400 plates scanned in a day, to populate a MySQL database with photometric values and errors for each of the resolved stellar images; typically there are ∼50,000 on a standard plate but there are more on "A" plates. Light curves are generated very rapidly for any object by extracting from the database the magnitudes thus determined from all plates, or only those with magnitude measures meeting a set of user-selected criteria. Variability measures and tests of their validity can then be derived readily. Additional variability analysis tools are being developed, and will be made available when the full database becomes public.

The full *DASCH* output database of ∼450,000 plate images and derived magnitudes for each resolved object (∼1Pb in total!) will be made available for world access as it is completed incrementally. The present plans are to digitize the northern sky at Galactic latitudes $|b| > 10°$ first, to allow comparisons with existing surveys such as SDSS, PTF, Pan-STARRS-1 and the CRTS. That stage could be completed by mid-2013. Given the difference in limiting magnitudes (up to 14–18 for *DASCH* against 20–22 for the modern surveys), such comparisons will be mainly for context or extreme transients. The southern sky above/below the Galactic plane will come next, with expected completion in early 2015, followed by the Galactic Plane by mid-2016—or still well before LSST. The reason for doing the Galactic Plane and Bulge last is to allow time to develop analysis of crowded-field photometry further by invoking point-spread function (psf) and image subtraction techniques in order to improve the present isophotal photometry used for SExtractor. Experiments have recently been undertaken to optimise the use of PSFEx+SExtractor for magnitude- and position-dependent fitting of the plate psf.

Acknowledgements

We thank the *DASCH* team and gratefully acknowledge support for *DASCH* by the HCO, the NSF (grants AST0407380 and AST0909073), and by the *Cornel and Cynthia K. Sarosdy Fund for DASCH*.

References

Courvoisier, T. J. L., *et al.* 1988, *Nature*, 335, 330
Grindlay, J. E., *et al.* 2009, *ASPC*, 410, 101
Johnson, J. A., & Winn, J. 2004, *AJ*, 127, 2344
Johnson, J. A., *et al.* 2005, *AJ*, 129, 1978
Laycock, S., *et al.* 2010, *AJ*, 140, 1062
Los, E., Grindlay, J., Tang, S., Servillat, M., & Laycock, S. 2010, *ASPC*, 442, 269
Pojmanski, G. 2002, *Acta Astronomica*, 52, 397
Servillat, M., Los, E., Grindlay, J., Tang, S., & Laycock, S. 2011, *ASPC*, 442, 273
Simcoe, R., *et al.* 2006, *Proc. SPIE*, 6312, 17
Tang, S., Grindlay, J., Los, E., & Laycock, S. 2010, *ApJ*, 710, L77
Tang, S., Grindlay, J., Los, E., & Servillat, M. 2011, *ApJ*, 738, 7
Tang, S., *et al.* 2012, *ApJ*, submitted

New Horizons in Time-Domain Astronomy
Proceedings IAU Symposium No. 285, 2011
R.E.M. Griffin, R.J. Hanisch & R. Seaman, eds.

© International Astronomical Union 2012
doi:10.1017/S1743921312000178

Spectroscopic Surveys

Francesca Primas

European Southern Observatory, D-85748 Garching, Germany
email: `fprimas@eso.org`

Invited Talk

Abstract. Surveys, surveys, and yet more surveys! During the last decade we have all witnessed a flourishing of imaging and spectroscopic surveys, of different sizes and over different areas of the sky. Although initially set-up for specific scientific goals, they should all share a multi-purpose flavour that can boost their impact and their exploitation by the scientific community. There are, however, fields that need more dedicated observing strategies, coordination and possibly data infrastructure in order to exploit fully these huge datasets. Time-domain astronomy is one of them. In the following, I will review the very recent developments in spectroscopic surveys, and I will report on what ESO has been involved in and has committed itself to do.

Keywords. surveys, instrumentation: spectrographs, techniques: spectroscopic, Galaxy: general, Galaxy: abundances, Galaxy: kinematics and dynamics

1. Introduction

It is not really surprising that astronomy is shifting towards larger and larger projects, involving larger and larger teams. The community has voiced the need to "go big" in several instances. The projects are usually set up to answer specific astrophysical questions, they are granted large amounts of telescope time, and produce an almost overwhelming amount of data. While on the one hand this may mean that telescope time available for small-scale projects is shrinking, on the other hand it usually has the advantage of allowing for the exploitation of a lot of data for other scientific goals. Furthermore, be it a decadal survey or a roadmap exercise, the very top guidelines for future astronomical infrastructures always include some recommendation for surveys and survey facilities.

Back in 2008, the report from the 4th ESA/ESO Working Group† (WG for short) on "Galactic populations, chemistry and dynamics", delivered a set of recommendations heavily clustered around the need for survey facilities (Turon *et al.* 2008). The WG recommended to both organisations that they ensure a leading role for Europe in the exploitation of GAIA data. This was translated into a requirement for spectroscopy of selected samples of GAIA targets before the mission, and specific spectroscopic follow-ups after it. To ESO alone, the WG recommended the investigation of possible ways to build a blue multiplexed spectrograph (>100 fibres, R~20-30,000, on a 4-m/8-m telescope) and an IR multiplexed spectrograph (>500 fibres, similar resolution) on a 4-m telescope.

When it comes to time-domain astronomy, coordination and complementarity among different instruments/facilities is one of the keys for success. Telescopes and instruments need to be coordinated, to react rapidly to triggers, and to be complementary across the energy spectrum. Most large facilities already have a "Target of Opportunity" observing

† In 2003, following an agreement to cooperate on science planning issues, the European Space Agency and the European Southern Observatory decided to establish a number of working groups that would be tasked to explore synergies in important areas of mutual interest and to make recommendations to both organisations.

schema in place or even a "Rapid Response Mode"‡, and in the literature one can find more and more reports on how successful those multi-site, multi-instrument campaigns are. However, when looking at large-scale surveys, one immediately realises that the observing strategies implemented and/or the access to the survey data may not always be suited to catching the most extreme and unusual phenomena in the universe.

2. On-Going and Future Spectroscopic Surveys

Even if the scientific goals are clear, implementing and carrying out a survey is always a challenging endeavour. The team of proposers/investigators must define well both observing and the analytical strategies: Do we need a dedicated telescope? Can we execute a survey as a large programme (thus applying for general open time at one or more facilities)? Do we have enough manpower to exploit the data set we will be collecting? On the other hand, especially when the survey is executed by Observatory staff on behalf of the team ("Service Mode" in ESO terminology), one faces a different set of challenges: How and when do we schedule it? How do we optimise its execution, especially when competing with other normal programmes scheduled on the same telescope if not the same instrument? Do we have the tools to monitor its progress properly? Which data products do we deliver, and how? Which (advanced) data products do we want to receive back from the team?

Notwithstanding those challenges, many large spectroscopic surveys have already started, and more will start in the near future. The following summary is meant to give a flavour of what has been going on in recent years, but it is certainly far from complete. Observing strategies that provide easier access to time-domain events (such as multi-epoch observations) are highlighted in *italics*.

- SEGUE: the Sloan Extension for Galactic Understanding and Exploration (parts 1 and 2) was carried out at Apache Point Observatory. SEGUE-1 (Yanny *et al.* 2009) obtained spectra of almost 240,000 stars covering a range of spectral types; SEGUE-2 targeted mostly blue horizontal-branch stars, K giants and M giants (~120,000). Both surveys have been completed; they cover a total of ~4500 deg^2 in the northern hemisphere and include 360,000 stars in the magnitude range $g = 14.5$–23.5) at a resolving power $R \sim$2,000.

- ARGOS: this is a survey targeting 28,000 K giants in 28 "bulge" fields (PIs: Freeman and Bland-Hawthorn). It uses the multi-fibre AAOmega 2dF spectrograph at the Anglo-Australian Telescope and covers two regions of spectra: 5000–5600 Å (at $R \sim$3000) and 8400–8800 Å (at $R \sim$11000). It is close to completion (if not now already completed).

- BRAVA: the Bulge RAdial Velocity Assay (already completed) is a large-scale radial-velocity survey of the Galactic bulge. It has delivered spectra of ~9,000 M-dwarf stars at $R \sim$4,000 (Howard *et al.* 2008; 2009).

- RAVE: the Radial Velocity Experiment (PI: Steinmetz) started in 2003 and will run until 2012 with the 6dF multi-fibre spectrograph at the 1.2-m UK Schmidt Telescope of the Anglo-Australian Observatory. The survey is dedicated to the measurement of

‡ See http://www.eso.org/sci/observing/phase2/SMSpecial/TOOObservation.html or http://www.eso.org/sci/observing/phase2/SMSpecial/RRMObservation.html to learn more about how ESO implements those requests.

radial velocities, metallicities and abundance ratios, thus providing a vast stellar kinematic database for all major components of the Galaxy. As of mid-2011 it had obtained 511,000 observations of stars, *20,000 of which have been observed at different epochs*. The complete survey is supposed to cover 20,000 deg.^{2} in the southern hemisphere. To date, three major data releases have been released (Steinmetz *et al.* 2006, Zwitter *et al.* 2008 and Siebert et al. 2011).

• MARVELS: the Multi-object APO Radial Velocity Exoplanet Large-area Survey (PI: Ge) started in 2008, and is one of the four approved SDSS III large surveys. Its main goal is to monitor the radial velocities of 11,000 bright stars with the precision and frequency needed to detect gas giant planets that have orbital periods ranging from several hours to two years. *All stars will be surveyed 15-20 times/year*. The survey is expected to run until 2014.

• APOGEE: the SDSS III Apache Point Observatory Galactic Evolution Experiment (PI: Majewski) uses the high-resolution, near-infrared multi-fibre spectrograph APOGEE to make a detailed survey of the dynamics and chemistry of the Milky Way, especially at low Galactic latitudes. The survey has just started, and plans to target 100,000 stars. Observations cover the *H*-band NIR at *R* ~25–30,000, and *each field is to be surveyed three times to detect binaries and possibly other time-domain objects and events*. There is a strong interest in extending it to the southern hemisphere as well, which may prolong the lifetime of the survey to 2019.

• GALAH: The GALactic Archeology with Hermes survey (PIs: Freeman and Bland-Hawthorn) is due to start in 2013 on the Anglo-Australian Telescope, using the HERMES spectrograph. It will observe high-resolution (*R* ~28,000) spectra of a million Galactic-disk stars (V<14 mag), with S/N ~ 100. It is planned to run for 6 years, to cover the full Galaxy.

3. ESO and Surveys

The European Southern Observatory and its facilities in Chile have also been involved in, and have supported, different types of surveys, both imaging and spectroscopic, but there has clearly been an increased emphasis on this specific activity in recent years.

The precursor of today's on-going ESO Public Surveys (PS) is probably the ESO Imaging Survey (EIS; Project Scientist: Luiz da Costa), which was carried out from 1997 to 2002 on the 2.2-m telescope with the Wide Field Imager. That survey was possibly the first attempt by ESO to run large-scale imaging surveys as a service to the community, i.e, planning the observations, carrying them out, assessing the quality of the images, and reducing and releasing them and preparing the associated catalogues.

Since then, ESO has executed several "Large Programmes", i.e., programmes that are submitted to the time-allocation committee (aka OPC, Observing Proposals Committee) and which require a minimum (though large) number of nights and run over multiple semesters. Apart from sample sizes, the main difference between a Public Survey and a Large Programme is the time-scale over which the raw data are released to the community at large; in the case of a Public Survey the data release is immediate, while in the case of Large Programmes it is usually 12 months after the observations were made. Most of the latter have so far used the FORS and VIMOS instruments, thus combining imaging and multi-object spectroscopy capabilities, but it is certainly worth mentioning the HARPS

Table 1. ESO Public Surveys with a potential for time domain science.

Survey	Area $(\deg^2)$	Depth Measure	Filters and depth
VVV	520	5σ, Vega	$Z=21.9$, $K_s=18.1$
VMC	184	10σ, Vega	$Y=21.9$, $K_s=20.3$
VHS	20,000	5σ, AB	$Y=21.2$, $K_s=20.0$
VPHAS+	1800	10σ, AB	$u'/i'=21.8$, $g'/r'=22.5$, $H\alpha=21.6$

spectrograph (Mayor *et al.* 2003), which is almost fully dedicated to the search for extra-solar planets.

3.1. *Public Surveys on Survey Telescopes*

As of 2010/11, ESO is operating two survey telescopes on Cerro Paranal: VISTA and VST, equipped respectively with VIRCAM and OmegaCAM imagers. Although these facilities do not offer any spectroscopic capability at the moment, a couple of the public surveys that have been approved for execution over the next years† may have some relevance for exploiting time-domain phenomena. They are the the VISTA Variables in the Via Lactea (VVV) survey (PI: Minniti), and the VISTA Magellanic Survey (VMC, PI: Cioni). VVV targets mainly the Galactic bulge and a part of the adjacent plane, and is *multi-epoch: each "tile" will be covered more than 100 times, spaced over the entire lifetime of the survey.* Its product will be a catalogue of 10^9 objects, of which 10^6 are expected to be variables. VMC focuses on the Magellanic Clouds system, and its observing strategy also includes some *multi-epoch observations to determine the mean magnitude of short-term variables.*

Among the others, the VISTA Hemisphere Survey (VHS, PI: McMahon) and the VST Photometric H-α Survey of the Southern Galactic Plane (VPHAS+, PI: Drew) should be mentioned because they represent an important source of information. VHS will cover the entire Southern Sky, reaching 4 mag—deeper than DENIS and 2MASS. Its final catalogue will offer the possibility for studying low-mass and nearby stars and the merger history of the Galaxy. In contrast, VPHAS+ will map the entire Southern Galactic Plane in order to reconstruct the star-formation history and detailed structure of the Galactic disk. It will produce a catalogue of around 500×10^6 objects, which will include greatly enhanced samples of rare evolved massive stars, Be stars, Herbig and T Tau stars, post-AGB stars, compact nebulæ, white dwarfs and interacting binaries. The main characteristics of these surveys in terms of area and depth of measurement are summarised in Table 1.

3.2. *Public Spectroscopic Surveys*

Owing to a strong demand from the astronomical community, the ESO Scientific and Technical Committee made a clear recommendation to ESO back in 2010 April to explore possible ways to implement spectroscopic surveys. In 2010 August, ESO published a Call for Letters of Intent (LoI) regarding all instruments available at the La Silla Paranal Observatory. 24 proposals were received by the deadline (mid-October 2010). A Public Spectroscopic Survey Panel composed of international experts evaluated the 24 LoI, and recommended two of them, which were then invited to submit a full and detailed observing proposal at the Period 88 deadline (end of March, covering the semester 2011 October to 2012 March). After an eventual approval by the Observing Programme Committee, the finalisation of the implementation of these two spectroscopic surveys could commence. The two projects are:

† http://www.eso.org/sci/observing/policies/PublicSurveys/sciencePublicSurveys.html

• The GAIA–ESO survey (PIs: Gilmore and Randich), approved on the VLT Kueyen telescope. It will use the multiplex FLAMES instrument (GIRAFFE and FLAMES/UVES modes) for 4 years, with approximately 30 nights allocated per semester. The survey will produce a large database of high-resolution spectra of more than 10^5 stars across the Milky Way galaxy. Among the deliverables, all the spectra will be classified, not only in terms of spectral type and stellar population but also as far as their stellar parameters and chemical characteristics are concerned. The survey will start in January 2012, and will be carried out in classical Visitor Mode.

• The Transient Universe survey (PI: Smartt), approved on the La Silla New Technology Telescope (NTT). It will make use of EFOSC2 and SOFI. The survey aims to *follow-up ~150 optical transients by securing high-quality, time-series optical and near infrared spectra.* It will cover the full range of parameter space: luminosity, host metallicity and explosion mechanisms. One of the survey deliverables is the *ESO Transient Database* (ETABASE), which will allow the astronomical community to access these extensive datasets of transients.

3.3. *Forward Look*

Today's best strategy for tackling at least some of the top open questions in astrophysics seems to require the availability of moderate to large multi-object spectroscopic facilities (MOS), especially when the requirements are for fundamental Galactic astronomy. Both the ASTRONET Roadmap (Bode & Monnet 2008) and the ESO/ESA Working Group #4 (Turon *et al.* 2008) explicitly listed the need for MOS capabilities among their top recommendations. It is therefore not surprising that ESO was recently asked by its Scientific and Technical Committee to look into future MOS capabilities. ESO followed the same outline as in the case of the launch of Public Spectroscopic Surveys (Call for Letters of Intent, evaluation of the proposals received and pre-selection of most interesting projects), and selected two projects which are now carrying out their Phase A studies:

• 4MOST, the 4-meter Multi-Object Spectroscopic Telescope, which focuses on the follow-up of very large surveys such as GAIA and eROSITA, and is expected to observed 7×10^6 objects over more than 10,000 deg². The initial proposal plans for an installation on the VISTA or NTT telescopes. The facility would allow a wavelength coverage of 3700–10000 Å (with some gaps in between), for a multiplex capability of at least 1,500 fibres and a resolution of 3000–5000 or 20,000 (the latter, of course, achievable only with 10-20% of the total fibres).

• MOONS, the Multi-Object Optical and Near-infrared Spectrograph, which focuses on galaxy formation and evolution, large scale structures, high-redshift and Galactic archaeology, though its rather small field of view may be not efficient for GAIA follow-ups. The original proposal is for installation on any of the VLT Unit Telescopes, with a wavelength coverage of 0.8–1.8μ, a multiplex capability of 250 (up to 500) fibres and a resolution of 3000–5,000 or 20,000.

The final decision is expected in 2013.

4. Concluding Remarks

It is hard to believe that today's astronomical community will ever suffer from lack of data! Several large-scale surveys have already been completed, or are on-going, or will start soon. They are usually tailored to answer specific astrophysical questions, but we all agree that it would be a big loss if we were not given the opportunity to exploit those immense datasets for other scientific goals besides the original ones.

In order to achieve that, the fundamental requirement for realising such a goal is a user-friendly interface which makes the data available to the public at large, preferably in a co-ordinated way (e.g. through the Virtual Observatory). Some of the surveys mentioned in this report were (or are) indeed *not* public, but were (or are) carried out by a consortium or collaboration of institutes (such as SDSS). However, in those cases a huge effort has been already made to ensure releases of data to the public at regular intervals.

ESO's approach is slightly different, because its surveys are public, meaning that raw data are made available to the community as soon as observations are made. Furthermore, ESO has also fully recognised the importance of making ready-to-use (advanced) data products (such as calibrated catalogues) available, to ensure maximum involvement by its community. ESO has recently introduced the concept of Phase 3, to follow naturally after Phases 1 (submission and evaluation of observing proposals) and 2 (implementation and optimisation of strategies and programme execution). Phase 3 defines the "closing the loop" phase, during which high-level data products are submitted to ESO by the PIs of Large Programmes and Public Surveys (i.e., those with the best expertise to provide them), and ESO then releases them to the community after some detailed consistency checks. For that purpose, a new Phase 3 query interface has recently been released (Arnaboldi *et al.* 2011). More information can be found at: `http://www.eso.org/sci/observing/phase3.html`

Yet this may not be sufficient for time-domain astronomy. The large volume of data to be searched for transients and the associated real-time decisions that need to be taken argue strongly in favour of a more automated retrieval and distribution of time-critical information. It is hoped that the concept of a common facility like the Virtual Observatory for integrating a vast variety of astronomical archives and computational tools around the world will offer a robust infrastructure for this field of research. But a lot of work still lies ahead of us.

References

Arnaboldi, M., *et al.* 2011, *ESO Messenger*, 144, 17
Bode, M. & Monnet, G. 2008 *The Messenger*, 134, 2
Howard, C. D., *et al.* 2008 *ApJ*, 688, 1060
Howard, C. D., *et al.* 2009 *ApJL*, 702, L153
Mayor, M., *et al.* 2003 *The Messenger*, 114, 20
Siebert, A., *et al.* 2011 *AJ*, 141, 187
Steinmetz, M., *et al.* 2006 *AJ*, 132, 1645
Turon, C., *et al.* 2008 *The Messenger*, 134, 46
Yanny, B., *et al.* 2009 *AJ*, 137, 4377
Zwitter, T., *et al.* 2008 *AJ*, 136, 421

New Horizons in Time-Domain Astronomy
Proceedings IAU Symposium No. 285, 2011
R.E.M. Griffin, R.J. Hanisch & R. Seaman, eds.

© International Astronomical Union 2012
doi:10.1017/S174392131200018X

Time-Domain Astronomy
with Swift, Fermi and Lobster

Neil Gehrels[1], Scott D. Barthelmy[1], and John K. Cannizzo[2]

[1] Astroparticle Physics Laboratory, NASA Goddard Space Flight Center,
Greenbelt, MD 20771, USA
email: `neil.gehrels@nasa.gov`

[2] Astroparticle Physics Laboratory, CRESST/UMBC/Goddard Space Flight Center,
Greenbelt, MD 20771, USA

Invited Talk

Abstract. The dynamic transient gamma-ray sky is revealing many interesting results, largely due to findings by Fermi and Swift. The list includes new twists on gamma-ray bursts (GRBs), a GeV flare from a symbiotic star, GeV flares from the Crab Nebula, high-energy emission from novae and supernovae, and, within the last year, a new type of object discovered by Swift—a jetted tidal disruption event. In this review we present highlights of these exciting discoveries. A new mission concept called Lobster is also described; it would monitor the X-ray sky at order-of-magnitude higher sensitivity than current missions can.

1. Overview

The highly variable nature of the gamma-ray sky revealed in the last few years, as well as the detection of very high energy emission from unlikely sources, have surprised the high-energy community and sent theorists scrambling back to the drawing board. Giant flares have been discovered in a source that was once thought to represent a standard measure of constant flux, and fairly benign semi-detached interacting binaries such as symbiotic stars with a WD accretor have shown > 100 MeV emission. Such high energy emission from a WD accretor (in the galactic transient V407 Cyg, a symbiotic star) is highly unexpected given that $GM_{\mathrm{wd}}m_{\mathrm{electron}}/(R_{\mathrm{wd}}k_B) \simeq 100$ eV, and shows immediately that such emission cannot be associated with gravitational release of energy, either thermal or nonthermal. At the largest end of the accretor mass scale, supermassive black holes, in March 2011 Swift discovered Sw 1644+57 (initially GRB 110328A) which is believed to be a jetted tidal disruption event (TDE). Perhaps the greatest surprise came with the recent discovery of GeV flares from the Crab nebula, previously considered to be a canonical standard candle powered primarily by simple pulsar spin-down.

2. High Energy Variability in Swift and Fermi

Swift (Gehrels *et al.* 2004) carries three instruments: a wide-field Burst Alert Telescope (BAT; Barthelmy *et al.* 2005) which detects GRBs and positions them to arc-minute accuracy, the narrow-field X-Ray Telescope (XRT; Burrows *et al.* 2005), and the UV-Optical Telescope (UVOT; Roming *et al.* 2005). BAT is a coded-aperture hard X-ray ($15 - 350$ keV) imager with $0.5\,\mathrm{m}^2$ of CdZnTe detectors (32,000 individual sensors; $\sim$2400 cm^2 effective area at 20 keV including mask occultation) and a 1.4 sr half-coded field of view. XRT is a Wolter 1 grazing-incidence, imaging X-ray telescope with a $0.2 - 10$ keV energy range, 120 cm^2 effective area at 1.5 keV, field of view of $23'.6 \times 23'.6$, point spread function (PSF) half-power diameter of $18''$ ($7''$ FWHM), and sensitivity of

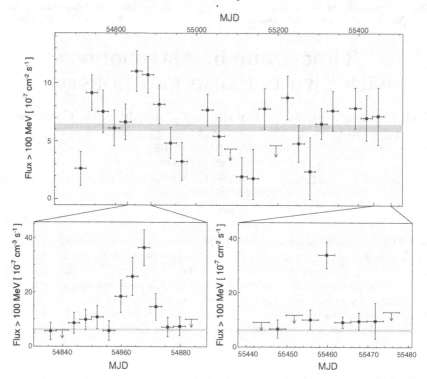

Figure 1. A light curve of the Crab nebula for photons with $E > 100$ MeV as observed by FERMI/LAT (Abdo *et al.* 2011).

approximately 2×10^{-14} erg cm^{-2} s^{-1} in 10^4 s. The UVOT is a modified Ritchey-Chrétien reflector with a 30-cm aperture, $170 - 650$ nm wavelength range, field of view $17' \times 17'$, PSF FWHM of 1.9 arcsec at 350 nm, and sensitivity of 23rd magnitude in white light in 10^3 s. The SWIFT mission was built by an international team from the US, UK, and Italy. After five years of development it was launched from Kennedy Space Center on 2004 November 20. Full normal operations commenced on 2005 April 5.

FERMI has two primary high-energy detectors: the Large Area Telescope (LAT), which operates between 20 MeV and $\gtrsim 300$ GeV (Atwood et al. 2009) and the Gamma-Ray Burst Monitor (GBM), which operates between 8 keV and 40 MeV (Meegan *et al.* 2009). The LAT is a 3000-kg instrument consisting of 77 m^2 of silicon detectors. It is a pair conversion telescope with silicon detector tracker and caesium iodide calorimeter. The LAT scans the full sky every 3 hr; all data are made public immediately. The GBM is a scintillator-based instrument with 12 sodium iodide detectors and two bismuth germanate detectors. FERMI was launched into low-Earth orbit in 2008 June.

There are many types of objects for which variability has been observed. Galactic black-hole transients are seen to have major outbursts thought to be caused by limit cycle oscillations of the accretion disk. Those outbursts can last for hundreds of days. Novae—thermonuclear explosions on the surfaces of WDs—can be quite powerful at energies of several hundred MeV and are thought to be providing information about the dynamics of the expanding ejecta. Flare stars are seen to vary dramatically in the $\sim$10–100 keV bandpass, providing information about changes in the magnetic field configuration of the star. Blazars are AGN in which we view the central BH directly down the jet. There can be large flares presumably due to the introduction of accreted material near the BH. GRBs are important sources for SWIFT and FERMI.

Both missions continue to enlarge the data-base of known and well-categorized GRB behaviour, as well as expand our knowledge by making unexpected discoveries (Gehrels *et al.* 2009). Long GRBs are thought to be due the core collapse of massive, low-metallicity stars in the early Universe, whereas short GRBs are due to NS–NS mergers. Finally, one of the greatest surprises of the past several years in high-energy astrophysics has been the discovery of major, high-energy flares in the Crab nebula.

3. FERMI/LAT and AGILE Crab Flares

The Crab Nebula is the result of a supernova. It was observed by the Chinese in 1054 AD; it lies at a distance $\sim$2 kpc, and has a physical extent of $\sim$ 3 pc. The observed power output $\dot{E}_{out}$ from the nebula, kinetic plus radiation, is $\sim 5 \times 10^{38}$ erg s^{-1}. The expanding ejecta which make up the nebula are thought to be powered primarily by the spin-down energy from the pulsar, $\dot{E}_{loss} = I\Omega\dot{\Omega}$ (Shapiro & Teukolsky 1983), where I is the neutron star moment of inertia and Ω the rotational rate ($\sim$30 Hz). Evaluating the loss rate assuming neutron star parameters $M = 1.4M_{\odot}$, $R = 12$ km, and $I = 1.4 \times 10^{45}$ g yields $\dot{E}_{loss} = 6.4 \times 10^{38}$ erg s^{-1}, in line with the total nebula power output. On the basis of those simple considerations, the recent finding of large GeV flares coming from the Crab Nebula was totally unexpected. Those flares (Fig. 1) have been seen with AGILE (Tavani *et al.* 2011) and by FERMI/LAT (Abdo *et al.* 2011). The synchrotron nebula increased and then decreased its power output by a factor $\sim$30 over observations of several weeks. During those high-energy observations there was also low-energy coverage by CHANDRA, SWIFT, RXTE, and MAXI. Little or no variability was seen at other wavelengths. At the smallest temporal resolution of the observations, the $\sim$1-hr variability time-scales imply a $\sim$milliarcsec scale at the distance of the Crab. The luminosity of the brightest flare was $\sim 10^{39}$ erg s^{-1}. It may be that the flares are due to magnetic reconnection in small knots within the nebula, and therefore represent localized but large perturbations in the overall energy budget. The Crab Nebula has been recently found to be variable in the low-energy gamma-ray band too (Wilson-Hodge *et al.* 2011). The variability lasts over time-scales of a year—much longer than the impulsive flare at high energies.

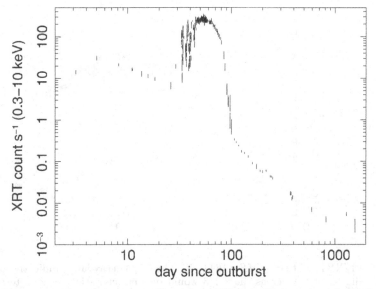

Figure 2. The 2006 nova outburst from RS Oph (Osborne *et al.* 2011).

4. Nova Studies with SWIFT

Novae can occur in interacting binaries containing a WD accretor, and are caused by the thermonuclear detonation of accreted material on the surface of a WD (Gallagher & Starrfield 1978). That can occur if the temperature and pressure at the base of the accumulated layer of accreted matter are in the appropriate regime. With its flexible scheduling and X-ray telescope, SWIFT has opened a new window on nova studies. To date SWIFT has observed 28 novæ. It has detected keV emission from shocked ejecta and supersoft (SS) emission from the WD surface. Extensive observations ($\sim$400 ks) of the 2006 nova outburst from RS Oph (Fig. 2) found an unexpected SS state, and 35-s QPO (Osborne et $al.$ 2011). Detailed analyses of SWIFT observations revealed a mass ejection of $\sim$3 $\times$ $10^{-5} M_\odot$ at $\sim$4000 km s^{-1} into the wind of the mass-losing red giant companion in the system.

5. Swift J1644+57 – The First Jetted TDE

Sw 1644+57 was triggered as GRB 110328A (Burrows et $al.$ 2011). Evidence connecting the X-ray/gamma-ray source with a galactic nucleus (Berger et al. 2011) was confirmed by precise HST and CHANDRA localizations (Fruchter et $al.$ 2011; Levan et $al.$ 2011). The host dwarf (non-AGN type) galaxy lies at $z = 0.35$ (Levan et $al.$ 2011). The strong circumstantial evidence associating the X-ray emission with the centre of the dwarf galaxy suggests a TDE (Bloom et $al.$ 2011). Long term monitoring showed its X-ray light curve to be quite different from a normal GRB (Shao et $al.$ 2011). Fig. 3 reproduces the long-term light curve. The bottom panel shows the flux binned in time as 0.125 dex. The two superposed decay segments indicate the putative theoretical decay laws for stellar debris accretion and a long-term accretion disk (Cannizzo et $al.$ 2011). It is surprising that such events have not been identified earlier, but now that we have some experience to guide us, a second such event has already been tentatively identified (Cenko et $al.$ 2011).

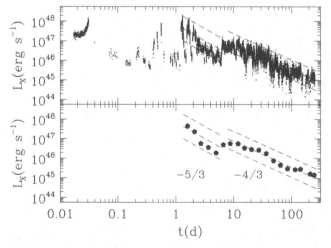

Figure 3. SWIFT/XRT light curve of Sw 1644+57. The $-5/3$ decay law is indicative of the direct accretion of stellar debris, whereas the $-4/3$ would be characteristic of a long term accretion disk (after Cannizzo et $al.$ 2011).

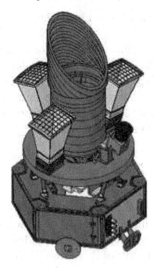

Figure 4. LOBSTER observatory.

6. LOBSTER **Mission Concept**

The tantalizing glimpses afforded us to date by serendipitous sources such as Sw 1644+57 have underscored the need for more efficient discovery techniques in order to increase the future rate of discovery of such objects. To that end, LOBSTER (Fig. 4) is a mission concept led by Goddard Space Flight Center, MIT, the University of Leicester and the University of Arizona to combine an X-ray wide-field imager (WFI) with a narrow-field IR telescope (IRT). The underlying strategy is similar to that of SWIFT: a detection at high energies, followed up by more detailed observations. The WFI has a combined 0.5-sr field of view that covers ~50% of the sky every 3 hours in multiple pointings. It is based on CCD technology and lobster-eye microchannel optics (Angel 1979). The IRT has a 40-cm diameter mirror, a wavelength range of $0.6 - 2.1\mu$, and is capable of multiband photometry and $\lambda/\delta\lambda = 30$ slit spectroscopy. An estimation of the rate for jetted TDEs in an instrument like LOBSTER, based on Sw 1644+57, is $\sim 1 - 2$ yr^{-1}

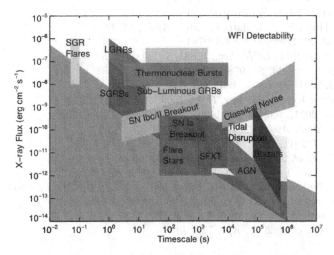

Figure 5. LOBSTER/WFI transient detectability.

(Cannizzo *et al.* 2011). Fig. 5 indicates in more generality the parameter space for LOB-STER transients.

LOBSTER will also provide capabilities for a major step forward in studies of the high-redshift universe using GRBs. The prompt emission is detected by the WFIs and observed in X-ray afterglow by the WFIs, and in infrared afterglow by the IRT. Improvements over SWIFT include: (1) a more sensitive wide-field instrument, (2) an infrared follow-up telescope and (3) slit spectroscopy by the follow-up telescope. The mission is expected to detect more than 25 bursts per year at redshift $z > 5$, and to extend detections to $z > 12$.

7. Summary

The high-energy sky is dominated by transient sources, which is somewhat surprising given our prior expectations. This is especially true with the unexpected discovery of large flares in the Crab Nebula, which has been considered historically as a fiducial mea-suring stick, giving rise to its own unit of flux, "the Crab". FERMI and SWIFT continue to monitor the sky, continuously and with sensitivity. Those missions have yielded impor-tant results: in the SWIFT era it has been definitively established, through localizations based on the prompt SWIFT/BAT discoveries, that long and short GRBs have different origins. That result was reached by identifying differences in their progenitor galaxies, and through studying offsets within those hosts. In 2011 March a new type of object was discovered: a tidal disruption event viewed down the jet. Nova and supernova obser-vations are also providing important clues about their high-energy transient behaviour, thereby helping to disclose their true nature.

References

Abdo, A. A., *et al.* 2011, *Science*, 331, 739
Angel, J. R. P. 1979, *ApJ*, 233, 364
Atwood, W. B., *et al.* 2009, *ApJ*, 697, 1071
Barthelmy, S. D., *et al.* 2005, *Space Sci. Rev.*, 120, 143
Berger, E., *et al.* 2011, *GCN Circ.*, 11854, 1
Bloom, J. S., *et al.* 2011, *Science*, 333, 203
Burrows, D. N., *et al.* 2005, *Space Sci. Rev.*, 120, 165
Burrows, D. N., *et al.* 2011, *Nature*, 476, 421
Cannizzo, J. K., Troja, E., & Lodato, G. 2011, *ApJ*, 742, 32
Cenko, S. B., *et al.* 2011, *arXiv*, 1107.5307v1
Fruchter, A. F., *et al.* 2011, *GCN Circ.*, 11881, 1
Gallagher, J. S. & Starrfield, S. 1978, *ARAA*, 16, 171
Gehrels, N., *et al.* 2004, *ApJ*, 611, 1005
Gehrels, N., Ramirez-Ruiz, E., & Fox, D. B. 2009, *ARAA*, 47, 567
Levan, A. J., *et al.* 2011, *Science*, 333, 199
Meegan, C., *et al.* 2009, *ApJ*, 702, 791
Osborne, J. P., *et al.* 2011, *ApJ*, 727, 124
Roming, P. W. A., *et al.* 2005, *Space Sci. Rev.*, 120, 95
Shao, L., Zhang, F.-W., Fan, Y.-Z., & Wei, D.-M. 2011, *ApJ*, 734, 33
Shapiro, S. L. & Teukolsky, S. A. 1983, *Black Holes, White Dwarfs, and Neutron Stars: The Physics of Compact Objects* (New York: John Wiley & Sons)
Tavani, M., Bulgarelli, A., & Vittorini, V., *et al.* 2011, *Science*, 331, 736
Wilson-Hodge, C. A., *et al.* 2011, *ApJ*, 727, 40

Day 2:

Explosive or Irreversible Changes

New Horizons in Time-Domain Astronomy
Proceedings IAU Symposium No. 285, 2011   © International Astronomical Union 2012
R.E.M. Griffin, R.J. Hanisch & R. Seaman, eds.   doi:10.1017/S1743921312000208

The Dynamic Radio Sky

James M. Cordes

Astronomy Department, Cornell University, Ithaca, NY 14853, USA
email: cordes@astro.cornell.edu

Invited Talk

Abstract. The radio band is known to be rich in variable and transient sources, but exploration of it has only begun only in the last few years. Relevant time scales are as small as a fraction of a nanosecond (giant pulses from the Crab pulsar). Short transients (less than one second, say) have signal structure in the time-frequency plane at the very least because of interstellar plasma propagation effects (dispersion and scattering), and in some cases due to emission structure. Optimal detection requires handling a range of signal types in the time-frequency plane. Short bursts by necessity have very large effective radiation brightness temperatures associated with coherent emission processes. This paper surveys relevant source classes and summarizes propagation effects that must be considered to optimize detection in large-scale surveys. Scattering horizons for the interstellar and intergalactic media are defined, and the role of the radio band in panchromatic and multimessenger studies is discussed.

1. Introduction

Transients across the electromagnetic spectrum as well as in neutrinos, cosmic rays and gravitational waves are of special interest because they carry information both about sources, particularly their energetics and environments, and also about intervening media. Comprehensive all-sky surveys of transients have been made for decades in the X-ray and gamma-ray bands. Detectors in those bands and in other tracers (cosmic rays, neutrinos and gravitational waves) inherently have wide fields of view (FoV). Telescopes that involve mirrors have relatively small FoVs; that is especially true at radio frequencies, where large antennæ have traditionally been used to achieve requisite sensitivities (i.e. $A\Omega = \lambda^2$). For that reason, much of the phase space for radio transients is unexplored, especially those that have low rates. It is possible that very bright but rare radio transients occur and can be detected with even small collecting areas.

The current situation is being rectified at radio wavelengths under the paradigm where a large number of small antennæ ("large-N/small-D") are used to supply adequate collecting area. At centimetre wavelengths a large number of small dish antennæ provides both large total area (NA) and large FoV ($\Omega \sim N_{\mathrm{pix}}\lambda^2/A$). Focal-plane arrays are increasingly being used to provide multiple-pixel (N_{pix}) systems. Feed clusters are used at higher frequencies and phased-array feeds are in an R&D phase for frequencies ~ 0.5 to 1.5 GHz. At metre wavelengths a similar approach uses phased-arrays comprised of dipole antennæ instead of dish concentrators, yielding radian-like FoV (e.g. LOFAR, Long-wavelength Array [LWA]).

Synoptic surveys in the radio band are essential for several reasons. First, the Venn diagram of transients includes some overlap of source classes (e.g. gamma-ray burst [GRB] afterglows) between radio, optical-infrared (OIR), and X-rays along with significant *non-overlap* (e.g. coherent radio bursts, optical novæ). Comprehensive surveys can therefore use radio and OIR transient detections in source classification. On the other hand, radio and gamma-ray transients show a great deal of commonality in blazars and pulsars.

While prompt radio bursts from GRBs have yet to be seen, in terms of energy it is very plausible for such bursts to exist and yet to have been missed so far because of insufficient temporal and sky coverage. In the rest of this paper, I discuss source types, propagation effects and some of the observing programmes that are likely to take place this decade.

2. Radio Transients

Nature can be profligate in producing radio photons and thus make sources detectable. It does so in many cases via coherent radiation processes that give N^2 rather than N scaling from a collection of N particles. In most sources radio emission is not the dominant energy channel for dumping free energy, but it is an extremely significant information channel. Examples include pulsars and radio pulses from ultra high-energy cosmic rays.

2.1. Fast and Slow Transients

It is useful to classify the wide range of time-scales in terms of how they are best sampled empirically. For radio telescopes with conventional fields of view, such as those provided by paraboloids with single-pixel feeds, slow transients are those that can be sampled in a raster scan survey because they stay "on" for at least as long as it takes to scan the relevant sky region (e.g. the Galactic plane, the Galactic centre or the entire sky). Depending on survey speed, transients of days or more may be considered "slow".

Fast transients, conversely, are those that would be missed in the time it takes to scan the sky, leading to incompleteness of the survey. Sub-second transients are linked to coherent radiation and, by making a simple light travel-time argument, to sources in extreme matter states. They are also affected by plasma propagation effects such as dispersion and multipath scattering, which can distort pulses and inhibit detection. By the same token, such transients are excellent probes of the intervening interplanetary, interstellar and intergalactic media.

2.2. Examples

All known, convincing, fast radio transients appear to be Galactic in origin because the column density of electrons toward them can be accounted for by the Milky Way. No firm statements about slow transients can be made on that basis because propagation effects do not affect the transient burst shape, although interstellar scintillation can modify the apparent flux density (see below). Examples of fast transients include giant pulses from the Crab pulsar, modulations of pulsar signals, rotating radio transients (RRATs; McLaughlin et al. 2006) and some solar bursts. Transients with long time-scales include intermittent pulsars that turn on and off on time-scales of weeks to months, magnetars that are pulsed at rotation rates ~ 5 to 10 s but are on or off over time-scales of years (Camilo et al. 2006), flare stars (Osten & Bastian 2008), brown dwarfs, GRB afterglows (with and without gamma-ray emission), and tidal disruption events. There are, of course, long-duration transients seen in imaging surveys that are not yet identified definitively (Hyman et al. 2006, Bower et al. 2007, Frail et al. 2011, Ofek et al. 2011).

2.3. Phase Space

In the two-dimensional space of peak flux density (S_{pk}) and time-scale, known radio transients range from 100 μJy to > 1 MJy (more than 10 decades) and in duration W from less than 1 ns to greater than weeks (more than 15 decades). Expressed as a phase space in "pseudo-luminosity" ($S_{\mathrm{pk}} D^2$, $D = $ distance) and a dimensionless time νW (with $\nu = $ radio frequency), the large ranges translate into effective radiation temperatures from non-thermal incoherent radiation up to $\sim 10^{12}$K for some AGNs, to much higher

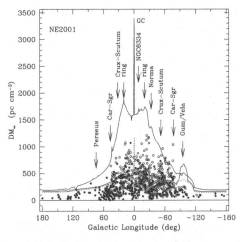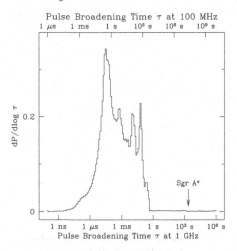

Figure 1. Left: Maximum dispersion measures DM_∞ predicted by the NE2001 electron density model for lines of sight integrated through the plane of the Milky Way. Right: Histogram of the pulse-broadening time expected from sources distributed throughout the disk of the Milky Way, defined as a disk of radius 10 kpc and thickness 1 kpc. The scattering measure and pulse broadening were calculated from the NE2001 model (Cordes & Lazio 2002). The bottom horizontal scale gives values for a radio frequency of 1 GHz and the top axis for 100 MHz.

temperatures for coherent sources such as molecular and plasma masers and pulsars. The record holder is a 0.4 ns shot pulse from the Crab pulsar with an effective temperature of 10^{42} K (Hankins & Eilek 2007). These large values illustrate the "bang for the buck" in creating radio photons with low amounts of total energy and the possibility that bright beacons await discovery from cosmological as well as Galactic sources.

3. Plasma Propagation Effects

Intervening plasmas dominate the effects that modify radio emission from both steady and time-dependent sources. Of course plasma effects are accompanied by gravitatational lensing and redshift-dependent phenomena on cosmological lines of sight.

3.1. *Dispersion*

Fast transients have signal shapes modified by the strongly frequency-dependent travel time. When a pulse propagates dispersively its arrival time varies with frequency as $\Delta t = 4.15$ ms $DM\nu^{-2}$ for dispersion measure DM (the line-of-sight integral of electron density) in standard units (pc cm^{-3}) and ν in GHz. Equivalently, the sweep rate in frequency is $\dot{\nu} \propto \nu^3 DM^{-1}$. If uncompensated, a pulse measured across a finite bandwidth is smeared out. However, de-dispersion techniques are well developed, and enable this deterministic effect to be removed. DM is not known *a priori*, so trial values must be used from a set of plausible values. For Galactic sources DM ranges between 2.4 and ~ 1500 pc cm^{-3} among known pulsars. The NE2001 model for free electrons in the Galaxy (Cordes & Lazio 2002) predicts dispersion measures up to about 3400 pc cm^{-3} for a direct sight-line through the Galactic centre. However, much larger values will obtain for lines of sight that pierce unmodelled dense HII regions. Fig. 1 (left) shows measured and modelled values of DM vs. Galactic longitude for directions through the plane of the Galaxy. The upper curve shows the maximum DM obtained from the NE2001 model.

Extragalactic sources will show contributions to DM from foreground Galactic electrons as well as from a host or intervening galaxy (if relevant) and from the intergalactic

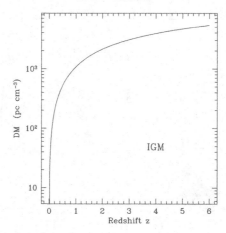

Figure 2. Dispersion measure from the intergalactic medium for a ΛCDM cosmology with
$\Omega_{M\,0} = -0.27$, $\Omega_{b\,0} = 0.046$, and $H_0 = 70.4$ km s^{-1} Mpc^{-1}.

medium (IGM). The IGM contribution is

$$\mathrm{DM_{igm}}(z) = cH_0^{-1}n_{e0} \int_0^z dz' \frac{(1+z')}{E(z')},\qquad (3.1)$$

where n_{e0} is the electron density from the baryonic fraction of the total closure density.
A fiducial value for intergalactic dispersion is $\mathrm{DM_{igm\,0}} = cH_0^{-1}n_{e0} \approx 10^3$ pc cm^{-3}.

3.2. Scattering and Scintillation

Multipath propagation along Galactic lines of sight occurs because there is microstructure
in the free electron density down to scales $\sim 10^3$ km (e.g. Rickett 1990) which diffracts
and refracts radio waves. Relevant effects include angular broadening ("seeing"), pulse
broadening and intensity scintillations.

Measured pulse broadening in Galactic lines of sight is as large as ~ 1 s but scales very
strongly with frequency, $\tau_d \propto \nu^{-4}$. There are lines of sight with much larger predicted
scattering times. Fig. 1 (right-hand panel) shows the distribution of τ_d expected from the
electron-density model NE2001. On lines of sight to the Galactic centre, τ_d is thousands of
seconds at 1 GHz and prohibits the detection of pulsars around Sgr A* at the conventional
frequencies used in pulsar searches. Searches for fast transients from the Galactic centre
region must therefore be done at high frequencies $\gtrsim 10$ GHz, where τ_d is $\sim 10^4$ times
smaller.

Diffractive interstellar scintillation (DISS) results from interference between scattered
wavefronts, and produces structure in both time and frequency with representative scales
of 100 s and 0.1 MHz but with ranges in each of many orders of magnitude. Refractive
scintillation (RISS) is caused by focusing and defocusing of radiation from scales much
larger than those responsible for DISS in the strong-scattering regime. RISS is broadband,
$\Delta\nu/\nu \sim 1$, and has time-scales of hours and longer, depending on the line of sight. RISS
from Galactic plasma appears to underly the "intraday variability" (IDV) of some active
galactic nuclei (e.g. Koay *et al.* 2011).

Broadening of pulses from extragalactic sources by intergalactic or extragalactic plasma
has not yet been measured definitively. Propagation through a face-on intervening galaxy
like the Milky Way would scatter radiation into ~ 1 mas at 1 GHz and broaden a pulse by
a fiducial broadening time $\Delta t_{\mathrm{face\ on}} = \theta_s^2/2H_0 = 5\theta_{s,\mathrm{mas}}^2$ sec. Pulse broadening from any
host galaxy of a transient source is deleveraged from this value by a geometrical factor

$\sim H/D$, where H is the path length through the host galaxy and D is the distance. Pulse broadening could be much larger, however, from edge-on and starburst galaxies. If the IGM is turbulent (as is plausible) like the interstellar medium, then that too could make a sizable contribution. The detection of pulses from cosmological sources would represent a very powerful tool for probing the IGM, as would relatively nearby sources in local-group galaxies.

3.3. *Pulse Broadening Horizon*

Pulse broadening causes the detectability of fast transients to degrade over and above the effects of the inverse-square law. Pulses will not be selected if scattering broadening is larger than the intrinsic pulse width, W. From measurements of pulse broadening for pulsars seen at different distances we can define the *Galactic horizon* in terms of DM by requiring that $\tau_d \lesssim W$. For 1 ms pulses the horizon is $\sim$5 kpc at 1 GHz, while 1 μs pulses can be seen to only 2.4 kpc at 1 GHz. These values apply only to sources within the Galactic disk. For a sight-line perpendicular to the disk, the seeing distance "breaks out" if it is more than $\sim$1 kpc. At low frequencies such as 100 MHz, a $1-\mu s$ pulse can be seen only to $\sim$100 pc.

The *cosmological horizon* can be calculated from the variation of DM with redshift z assuming that the IGM is completely ionized and that the relationship of pulse-broadening time to DM is the same as for Galactic sources—though that is at best a very crude approach. We find that a 1-ms pulse can be detected to $z \approx 0.2$ and that the broadening is $\tau_d \sim 100$ ms for transients originating at $z = 1$.

4. How Bright Can Fast Transients Be?

We already know that giant pulses from certain pulsars, like the Crab, are emitted frequently enough to be detectable at plausible rates out to 1.5 Mpc with the Arecibo Telescope. That number results from scaling 0.43 GHz pulses of amplitude $S_{\rm pk} \sim 200$ kJy that are pulse-broadened to ~ 100 μs duration and occur at a rate ~ 1 hr^{-1}. They correspond to a "pseudo-luminosity" $S_{\rm pk}D^2 \sim 10^{5.8}$ Jy kpc^2. Surely there are more luminous, giant-pulse-emitting pulsars that are detectable even further.

The case can be made that there are other burst sources that can tap larger sources of free energy than are available in pulsars like the Crab. They include:

- **Hyperfast rotators:** Pulsars born near the break-up limit (~ 1 ms) with canonical or magnetar-like magnetic fields (10^{12} to 10^{15} Gauss) can rapidly dump their rotational energy (using a moment of inertia of 10^{45} gm cm^2),

$$\frac{1}{2}I\Omega^2 = 2\pi^2 I P^{-2} \sim 10^{51.3} \text{ erg } I_{45}P_{\rm ms}^{-2}, \tag{4.1}$$

which is comparable to the non-neutrino energy released in a supernova. That energy will be released in a short amount of time as the pulsar rapidly spins down, but along the way it can drive giant-pulse emission much larger than seen from the Crab, perhaps by a factor of 10^6, e.g $S_{\rm pk}D^2 \sim 10^{12}$ Jy kpc^2. Such pulses could be seen out to ~ 1 Gpc.

- **Prompt GRB Counterparts:** The magnetospheres of neutron stars may reactivate when they merge, with orbital motion substituting for spin in generating voltage drops that accelerate particles and create electron-positron pairs (Hansen & Lyutikov 2001). NS–NS mergers are strong candidates for short-duration GRBs. The energy involved is similar to, or exceeds, that for extreme giant pulse-emitting objects. The GRB peak luminosity is fiducially $L_\gamma = 10^{51}L_{\gamma,51}$ erg s^{-1}. We assume that the true radio luminosity is a multiple ϵ_r of the gamma-ray luminosity, $L_r = \epsilon_r L_\gamma$. Over many kinds of astrophysical objects (stars, pulsars, AGNs), ϵ_r ranges from about 10^{-8} (Crab pulsar) to

10^{-3} (blazars). We estimate the pseudo-luminosity L_p in units of Jy kpc^2 by calculating the peak radio flux from the GRB assuming (only fiducially) isotropic emission. Beaming may influence the radio luminosity estimate as it does the gamma-ray luminosity. This approach gives a stupendously large pseudo-luminosity and peak flux density,

$$L_p = 10^{15.9} \text{ Jy kpc}^2 \left(\frac{\epsilon_{r,-5} L_{\gamma,51}}{\Delta\nu_{r,\text{GHz}}} \right), \qquad S_{\text{pk}} = 10^{2.9} \text{ Jy } \left(\frac{\epsilon_{r,-5} L_{\gamma,51}}{\Delta\nu_{r,\text{GHz}}} \right) \left(\frac{3 \text{ Gpc}}{d_L} \right)^2 . \quad (4.2)$$

where we use a fiducial value $\epsilon_r = 10^{-5} \epsilon_{r,-5}$ and a radio emission bandwidth $\Delta\nu_r = 1 \text{ GHz} \Delta\nu_{r,\text{GHz}}$.

5. Transient Surveys: Fast and Slow

Pulsar Survey pipelines with time sampling of 50 to 100 μs now routinely include analysis for single bursts, which led to the discovery of RRATs. Analysis of archival data, such as Very Large Array data, has aimed to identify slow transients. Planning for new telescopes and those coming on-line—such as the EVLA, LOFAR, ASKAP, LWA, MeerKAT and the SKA—now includes transients as key science.

6. Energetics vs. Information

In many circumstances radio luminosities represent a trivial amount of the total power budget of a source. However, there are exceptions; in long-period pulsars the luminosity is in fact more than a few per cent of the spindown loss rate $\dot{E} = I\Omega\dot{\Omega}$. More importantly, in many cases radio wavelengths have provided crucial, unique information about high-energy and potential gravitational-wave sources, in part because radio photons are easily produced. The discovery of "intermittent" pulsars (Kramer *et al.* 2006) indicates that radio emission traces the overall spindown energy losses from spin-driven pulsars. Intermittent pulsars include cases where radio emission shuts off for long periods (10^6 s) and is on for only a fraction of that time. During the "on" period, $\dot{E}$ can be tens of percent larger than during the "off" period. The situation may be similar in a host of other objects which, owing to beaming, may be quiet in high-energy bands but quite loud in others. GRB orphan afterglows have been seen (e.g. Soderberg *et al.* 2010), but prompt emission may also occur without any counterpart GRB (Cordes 2007).

References

Bower, G. C., *et al.* 2007, *ApJ*, 666, 346
Camilo, F., *et al.* 2006, Nature, 442, 892
Cordes, J. M. & Lazio, T. J. W. 2002, astro-ph/0207156
Cordes, J. M. 2007 SKA Memo 97, http://www.skatelescope.org, 2007
Frail, D. A., *et al.* 2011, arXiv:1111.0007
Hallinan, G., *et al.* 2007, ApJL, 663, 25
Hankins, T. H. & Eilek, J. A. 2007, ApJL, 670, 693
Hansen, B. M. S. & Lyutikov, M. 2001, MNRAS, 322, 695
Hyman, S. D., *et al.* 2006, ApJ, 639, 348
Kramer, M., *et al.* 2006, *Science*, 312, 549
Koay, J. Y., *et al.* 2011, *AstAp*, 534, L1
McLaughlin, M. A., *et al.* 2006, Nature, 439, 817
Ofek, E. O., *et al.* 2011, ApJ, 740, 65
Osten, R. A. & Bastian, T. S. 2008, ApJ, 674, 1078
Rickett, B. J. 1990, *ARAA*, 28, 561
Soderberg, A. M., *et al.* 2010, Nature, 463, 513

New Horizons in Time-Domain Astronomy
Proceedings IAU Symposium No. 285, 2011
R.E.M. Griffin, R.J. Hanisch & R. Seaman, eds.

© International Astronomical Union 2012
doi:10.1017/S174392131200021X

Cosmic Explosions (Optical)

S. R Kulkarni

Caltech Optical Observatories, Pasadena, CA 91125, USA
email: prkulkarni@mac.com

One of the principal motivations of wide-field and synoptic surveys is the search for, and study of, transients. By transients I mean those sources that arise from the background, are detectable for some time, and then fade away to oblivion. Transients in distant galaxies need to be sufficiently bright as to be detectable, and in almost all cases those transients are catastrophic events, marking the deaths of stars. Exemplars include supernovæ and gamma-ray bursts. In our own Galaxy, the transients are strongly variable stars, and in almost all cases are at best cataclysmic rather than catastrophic. Exemplars include flares from M dwarfs, novæ of all sorts (dwarf novæ, recurrent novæ, classical novæ, X-ray novæ) and instabilities in the surface layers of stars such as S Dor or η Carina. In the nearby Universe (say out to the Virgo cluster) we have sufficient sensitivity to see novæ. In §1 I review the history of transients (which is intimately related to the advent of wide-field telescopic imaging). In §2 I summarize wide-field imaging projects, and I then review the motivations that led to the design of the Palomar Transient Factory (PTF)†. Next comes a summary of the astronomical returns from PTF (§3), and that is followed by lessons that I have learnt from PTF (§4). I conclude that, during this decade, the study of optical transients will continue to flourish (and may even accelerate as surveys at other wavelengths—notably radio, UV and X-ray—come on-line). Furthermore, it is highly likely that there will be a proliferation of highly-specialized searches for transients. Those searches may well remain active even in the era of LSST (§5). I end the article by discussing the importance of follow-up telescopes for transient object studies—a topical issue, given the Portfolio Review that is being undertaken in the US.

1. History: Phase Space

One could say that large surveys of the sky for star positions, stellar photometry and stellar classification (via low-resolution spectroscopy) effectively define the start of the modern era of astronomy. The early returns resulting from the discovery and study of variable stars were stunning. RR Lyrae and Cepheid variables proved to be rather precise yardsticks, and astronomers came to appreciate the physical scale of our Galaxy and eventually the physical scale of the Local Universe.

All transients were initially classified as "nova stella" or new stars, and the abbreviation "novæ" came to be applied specifically to just classical novæ. One of the main goals of the study of novæ was to determine a distance to the Andromeda galaxy using the fact that fast novæ are, on average, brighter than the slow ones. Thus luminosity is converted to time-scale (which, for modest velocities, can be regarded as independent of frame). A large amount of Mt. Wilson 60-inch telescope time was spent by E. Hubble studying novæ in M31.

† PTF is a collaboration of the following entities: California Institute of Technology, Lawrence Berkeley National Laboratory (LBNL), Weizmann Institute of Sciences, Las Cumbres Observatory Global Telescope, Columbia University, Oxford University, Infrared Processing & Analysis Center (IPAC) and UC Berkeley. LBNL is responsible for the image subtraction pipeline, IPAC for the photometric pipeline and U. C. Berkeley for the classification engine.

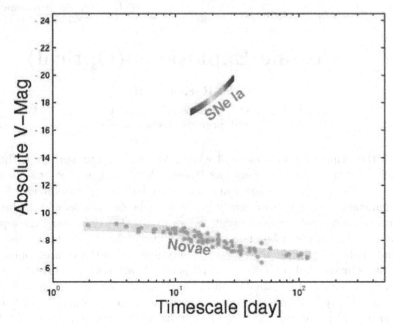

Figure 1. Phase space of classically known non-relativistic explosions in the Zwicky era: novæ and Type Ia supernovæ. The horizontal axis is the decay time-scale and the y-axis is the peak luminosity but shown as absolute magnitude in the V-band. Type II explosions have not been included because the explosion physics is mediated and hence masked by the envelope. Notice the wide gap between novæ and supernovæ.

The modern era of transients, with a focus on controlled cadence and a physics-based enquiry, began with F. Zwicky and W. Baade. Recognizing the importance of the then newly-invented "Schmidt" type wide-field telescope† Zwicky obtained funds from a wealthy family in Pasadena and had constructed an 18-inch telescope using the Schmidt-camera design. The "P18" was the first telescope on Palomar mountain‡. The first major result was the recognition of two distinct families: classical novæ and "super-novæ" (Baade & Zwicky 1934); see Fig. 1. In their very next paper the authors made the bold conjecture that supernovæ mark the transmutation of an aging star into a neutron star, a very compact object (itself a novel hypothesis first proposed by L. Landau in 1931). The resulting enormous release of gravitational binding energy would accelerate some particles to relativistic velocities or cosmic rays. Next, thanks to the systematic survey carried out by Zwicky, families of supernovæ were recognized (Type I and Type II have certainly survived the passage of time)¶.

The success of P18 motivated Zwicky to seek a larger Schmidt-type telescope, and that led to the 48-inch telescope ("P48")†. It saw first light at about the same time as the Palomar 200-inch (circa 1948). P48 undertook the ambitious and comprehensive Northern hemisphere sky surveys (POSS-I and POSS-2). The photographic plates were

---

† Alerted to Zwicky by W. Baade, who knew the inventor Bernhard Schmidt.

‡ The telescope still exists and will soon be moved to the Palomar Museum; the P18 dome is now our aeronomy centre, housing a polar telescope for seeing monitoring.

¶ It would be interesting to revisit types III, IV and V that Zwicky had proposed.

† Currently known as the Samuel Oschin telescope after a benefactor whose gift allowed the telescope to be rejuvenated.

replaced by CCDs with ever increasing sophistication: "3-banger", QUEST (Baltay *et al.* 2007), CFH12K (Cuillandre *et al.* 2001).

2. The Era of (Optical) Synoptic Surveys

It is clear that we are now well into the era of synoptic and/or wide-field astronomy. In the 1-m to 2-m category we have the Catalina Sky Survey, Palomar Transient Factory (PTF), Pan-STARRS-1, SkyMapper and La Silla-QUEST. In the 2-m to 4-m category we have the magnificent Canada-France-Hawaii MegaCam, the VLT Survey Telescope (VST), the One Degree Imager on WIYN (under construction) and Dark Energy Camera on the Blanco 4-m telescope (and expected to be commissioned during 2012). Finally, in the behemoth category we have Suprime-Cam (and soon Hyper Suprime-Cam, HSC) on the Subaru 8.3-m telescope (also expected to be commissioned during 2012). The ultimate in this category will be the Large Synoptic Survey Telescope, an 8.3-m telescope but equipped with an imager with a field of view about five times larger than that of HSC. All in all we have an impressive list of facilities and projects. Given the large to gigantic formats of the imagers (0.1–1 gigapixel) some of these projects have the capacity to keep the entire global astronomical community busy for this entire decade simply with undertaking follow-up or follow-along observations.

3. The Palomar Transient Factory

The Palomar Transient Factory (PTF) was motivated by two considerations (Rau *et al.* 2009). First was the exploration of transients in the sky. A phase-space diagram is useful to summarize the known phenomenology graphically. An explosion has several basic parameters: the energy of the explosion, the mass of ejecta, the velocity of ejecta, the rise time of the explosion, the peak luminosity, and the decay time of the explosion. Peak luminosity and decay time-scales are easily measured, and as such are favourable for the principal axes of the phase space of transients. Fig. 1 shows the "classical" phase space of explosive transients. We were keen to find objects in the nova-supernova "gap".

A second motivation was the great promise of entirely new areas of astronomy:

(*a*) high-energy cosmic rays,
(*b*) high-energy neutrinos,
(*c*) highest-energy photons, and
(*d*) gravitational wave astronomy.

What is common to all these fields (apart from the fact that they represent new frontiers for astronomy and are quite expensive) is that the sources of interest to these facilities are associated with truly spectacular explosions. Furthermore, the horizon (radius of detectability), for reasons either of optical depth (GZK cutoff; $\gamma\gamma \rightarrow e^{\pm}$) or of sensitivity, is limited typically to 100 Mpc. Moreover, these facilities provide relatively poor localization. Thus, for the fields in question the study of explosions in the Local Universe ($d \lesssim 200$ Mpc) is critical for two reasons: for improving the localization via low-energy observations (which usually means in the optical), and for identifying and filtering out the torrent of false positives (because the expected rates of sources of interest is a tiny fraction of the known transients).

In Fig. 2 we display the phase space informed by theoretical considerations and speculations. Basing our surmises on the history of our subject we should not be surprised to find, say a decade from now, that we were not sufficiently imaginative.

PTF (see Law *et al.* 2009) consists of two dedicated telescopes: the P48, with a large field of view (7.2 deg^2), acting as the Discovery Engine, and the Palomar 60-inch (P60) acting as the Photometric Engine. The former is equipped with the re-engineered

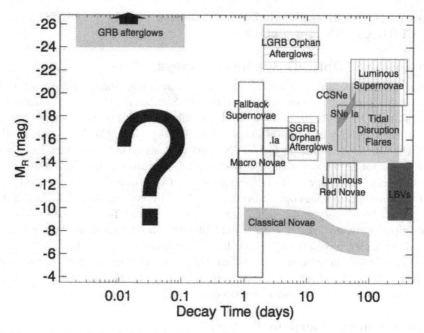

Figure 2. Theoretical and physically plausible possibilities are marked in the explosive transient phase space. The original figure is from Rau *et al.* (2009). The updated figure (to show the unexplored sub-day phase space) is from the LSST Science Book (v2.0).

CFH12K 96 Megapixel CCD mosaic, and the latter with a single 4 Megapixel CCD. The clarity afforded by the singular focus allowed us to optimize PTF for transient studies. Specifically, unlike some projects such as SDSS SN search, PS-1 and SkyMapper, we undertake the search for transients in a single band (R band during most of the month and g band during the darkest period). This simplification alone gives us a factor of five more target throughput.

Next, we adopted a "No Transient Left Behind" strategy. Specifically, a transient without any additional information (such as low-resolution spectroscopy) does not in most instances represent a meaningful advance. After all, it is relatively easy to discover ageing supernovæ, the occasional flare from the numerous dwarf stars, a burp from an accreting system and so on. To address this important issue we employ the P60 to obtain 3-colour photometry to classify crudely the transient and use a variety of criteria for relatively prompt low-resolution spectroscopy on an armada of larger telescopes (the Palomar 200-inch, the KPNO 4-m, the WHT 4.2-m and the Lick 3-m). As a result we have amassed a set of nearly 1500 *spectroscopically classified* supernovæ of which a good fraction was detected prior to maximum.

Given that follow-up is at premium, having a small sample of transients with desired properties is more valuable than a large sample of transients with a pot-pourri of properties. Thus, the choice of pointings and cadence control are critical. We have scoured around the sky for regions with large local over-densities ($d \lesssim 200\,\mathrm{Mpc}$). The special hundred PTF pointings contain 4 times more light than randomly chosen pointings (Kasliwal 2011). Cadence control is even more important. For instance, the logistics of obtaining UV spectra of Type Ia supernovæ require that the supernovæ be identified 10 to 14 days prior to peak. In 2010 we focused on finding such supernovæ, and we now have three dozen HST UV spectra of nearby Type Ia's—an order of magnitude increase in the sample size. Our sample enables us to study rest-frame spectra of both nearby and cosmological Type Ia supernovæ (Cooke *et al.* 2011).

Cadence control is essential for decreasing the latency between *detection* (by which I mean that a transient has been identified with a reliable degree of certainty) and *discovery* (by which I mean that the astronomer has a useful idea of the nature of the transient). At the very minimum we should know if the source is Galactic or extra-galactic. At the next level, it would be useful to have the first estimate of sub-typing (e.g., flare star/DN/CV; Type Ia/Ibc/II SN)). During 2011 we made efforts to decrease the latency and were (very luckily) rewarded with the discovery of a Type Ia supernova in the very local Universe (PTF11kly in Messier 101) just 10 hours after the explosion (Nugent *et al.* 2011). The proximity and near natal discovery allowed us and other astronomers to shed valuable new light on the progenitor of a Type Ia supernova.

With 38 refereed publications to date† (a mean rate of 1 publication per month) PTF has been productive. PTF supported the thesis projects of three students. In addition to the two results summarized above, key discoveries and findings include the clear identification of a class of luminous supernovæ whose spectra show no hydrogen (Quimby *et al.* 2011), the gradual "colouring of the phase space", as shown well in Fig. 1 of Kasliwal (p. 62), and an apparent new type of nuclear event (PTF10iya; Cenko *et al.* 2011).

4. Lessons Learnt

• First and foremost, a clear vision for the project must be articulated. The statement that it behooves astronomers to explore phase space (wrongly attributed to Zwicky) is frequently heard at august meetings as the one we have just had. Even a careless reading of Zwicky's morphological approach to problems stresses the importance of a sensible exploration of phase space and consistency with physics. In this regard, Fig. 1 on page 62 was an essential exercise in motivating PTF.

• Merely identifying a transient does not constitute a discovery. We are now operating in a régime in which transients are a dime a dozen. Going forward, culling either by brute force (spectroscopy) or by cadence control is essential.

• A horizontal structure is best suited to the pursuit of transients. Such a structure allows for key project teams that are in effect independently operated to undertake follow-up efficiently and rapidly. However, nature is rarely so neatly organized. Undoubtedly, there will be overlap by the way of objects and interests. A strong leadership is essential to resolve such situations.

• Software pipelines have to be fully functional before the searches begin. This statement is a truism, but apparently is not always satisfied in practice.

• The collaboration should be flexible enough to change search strategies in order to respond to new transients discovered at the edge of a currently studied phase space. In 2009 and 2010 we focused on a standard SN programme, planning for a five-day cadence. On crossing the 1,000-SN mark we increased the cadence; with each ratcheting down of the latency we had interesting finds, and the final prize was PTF11kly in M101 (Nugent *et al.* 2011).

5. Future: Beetles and Behemoths

The following question inevitably arises at meetings such as this: is there a need for more synoptic surveys since even a modest-aperture synoptic survey can generate more

† http://www.astro.caltech.edu/ptf

data than can be followed up? This is a meritorious question and my answer is that there is considerable room for highly-focused projects. I list three examples. (1) The Catalina Sky Survey, a survey based on 1-m telescope equipped with a routine CCD detector, is remarkable for its harvest of NEOs and pinpointing (with considerable heads-up) the fiery entry of 2008TC3 (Boattini *et al.* 2009). (2) PTF, with its laser-like focus on transients has, as remarked in §3, been very productive and has also made noticeable contributions to our field. Finally, (3) the CFHT SN Legacy Survey (CFHT; Conley *et al.* 2011) has made deep contributions to Ia cosmology and the SDSS Stripe 82 project has made contributions across board—SN cosmology, AGN variability, stellar variability, tidal disruption events (e.g., van Velzen *et al.* 2011). Looking at these gains I would venture to say that focused projects have the ability to trump much larger facilities which are subject to competing cadence demands.

The second reason why I remain bullish on the continued value of the current portfolio of synoptic surveys is that the systematic and sensible exploration and study of the phase space has only just begun. The phase space with decay time of less than a day (for which there are several plausible models) and the entire physics of the rise time of explosive transients is in effect wide open.

Finally, the transient game is currently dominated by optical synoptic facilities and projects. The next to join this club would be radio facilities at both metre and decimetre wavelengths (EVLA, LOFAR, MWA, upgraded GMRT, APERTIF, MeerKAT). With some luck, synoptic space-based projects may also happen during this decade (e.g., LIM-SAT† in the UV, reuse of WISE for thermal IR searches and a Lobster-type mission in the X-rays). One can easily imagine joint studies with optical facilities.

Projects such as PTF, CSS and PS-1 demonstrate that there is ample scope for 1-m to 2-m-class surveys to continue well into this decade. The magnitude limit of 21 is ideally suited to classification spectroscopy (from which I do not see a simple escape). Going fainter is only an advantage if one has the ability to discriminate between the "unknown unknowns" against the dense fog of known transients. Recall that follow up at the 22-mag requires 6 times as much as follow up time (on the same telescope) as a 21-mag event.

However, as noted several times, detecting a transient is merely the first step. Progress only happens when follow-up observations can be undertaken. Thus, follow-up telescopes are *an essential requirement* for transient studies. Currently, in the US there is a debate over the future of existing facilities. It is easy (trivial) to make an argument for ever larger telescopes. However, for many fields consolidation of resources (with strong focusing of the same) is more important. Indeed, some of the great discoveries over the past two decades have come from concerted efforts on modest-sized telescopes (e.g., planets around normal stars). PTF is an example of repurposing existing telescopes for singular goals at a modest cost.† Both discovery (usually arising from focused or smaller facilities) and depth of understanding (usually requiring larger facilities) are essential. However, the balance has tipped far too much for larger telescopes.

So enthused am I with the prospects and promises of focused programmes (along with appropriate follow-up) that, along with my colleagues, I am now proposing a second-generation of PTF. We propose to equip fully the entire focal plane of P48 (40 deg^2), build an efficient IFU ultra-low resolution ($R = \lambda/\Delta\lambda \sim 100$) spectrograph‡ optimized for classification, and a Rayleigh-scattering AO system¶ for sensitive photometry even in

---

† This is a cluster of UV telescopes with a total of 1,000 square degrees and promoted by an ad hoc group of astronomers from Israel, Caltech, India and Canada.
† The capital and 4-year operating cost of PTF is under $3M.
‡ http://sites.google.com/site/nickkonidaris/sed-machine
¶ http://www.astro.caltech.edu/Robo-AO

crowded host galaxy fields. In addition to P60 we are seeking another telescope so that the P48 along with these two telescopes will be one of the first new-generation transient facilities. The primary focus of this facility will be to probe the sub-day phase space (marked by a big "?" in Fig. 2). I propose to name it the "Zwicky Transient Facility" (ZTF) after the founder of our field. With some effort it should be possible for this facility to see first light by 2015. Should interesting sub-minute bursts be discovered, then one can imagine upgrading ZTF with CMOS. It is thus hoped that ZTF, a mere beetle, will prove productive and interesting even as general-purpose behemoths come on line.

Acknowledgements

I would like to acknowledge the National Astronomy Observatory of Japan and the Japan Society for Promotion of Science for hosting my sabbatical stay in Japan during which period this article was completed.

I would like to thank all the members of the PTF experiment. It is their hard work and creativity that has made PTF productive.

I would further like to acknowledge the key role played by the following in the PTF: L. Bildsten, J. Bloom, S. B. Cenko, A. Gal-Yam, M. Kasliwal, R. Laher, N. Law, P. Nugent, E. Ofek, R. Quimby and J. Surace. The hard-working and knowledgeable staff of Caltech Optical Observatories made it possible for rapid re-engineering of the focal plane of P48 and refurbishment of P48. The very low down-time, despite the age of P48 and P60, speaks superbly of the deep knowledge and hard work of the Palomar Mountain staff. D. Fox, B. Cenko and M. Kasliwal were in charge of the automation, robotization and optimization of P60, and E. Ofek was (and continues to be) the sequencer for P48.

PTF got a great start with CFH12K, and I sincerely appreciate the generosity of the builders of CFH12K, Christian Veillet (the then Director of CFHT) and the CFHT Corporation. I would like to thank D. Frail, G. Helou and T. Prince for helping me debate strategies in this rapidly evolving field. Finally, I would like to acknowledge and thank W. Rosing, whose initial "investment" in PTF allowed me to garner funds from other parties and complete the funding for PTF in record time.

References

Baade, W., & Zwicky, F. 1934, *PNAS*, 20, 254
Baltay, C., *et al.* 2007, *PASP*, 119, 1278
Boattini, A., *et al.* 2009, *AAS-DPS*, 41, #9.02
Cenko, S. B., *et al.* 2011, arXiv1103.0779
Cooke, J., *et al.* 2011, *ApJ*, 727, L35
Conley, A., *et al.* 2011, *ApJS*, 192, 1
Cuillandre, J.-C., Starr, B., Isani, S., & Luppino, G. 2001, *ExA*, 11, 223
Kasliwal, M. M. 2011, PhD Thesis, Caltech
Law, N. M., *et al.* 2009, *PASP*, 121, 1395
Nugent, P. E., *et al.* 2011, *Nature*, 480, 344
Quimby, R. M. *et al.* 2011, *Nature*, 474, 487
Rau, A., *et al.* 2009, *PASP*, 121, 1334
van Velzen, S., *et al.* 2011, *ApJ*, 741, 73

New Horizons in Time-Domain Astronomy
Proceedings IAU Symposium No. 285, 2011
R.E.M. Griffin, R.J. Hanisch & R. Seaman, eds.

© International Astronomical Union 2012
doi:10.1017/S1743921312000221

Systematically Bridging the Gap between Novæ and Supernovæ

Mansi M. Kasliwal[1,2] (on behalf of the Palomar Transient Factory Collaboration)

[1] Observatories of the Carnegie Institution for Science, Pasadena, CA 91101, USA
[2] Dept. of Astrophysical Sciences, Princeton University, Princeton, NJ 08544, USA
email: mansi@astro.caltech.edu

Abstract. Until recently, the venerable field of cosmic explosions has been plagued with a glaring six-magnitude luminosity gap between the brightest novæ and the faintest supernovæ. A key science driver of the Palomar Transient Factory was a systematic search for optical transients that are fainter, faster and rarer than supernovæ. Theorists predict a variety of mechanisms to produce transients in that "gap", and observers have the best chance of finding them in the local universe. The talk presented the discoveries and the unique physics of cosmic explosions which bridge that gap between novæ and supernovæ. As Fig. 1 illustrates, there is now evidence of multiple, distinct populations of rare transients in the "gap".

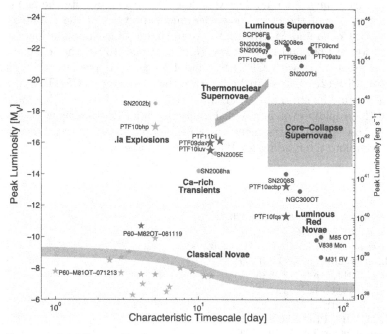

Figure 1. Framework of Cosmic Explosions expressed as peak luminosity versus characteristic time-scale. In 2005 we only knew about classical novæ and supernovæ (grey bands). In the past 5 years systematic searches, serendipitous discoveries and archival searches have uncovered a plethora of novel, rare transients. Several new classes are emerging, and the governing physics is being widely debated: luminous red novae (the electron-capture induced collapse of rapidly rotating O–Ne–Mg white dwarfs?), luminous supernovæ (magnetars or pair instability explosions?), Type Ia explosions (helium detonations in ultra-compact white-dwarf binaries), calcium-rich halo transients (helium deflagrations or fallback onto a black hole?) (from Kasliwal, M.M., 2011, PhD Thesis, California Institute of Technology).

New Horizons in Time-Domain Astronomy
Proceedings IAU Symposium No. 285, 2011
R.E.M. Griffin, R.J. Hanisch & R. Seaman, eds.

© International Astronomical Union 2012
doi:10.1017/S1743921312000233

Supernovæ and Transients with Euclid and the European ELT

Isobel Hook[1,2]

[1] Oxford Astrophysics, University of Oxford, Oxford, OX1 3RH, U.K.
email: imh@astro.ox.ac.uk

[2] INAF–Osservatorio Astronomico di Roma, Monte Porzio Catone, Italy

Abstract. The prospects are described for studies of large samples of supernovæ and other variable objects with two proposed future facilities: (1) the European Extremely Large Telescope, a general-purpose 40-m-class ground-based optical-IR telescope, and (2) Euclid, an M-class mission within ESA's Cosmic Vision programme, primarily for cosmology. The capabilities and status of the two facilities are briefly described. Their suitability for the study of time-varying objects in general, and of supernovæ in particular, is discussed. It is shown that Euclid has the potential for NIR imaging of a few thousand Type Ia supernovæ to intermediate z, while the E-ELT will be capable of spectroscopic and classification measurements of Type Ia supernovæ to $z = 4$.

Keywords. Telescopes

---

1. The European Extremely Large Telescope

1.1. *Overview*

The European Extremely Large Telescope (E-ELT) project aims to design and construct a 40-m class optical-IR telescope, which will be the largest such telescope in the world. The project is being led by ESO on behalf of its member states. It has been identified by the ASTRONET process as a top priority of European ground-based astronomy (Bode, Cruz & Molster 2008).

The E-ELT has a very broad science case. The scientific drivers for the telescope have been developed over a period of many years in collaboration with the project's Science Working Group. The wider astronomical community has also provided input through the Design Reference Science Plan (DRSP, Kissler-Patig, Yoldas & Liske (2009)) and through a series of ESO and OPTICON-sponsored workshops. The workshops have highlighted synergies with ALMA, JWST, SKA and survey facilities in general.

The key driving themes that have emerged range from extra-solar planets to studies of fundamental constants, dark matter and dark energy. There are many science cases that involve the time domain, some examples of which are listed below:

- Identification of variable sources from other facilities
- Solar system: monitoring weather and volcanic activity
- Exo-planets: radial velocities, direct imaging and transit measurements
- The Galactic Centre: orbits of stars close to the central black hole
- Supernovæ
- Gamma-Ray Bursts
- Expansion of the Universe: measuring the effects of Dark Energy in real time

In addition, if fast detectors capable of sub-minute and sub-second resolution are installed, studies of extreme physics (pulsars, neutron stars, black holes) stellar phenomena, GRBs, transits and occultations also become possible (Shearer *et al.* 2010). Finally, future unknown science programmes that we cannot imagine today, may well turn up the most exciting discoveries.

1.2. *Project Status*

The telescope optical design is a 5-mirror design with a 39-m diameter primary mirror. In April 2010 Cerro Armazones in Chile was adopted as the baseline site. As well as being an excellent site for astronomical observations, Cerro Armazones is about 20 km from the VLT site at Cerro Paranal, providing significant advantages of joint operations of the two facilities. The instrument plan is under development. Phase A studies for 8 instruments and 2 adaptive optics (AO) systems have been completed and two "first light" instruments have been selected: a diffraction-limited near-IR camera that will be fed by AO, and an integral field spectrograph capable of operating with or without AO. The full instrument suite for the E-ELT will be built up over first decade.

Phase B of the E-ELT project was completed in December 2010, following the successful completion of an external design review. Since then the project has been through a cost and risk reduction exercise, resulting in the diameter of the primary mirror being reduced to 39-m (from 42-m, although the precise definition of the diameter also changed). The construction cost is approximately 1 billion Euros, to be shared by the ESO member states. The construction proposal will be presented to ESO Council in December 2011. If approved, construction could start in 2012, leading to first light early in the next decade.

2. Overview and Status of EUCLID

EUCLID is a proposed 1.2-m optical-IR space telescope within ESA's Cosmic Vision 2015–2025 programme. Its primary science goal is precison measurements of cosmological parameters via weak lensing and galaxy clustering techniques. The instrumentation consists of an optical imager and a near-infrared (NIR) imager and spectrograph, all with a field of view of 0.5 deg.^{2} The satellite will be launched by a Soyuz rocket and will operate at the L2 Lagrange point for a 6-year mission duration.

EUCLID will carry out two main surveys, Wide and Deep. The Wide survey will cover > 15000 deg.^{2}, with imaging in a single broad optical (R+I+Z) filter to a depth of H(AB)=24.5 (10σ for a point source), imaging in three near infrared bands (Y, J, H) to a depth of H(AB)=24 (5σ, extended source) and NIR slitless spectroscopy to a depth of $3 \times 10^{-16} \mathrm{cm}^{-2}\mathrm{s}^{-1}$ (3.5σ unresolved line flux). EUCLID also has a potential extension to carry out a Galactic-plane survey.

The Deep survey will cover > 40 deg.^{2}, reaching 2 mags deeper than the Wide survey in both optical and NIR imaging and NIR Spectroscopy. The Deep field is primarily for calibration purposes but will enable a vast range of additional science. The depth will be built up by 40 repeat visits, each of which will reach the same depth as the Wide survey. The exact strategy and timing of these observations has yet to be optimised. In the context of this Symposium, the repeat visits to the Deep field are particularly interesting in that they will facilitate the detection of variable and transient objects.

EUCLID has been proposed as an M-class mission, with the Definition Study report ("Red book") submitted in July 2011. ESA's downselection process in October 2011 resulted in selection of EUCLID for the second M-class launch slot. The mission is now working towards the important milestone of its adoption by ESA in June 2012. Launch would then take place in 2019.

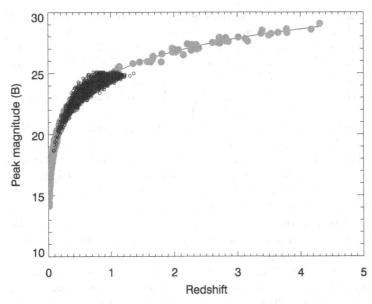

Figure 1. Projected Hubble diagram for Type Ia supernovæ including simulated data from E-ELT and EUCLID. The grey circles with $z < 1$ represent the sample that already exists, or will exist within the next few years. These are almost entirely optical observations. The potential sample of SNe with NIR measurements from EUCLID are shown as black dots (simulated data provided by K. Maguire). The circles with $z > 1$ represent the sample that could be obtained with a combination of JWST imaging and E-ELT spectroscopic follow-up.

3. Example of Science Case: Type Ia Supernovæ

One of the biggest problems facing modern physics is the lack of understanding of Dark Energy which appears to be causing the acceleration of the Universe. Measurements from different techniques including Type Ia supernovæ (SNe Ia), galaxy clustering and the Cosmic Microwave Background are consistent with dark energy being equivalent to the cosmological constant (see, for example, Sullivan *et al.* 2011), and yet there is no natural explanation for the size of the measured value of its energy density.

SNe Ia are one of the best-known distance indicators; their high luminosity enables them to be used to measure the geometry of the Universe to cosmological distances. Currently there are over 500 well-measured SNe with $z < 1$. The main limiting factors in SN cosmology are now the quantity and quality of the nearby SN Ia sample and the related issues of calibration (partially caused by the fact that the nearby and distant SN sets have traditionally been measured through different filter sets) and extinction by dust (in particular disentangling this effect from intrinsic variations in SN luminosity and colour).

Improvements in the nearby sample are now being made by searches such as the Palomar Transient Factory, Skymapper and PanSTARRS. The question of dust extinction and intrinsic luminosity/colour variations is being, or will be, addressed by observing at longer wavelengths where the effects of dust are less pronounced. In total about 80 distant SNe have been measured in the NIR (Freedman *et al.* 2009, Nobili *et al.* 2009, Riess *et al.* 2000). In the next few years we can expect of the order of 100 more to be measured by VISTA combined with optical data from the Dark Energy Survey.

EUCLID has the potential to make more than an order of magnitude improvement in the statistics of SNeIa measured in the near-IR, up to $z \sim 1$, provided a suitable survey strategy is implemented for the repeat visits of the Deep field. Simulations have been carried out in order to estimate the number and redshift distribution of SNe observed in the EUCLID deep field. One example is shown in Fig. 1, in which the Deep survey of 6 months total time is divided into periods with $\sim$ 4-day and $\sim$ 10-day cadences. In this example EUCLID could measure good NIR light curves (at least 4 points per curve) and colours of about 2800 Type Ia SNe to $z \sim 1.2$, of which the $\sim$ 1700 with $z < 0.8$ would form an unbiased sample for cosmology (Fig.1). We stress that this is just one example, and that the final strategy for the deep field survey has yet to be defined.

At even higher redshifts ($z > 2$) the SNIa Hubble diagram is completely unmeasured. A basic check of the behaviour of the expansion history of the Universe in the matter-dominated regime could be made by observing SNeIa at higher redshifts than is currently possible. Such SNe would be within reach of discovery by JWST (again, assuming sufficient time were dedicated to a search). SNe are point sources and so would benefit from the AO correction and the full spatial resolution of the E-ELT. Simulations show that the E-ELT would be capable of obtaining spectra of SNeIa out to $z = 4$ (Hook (2010)), using an AO-assisted integral field spectrograph such as that selected as one of the first-light instruments for E-ELT (based on the HARMONI concept, Tecza et al. 2009). The simulations show that the E-ELT spectra will be of sufficient quality to determine the host-galaxy redshift and, crucially, to determine the type of supernova by detecting the presence of the Si II 4000 Å feature (in the observed K band spectrum at redshift of 4). Observations of a sample of 50 SNeIa in the redshift range $1 < z < 4$ requires 400 E-ELT hours spread over a period of 5 years.

4. Conclusions

Two major new facilities are expected to be in operation around the turn of the decade: E-ELT, a general purpose giant telescope, and EUCLID, a survey machine for cosmology and legacy science. Both will operate at optical and near-infrared wavelengths and will be immensely powerful for a wide range of science, including time-domain astronomy. That has been illustrated with the example of observations of distant supernovæ for cosmology, where EUCLID has the potential for NIR imaging of a few thousand SNeIa to intermediate z, and E-ELT will be able to make spectroscopic measurements of SNeIa to $z = 4$.

References

Bode, M., Cruz, M. J., & Molster, F. J. (Eds.), 2008, *The ASTRONET Infrastructure Roadmap*, www.astronet-eu.org/IMG/pdf/Astronet-Book.pdf

Freedman, W., et al., 2009, *ApJ*, 704, 335

Hook I., 2010, in: A. Shearer (ed.), *High Time Resolution Astrophysics IV: The Era of Extremely Large Telescopes, Proceedings of Science*, 108

Kissler-Patig, M., Kupcu Yoldas, A., & Liske, J. 2009, *ESO Messenger* 138, 11

Nobili, S. et al., 2009, *ApJ*. 700, 1415

Riess, A., et al., 2000, *ApJ*, 536, 62

Shearer, A., et al.. 2010, arXiv:1008.0605v2

Sullivan, M., et al., 2011, *ApJ*, 737, 102.

Tecza, M., Thatte, N., Clarke, F., & Freeman, D., 2009, in: A. Moorwood (ed.), *Science with the VLT in the ELT Era*, (Netherlands: Springer), p. 267

New Horizons in Time-Domain Astronomy
Proceedings IAU Symposium No. 285, 2011
R.E.M. Griffin, R.J. Hanisch & R. Seaman, eds.

© International Astronomical Union 2012
doi:10.1017/S1743921312000245

Search for Electromagnetic Counterparts to LIGO-Virgo Candidates: Expanded Very Large Array† Observations

Joseph Lazio[1], Katie Keating[2], F. A. Jenet[3] and N. E. Kassim[4]

[1] Jet Propulsion Laboratory, California Institute of Technology, Pasadena, CA 91109, USA.
email: Joseph.Lazio@jpl.nasa.gov

[2] National Research Council, Naval Research Laboratory, Washington, DC 20375, USA.

[3] Center for Gravitational Wave Astronomy, University of Texas, Brownsville, TX 78520, USA

[4] Remote Sensing Division, Naval Research Laboratory, Washington, DC 20375, USA

Abstract. This paper summarizes a search for radio-wavelength counterparts to candidate gravitational-wave events. The identification of an electromagnetic counterpart could provide a more complete understanding of a gravitational-wave event, including such characteristics as the location and the nature of the progenitor. We used the Expanded Very Large Array (EVLA) to search six galaxies which were identified as potential hosts for two candidate gravitational-wave events. We summarize our procedures and discuss preliminary results.

Keywords. Gravitational waves, methods: observational, radio continuum: general

---

1. Gravitational-Wave Astronomy and the Time Domain

Gravitational waves (GWs) are fluctutions in the space-time metric, equivalent to electromagnetic waves resulting from fluctuations in an electromagnetic field. Because of the weakness of the gravitational force, however, a laboratory demonstration of GWs comparable to Hertz's demonstration of electromagnetic waves is not possible. Indeed, the characteristic scale for the luminosity of a GW source is $L_0 = 2 \times 10^5 \, M_\odot \, c^2 \, s^{-1}$, indicating immediately that the generation of GWs will occur in astrophysical environments in which large masses are moving at high velocities.

Precise timing of pulses from the radio pulsar PSR B1913+16, which is one member of a double neutron star system, has already revealed indirect evidence for GWs (Hulse & Taylor 1975; Weisberg *et al.* 2010). In this system, the rate of orbital-period decay as predicted by general relativity is consistent with that observed from the pulsar timing measurements. Since the discovery of that system, other pulsars in neutron star–neutron star binaries have also been discovered, and the predicted levels of orbital-period decay from general relativity remains consistent with those from the measurements.

From the standpoint of time-domain astronomy, many of the other predicted sources of GWs are rapidly time-varying phenomena. They include the mergers of compact objects (neutron star–neutron star, neutron star–black hole and black hole–black hole mergers), asymmetric supernovae, rapidly rotating asymmetric neutron stars, and exotic objects such as oscillating cosmic strings.

While evidence for GWs remains indirect, the promise of direct detection of GWs has excited considerable international interest. In Europe, the AstroNet *A Science Vision*

---

† The National Radio Astronomy Observatory is a facility of the National Science Foundation operated under cooperative agreement by Associated Universities, Inc.

for European Astronmy posed "Can we observe strong gravity in action?" as a key question for this decade, while in the U.S., gravitational-wave astronomy was identified as a scientific frontier discovery area in the *New Worlds, New Horizons in Astronomy and Astrophysics* Decadal Survey. In this respect, GW astronomy is similar to the experience in opening up new spectral windows in the electromagnetic spectrum. As each new spectral window has been opened, entirely new classes of sources have been discovered. Indeed, one of the most surprising results of GW astronomy would be if *no* new classes of sources were discovered as this new window on the Universe is opened.

2. Electromagnetic Counterparts to Gravitational Wave Events

Supernovæ have a well-known electromagnetic signature, and asymmetric supernovæ are predicted to be gravitational wave emitters. It is therefore natural to anticipate that other gravitational-wave emitters might also display electromagnetic counterparts. Moreover, the identification of electromagnetic counterparts to GW events would have a number of benefits:

• Precise localization of the event, potentially at the sub-arcsecond level, is possible via electromagnetic observations. Such localization may be crucial in understanding the nature of the event (for instance, whether in the nucleus of a galaxy or in its outskirts).

• The characteristics of the electromagnetic counterpart, such as its spectrum, are likely to constrain the environment or progenitor, or both, of the GW event.

In many respects, determining the electromagnetic counterpart to a GW event is analogous to determining the (electromagnetic) spectrum of a transient discovered in one band and then following up at others (e.g., a gamma-ray burst discovered at gamma-ray wavelengths and then followed up at X-ray, optical and radio wavelengths).

Our focus on radio-wavelength counterparts is motivated by several considerations:

• Non-thermal, high-energy particles often produce radio emission easily, particularly in the presence of a magnetic field (e.g. cyclotron and synchrotron emission).

• Radio observations can yield precise astrometry, in the best cases obtaining positions at the milliarcsecond level.

• Radio wavelengths are unaffected by dust obscuration, either from the immediate environment of the event or from intervening objects.

• Radio telescopes can observe during the day, offering rapid follow-up.

• If a GW event also produces a radio burst or pulse, the propagation of this pulse will be delayed by its propagation through the ionized interstellar (and intergalactic) medium. Such delays can be minutes to hours, depending upon the electron column density along the line of sight, but they potentially allow for detailed follow-up of the burst.

3. LIGO-Virgo Observations

The radio-wavelength observations that we describe below are based on coordinated observations between the Laser Interferometric Gravitational-wave Observatory (LIGO) and the Virgo that occurred during the Autumn of 2010 (LIGO Scientific Consortium & Virgo Collaboration 2011). LIGO has two elements, one located in Hanford, WA, USA, and the other in Livingston, LA, USA, while Virgo has one element located near Pisa, Italy. Together they form a 3-element interferometer.

The LIGO-Virgo interferometer measures differences of arrival times. Analyses of test waveforms injected into the LIGO-Virgo processing pipeline indicate that a candidate's position can be localized only to a region of order 10 deg^2. Ordinarily, such a large uncertainty region could not be useful in searches for a radio counterpart with the current generation of telescopes, because their fields of view are too small. However, the most likely sources that the LIGO-Virgo interferometer could detect at reasonable signal-to-noise ratios would have occurred within 40 Mpc. If one assumes that a candidate event is associated with a galaxy, then the typical number of galaxies within the certainty region of a GW candidate is only three. This small number of galaxies can be usefully searched.

4. Expanded Very Large Array Observations

The Expanded Very Large Array (EVLA) is a 27-element radio interferometer operating between 1 and 50 GHz. It has been the focus of a recent major upgrade (which is nearly complete), and it is being commissioned with science programmes that are now well established. For the purpose of these observations, the EVLA offers a number of attractive features:

• The wide frequency (wavelength) coverage potentially allows "tuning" of the observations to a frequency well matched to the expected physics. In this case, we observed at 5 GHz (λ6cm), at which both expanding synchrotron fireballs and relativistic jets are likely to be detectable.

• At our observational frequency, the nominal field of view (approximately 7') is well matched to the size of most local galaxies. In practice, the field of view is usually defined as the region over which the antenna response is at least half its peak value; sources outside the nominal field of view can still be detected, provided that they are sufficiently strong to compensate for the decreased antenna response. Accordingly, in order not to miss a potential candidate, we imaged a much larger region—typically 30'.

• The angular resolution of the EVLA can be adjusted by moving the individual elements. During most of our observations the angular resolution achieved was about 4″, which provides about 0.″4 localization (equivalent to 20 pc at 10 Mpc) for a reasonable signal-to-noise ratio.

We have now conducted three epochs of observations for each of the two LIGO-Virgo candidates, observing all of the nearby galaxies within the uncertainty region. Fig. 1 (left) shows the field around one of the galaxies; Fig. 1 (right) displays the flux curves of the detected sources. Typically we detected ∼6 sources in the field of each galaxy. That small number of sources is consistent with the number of extragalactic sources expected (Windhorst 2003), but does not exclude the possibility that one of the radio sources is a counterpart to the LIGO-Virgo candidate.

Data acquisition and reduction of all three epochs for both candidates has only recently been concluded. In assessing the reality of any potential radio-wavelength counterpart to either LIGO-Virgo candidate, other potential sources of variability must also be considered. On the time-scales and cadence of our observations, it is unlikely that intrinsic variability of any active galactic nuclei (AGN) in the field of view would represent a source of contamination. However, at these wavelengths, refractive interstellar scintillation due to the Galaxy's interstellar medium is a potential source of contamination (for instance, an AGN unrelated to the LIGO-Virgo candidate might show variability on the time-scales of our observations).

5. Future

LIGO and Virgo are currently being upgraded. When that is completed (∼2014) they will be more sensitive and able to probe to larger distances. We expect a much larger number of galaxies to be identified as hosts, and as a search of all of them with the EVLA

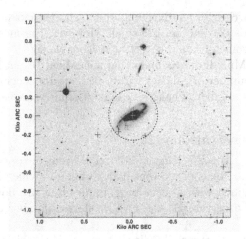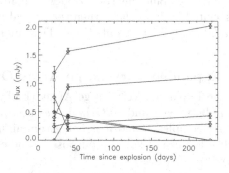

Figure 1. Left: Radio wavelength (λ6cm) sources detected in the field of view of one of the galaxies representing a potential host of a candidate LIGO-Virgo gravitational-wave event. Plusses represent the detected sources, and the circle represents the approximate "field of view," defined as the half-power point of the antenna response. Right: Radio flux curves for those sources.

would then be very time consuming (see also Metzger & Berger 2011), a change in observing strategy will be needed. An approach similar to current follow-ups of supernovae could be profitable: if an optical counterpart is found, the EVLA could used to assess the radio-wavelength properties of the counterpart.

Later in the decade other radio-wavelength facilities are likely to be available, thus presenting additional opportunities. Among them are (1) the Low Frequency Array (LOFAR), which could conduct wide-field "blind" searches at metre wavelengths (30–240 MHz) for northern-hemisphere counterparts, (2) the Karoo Array Telescope (MeerKAT), which could conduct southern-hemisphere follow-up observations similar to those of the EVLA, and (3) the Australian Square Kilometre Array Pathfinder (ASKAP), which could conduct wide-field "blind" searches at decimetre wavelengths ($\sim$ 1 GHz) for southern-hemisphere counterparts.

Acknowledgements

Part of this research was carried out at the Jet Propulsion Laboratory, California Institute of Technology, under a contract with the National Aeronautics and Space Administration. Support for basic research in radio astronomy at the NRL comes from 6.1 Base funding. An NRC-NRL Research Associateship to K.K. and a CAREER grant from the NSF (award AST 0545837) to F.A.J. are also acknowledged.

References

Hulse, R. A. & Taylor, J. H. 1975, *ApJ*, 195, L51
LIGO Scientific Collaboration & Virgo Collaboration, 2011, arXiv:1109.3498
Metzger, B. D. & Berger, E. 2011, *ApJ*, in press; arXiv:1108.6056
Weisberg, J. M., Nice, D. J., & Taylor, J. H. 2010, *ApJ*, 722, 1030
Windhorst, R. A. 2003, *New Astron. Rev.*, 47, 357

New Horizons in Time-Domain Astronomy
Proceedings IAU Symposium No. 285, 2011
R.E.M. Griffin, R.J. Hanisch & R. Seaman, eds.

Explosions on a Variety of Scales

Lars Bildsten

Kavli Institute for Theoretical Physics, University of California,
Santa Barbara, CA 93106, USA

email: Bildsten@kitp.ucsb.edu

Invited Talk

Summary. The theoretical community is beginning to appreciate (and predict) the potential diversity of explosive outcomes from stellar evolution, while the supernovæ surveys are finding new kinds of supernovæ. This talk described two such new supernovæ. The first are ultraluminous core collapse supernovæ with radiated energies approaching 10^{51} ergs. The talk went on to present our recent work that explains these events with late-time energy deposition from rapidly rotating, highly magnetized neutron stars: magnetars. It concluded with our theoretical work on helium shell detonations on accreting white dwarfs that predict a new class of supernovæ called ".Ia's". The first such candidate may well have been found by the Palomar Transient Factory.

Transients with Pan-STARRS–1

Stephen Smartt (and the PSI Science Consortium)

Astrophysics Research Centre, School of Mathematics and Physics,
Queen's University Belfast, Belfast, BT7 1NN, UK

email: s.smartt@qub.ac.uk

Invited Talk

Summary. The Pan-STARRS1 sky survey began operations in May 2010. Since then we have discovered more than 4000 optical transients in the Medium Deep Fields (11 fields of approximately 7 square degrees). Several hundred have been confirmed with optical spectroscopy. The talk discussed the challenges involved in automating reliable difference imaging, object detection and classification. The science part of the talk then focussed on the discovery of high-redshift, very luminous supernovæ which appear to arise in faint hosts.

New Horizons in Time-Domain Astronomy
Proceedings IAU Symposium No. 285, 2011
R.E.M. Griffin, R.J. Hanisch & R. Seaman, eds.

© International Astronomical Union 2012
doi:10.1017/S1743921312000269

Light Echoes of Transients and Variables

Armin Rest

Space Telescope Science Institute, 3700 San Martin Drive, Baltimore, MD 21218, USA
email: `arest@stsci.edu`

Summary. Tycho Brahe's observations of a supernova in 1572 challenged the contemporaneous European view of the cosmos that the celestial realm was unchanging. 439 years later we have once again seen the light that Tycho saw, as some of the light from the 1572 supernova is reflected off dust and is only now reaching Earth. These light echoes, as well as ones detected from other transients and variables, give us a very rare opportunity in astronomy: direct observation of the cause (the supernova explosion) and the effect (the supernova remnant) of the same astronomical event. Furthermore, in some cases we can compare light echoes at different angles around a supernova remnant, and thus investigate possible asymmetry in the supernova explosion. In addition, in cases where the scattering dust is favorably positioned, the geometric distance to the SN remnant can be determined using polarization measurements. These techniques have been successfully applied to various transients in the last decade, and the talk gave an overview of the scientific results and techniques, with a particular focus on the challenges we will face in the current and upcoming wide-field time-domain surveys.

A New Class of Relativistic Outbursts from the Nuclei of Distant Galaxies

S. B. Cenko[1], S. R. Kulkarni[2], D. A. Frail[3], and J. S. Bloom[1]

[1] University of California, Berkeley, CA 94720, USA
email: `cenko@berkeley.edu`

[2] California Institute of Technology, Pasadena, CA 91125, USA

Summary. The recent discovery of the transient source Swift J164449.3+573451 (Swift J1644) has revealed a potentially new class of high-energy outbursts. Like long-duration gamma-ray bursts, these sources exhibit prompt, catastrophic energy release which drives relativistic outflows. However, the central engine powering those events is the supermassive black hole at the centre of a normal galaxy. While not unequivocal, the data can best be explained by the tidal disruption of a star which passes too close to the nuclear black hole, creating an episode of hyper-critical accretion. Motivated by this fascinating discovery, we have searched for new examples that have the necessary properties (luminous X-ray and/or radio, long-lived high-energy emission, evidence for beaming) and have found Swift J2058 (2011 May 18) and PTF 11agg (2011 Jan. 30). The talk discussed the properties of these sources, what may be learned from them in the future, and the detection rate for future transient surveys.

The details which the talk disclosed can be found at
`http://adsabs.harvard.edu/abs/2011Sci...333..199L`, —/2011Sci...333..203B
and —/2011arXiv1107.5307C.

Day 3:

Things That Tick

New Horizons in Time-Domain Astronomy
Proceedings IAU Symposium No. 285, 2011
R.E.M. Griffin, R.J. Hanisch & R. Seaman, eds.

© International Astronomical Union 2012
doi:10.1017/S1743921312000282

Spectroscopic Binaries: Towards the 100-Year Time Domain

R. F. Griffin

Institute of Astronomy, The Observatories, Madingley Road, Cambridge, CB3 0HA, UK
email: rfg@ast.cam.ac.uk

Invited Talk

Abstract. Good measurements of visual binary stars (position angle and angular separation) have been made for nearly 200 years. Radial-velocity observers have exhibited less patience; when the orbital periods of late-type stars in the catalogue published in 1978 are sorted into bins half a logarithmic unit wide, the modal bin is the one with periods between 3 and 10 days. The same treatment of the writer's orbits shows the modal bin to be the one between 1000 and 3000 days. Of course the spectroscopists cannot quickly catch up the 200 years that the visual observers have been going, but many spectroscopic orbits with periods of decades, and a few of the order of a century, have been published. Technical developments have also been made in 'visual' orbit determination, and orbits with periods of only a few days have been determined for certain 'visual' binaries. In principle, therefore, the time domains of visual and spectroscopic binaries now largely overlap. Overlap is essential, as it is only by combining both techniques that orbits can be determined in three dimensions, as is necessary for the important objective of determining stellar masses accurately. Nevertheless the actual overlap—objects with accurate measurements by both techniques—remains disappointingly small. There have, however, been unforeseen benefits from the observation of spectroscopic binaries that have unconventionally long orbital periods, not a few of which have proved to be interesting and significant objects in their own right. It has also been shown that binary membership is more common than was once thought (orbits have even been determined for some of the IAU standard radial-velocity stars!); a recent study of the radial velocities of K giants that had been monitored for 45 years found a binary incidence of 30%, whereas a figure of 13.7% was given as recently as 2005 for a similar group.

Keywords. stars:binaries:spectroscopic

After all that we have been hearing about big, and in many cases futuristic, projects, involving huge expense and large numbers of people, I am afraid that this talk will seem rather low-key. The truth is that, as far as spectroscopic binaries are concerned, it is inevitable that for a long time to come the 100-year time domain must be largely *retrospective*, because, no matter how frequent and sensitive future observations may be, they can not retrieve time that is already past. Plenty of good visual orbits with periods of 100 years or more have been determined, so it might be argued that even if spectroscopic orbits with 100-year periods could be derived, that would not be much of an achievement, because binary systems are just binary systems and it is only a question of what observational technique you use to observe them that decides whether you call them visual binaries or spectroscopic ones. But that is not quite so. In order to obtain stellar masses, which I suppose are the most critical data of all about stars, binaries need to be observed both 'visually'—by which I mean in angular measure on the sky, whether literally by eye or by interferometric or other such procedures—*and* spectroscopically. The visual orbit has a scale only in angular measure and not in kilometres, while the scale of a spectroscopic orbit by itself includes the undetermined factor sin i, i being the orbital inclination; the factor even enters *cubed* in the expression for the mass. Unfortunately for the progress of astrophysics, there has been remarkably little overlap between the period distributions of visual and spectroscopic binaries, although I have been doing my best to remedy that from the spectroscopic side.

Visual binaries have been carefully measured for position angle and angular separation since the time of Wilhelm Struve nearly 200 years ago. The observers have clearly realised that in most cases they could not hope to obtain orbits within their own lifetimes, but they nevertheless selflessly accumulated good measurements for posterity and we are certainly reaping the benefits from them now. Spectroscopists have not been so altruistic. They have seized on the short-period systems and tended to ignore the others. There is, perhaps, an extenuating circumstance, inasmuch as among the visual binaries it is the wide—and therefore usually slow—ones that are the easiest to resolve and measure. Among spectroscopic binaries it is the ones with the largest amplitudes—and therefore usually the shortest periods—that are the easiest. The fierce observational selection that arises in that way has even been exacerbated in recent years by the parcelling out of observing time by time-allocation committees which lack patience and require to see results published before they will grant any further time.

Soon after the methods of measuring accurate radial velocities were developed at Lick and Bonn in the 1890s, Lick Observatory started the great project of measuring repeatedly the velocities of all stars brighter than magnitude $5\frac{1}{2}$. After that project was concluded in 1925, however, the bright stars were almost entirely ignored by radial-velocity observers. Although one might hope that now, after more than 100 years, 100-year orbits could be derived, in actuality that is not so, because there is a gap of 50 years in the data. I can illustrate the situation that pertained not much more than 30 years ago by this quotation from the sixth edition of Smart's book, published in 1977: "The orbital periods of spectroscopic binaries are generally several days only . . . ", and by a histogram (Fig. 1) showing the distribution in half-dex period bins of the 204 late-type orbits that were listed with quality c or better in the 1978 Catalogue of Spectroscopic Binary Orbits. [The modal bin is the 3–10-day one.]

When I was a comparatively young man I was somewhat of an instrument-building enthusiast, and in that capacity I developed the method of measuring radial velocities by cross-correlation of spectra, first on the Cambridge 36-inch telescope and then, in collaboration with Jim Gunn, on the 200-inch at Palomar. I felt obliged to obtain a good lot of astronomical results at Cambridge, just to show that the method worked—a conclusion that was strongly resisted by the great majority of those who saw themselves as experts at the time. But then I got interested in binary stars, and the absence of any demand other than my own for observing time on the 36-inch reflector led me to see that, despite the disadvantages of site and climate, the constant access to the telescope provided freedoms not enjoyed by many other astronomers to pursue a project on spectroscopic binaries.

Partly with a view to providing a regular input of binary orbits into the literature, but also partly to advertise the seemingly despised method of cross-correlation that I had developed, in 1975 I started in *The Observatory* Magazine a series of orbit papers, featuring the words *PHOTOELECTRIC RADIAL VELOCITIES* conspicuously in the series title. Of course that expression appears tautological now that radial velocities are all measured photoelectrically, but at the time it was very distinctive. At the start of the series I had already been observing for nine years and felt that I had enough data 'in the bank' to put a paper of that series in every issue of *Observatory* (it is published every two months) indefinitely, as long as my own interest held up and the Editors were willing

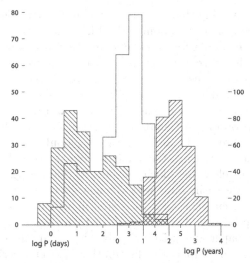

Figure 1. Period distribution of three sets of binary orbits. The ordinates on the left refer to the left-hand (diagonally shaded) distribution and the central one (no hatching), while the right-hand scale refers to the right-hand shaded distribution. From left to right: *a*) The 204 spectroscopic orbits of tolerable quality (qualities *a − c*) for stars with at least one component of type F5 or later, listed in the *Seventh Catalogue of the Orbital Elements of Spectroscopic Binary Systems* (*Publ. Dominion Astr. Obs.*, vol.17, 1978); that was the total number of such orbits known at that time, soon after my series of papers giving photoelectrically determined spectroscopic orbits started. (*b*)The 291 spectroscopic orbits published in Papers 1–200 in my series of papers in *The Observatory* in 1975–2009. (*c*) The 294 'visual' orbits published in IAU Commission 26 *Circulars* 165–174 (2008 June–2011 June), which could be expected in principle to include *all* such orbits published in the three-year interval shortly before those dates. The point of this diagram is to illustrate the disappointingly small overlap between the distributions of the spectroscopic orbits and the published visual ones, notwithstanding that it has been demonstrated that modern interferometers can produce orbits for binary systems with periods little longer than log *P* (days) = 0.

to accept the papers. In fact the series is still going!—the current issue of the Magazine carries the paper with the serial number 219.

Another programme that I started in the very first observing season, 1966, was on the 'Redman K stars'. There was a stage in the development of the cross-correlation method when the faintest stars that were readily observed with the 36-inch telescope were of the seventh magnitude. Redman, then Director of The Observatories at Cambridge, pointed out the existence of hundreds of seventh-magnitude K stars whose radial velocities he himself had measured from 90-Å/mm spectrograms taken with the 72-inch reflector at the Dominion Astrophysical Observatory, with which the new photoelectric velocities could be compared. I did indeed observe 80-odd of those stars in 1966 and again in 1969, and published the results. Whereas Redman had needed to give two-hour exposures on the 72-inch reflector, my observations took only a few minutes on the 36-inch and were several times more accurate! Nine of the stars appeared, with varying degrees of certainty, to change their velocities between 1966 and 1969. In 1973 I obtained another round of measurements of the Redman stars, because it was beginning to become apparent that stars were liable to change their velocities on longer time-scales than those with which most spectroscopists have usually concerned themselves; since then the same set of stars has been re-observed in most seasons, with just one observation per season except for stars that have proved to vary in velocity. The systematic coverage of one observation in each of 30 seasons over a total interval of 45 years has just been completed and a paper

giving the results hs been submitted. The proportion of binaries among that sample of
K giants amounts to 30%, more than double the 13.7% that was asserted (to a precision
of 0.1%!) from a much larger but much less careful observing programme in 2005.

A third project that has recently been published refers to spectroscopic binaries in
the field of the Hyades star cluster. Inasmuch as there was no history of interest in
stellar radial velocities at Palomar Observatory, there was no entrenched opposition to
the powerful new method, and as early as 1971 Jim Gunn and I were allowed to make
a (much improved) version of the Cambridge spectrometer for the 200-inch telescope.
Initially it was to investigate the dynamics of certain globular star clusters—it had never
previously been possible to measure the velocities of the individual red giants in such
clusters with an accuracy below the velocity dispersion that the virial theorem might lead
one to expect, but that was easily done with the new spectrometer. The instrument was
also used on certain open clusters, and discovered many binaries in the Hyades. Many
of them were written up in several papers published long ago, but I have been following
others ever since, and a paper giving 54 new orbits is now in press.

It is of no small interest to compare the distribution of orbital periods of the binaries in
the three programmes I have described—field stars in the *Observatory* series Fig. 1, the
Redman K giants (Fig. 2), and the Hyades (Fig. 3)—with the histogram I showed from
the 1978 orbit catalogue. My choice of stars is still affected by observational selection but
not nearly as seriously as the *Catalogue* distribution. You will see that the modal 'bin'
for the *Observatory* and Redman stars is the 1000–3000-day one, while in the Hyades it
is the 3000–10000-day one—fully three orders of magnitude longer than the one in the
Catalogue! I would point out that periods above 10^4 days are still heavily discriminated
against by observational selection even in my own work; 10^4 days is some 27 years, and
$10^{4.5}$ days is 86 years and so is far beyond a working lifetime. Within that limitation,
the distribution of orbital periods (on a logarithmic scale) may be considered to be
monotonically rising even to the longest periods currently accessible.

I should perhaps point out that there is nobody to identify for you, on Day 1 of your
professional lifetime, all of the stars whose periods are of the maximum length that you
can hope to document, so that you can get started on them straight away. Inevitably
you learn about many of the long-period stars only when it is really too late! For myself,
before I could start observing it was necessary to develop the procedure without which

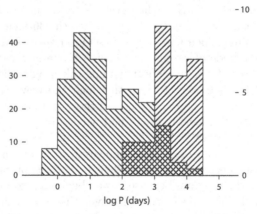

Figure 2. The *Seventh Catalogue* distribution of spectroscopic binary periods (Fig. 1(*a*)) is contrasted here with the distribution of periods for the binaries found among the 'Redman K stars'
(R.F. Griffin, JA&A, in press, 2012). Since the selection criteria for those stars discriminated
against dwarfs there are no very short periods in the Redman sample.

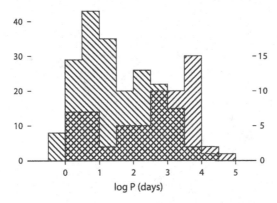

Figure 3. The *Seventh Catalogue* distribution of spectroscopic binary periods (Fig. 1(a)) is contrasted here with the distribution of periods for Hyades spectroscopic binaries with known orbits (mostly from R.F. Griffin, JA&A, in press, 2012).

it would have been impossible to follow a large number of low-amplitude binaries year after year and decade after decade. My successors can start observations on Day 1, on instrumentation (better, actually, than any that I have ever had) that is already provided for their use by others.

It is obvious not only that I personally am not in a position to extend the period distribution of spectroscopic binaries much further towards longer periods, but also that progress in that direction is in any case going to be extremely slow, a task for generations yet to come. A more practical possibility for increasing the much-to-be-desired overlap between the periods of spectroscopic and visual binaries would seem to be from the 'visual' side—and indeed the instrumental enthusiasts among the visual-binary people have by no means been idle. They have developed first of all the speckle method and then interferometry with widely separated apertures, that have given resolutions down to a single millisecond of arc and allowed good 'visual' orbits to be produced for 'spectroscopic' binaries with periods down to a few days, such as that of β Aurigae. In principle, therefore, there is now a huge overlap between the domains of visual and spectroscopic binaries, which ought to allow stellar masses to be derived on a wholesale scale for stars of all different types and luminosities. Unfortunately that does not seem to be happening. I have looked at the distribution of orbital periods whose publication has been documented in the last ten issues of the IAU Commission 26 Circulars, which in principle record all the newly determined 'visual' orbits, and I have added that distribution to Fig. 1 for comparison with two of the distributions I have already given for spectroscopic orbits. The modal bin of the visual orbits is the 100–300-year bin, and there are more than 20 orbits with altogether ludicrous periods of more than a thousand years, no more than a sixth of whose periods can possibly be covered by observation! Only three of the 300 orbits have periods as 'short' as 10 years. We really need someone to light a fire under the visual people and explain to them what they ought to be doing!

The observation of spectroscopic binaries has not proved to be such an arid and boring exercise as you might imagine, partly because many of the stars concerned prove to be interesting objects in their own right. That is particularly the case with the whole class of composite-spectrum binaries, in which the late-type star whose radial velocity I can measure has a companion of early type. The spectra of many such pairs have been skilfully disentangled by my colleague and co-author, and the systems have been revealed as having remarkable diversity.

I would like to finish by referring to some of the results, and challenges, of orbit deter-

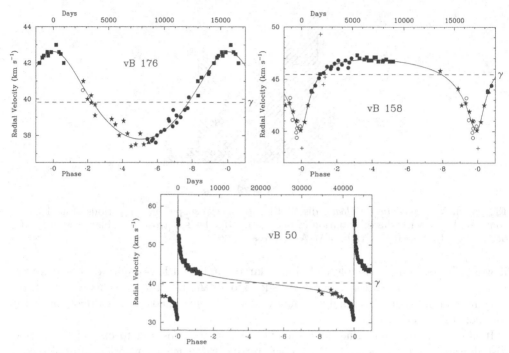

Figure 4. Orbits of Hyades binaries van Bueren 176, 158, and 50. The different symbols represent observations obtained with different instruments.

mination of long-period spectroscopic binaries. First, δ Andromedae was quite definitely shown by both Lick and Bonn measurements to have had a maximum of velocity in the early years of the last century. So in due course it was bound to have another one. For about 25 years after I put it on my observing programme in 1969 it went absolutely nowhere, but then it started to move, and now it appears to be right on the crest of a fresh maximum. Another bright star with a similar period, that I shall be able to write up if I can keep going just a few more years, is τ Piscium. The late-type primary star in the beautiful visual pair Albireo (β Cygni) remains a conundrum. It definitely showed a small rise in velocity during the early part of the last century, and it has equally definitely shown a small decline in the forty years that I have been observing it. Its case well illustrates the difficulties that are caused by the half-century of neglect that I mentioned previously. The two intervals of observation may be linked directly, giving an orbital period of a little over 100 years, but there could equally well be an intervening cycle, in which case the period would be about 56 years. The visual orbit by Hartkopf is not well determined and has a period of 97 years, but the radial velocities cannot be made to fit it well. Fig. 4 shows the orbits of three Hyades binaries: van Bueren 176, which has been seen almost round its 43-year orbit, van Bueren 158, whose orbit of about 48 years still has some way to go and meanwhile is less well determined, and finally van Bueren 50, the *pièce de résistance*, whose period of 115 years is determined within about 10% because although there is an enormous gap in phase coverage we have seen both nodes of its extremely eccentric orbit. It also has an interferometric orbit which has very similar elements and a period of 112 years.

New Horizons in Time-Domain Astronomy
Proceedings IAU Symposium No. 285, 2011
R.E.M. Griffin, R.J. Hanisch & R. Seaman, eds.

© International Astronomical Union 2012
doi:10.1017/S1743921312000294

On the Sensitivity of Period Searches

A. Schwarzenberg-Czerny[1,2]

[1] Nicolaus Copernicus Astronomical Centre, 00-716 Warsaw
[2] Adam Mickiewicz University Observatory, PL 60-286, Poznań, Poland
email: alex@camk.edu.pl

Invited Talk

Abstract. Astronomical time series are special in that time sampling in them is uneven yet often with periodic gaps due to daytime, moon and seasons. There is therefore a need for special-purpose time-series analysis (TSA) methods. The emergence of massive CCD photometric surveys from the ground and space raises the question of an automatic period search in $\gg 10^5$ light curves. We caution that already at the planning stage it is important to account for the effects of time sampling and analysis methods on the sensitivity of detections. We present a transparent scheme for the classification of period-search methods. We employ tools for evaluating the performance of those methods, according to the type of light curves investigated. In particular we consider sinusoidal and non-sinusoidal oscillations as well as eclipse or transit light curves. From these considerations we draw recommendations for the *optimum analysis of astronomical time series*. We present briefly the capability of an automatic period-search package TATRY. Finally we discuss the role of Monte Carlo simulations in the analysis of detection sensitivity. As an example, we demonstrate a practical method to account for the bandwidth (multi-trial) penalty in the statistical evaluation of detected periods.

Keywords. methods: data analysis, methods: statistical, (stars:) binaries: eclipsing, stars: oscillations (including pulsations), (stars:) planetary systems, (Galaxy:) globular clusters: general, (galaxies:) Magellanic Clouds

1. Introduction

We present a biased overview of methods for enhancing period searches in astronomical data, relying heavily on our own work. In this context, astronomical data are ones with uneven time sampling. Data having regular time sampling are best analysed using FFT-based methods and are not discussed here. The massive present-day photometric surveys yield $\gg 10^5$ light curves, so we concentrate on methods that permit a fully automatic period analysis. First we turn attention to the proper planning of observation sampling. We then discuss the orthogonal models of periodic signals as they permit analytical evaluation of statistical properties in period search. From statistics we employ the concept of test power in order to evaluate the sensitivity of period-search methods. On that basis we are able to derive recommendations for optimum period analysis. We continue with the discussion of two correction effects to be accounted for in calculating realistic probability distributions. Finally we turn our attention to pitfalls and profits of Monte Carlo simulations for statistical analysis of time series.

2. Sample Planning

A frequent fatal error in planning astronomical observations is observing objects at the same position with respect to the meridian, say on D consecutive nights. The corresponding sampling pattern may then be represented as a product $\Pi_D(t) \cdot \text{Ш}_1(t)$, where Π and

III denote the Heaviside top-hat and Dirac picket-fence functions in the time domain. Applying the convolution $*$ and its theorem to the Fourier transforms $\mathcal{F}$, one obtains the corresponding window function in the frequency domain, $W = |\mathcal{F}\Pi_D * \mathcal{F}III_1|^2 = |\text{sinc}(D\nu) * III_1(\nu)|^2$. It corresponds to an infinite series of peaks/aliasses of width $1/D$. Next, let us consider the same number of observations scattered uniformly over a fraction d of nights. Such a pattern may be represented by $\Pi_D(t) \cdot [III_1(t) * \Pi_d(t)]$ and $W = |\text{sinc}(D\nu) * [III_1(\nu) \cdot \text{sinc}(d\nu)]|^2$. Now all aliases beyond width $1/d$ of the $\text{sinc}(d\nu)$ function are greatly reduced in size. Such a reduction is obtained by reshuffling observations on different nights without changing their time span *per night*. The effect resembles the aperture synthesis obtained by shifting radio antennæ between observations.

3. Period Search by Quadratic Norms

3.1. *Statistical Principles of Detection: Periodogram Statistics and Distributions*

Our considerations are based on R.A. Fisher's statistical theory of the least squares (LSQ) fit of n observations $\mathbf{x} = (x_1, \cdots, x_n)$ with the orthogonal model $\mathbf{x}_{\parallel} = \sum_i c_i \mathbf{p}_i(\varphi)$, where vectors/functions $\mathbf{p}_i(\varphi) \equiv \mathbf{p}_i(2\pi t)$ are orthogonal with respect to the scalar product defined by observed phases: $0 = (\mathbf{p}_i, \mathbf{p}_j) \equiv \sum_\varphi w_\phi p_i(\varphi)p_j(\varphi)$ for $i \neq j$. Hereafter we assume that the average values $(1, \mathbf{x}) = (1, \mathbf{x}_{\parallel}) = 0$. By virtue of the Fisher lemma the model $\mathbf{x}_{\parallel}$ and residuals from fit $\mathbf{x}_{\perp} \equiv \mathbf{x} - \mathbf{x}_{\parallel}$ are orthogonal $(\mathbf{x}_{\parallel}, \mathbf{x}_{\perp}) = 0$. In consequence, an n-dimensional analogue of Pythagoras theorem holds:

$$\|\mathbf{x}\|^2 = \quad \|\mathbf{x}_{\parallel}\|^2 + \|\mathbf{x}_{\perp}\|^2 \quad \text{where}$$
$$n = \quad n_{\parallel} + \quad n_{\perp} \tag{3.1}$$
$$\text{observation} = \text{model} + \text{residuals}$$

$\|\mathbf{x}\|^2 \equiv (\mathbf{x}, \mathbf{x})$ and n, $n_{\parallel}$ and $n_{\perp}$ denote the number of observations, the number of model parameters and the number of degrees of freedom of the residuals. Because of the relation $\|\mathbf{x} - \mathbf{x}_{\parallel}\|^2 = \|\mathbf{x}\|^2 - 2(\mathbf{x}, \mathbf{x}_{\parallel}) + \|\mathbf{x}_{\parallel}\|^2$, where only the middle term depends on the frequency ν, our considerations for LSQ also apply to the case of the cross-correlation function (CCF) periodogram.

Suitable families of orthogonal functions $\mathbf{p}_i$ are either Szegö trigonometric polynomials or top-hat functions corresponding to the phase bins (Schwarzenberg-Czerny 1996, 1989). Nominally, Szegö polynomials follow from Gramm-Schmidt orthonormalization of Fourier harmonics, yet convenient recurrence formulæ also exist. Phase folding and binning of data is equivalent to LSQ fitting a step function composed of a linear combination of top-hats. The box function employed for planetary transit searches corresponds to two phase bins of unequal width (Schwarzenberg-Czerny & Beaulieu 2006), so the present considerations apply in this area too. Quite unique orthogonal functions were employed by MACHO (Akerlof *et al.* 1994).

A statistics $\Theta(\nu, \mathbf{x})$ is the merit figure indicating the quality of the fit. A periodogram is the plot of $\Theta(\nu, \mathbf{x})$ against ν. Patterns in the periodogram may relate to the presence in the data of oscillations with the corresponding frequency. The significance of those frequencies depends on the probability distribution of Θ for hypothetical data consisting of pure noise. In Statistics, that case is called a null hypothesis, H_0. Θ must be dimensionless, as no statistical conclusions may depend on units. There are three ways to construct dimensionless Θ statistics from the dimensioned $\|\mathbf{x}\|$, $\|\mathbf{x}_{\parallel}\|$ and $\|\mathbf{x}_{\perp}\|$ (Table 1). Because of Eq. (3.1), all these Θ's are uniquely related:

$$\Theta_{\perp} = 1 - \Theta_{\parallel} = \frac{1}{1 + \Theta_{AOV}}, \tag{3.2}$$

Table 1. Basic Classes of Period Statistics

Statistics Definition	Distribution	Name	Analogue
$\Theta_{AOV} = \frac{\|\mathbf{x}_\||\|}{\|\mathbf{x}_\perp\|}$	$F(n_\||, n\perp; \Theta_{AOV})$	Fisher-Snedecor	AOV[1], mhAOV[2]
$\Theta_\|| = \frac{\|\mathbf{x}_\||\|}{\|\mathbf{x}\|}$	$\beta(n_\||, n\perp; \Theta_\||)$	β distribution	Power[3], L-S[4,5]
$\Theta_\perp = \frac{\|\mathbf{x}_\perp\|}{\|\mathbf{x}\|}$	$\beta(n\perp, n\||; \Theta_\perp)$	β distribution	χ^2, PDM* [6,7]

References: (1)- Schwarzenberg-Czerny (1989), (2)- Schwarzenberg-Czerny (1996), (3)- Deeming (1975), (4)- Lomb (1976), (5)- Scargle (1982), (6)- Stellingwerf (1978), (7)- Schwarzenberg-Czerny (1997)

so the corresponding F and β distributions may be obtained from each other by suitable changes of variable. From this we find that conclusions drawn from the Θ_{AOV}, $\Theta_\||$ and $\Theta_\perp$ periodograms must all be identical *if and only if* the model $\mathbf{x}_\||$ remains the same. In other words, what counts is not the shape of the periodogram peak but its probability (Schwarzenberg-Czerny 1998). Turning that argument *ad absurdum*, one may state that obtaining a clean, single-peak periodogram is sufficient to raise any periodogram to the power of 1000 or so. As no additional information is supplied, such a nice view has spurious meaning. In practical terms it is sufficient to discuss periodograms in which oscillations correspond to peaks. The results would also apply to the periodograms showing through at the corresponding frequencies.

However, to the human eye equivalent periodograms may look deceptively different. For a high S/N and χ^2 periodogram, an alias minimum of $\Theta_\perp$ that is twice as high as the true minimum would not look significant. At the same time the corresponding alias peak power $\Theta_\|| = 1 - \Theta_\perp$ would almost match the true peak, pretending to be significant. For the human eye it is therefore better to plot a $\log \Theta_{AOV}$ periodogram, as the probability distribution of its values is close to the normal one. Then a twice-higher peak has twice the σ significance.

3.2. *Sensitivity of Detection: Test Power*

To evaluate the sensitivity of detection we must consider two different hypothetical data sets: for a pure noise with standard deviation 1, and for noise plus a periodic signal of amplitude A (same units). In Statistics these two cases are called the "null" and the "alternative" hypotheses, H_0 and H_1, respectively. Accordingly, for H_0 and H_1, Θ obeys different probability distributions, $P_0(\Theta)$ and $P_1(\Theta)$. Ideally the two distributions are separated by a critical value, Θ_c. We could say that $\Theta < \Theta_c$ corresponds to a pure noise and $\Theta > \Theta_c$ to the detected of a signal. However, in reality the two distributions overlap for a range of Θ. Two kinds of errors thus arise: one claims detection while in reality H_0 is true (*false positives*), and conversely one claims no detection while in reality H_1 remains true (*misses*). In classical statistics we fix Θ_c so that false positives seldom occur, i.e. the significance level $\alpha = P_0(\Theta < \Theta_c)$ is close to 1. Then the *test power* of the criterion Θ_c is defined as $\beta = P_1(\Theta < \Theta_c)$, where the probability of misses is $1 - \beta$. Thus, for a fixed Θ_c, large β corresponds to good detection sensitivity. The analytical formulæ for P_0 are listed in Table 1. No corresponding formulæ are known for P_1, as they depend in a complex way on signal shape. However, for small signal-to-noise, $A/1 \ll 1$, it is possible to derive approximate asymptotic formulæ for P_1 (Schwarzenberg-Czerny 1999). In that approximation, P_0 and P_1 retain the same shape yet are shifted, in units of their standard deviation, by

$$\Delta\Theta/\sqrt{Var\{\Theta\}} = A^2 n \frac{\|s_\||\|^2}{\sqrt{2n_\||}} \quad \text{where} \quad (3.3)$$

$$\|s_\|\|^2 = \frac{(x_{\|signal}, x_{\|model})^2}{\|x_{\|signal}\|^2 \|x_{\|model}\|^2}. \tag{3.4}$$

$x_{\|signal}$ and $x_{\|model}$ denote the shapes of the real signal and of the fitted model, respectively. The bigger $\Delta\Gamma$ is, the more sensitive is our method/model for a given signal.

After feeding into Eqs. (3.3, 3.4) the von Mieses function $e^{-\kappa \cos^2 \varphi}$ as a signal and Fourier harmonics or top-hat functions as model functions, our calculations yield the test power for the most popular Fourier and binning periodograms. Small κ values corespond to near sinusoidal input signals, and large ones to narrow gaussian spikes of width $\sqrt{2\kappa}$. These calculations reveal that excessively crude or excessively fine models both yield decreased sensitivity because (respectively) of the factors $\|s_\|\|^2$ and $1/\sqrt{n_\|}$ in Eq. (3.3). Thus, for optimum sensitivity, the resolution of the model should just resolve the features in the signal.

4. Related Issues

4.1. *Strength and Pitfalls of Monte Carlo TSA*

Monte Carlo (MC) simulations have venerable origins, and date back to von Neumann's work in 1940 at Los Alamos. MC constitutes a powerful method for studying likely events and their expectation integrals, $E\{F(x)\} \equiv \int F(x)f(x)dx$. The application of MC methods follows rules of Statistics as a branch of Mathematics. Current prevailing editorial policies seem to expect any observer or referee to be able to analyse the reliability of conclusions by MC simulations. In my opinion it is as justifiable as expecting an observer to perform state-of-the-art hydrodynamic simulations and/or to support observations with quantum mechanical calculations of involved atoms and transition probabilities. It is doable, but by no means by all. The policy results in the emergence of many poor simulations at best, and in the publication (in otherwise respectable journals) of chains of logically linked wrong papers based on shabby simulations, at worst. In particular, the application of MC methods for rare events, such as in significance analysis, is always uneconomic and often risky, because of untested statistical properties of random number generators and discrete computer arithmetic for these rare events.

4.2. *Corrected Significance: Bandwidth Penalty*

From Table 1 one may derive the analytic tail probability of large Θ for a single frequency: $Q_1(\Theta > \Theta_0) \equiv 1 - P_1(\Theta < \Theta_0)$. As more and more frequencies are examined in the periodogram, the probability of a spurious occurrence of a peak due to pure noise increases, in the same way as the probability of winning a lottery increases with the number of trials. This increased probability, called *bandwidth penalty*, has to be accounted for any realistic statistical evaluation. Because the aliasing values of a periodogram at different frequencies may be strongly correlated, out of N investigated frequencies only $N_{eff} \leqslant N$ may be independent. In that case the postulated tail probability, according to Horne & Baliunas 1986, could be:

$$Q_N(\Theta_0) = Q_1(\Theta_0)^{N_{eff}}. \tag{4.1}$$

The hitch is in the unknown value of N_{eff}. Paltani (2004) proposed a useful method to estimate N_{eff} by MC simulations, though relying on their mean, rather than extreme, values. The Paltani procedure, improved by us in steps (e) and (f), may be summarized in the following way:

(a) Replace observations x with a simulated white noise;
(b) Calculate the periodogram Θ for a given frequency grid;
(c) Find the extreme value Θ_s of the simulated periodogram;
(d) Repeat steps (a)–(c) as many times as desired for accuracy;
(e) Find median value Θ_m of Θ_s, where $Q_N(\Theta_m) = 0.5$;
(f) Solve Eq. (4.1) for $N_{eff} = \frac{\ln 0.5}{\ln[Q_1(\Theta_m)]}$;
(g) Calculate $Q_N(\Theta_0)$ from Eq. (4.1) for Θ_0.

In this way one makes use of likely events $Q_N(\Theta_m) \gg Q_N(\Theta_0)$, which would be rejected in the brute-force simulations.

4.3. *MC Study of Bandwidth Correction*

The modified Paltani method enables a convenient study of N_{eff} by MC simulations for several realistic, uneven sampling patterns. For illustration, we included a case with a large time gap in the middle of the observations. The simulations depend on several parameters: the number of calculated frequencies N, the number of observations N_{obs}, the maximum number of resolved frequencies $N_{max} = \Delta t \Delta \nu$ where Δt and $\Delta \nu$ denote the ranges of time and frequency spanned by the observations and their periodogram, respectively, and the number of model parameters $N_{\parallel}$.

In all simulations the condition $N_{eff} < N$ held strictly. However, another condition, $N_{eff} \leqslant N_{max} \sqrt{N_{\parallel} - 1}$, held only approximately. It is expected that a finer model would yield a more precise phase determination, so some factor involving $N_{\parallel}$ seems in place. Surprisingly, the condition $N_{eff} \leqslant N_{obs}$ does not hold in our simulations. To our knowledge, that effect was not mentioned in the literature. At this stage no definite explanation seems available. However, one could suspect that periodogram values depend on observations in such a complicated way that effectively they become chaotic. In the same sense, the consecutive values of a random number generator show no correlation despite them all depending on just one seed value. An alternative explanation would be that Eq. (4.1) never held, i.e. that in a periodogram there were no truly independent frequencies.

4.4. Corrected Significance: Correlation or Red Noise Effect

The presence of a *correlation (red noise)* in observations may ruin simplistic statistical estimates. For example, the LSQ fit of a sine to the solar spot Wolfer numbers spanning 100 years yields the nominal period of $P \approx 11$ y with an error of the order of $0.002P$. However, the propagation of such an ephemeris for the next 50 years demonstrates that a realistic period error was $\approx 0.1P$. This has happened because consecutive residuals from the fit are correlated (they keep the same sign for decades), while the standard LSQ error estimates implicitly *assume* that residuals are *(uncorrelated) white noise*. Conversely, for simulated data consisting of white noise plus the oscillation of the same variances or amplitudes as above, the $0.002P$ error estimate proves realistic.

This remarkable correlation effect is seldom discussed in texts on LSQ. In fact, the correlation of every N_{corr} consecutive observations decreases the effective number of observations by roughly a factor of N_{corr}, and hence increases the real LSQ errors by a factor of $\sqrt{N_{corr}}$ (Schwarzenberg-Czerny 1991). A simple way to estimate N_{corr} is by counting the number of sign changes in the residuals from the fit (the *post mortem* analysis). For white noise, one expects $N_{obs}/2$ changes of sign (every second residual should change sign, on average). If the observed number of sign changes in the residuals is $N_{sign} < N_{obs}/2$, then the number of consecutive correlated observations is $N_{corr} \approx N_{obs}/(2N_{sign})$.

Conclusion: Statistics does work for planning and analysis of astronomical time series observations, though care is needed.

References

Akerlof, C., Alcock, C., Allsman, R., *et al.* 1994, *AJ*, 436, 787
Deeming, T. J. 1975, *Astrophys. Sp. Sc.*, 36, 137
Gieronimus, Ya.L., 1958, *Orthogonal polynomials of circle and real intervals*, Moscow, GIFML (in Russian)
Horne, J. H. & Baliunas, S. L. 1986, *ApJ*, 302, 757
Lomb, N. R. 1976, *Atrophys. Sp. Sc.*, 39, 447
Paltani, S. 2004, *A&A*, 420, 789
Scargle, J. D. 1982, *ApJ*, 263, 835
Schwarzenberg-Czerny, A. 1989, *MNRAS*, 241, 153
Schwarzenberg-Czerny, A. 1991, *MNRAS*, 253, 198
Schwarzenberg-Czerny, A. 1996, *ApJ*, 460, L107
Schwarzenberg-Czerny, A. 1997, *ApJ*, 489, 941
Schwarzenberg-Czerny, A. 1998, Baltic Astronomy, 7, 43
Schwarzenberg-Czerny, A. 1999, *ApJ*, 516, 315
Schwarzenberg-Czerny, A. & Beaulieu, J.-P. 2006, *MNRAS*, 365, 165
Stellingwerf, R. F. 1978, *ApJ*, 224, 953

Appendix

The following recurrence yields Szegö polynomial expansion coefficients $c_n, n = 0, 1, \cdots, 2N$ for observations $t_m, x_m, m = 1, \cdots, M$ (Gieronimus 1958, Schwarzenberg-Czerny 1996):

$$\mathbf{p}_0(\mathbf{z}) = 1 \qquad c_n = \frac{(\mathbf{f}, \mathbf{p}_n)}{\|\mathbf{p}_n\|^2} \qquad \alpha_n = \frac{(\mathbf{z}\mathbf{p}_n, 1)}{\|\mathbf{p}_n\|^2} \tag{1}$$

$$\mathbf{p}_{n+1}(\mathbf{z}) = \mathbf{z}\mathbf{p}_n(\mathbf{z}) - \alpha_n \mathbf{z}^n \overline{\mathbf{p}_n(\mathbf{z})} \tag{2}$$

where $z_m = e^{2\pi i \omega t_m}$, $\mathbf{f} = \mathbf{z}^N \mathbf{x}$, i.e. component-by-component product of vectors $\mathbf{z}^N$ and $\mathbf{x}$. The modified Eq. (1) for α_n remains valid for the FFT limit case $\mathbf{p}_n(\mathbf{z}) \to \mathbf{z}^n$.

New Horizons in Time-Domain Astronomy
Proceedings IAU Symposium No. 285, 2011
R.E.M. Griffin, R.J. Hanisch & R. Seaman, eds.

© International Astronomical Union 2012
doi:10.1017/S1743921312000300

Sines, Steps and Droplets: Semi-parametric Bayesian Modelling of Arrival Time Series

Thomas J. Loredo

Department of Astronomy, Cornell University, Ithaca, New York, USA
email: loredo@astro.cornell.edu

Abstract. I describe ongoing work developing Bayesian methods for flexible modelling of arrival-time-series data without binning. The aim is to improve the detection and measurement of X-ray and gamma-ray pulsars and of pulses in gamma-ray bursts. The methods use parametric and semi-parametric Poisson point process models for the event rate, and by design have close connections to conventional frequentist methods that are currently used in time-domain astronomy.

Keywords. methods: statistical, methods: data analysis, pulsars: general, gamma rays: bursts

---

Measuring the arrival times, directions and energies of individual quanta—photons or particles—potentially provides the finest possible resolution of dynamical astronomical phenomena, particularly for high-energy sources that produce low detectable fluxes. The simplest methods for signal detection and measurement bin the data for statistical or computational convenience (for example, to allow the use of asymptotic Gaussian approximations, or to enable fast Fourier decomposition with an FFT). But methods that instead directly analyze the event data without binning can detect weaker signals and probe shorter time-scales than can ones that require binning.

The ongoing work I briefly describe here was motivated by studies of X-ray and gamma-ray pulsars, which produce periodic signals, and gamma-ray bursts, which produce chaotic signals that are typically comprised of multiple overlapping pulses. In the former case there may be less than one event per period (particularly in energy-resolved studies); in the latter, time-scales as short as milliseconds are relevant, and detected photons are sparse at high energies. Both phenomena motivate the development of data analysis techniques that can milk every hard-won event for what it is worth.

For simplicity we focus here just on the arrival-time data (also known as time-tagged event data), and presume that the events being analyzed have been selected to have directions consistent with an origin from a single source, and moreover that energy dependence of any putative signal is not significant (so the signal's temporal signature is not corrupted by ignoring event energies). We can represent the data as points on a time-line, as in Fig. 1. The dots denote events at times t_i that have been detected within small time intervals (δt), which represents the instrumental time-resolution. The empty intervals, denoted Δt_j, are informative; seeing no events in an observed interval provides a constraint on the signal, in contrast to simply not observing during the interval.

Conventional approaches to detecting signals in such data (binned or unbinned) adopt an approach of frequentist hypothesis testing: one devises a test statistic that measures departure of the data from the predictions of an uninteresting "null" model, and uses it to see if the null model may be safely rejected (implying that an interesting signal is present). No explicit signal model is needed to define such a test.

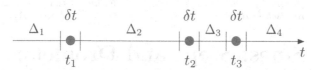

Figure 1. Arrival-time series depicted as points on a timeline, with detection and nondetection intervals noted.

Here I describe methods that were developed from the Bayesian approach, where one compares the null model to explicit alternative models that describe interesting signals. One motivation was to show how conventional "alternative free" test statistics arise in this framework (exactly or approximately). It illuminates implicit assumptions underlying conventional methods; more constructively, it provides a framework whereby generalizations of the implicit models may lead to new methods. Some other virtues of adopting a Bayesian approach to these problems, both pragmatic and conceptual, are outlined below; Loredo (2011) gives a more extensive, but still introductory, discussion.

We model the data with a non-homogeneous Poisson point process in time. A model, M, specifies an intensity function (event rate) $r(t; \mathcal{P})$ that depends on the model's parameters, $\mathcal{P}$. Parametric models have a parameter space of fixed dimension; for example, a periodic-signal model will typically have frequency, amplitude and phase parameters, and possibly additional parameters describing the light-curve shape. Non-parametric models have a parameter space whose (effective) dimension may grow with sample size, adapting to the data; it may formally be infinite-dimensional. Semi-parametric models have a parameter space with a fixed-dimension part (e.g. the frequency and phase of a periodic model), and a non-parametric part (e.g. an adaptive light-curve shape).

The data drive Bayesian inferences via the likelihood function, the probability for the data, $D = (\{t_i\}, \{\Delta t_j\})$, given values for the parameters. Referring to Fig. 1, we build the likelihood function by calculating the product of Poisson counting probabilities for zero counts in the empty intervals, and one count in each detection interval. The zero-count probabilities are of the form $\exp[-\int_{\Delta_i} dt\, r(t)]$, and the one-count probabilities are of the form $[r(t_i)\delta t] \exp[-r(t)\delta t]$ (presuming δt is small so that $r(t)\delta t \ll 1$). The likelihood function is thus:

$$\mathcal{L}_M(\mathcal{P}) \equiv p(D|\mathcal{P}, M) = \exp\left[-\int_T dt\, r(t)\right] \prod_{i=1}^{N} r(t_i)\delta t, \qquad (0.1)$$

where T denotes the full observing interval and N is the number of detected events. To go further, we must give specific rate models and priors for the model parameters. To fit a particular model, we use Bayes's theorem to calculate a posterior probability for the parameters, $p(\mathcal{P}|D, M) = p(\mathcal{P}|M)\mathcal{L}_M(\mathcal{P})/Z_M$, where Z_M is a normalization constant given by the integral of the product of the prior probability density, $p(\mathcal{P}|M)$, and the likelihood function. To detect a signal, we instead use Bayes's theorem on a hypothesis space including one or more models for interesting signals, and the null model (here, a constant-rate model with a single parameter, the amplitude, A, with $r(t) = A$). In this space, the normalization constant for a particular model, Z_M, plays the role of the likelihood for the model (as a whole); Z_M is often called the marginal likelihood for the model, where "marginal" refers to the integration over $\mathcal{P}$ used to calculate its value.

As a simple starting point, consider a model where the logarithm of the rate is proportional to a sinusoid plus a constant; the logarithm guarantees that the rate itself is non-negative. We may then write the rate as $r(t) = A \exp[\kappa \cos(\omega t - \phi)]/I_0(\kappa)$, where A is the time-averaged rate and $I_0(\cdot)$ denotes the modified Bessel function of order 0 (this normalizes the exponential factor so that A is the time-average). Light curves with that shape have a single peak per period; its width (equivalently, the duty cycle) is determined by the concentration parameter, κ, with large values corresponding to sharp peaks and $\kappa = 0$ corresponding to a constant rate. To estimate the frequency and concentration, we calculate the posterior for all four parameters, $p(A, \omega, \kappa, \phi | D, M)$, and marginalize (integrate) over A and ϕ. By adopting a flat prior for the phase, and nearly any prior for A that is independent of the other parameters, we find a marginal posterior probability density for frequency and concentration proportional to $I_0[\kappa R(\omega)]/[I_0(\kappa)]^N$, where $R(\omega)$ is the Rayleigh statistic given by

$$R^2(\omega) = \frac{1}{N}\left[\left(\sum_{i=1}^{N}\cos\omega t_i\right)^2 + \left(\sum_{i=1}^{N}\sin\omega t_i\right)^2\right]. \tag{0.2}$$

Estimation of ω alone, accounting for uncertainty in all other parameters, is found by further integrating over κ, which is easy to do numerically. Detection requires calculating the marginal likelihood, corresponding to a further integration over ω, and must be done numerically. It can be time-consuming for blind searches, but not significantly more so than the kind of frequency-grid searching employed by conventional tests.

The Rayleigh statistic was invented for the well-known (frequentist) Rayleigh test for periodic signals in arrival-time series; Lewis (1994) gives a review of the Rayleigh test and other frequentist period detection methods mentioned below. The quantity $2R^2(\omega)$ is called the Rayleigh power; it is the point process analogue of the periodogram or Fourier power spectral density. From a Bayesian point of view, the Rayleigh test implicitly assumes that periodic signals may be modelled well by log-sinusoid rate functions. Notably, there is no parameter corresponding to κ in the Rayleigh test; also, in practice it is known to work well only for smooth light curves with a single broad peak per period. Such light curves correspond to values of κ near unity—another implicit assumption of the Rayleigh test. These results indicate that one can implement Bayesian period searches using conventional computational tools already at hand for the Rayleigh test. They also indicate that explicit consideration of the κ parameter may lead to procedures which are more sensitive to sharply-peaked light curves than are the conventional Rayleigh tests.

Many pulsars have light curves with two or more peaks per period. That suggests generalizing the log-sinusoid model to a log-Fourier model, with the logarithm of the rate proportional to a sum of harmonic sinusoids. By adopting a finite sum of m harmonics with concentration parameters κ_k ($k = 1$ to m, with the fundamental corresponding to $k = 1$), we may proceed analogously to the above analysis. The larger number of phases and concentration parameters thwarts analytical integration; in an approximate treatment the posterior distribution for frequency and concentration is proportional to $\exp[S(\omega)]$, with $S(\omega) \equiv \sum_{k=1}^{m}\kappa_k R(k\omega)$. The frequentist test generalizing the Rayleigh test to multiple harmonics is the Z_m^2 test, with $Z_m^2(\omega) = 2\sum_{k=1}^{m}R^2(k\omega)$, a sum of Rayleigh powers at harmonics. Notably, the Bayesian analysis uses the sum of $R(k\omega)$ values ("harmonic Rayleigh amplitudes") rather than powers. Note that, for $\kappa_k = 1$, $S^2(\omega) = Z_m^2(\omega)/2 + \sum_k\sum_{j\neq k}R(j\omega)R(k\omega)$; that is, S^2 contains information not in Z_m^2. Roughly speaking, Z_m^2 corresponds to incoherently summing power in harmonics, but the quantity arising in the Bayesian treatment of a harmonic model instead sums amplitudes, accounting for phase information ignored by Z_m^2.

A popular frequentist period-detection method that aims to be sensitive to periodic signals of complex shape is χ^2 epoch folding (χ^2-EF). One folds the arrival times given by a trial period to produce a phase, θ_i, for each event, with θ_i in the interval $[0, 2\pi]$. For a constant signal, the phases should be uniformly distributed. The χ^2-EF method bins the phases into B bins and uses Pearson's χ^2 to test consistency of the binned phases with a uniform distribution. This motivates a Bayesian model with a piecewise-constant rate function and B steps per period. Gregory & Loredo (1992) (GL92) analyzed such a model, giving the rate in a bin as Af_k, where A is the average rate, and the shape parameters, f_k ($k = 1$ to B), specify the fraction of the rate attributed to each bin, with $\sum_k f_k = 1$. They assigned a constant prior to the shape parameters, from the intuition that that spreads probability across all possible shapes. After marginalizing over the shape parameters, the posterior for frequency and phase is inversely proportional to the multiplicity of the set of counts of events in phase bins. In a large-count limit, that is approximately $\exp(-\chi^2/2)$, providing a tie to χ^2-EF. The method has performed impressively, detecting an X-ray pulsar where the Rayleigh test failed, and performing well in a simulation study by Rots (1993) that compared it to other methods.

Despite those successes, there is room for improvement in the GL92 analysis, for a surprising reason. The constant prior adopted in GL92 does not in fact spread probability over all possible shapes. As the number of bins increases, the constant prior assigns ever larger probability to the neighborhood of flat models, making it harder than necessary to detect narrow peaks. The reason is a "curse of dimensionality" known as *concentration of measure*: a multi-dimensional distribution built out of the product of broad one-dimensional distributions with finite moments concentrates its probability in a decreasing volume of parameter space as dimension increases. Concentration can be avoided by letting the parameters of the one-dimensional component distribution vary with the target dimension. A theoretically appealing way to do that is to require *divisibility* of the prior; for example, the four-bin prior should reduce to the two-bin prior if we create a two-bin model out of the combination of bins 1 and 2, and 3 and 4. A divisible Dirichlet distribution prior accomplishes that, and improves the sensitivity to sharply peaked light curves so long as any constant background component is small (Loredo 2011).

Extending the construction to functions described with an infinite number of bins or points leads one to consider *infinitely divisible* priors for non-parametric functions: Gaussian process priors for curve fitting, Dirichlet process priors for modelling probability densities, and Lévy process priors for modelling Poisson intensities. With a team of statistician and astronomer colleagues, I am developing methods using priors built on Lévy processes for modelling pulses in gamma-ray bursts. This approach can quantify uncertainty even in a regime where pulses are highly overlapping. Its implementation involves compound Poisson processes, as arise in simple models of accumulation of rain, where drops with a distribution of sizes fall radomly over a region of space. And so Bayesian modelling of arrival-time series has led us from sines to steps to droplets.

Acknowledgements

Some of the research summarized here is currently supported by grant NNX09AK60G from NASA's Applied Information Systems Research Program. I am grateful to my collaborators on that grant, Mary Beth Broadbent, Carlo Graziani, Jon Hakkila, and Robert Wolpert, for their continuing contributions to this research.

References

Gregory, P. C. & Loredo, T. J. 1992, *ApJ*, 398, 146

Lewis, D. A. 1994, in: J. L. Stanford & S. B. Vardeman (eds.), *Statistical methods for physical science, Methods of Experimental Physics Vol. 28* (San Diego: Academic Press), p. 349

Loredo, T. J. 2011, in: J. M. Bernardo, M. J. Bayarri, J. O. Berger, A. P. Dawid, David Heckerman, A. F. M. Smith, & M. West (eds.), *Bayesian Statistics 9* (Oxford: Oxford Univ. Press), p. 361

Rots, A. H. 1993, in: P. J. Serlemitsos & S. J. Shrader (eds.), *BBXRT: a preview to astronomical X-ray spectroscopy in the 90's* (Maryland: NASA/GSFC), p. 363

New Horizons in Time-Domain Astronomy
Proceedings IAU Symposium No. 285, 2011
R.E.M. Griffin, R.J. Hanisch & R. Seaman, eds.

© International Astronomical Union 2012
doi:10.1017/S1743921312000312

Variable Stellar Object Detection and Light Curves from the Solar Mass Ejection Imager (SMEI)

R. A. Hounsell[1], M. F. Bode[1], M. J. Darnley[1], D. J. Harman[1], P. P Hick[2], A. Buffington[3], B. V. Jackson[3], J. M. Clover[3] and A. W. Shafter[4]

[1] Astrophysics Research Institute, Liverpool JMU, Birkenhead, CH41 1LD, UK
email: rah@astro.livjm.ac.uk

[2] San Diego Supercomputer Center, University of California, La Jolla, CA 92093, USA

[3] Center for Astrophysics and Space Sciences, Univ. of California, La Jolla, CA 92093, USA

[4] Department of Astronomy, San Diego State University, San Diego, CA 92182, USA

Abstract. With the advent of surveys such as the Catalina Real-Time Transient Survey, the Palomar Transient Factory, Pan-STARRS and GAIA, the search for variable objects and transient events is rapidly accelerating. There are, however important existing data-sets from instruments not originally designed to find such events. One example of such an instrument is the Solar Mass Ejection Imager (SMEI), an all-sky space-based differential photometer which is able to produce light curves of bright objects ($m \leqslant 8$) with a 102-minute cadence. In this paper we discuss the use of such an instrument for investigations of novæ, and outline future plans to find other variable objects with this hitherto untapped resource.

Keywords. stars: novæ, cataclysmic variables; astronomical data bases: miscellaneous; instrumentation: photometers

1. Introduction

The Solar Mass Ejection Imager (SMEI) is a high-precision white-light differential photometer (Buffington *et al.* 2006; Buffington *et al.* 2007) on board the CORIOLIS solar satellite, and has been in operation since January 2003 (Eyles *et al.* 2003). The instrument consists of three baffled CCD cameras, each with a $60° \times 3°$ field of view, combining to sweep out almost the entire sky every 102 minutes with each orbit of the spacecraft (Hick *et al.* 2007). The peak throughput of the instrument is at approximately 700 nm, with a FWHM ~ 300 nm, and it can reliably detect brightness changes in point sources down to approximately 8th magnitude.

SMEI was originally designed to map large-scale variations in heliospheric electron densities from Earth orbit by observing the Thomson-scattered sunlight from solar-wind electrons in the heliosphere (Jackson *et al.* 2004). In order to isolate the faint Thomson-scattered sunlight, the much larger white-light contributions from the zodiacal dust cloud, the sidereal background and individual point sources (bright stars and planets) must be determined and removed (Hick *et al.* 2007 give details of the method). Brightness determination of point sources is therefore a routine step in the SMEI data analysis. SMEI is thus capable of providing unparalleled light curves with 102-minute cadence for many bright variable objects, allowing valuable opportunities to explore their fundamental parameters.

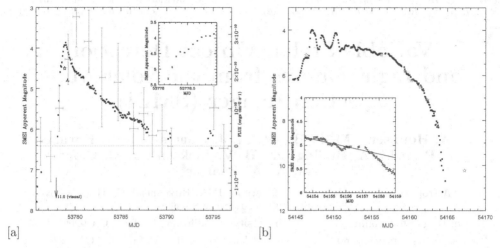

Figure 1. (a) SMEI light curve of RS Oph (black squares) in terms of "SMEI magnitude" (defined by Buffington *et al.* 2007) *vs.* time (left-hand y-axis). Over-plotted (grey) are the SWIFT BAT 14–25 keV data from Bode *et al.* (2006), right-hand y-axis. (b) SMEI light curve of V1280 Sco (black squares); superimposed (grey stars) are data from the "π of the Sky" project. The inset shows the region around the light-curve break which is associated with the onset of dust formation. The solid line indicates the fit to the pre-break SMEI light curve and its extrapolation. Images have been reproduced by permission of the AAS from Hounsell *et al.* 2010, which includes more details for each plot.

2. Examination of Novæ

Previous analysis of the SMEI data set (Hounsell *et al.* 2010) focused on known bright novæ, in which three classical novæ (KT Eridani, V598 Puppis, V1280 Scorpii) and two recurrent novæ (RS Ophiuchi and T Pyxidis; see Hounsell *et al.* 2011 for the latter) were examined. The light curves produced by SMEI are unprecedented in their detail, and reveal (among other features) the epoch of the initial explosion, the reality and duration of any pre-maximum halt (found for all fast novæ in our sample, see Fig. 1b and Fig. 2a and b), the presence of secondary maxima (V1280 Sco, Fig. 1b), quasi-periodic variations (T Pyx, Fig. 3), the speed of decline of the initial light curve, plus precise timing of the onset of any dust formation (V1280 Sco, Fig. 1b). The SMEI data have also confirmed that two of the novæ (KT Eri and V598 Pup), which reached naked-eye brightness at peak, were undetected at that point and were only later re-discovered.

3. Current and Future Work

With approximately 8.5 years of all-sky photometry the SMEI archive contains a wealth of untapped data on many bright variable objects, many of which may be previously unidentified. Basing their work on an extrapolation of the observed nova density in M31, Darnley *et al.* (2006) have estimated a Galactic nova rate of $34^{+15}_{-12} yr^{-1}$, which is consistent with the previous estimate of 35 yr^{-1}± 11 derived by Shafter (1997). Of that subset Shafter (2002) estimates that roughly five novæ per year should reach $m_V = 8$ or brighter. However, in reality only one or two are actually detected, but nevertheless observations of many of those missed novæ must reside within SMEI data. To find such missing novæ a local point maximum search on four years' worth of SMEI data from camera 2 has been conducted. Initial results of the search have revealed the presence of over 1000 bright variable-object candidates, several of which are not found in the General

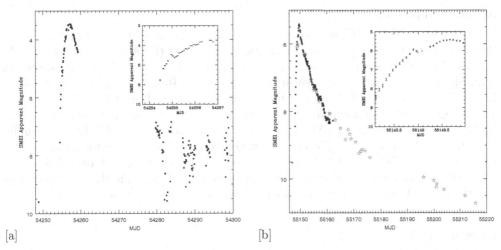

[a]   [b]

Figure 2. (a) SMEI light curve of V598 Pup. The inset shows the rising portion of the light curve with an expanded time scale. (b) The black squares show the SMEI light curve of KT Eri, observed with Liverpool Telescope SkyCamT (LTSCT, described by Steele *et al.* 2004 and also in **http://telescope.livjm.ac.uk/Info/TelInst/Inst/SkyCam/**) data superimposed (grey stars). These images have been reproduced by permission of the AAS from Hounsell *et al.* 2010, which also gives more details for each plot.

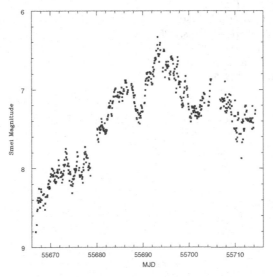

Figure 3. SMEI light curve of the 2011 outburst of T Pyx. The light curve clearly shows the quasi-periodic variations, and ends as the nova approached the Sun-wards exclusion zone (see Hounsell et al. 2011 for further details).

Catalogue of Variable Stars. We are currently conducting a search of the remaining years of data from all SMEI cameras. We hope to classify all variables found whilst also looking for novæ and other transients amongst them.

This work has demonstrates the importance of data sets like SMEI, and what a wealth of information can (unexpectedly) be held within such archives. SMEI data by themselves can thus provide significant information about many bright variables, but in combination with other all-sky data sets such as STEREO and potentially GAIA, SMEI can become an even more powerful tool.

Acknowledgements

The USAF/NASA SMEI is a joint project of the University of California San Diego, Boston College, the University of Birmingham (UK), and the Air Force Research Laboratory. The Liverpool Telescope is operated on the island of La Palma by Liverpool John Moores University in the Spanish Observatorio del Roque de los Muchachos of the Instituto de Astrofisica de Canarias with financial support from the UK Science and Technology Facilities Council. We thank Gerry Skinner for provision of the BAT data on RS Oph and pointing us to the AFOEV observations of the outburst. P.P. Hick, A. Buffington, B.V. Jackson, and J.M. Clover acknowledge support from NSF grant ATM-0852246 and NASA grant NNX08AJ11G. A.W. Shafter acknowledges support from NSF grant AST-0607682. R. Hounsell acknowledge support from STFC postgraduate studentships and the Royal Astronomical Society for providing a supportive grant.

References

Bode, M. F., *et al.* 2006, *ApJ*, 652, 629
Buffington, A., *et al.* 2007, *SPIE*, 6689,
Buffington, A., Band, D. L., Jackson, B. V., Hick, P. P., Smith, A. C., 2006, *ApJ*, 637, 880
Darnley, M. J., *et al.* 2006, *MNRAS*, 369, 257
Eyles, C. J., *et al.* 2003, *Solar Phys*, 217, 319
Hick, P., Buffington, A., Jackson, B. V., 2007, *SPIE*, 6689,
Hounsell, R., *et al.* 2011, *ATel*, 3373, 1
Hounsell, R., *et al.* 2010, *ApJ*, 724, 480
Jackson, B. V., *et al.* 2004, *Solar Phys*, 225, 177
Shafter, A. W., 1997, *ApJ*, 487, 226
Shafter, A. W., 2002, *AIPC*, 637, 462
Steele, I. A., *et al.* 2004, *SPIE*, 5489, 679

New Horizons in Time-Domain Astronomy
Proceedings IAU Symposium No. 285, 2011
R.E.M. Griffin, R.J. Hanisch & R. Seaman, eds.

© International Astronomical Union 2012
doi:10.1017/S1743921312000324

Surveying the Bright Sky

Arne A. Henden

AAVSO, 49 Bay State Road, Cambridge, MA 02138
email: arne@aavso.org

Abstract. The AAVSO is initiating a new survey of the sky. It will cover the entire visible sky, both north and south, on a daily basis, in two colours, and with a limiting magnitude of $V = 17$. This will be a perfect complement to LSST, but will be available years earlier and will continue into the indefinite future. The photometry will be publicly available within 24 hours through our website. Some details of the hardware and operations are described.

Keywords. surveys, techniques: photometric, stars: variables: other

1. Introduction

It would be wonderful if we had continuous coverage of the night sky, with multi-spectral information about every star visible to the faintest limits and with nearly infinite time-sampling. This utopian goal cannot be reached with modern technology, so any survey will only partially fill the parameter space. Perhaps not surprisingly, not many surveys have so far attempted to pursue that goal. Tycho, the star-mapper instrument on-board the HIPPARCOS satellite (ESA 1997), covered the entire sky during the 4-year mission, with B_t and V_t photometry down to 9th magnitude per observation (3σ). An average of 130 data points were acquired over the 3.5-year mission for 481553 objects.

The All-Sky Automated Survey (ASAS; Pojmanski 2002) started as a southern system at Las Campanas Observatory in 2000. It covers the southern sky (to Dec $+28°$) from $6.5 < V < 14$ mag; it has $15''$ pixels and makes approximately one visit every 2–3 days. The survey states that it is being done at V and I_c, though the I_c measurements have never been released publicly. A northern site has been operational since 2006, but only one paper (on KEPLER field variables) has been released (Pigulski *et al.* 2009), and no public data release has been made. No data have been available from the southern site since 2009, and a disk failure a few years ago resulted in a total loss of data for some time.

All other surveys only cover a portion of the night sky, or have other limitations. The wide-field exoplanet surveys like XO or SuperWASP use unfiltered CCDs. PTF and Pan-STARRs only cover either small sections of the northern sky, or only one filter, or only every few nights, and have saturation magnitudes around 14. The Catalina Sky Survey uses a northern and a southern telescope (though most of the current data come from the north), it stays away from the Galactic plane, and is only unfiltered. Future proposed surveys like LSST have saturation limits of 17th magnitude, only cover the sky in one passband at a time, return to a given field on perhaps a weekly basis, and have fixed locations (LSST at Cerro Pachon). The costs of these projects range from $100K to $400M just for the hardware, not counting the many man-years of programming effort and the continued operations cost at each site.

The AAVSO has begun an ambitious survey of the night sky to fill in the parameter space between wide-field telephoto lens surveys like ASAS, and the $100M-class surveys using fixed telescopes in the north and south. This survey, the 2nd Generation All-Sky Survey (2GSS), is described below.

2. Overview

2GSS is designed to fill a limited parameter space, namely:
- all-sky coverage
- cadence once per night
- dynamic range $10 < V < 17$
- spatial resolution around $2''$
- at least two simultaneous filters
- continuous operation for the indefinite future
- immediate public access to the database

All-sky cover is accomplished through the use of multiple sites around the world. A minimum of two sites would be required for north/south coverage. Owing to expected weather problems and the actual single-system coverage, we will install several systems in both hemispheres. We feel that all-sky is an important aspect, as we are primarily interested in variable-star monitoring, and several rare classification prototypes are found in each hemisphere. At the same time, transient discoveries are indiscriminant regarding their location.

While previous surveys have provided weekly or daily cadence over a small region of sky, we wish to cover the available night sky every 24 hours. Allowing for the solar and lunar avoidance regions, that amounts to about 30,000 square degrees to be covered nightly. It ensures that any variable stars with periods much longer than one day will have their light curves well covered; shorter-period stars with known period will have good coverage over multiple cycles; transient objects will be detected within one day of outburst or decline.

Tycho covered most stars down to about 9–10mag. ASAS covers down to 14 mag, though with blending through the Galactic Plane. LSST saturates at 17 mag. We've decided that the optimal magnitude range is 10–17, to mesh with the faint end of Tycho and the bright end of LSST, and also to provide 3 magnitudes of extended range beyond ASAS.

One of the biggest problems with ASAS is the $15''$ pixel size, which generally means that objects closer than $30''$ are measured as one. This blending causes problems in the Galactic Plane, when several stars are often contained within the aperture, and deciding which is the variable, and what is its true amplitude, is difficult. It only gets worse as you go fainter and more stars become visible. For that reason, we have opted to go with finer pixel resolution at the expense of field coverage per image, and to increase the number of systems and the cadence to recover the field coverage.

For deriving knowledge of the type of object being studied, having two or more simultaneous pass-bands of information helps enormously. You then know whether an object is red or blue, you know whether the variation has the same amplitude at both wavelengths, and you can discriminate against cosmic ray and other defects by imaging the same field with two separate systems. These are particularly important for transients, and such a survey will give valuable hints as to how to conduct the far more ambitious LSST survey.

We are committed to providing ready access to all of the data from the 2GSS survey. The standard photometry will be available within 24 hours, and we are planning to release the transient events within a few minutes after an exposure is completed. Longevity is equally important. The AAVSO database of observations dates back over 100 years, and our new survey is a key element of our future plans. We intend it to be a dynamic project, changing as technology improves but continuing into the indefinite future.

3. Hardware Description

We are pushing commercial technology as much as possible, while remaining cost-effective. We anticipate 5 "nodes" in the final configuration: 3 in the north as a hedge against poorer weather, and 2 in the south. We are using seed funding from the Robert Martin Ayers Sciences Fund to develop the first of the northern nodes, to be sited in southern Arizona. Each node will cover about 8,000 deg.² per night, and consists of two telescope/camera systems, either on a single mount or on two smaller mounts. Since this one is a prototype system, we are trying out several configurations of telescopes and mounts to find the optimal solution. Each node will have, inside a single enclosure,

- telescope: (2) APM305 305-mm aperture, f/2.8 focal ratio; corrected prime focus
- camera: (2) FLI ML16803; liquid cooled, 4096 × 4096 pixels, 9-μ pixels.
- filter holder; one filter per telescope, either g' or i'
- Paramount MX mount, one per telescope
- single linux computer running two Virtual Machines, one for each telescope system.

The telescope (seen in Fig. 1) is a unique design that uses a Keller corrector to cover the entire sensor with minimal vignetting and distortion. Since it is prime focus, we chose the Microline camera body for its small size and minimal obscuration. The FLI cameras also download in just a few seconds, minimizing the dead-time between exposures. Such a node, consisting of two telescopes and their mounts, costs $120K. Since they are small installations, we anticipate being able to locate the nodes on the grounds of existing observatories with minimal support costs. They are also small enough that they can be moved to other sites if necessary in later years.

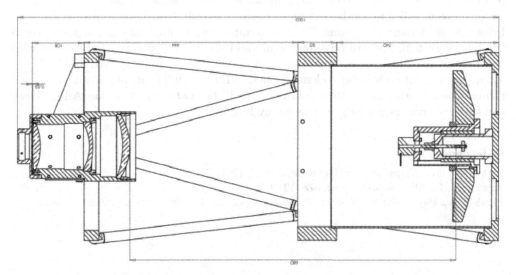

Figure 1. CAD drawing of the APM305 telescope.

4. Planned Operation

According to experimental data from our existing APASS system, which uses 20-cm telescopes and the same CCD sensor, we should reach 17th magnitude in about 15 seconds with these telescopes. We then add 5 seconds for download and 10 seconds to move to the next field, so approximately 1000 fields per night can be covered. The system has a focal length of 85.4 cm, a plate scale of $2''.17$ per pixel, and a field of view of $2°.47 \times 2°.47$ (6.12 deg.^{2}). To cover the $\sim$30,000 deg.^{2} of sky visible each night requires 4744 fields. Since each site covers 1000 fields, a total of 5 nodes is required for daily coverage, with the understanding that some fields may miss a night or two depending on weather. If poor weather persists, another node may perform double-duty, giving temporarily lower cadence but preventing long lapses in coverage.

The images are acquired on a linux computer. At the end of the night, a script is run to perform dark subtractions and flat-field each image using master calibration frames. As with our APASS pipeline, the images are then analyzed using Sextractor and DAOPHOT to obtain good aperture photometry, passed through WCS software to obtain good astrometry, and then stored on an external hard drive. Each star-list is then compared with the APASS master catalogue to obtain zero-points, and mean transformation coefficients are used to transform the data. The observations are then stored in a relational database for public retrieval. From our experiences with APASS, we anticipate this processing to take about 8 hours, so it will be completed during daylight before the next observing session.

Only calibrated star-lists will be transmitted from the local node back to a central site (which is currently planned to be an Amazon Cloud instance). The database there will be accessible through a web-based interface. Once a month local technicians will exchange the external hard drive and send the images back to AAVSO headquarters to be archived.

5. Current Status

The APM telescope is due to arrive during 2012 January. The FLI camera is already in-house and we have tested it in our electronics lab. We have filters and a Paramount, so by early 2012 we expect to be doing initial tests of a complete system. A 2-telescope node will be installed in southern Arizona during early 2012, with initial results available before the AAS meeting in June. We are submitting grant applications for purchasing a full 5-node system, and are pursuing Memorandum of Understanding agreements with appropriate sites. Our expectations are that we will be covering approximately 20,000 square degrees with a few-day cadence by mid-2012, with full configuration installed and commissioned during 2013. More information will be posted on the main AAVSO web site (http://www.aavso.org) as the survey begins.

References

ESA. 1997, *The Hipparcos and Tycho Catalogues*, ESA SP-1200

Pojmanski, G. 2002, *Acta Astronomica*, 52, 397

Pigulski, A,. Pojmański, G., Pilecki, B., & Szczygieł, D. M. 2009, *Acta Astronomica*, 59, 33

New Horizons in Time-Domain Astronomy
Proceedings IAU Symposium No. S285, 2011
R.E.M. Griffin, R.J. Hanisch & R. Seaman, eds.

High Time-Resolution Astronomy
on the 10-m SALT

Barry Welsh[1], David Anderson[1], Jason McPhate[1], John Vallerga[1],
Oswald H. W. Siegmund[1], David Buckley[2], Amanda Gulbis[2],
Marissa Kotze[2] and Stephen Potter[2]

[1] Experimental Astrophysics Group, Space Sciences Laboratory, UC Berkeley, CA 94720, USA
email: bwelsh@ssl.berkeley.edu

[2] SALT, South African Astronomical Observatory, Cape Town, South Africa

Abstract. We present a brief description of the Berkeley Visible Imaging Tube (BVIT) detector system, which is a user instrument on the 10-m Southern African Large Telescope (SALT), and include some preliminary observational results gained in mid-2011. The data show that BVIT is capable of revealing emission features occurring on time-scales of < 0.1 sec, thus opening up for the general user a window of high time-resolution astronomy at visible wavelengths.

Keywords. detector, transients

1. Introduction

The recording of emission from astronomical objects on time-scales of < 1 sec has generally not been possible at frequencies other than gamma-ray, X-ray or radio. (However, we note the exception of the ULTRACAM instrument: Dhillon & Marsh 2001). Although the next decade will be devoted to many surveys of the sky recorded (like PSST or Pan-STARRS) on time-scales of days to weeks, sparse attention has been focused on the visible variability of objects over very short time-periods. However, the Berkeley Visible Image Tube (BVIT) fills this void by allowing users to perform differential photon-counting photometric observations of faint objects at visible wavelengths with time-resolutions of micro-seconds. The BVIT (Siegmund et al. 2008) is currently a user-instrument on the 10-m Southern African Large Telescope (SALT, Buckley et al. 2010), and is ideally suited for making observations of rapid astrophysical emission phenomena associated with flare stars, low-mass X-ray binary systems, optical pulsars and cataclysmic variables. In addition, because of the high cadence of observations, BVIT can also be used in the accurate timing of eclipses in AM Her systems, exo-planet transits and stellar occultations by small solar-system bodies such as Kuiper Belt Objects. Details about both the instrument and its observational performance can be found at http://bvit.ssl.berkeley.edu.

2. The BVIT Instrument

BVIT is a simple instrument with minimal observational setup requirements and a high degree of post-acquisition data flexibility. At its core is a 25-mm-diameter microchannel plate sealed tube detector, whose SuperGen-2 photocathode converts incident visible photons into photoelectrons that are amplified by gains of up to 10^8. The resultant charge cloud is accelerated onto a delay-line anode which enables the position (x,y) and time of arrival (t) to be recorded. Similar detection schemes are currently being flown on the NASA *GALEX* and *HST-COS* missions. In Fig. 1 we show the entire BVIT instrument in its self-contained enclosure on the SALT; it includes dual B, V, R, $H\alpha$

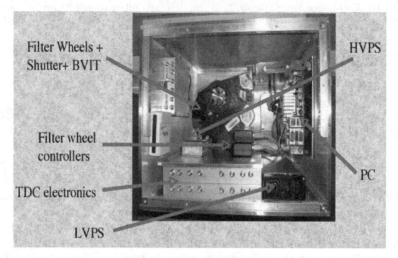

Figure 1. The BVIT detector system on SALT

and ND filter wheels, an adjustable field iris (1′.9 field of view), a field shutter and the support electronics for the detector system (HVPS, LVPS, Xilinx based time-to-digital board and a mini-ITX PC with 0.5 Tera-byte hard disk).

The 1′.9 field of view allows real-time monitoring of both the target and the sky-background time-tagged fluxes, thus enabling discrimination between fast transient events seen in the target and/or background signals. The ability to perform time-tagged photon flux monitoring of comparison stars lying within the detector field of view enables relative (differential) flux photometric comparison throughout the entire observation. Typically, sources of visible magnitude in the range $V = 12.5$ to 22.0 can be observed with high time-resolution on the SALT with BVIT.

3. BVIT Observations

BVIT observations on the SALT are currently funded by a 3-year NSF research grant. The main scientific objectives are simultaneous monitoring of low-mass X-ray binary systems with the RXTE satellite, AM Her-type eclipse monitoring, exo-planet transit timing and the search for optical counterparts to the Rotating Radio Transients (RRATs) (McLaughlin *et al.* 2006).

In this section we report on data recorded over a 4-night period (2011 June 2–6) on the SALT. Time-tagged photons were extracted in a selected radius around the position of the target, in addition to similarly selected time-tagged photons from a comparison star within the detector field of view and also photons from the sky background. The target data were then corrected for variations in the comparison star and sky background signals as a function of both time and position, to account for variable seeing/focus and any image tracking drifts. In Fig. 2 we show the results of that analysis process for the AM Her system of UZ For, displaying the egress from an eclipse at t ∼400 sec.

3.1. *GR Mus*

GR Mus (4U 1254-690) is a well-observed low-mass X-ray binary system (Barnes *et al.* 2007) that presently is in its "low-phase" of emission activity. In their "high" states these systems often show fast flickering on time-scales as short as 20 ms (50 Hz), together with longer-term eruptive events and quasi-periodic oscillation (QPO) events with frequencies less than 1 Hz. Those processes are linked to the transfer of mass from the companion

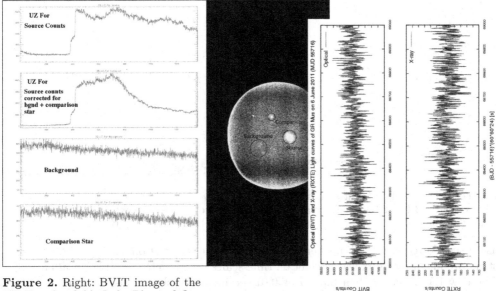

Figure 2. Right: BVIT image of the UZ For sky field. Left: Plots of flux versus time for the target star (raw and corrected), comparison star and sky background.

Figure 3. Plots of BVIT visible and RXTE X-ray flux versus time for GR Mus.

star to the accretion disk surrounding the central black hole (or neutron star). It has been shown by Shih *et al.* (2011) that the associated optical emission varies on the same time-scale as, and is correlated with, the soft X-ray emission, and demonstrates that X-ray reprocessing in such systems is mainly driven by soft X-rays. The presence of many QPOs and breaks in the frequency power-density spectrum of such X-ray binaries is thought to indicate the characteristic time-scale at some transition radius in the accretion disk.

In Fig. 3 we show flux versus time plots for thousands of data points recorded simultaneously with BVIT and the RXTE satellite on 2011 June 6. Both data sets confirm the very low state of activity of this source at present. We find no evidence for QPO activity in the power density spectrum of the BVIT data, and a cross-correlation analysis results in a null result (correlation $< \pm 0.1$) in terms of time-lags between emission in the visible and X-ray data. A full description of the treatment of these data is being prepared for publication.

3.2. *CN Leo*

CN Leo is a very active M6 flare star known for its recurrent flaring activity observed at radio, visible, UV and X-ray wavelengths (Fuhrmeister *et al.* 2008). In Fig. 4 we show B-band BVIT observations of a rarely seen double flare event occurring within a 500-sec time period on CN Leo. This double-flare takes ~1500 sec decay time to return to the flux level recorded immediately prior to the first flare. White-light flare emission is thought to consist primarily of blackbody radiation at the flare peak and hydrogen continuum during the decay time period (Zhilyaev *et al.* 2007), and as such, our B-band data are unable unambiguously to differentiate between those spectral components as a function of the flare's evolution with time. However, the high time-resolution capability of BVIT is able to reveal at least two prominent cases of pre-flaring at t = 63 sec and 74 sec (see Fig 5). In addition, radiative hydrodynamic models of short-lived (~15 sec) M-dwarf flares suggest that variations in the emitted flux should appear over time-frames as short

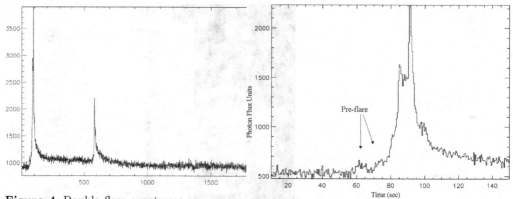

Figure 4. Double flare event seen on CN Leo with B-band BVIT observations; data binned at 1 sec.

Figure 5. Pre-flare activity on CN Leo revealed by BVIT.

as 0.1 sec (Allred *et al.* 2006), whereas previous visible observations suggested that no fine structure in emitted flux was evident on time-scales less than 0.3 sec (Beskin *et al.* 1990). The new BVIT data, when binned at 0.01-sec intervals, show emission structure at or below the 0.1-sec level. The BVIT data also reveals a small dip in the emitted flux at 1370 sec. Such diminutions in the stellar continuum after a flare event have been previously reported (Hawley *et al.* 1995) and may be due to the initial flare energy impacting on the photosphere, which leads to an increase in the hydrogen opacity and thus causing a small drop in the observed emitted flux.

4. Conclusion

Although BVIT has only been initially tested on the SALT, these early data and performance characteristics are encouraging in that it is clear that complex emission processes have signatures that can only be revealed through high time-resolved observations with cadences < 0.1 seconds. In addition, such high cadences are also important for the precise timing of various eclipse phenomena, especially exo-planet transits where variations in the eclipse ephemeris predictions could indicate the presence of other, unseen planetary objects.

References

Allred, J., *et al.* 2006, *ApJ*, 644, 484
Barnes, A., *et al.* 2007, *MNRAS*, 380, 1182
Beskin, G., *et al.* 1990, *BCrAO* 79, 67
Buckley, D., *et al.* 2010, *Proc. SPIE*, 7735, 174
Dhillon, V. & Marsh, T. 2001, *New. Astron. Revs*, 45, 91
Fuhrmeister, B., Liefke, C., Schmitt, J., & Reiners, A. 2008, *A&A*, 487, 293
Hawley, S., *et al.* 1995, *ApJ*, 453, 464
Kotze, M., *et al.* 2012 *MNRAS*, in preparation
McLaughlin, M., *et al.* 2006, *Nature*, 439, 817
Shih, I., Charles, P., & Cornelisse, R. 2011, *MNRAS*, 412, 120
Siegmund, O. H. W., *et al.* 2008, in: D. Phelan, O. Ryan & A. Shearer (eds.), *High Time Resolution Astrophysics. AIP Conf. Ser. Vol. 984,* (AIP), p. 103
Zhilyaev, B., *et al.* 2007, *A&A*, 465, 235

New Horizons in Time-Domain Astronomy
Proceedings IAU Symposium No. 285, 2011
R.E.M. Griffin, R.J. Hanisch & R. Seaman, eds.

© International Astronomical Union 2012
doi:10.1017/S1743921312000348

Pulsars

Benjamin W. Stappers

School of Physics and Astronomy, The University of Manchester, Manchester, M13 9PL,UK
email: Ben.Stappers@manchester.ac.uk

Invited Talk

Summary. Pulsars can be considered as the ultimate time-variable source. They show variations on time-scales ranging from nanoseconds to as long as years, and they emit over almost the entire electromagnetic spectrum. The dominant modulation is associated with the rotation period, which can vary from slighty more than a millisecond to upwards of ten seconds (if we include the magnetars). Variations on time-scales shorter than the pulse period are mostly associated with emission processes and are manifested as giant pulses, microstructure and sub-pulses (to name a few). On time-scales of a rotation to a few hundred rotations are other phenomena also associated with the emission, such as nulling, moding, drifting and intermittency.

By probing these and slightly longer time-scales we find that pulsars exhibit "glitches", which are rapid variations in spin rates. They are believed to be related to the interaction between the superfluid interior of the neutron star and the outer crust. Detailed studies of glitches can reveal much about the properties of the constituents of neutron stars—the only way to probe the physics of material at such extreme densities. Time-scales of about an hour or longer reveal that some pulsars are in binary systems, in particular the most rapidly rotating systems. Discovering and studying those binary systems provides vital clues to the evolution of massive stars, while some of the systems are also the best probes of strong-field gravity theories; the elusive pulsar-black hole binary would be the ultimate system.

Pulsars are tools that allow us to probe a range of phenomena and time-scales. It is possible to measure the time of arrival of pulses from some pulsars to better than a few tens of nanoseconds over years, making them some of the most accurate clocks known. Concerning their rotation, deviations from sphericity may cause pulsars to emit gravitational waves which might then be detected by next-generation gravitational-wave detectors. Pulsars themselves can be used as the arms of a Galactic-scale gravitational-wave detector. Measuring correlated deviations in the arrival times of pulses from a number of pulsars distributed throughout the Galaxy could give rise to a direct detection of the stochastic gravitational-wave background, which is associated with the astrophysics of the early Universe—most likely from supermassive black-hole binary systems, but potentially also from cosmic strings. While they are famed for their clock-like rotational stability, some pulsars—in particular the more youthful ones—exhibit modulation in pulse arrival times, often called timing noise. It was recently demonstrated that in at least some cases this variability is deterministic and is associated with modulations in the pulsar emission properties and the spin-down rate. This breakthrough may lead to further improvements in the precision which can be achieved with pulsar timing, and enhance still further the ability to test theories of gravity directly and to make a direct detection of gravitational waves.

I presented some of the history of what is known about the variations in pulsars on all these time-scales and reviewed some of the recent achievements in our understanding of the phenomena. I also highlighted how new transients associated with radio-emitting neutron stars are being discovered, and how other transient sources are being identified by the same techniques. These continued improvements have come about without new telescopes, but the next generation of very sensitive wide-field instruments will permit observational cadences which will reveal many new manifestations and will further revolutionise our understanding of this class of objects which have such high astrophysical potential.

Keywords. stars:pulsars:general, radiation mechanisms:general, gravitational waves, ISM:general

New Horizons in Time-Domain Astronomy
Proceedings IAU Symposium No. 285, 2011
R.E.M. Griffin, R.J. Hanisch & R. Seaman, eds.

Charting the Transient Radio Sky on Sub-Second Time-Scales with LOFAR

J. W. T. Hessels[1,2] (and the LOFAR Transients Key Science Project)

[1]Netherlands Institute for Radio Astronomy (ASTRON), 7990 AA Dwingeloo,
The Netherlands

[2]Astronomical Institute "Anton Pannekoek," University of Amsterdam, The Netherlands
email: J.W.T.Hessels@uva.nl

Summary. The LOw Frequency ARray (LOFAR) is a radio interferometric telescope that promises to open a largely unexplored window on transient sources in the "radio sky", from time-scales of nanoseconds to years. An important aspect of this will be the study of radio-emitting neutron stars in their various incarnations: slow pulsars, young pulsars, millisecond pulsars, magnetars, rotating radio transients, intermittent pulsars, et cetera. Pulsars and their brethren are the prototype of the more general "fast transients": sub-second, dispersed radio bursts which point the way to extreme, and potentially still unknown, phenomena. For instance, prompt radio bursts from supernovæ and other extra-galactic bursts have been hypothesized; these could prove to be powerful cosmological probes.

This talk discussed LOFAR's impressive ability to observe pulsars and to enlarge greatly the discovery space for (even rarer) fast transients. It also presented the latest pulsar observations made during LOFAR's commissioning period. These are demonstrating powerful observing techniques that will be crucial for the next generation of radio telescopes as well as the effort to increase our understanding of the dynamic nature of the Universe.

An expanded version of the talk can be found at http://adsabs.harvard.edu/abs/2011A

New Horizons in Time-Domain Astronomy
Proceedings IAU Symposium No. 285, 2011
R.E.M. Griffin, R.J. Hanisch & R. Seaman, eds.

© International Astronomical Union 2012
doi:10.1017/S1743921312000361

Probing the Physics of Planets and Stars with Transit Data

Suzanne Aigrain

University of Oxford, Department of Physics, Keble Road, Oxford, OX1 3RH, UK
email: Suzanne.Aigrain@astro.ox.ac.uk

Invited Talk

Summary. Virtually all exoplanet detection and characterisation methods are based on time-domain data. This invited talk gave an overview of some recent results in the field, highlighting some of the time-series-specific challenges encountered along the way. In particular it focussed on planetary transits: how to detect shallow, rare transits in noisy data, and how to model them with extreme accuracy to extract information about the transiting planet's atmosphere. Space-based transit surveys also constitute an extraordinary goldmine of information on stellar variability, and the talk touched briefly upon some recent statistical work in that field.

Asteroseismology

Don Kurtz

Jeremiah Horrocks Institute, University of Central Lancashire, Preston, PR1 2HE, UK
email: dwkurtz@uclan.ac.uk

Invited Talk

Summary. In 1926 in the opening paragraph of his now-classic book, *The Internal Constitution of the Stars*, Sir Arthur Eddington lamented, "What appliance can pierce through the outer layers of a star and test the conditions within?" While he considered theory to be the proper answer to that question, there is now an observational answer: asteroseismology. This talk introduced the concepts of asteroseismology, then looked at a selection of discoveries made for "ticking things" with the micromagnitude precision light curves of the KEPLER Mission.

Day 4:

Irregular and Aperiodic Changes

New Horizons in Time-Domain Astronomy
Proceedings IAU Symposium No. 285, 2011
R.E.M. Griffin, R.J. Hanisch & R. Seaman, eds.

© International Astronomical Union 2012
doi:10.1017/S1743921312000385

Variability in Active Galactic Nuclei

Erin Wells Bonning

Yale Center for Astronomy & Astrophysics, New Haven, CT 06520, USA
email: `erin.bonning@yale.edu`

Invited Talk

Abstract. This talk explored variability in active galactic nuclei (AGN) for a variety of scales across the time domain. From billion-year-scale intermittency to a quasi-periodic oscillation signal with a period of one hour, time-varying signals offer insights into a myriad of complex processes driven by the AGN central engine. Athough the era of time-domain observations of AGN across the spectrum has but just begun, already observations reveal the rich detail of phenomena associated with actively accreting black holes which challenge theoretical models.

An active galactic nucleus (a.k.a. AGN, Seyfert, quasar, etc.) is an accreting super-massive black hole (BH). A jet may or may not be present. Accreting stellar-mass black holes are well-studied in our Galaxy, and it can be instructive to compare the two cases. Stellar-mass black holes are observed as X-ray binaries (XRBs), where the BH accretes material from a companion star. The hot accretion-disk emission peaks in X-rays. In contrast, supermassive black holes have relatively cooler disks, peaking in the rest-frame UV. In both cases a jet may be produced by relativistic particles collimated by magnetic fields and ejected along a single axis.

1. Intermittency in AGN

Owing to the smaller mass of the BH in an X-ray binary, the associated time-scales are proportionally smaller and transient phenomena such as the formation and cessation of jets, quasi-periodic oscillations and disk variability are easily observed on human time-frames. For example, in XRBs, jets are observed to arise or vanish as a function of accretion state. In AGN, the relevant timescales are $\sim 10^7$–10^8 years, so the analogous transitions must be studied in large statistical samples. Nevertheless, some direct evidence of intermittency of AGN on long time-scales has been seen. For example, in the class of double-double radio galaxies (DDRGs), or X-shaped radio galaxies, discontinuous or cross-shaped radio emission is observed which is interpreted as due to the jet switching on and off. Another example is the case of Perseus A, a powerful radio galaxy at the centre of a massive cluster. The hot X-ray gas surrounding Perseus A shows distinct ripples around radio bubbles, evidence that its jet has been intermittently heating the cluster gas. In all of these examples, the inferred time-scale is on the order of 10^7–10^8 years, as expected if AGN behave as scaled-up accreting stellar-mass black holes.

2. Quasi-Periodic Oscillations

The X-ray power spectra of AGN and XRBs show a characteristic broken power law, with the break frequency scaling as the BH mass. X-ray binaries are also known for showing quasi-periodic oscillations (QPOs) in their light curves. The origin of these QPOs is not definitively understood, with a multitude of theoretical models proposed for their explanation. In general, they are thought to arise from some disturbance in the disk, with the frequency related to one of the disk dynamical, thermal, or other time-scales. The search for QPOs in AGN has been frustratingly difficult, with many claimed QPOs that are ultimately deemed statistically insignificant. There is one exception, a clear detection (5σ) of a one-hour period in the AGN RE J1034+396. This source is a high accretion-rate (possibly super-Eddington) narrow-line Seyfert 1. It is unknown why it has shown a QPO when so many others have not. Further study has revealed that the QPO was transient—it has not returned since the original observation. Long periods of decades or more have been searched for other sources in archives of historical observations; however, the significance of periodicity thus detected must remain low, as few cycles have ever been observed, and the number of historical observations of a given source is often sparse.

3. Measuring Black-Hole Masses

Time-domain studies of AGN prove very useful for determining BH masses from X-ray variability. BH masses in AGN are difficult to measure owing to the bright point source near the BH. Existing methods tend to rely on measurements of broad emission-line gas or invoke empirical black-hole–galaxy relations for narrow-line AGN. However, those relations may not necessarily hold for obscured AGN. The power spectra of AGN and XRBs are similar, and are characterized by a broken power-law with the break frequency proportional to the BH mass. However, measuring the X-ray power spectrum of AGN is very difficult owing to their low count rates. Alternatively, to determine the break frequency even for low-quality light curves for distant or less-luminous sources one can exploit the fact that the normalized excess variance of the light curve is equivalent to the integral under the high-frequency side of the broken power-law. A project that I carried out with James Kim successfully recovered known BH masses from AGN light curves for a small test sample, promising well for future studies of AGN in large surveys.

4. Blazar Observations with FERMI

One of the most recent advances in time-domain astrophysics came with the launch in 2008 of the FERMI gamma-ray space telescope. FERMI images the entire sky once every 3 hours, producing unprecedented coverage of transient and variable phenomena. This is central for studying blazars—relativistic jets from AGN which lie along the line of sight. These highly variable sources emit across the electromagnetic spectrum; the low-energy part of the SED is caused by synchrotron radiation from relativistic electrons in the jet, the high-energy (gamma-ray) peak mainly caused by inverse Compton scattering of ambient and synchrotron photons off those electrons. Numerous multi-wavelength campaigns, from radio to gamma-rays, have been carried out to image blazars synchronously, with results that show correlations between low and high energies (confirming the inverse Compton hypothesis); they link ejection of radio knots with large gamma-ray flares and changes in polarization (revealing the radio core to be a standing shock in the jet), and detect underlying disk emission from colour-magnitude relations. Unexpected phenomena have also been observed, such as a possible disk–jet connection related to changes in accretion state, and hysteresis in the spectral slope during flares.

5. Acknowledgments

I offer my thanks to colleagues at Yale in the blazar and AGN group, and in particular to James Kim. A Debra Fine fellowship is warmly acknowledged. I am also grateful for the opportunity to speak at this very enriching and rewarding symposium.

New Horizons in Time-Domain Astronomy
Proceedings IAU Symposium No. 285, 2011
R.E.M. Griffin, R.J. Hanisch & R. Seaman, eds.

© International Astronomical Union 2012
doi:10.1017/S1743921312000397

Variable Red Giants

Franz Kerschbaum and Walter Nowotny

University of Vienna, Department of Astronomy, A-1180 Wien, Austria
email: `franz.kerschbaum@univie.ac.at`

Invited Talk

Abstract. The longest-known class of pulsating variable stars, namely pulsating red giants, is also the one that involves the most complex physical processes. Pulsation, mass loss, nuclear synthesis, mixing, atmospheric and circumstellar chemistry and dust formation all interrelate with one another and make both the observational studies and the modelling efforts quite challenging. The paper outlines some of the current key questions, and recommends observational strategies.

Keywords. stars: late-type, stars: AGB and post-AGB, stars: variables: other, stars: oscillation, stars: winds, outflows

---

1. Introduction

Within a publication mainly devoted to Venus and Mercury transits, Hevelius (1662) added *Quibus accedit succincta Historiola, novæ illius, ac **miræ** stellæ in collo ceti, certis anni temporibus clare admodum affulgentis, rursus omnino evanescentis.*† That text gave the first report of the early observations of a very special star, discovered by David Fabricius on 1596 August 13 in the constellation Cetus. After fading during the ensuing months it was seen to brighten again about ten years later, which was rated by observers of the 16th or early 17th century as very unusual behaviour. From then on it received increasingly more attention, and in 1639 Johannes Holwarda was the first to derive a period of 11 months. *o Ceti*, or (following Hevelius) *Mira*, proved to be the first pulsating variable to be discovered, and thereby opened a new line of astronomical research.

Today we know Mira as red giant star on the Asymptotic Giant Branch (AGB, Habing & Olofsson 2004; Kerschbaum *et al.* 2011). This article is focused mainly on AGB stars but also addresses similar time-variable phenomena of Red Giant Branch (RGB) and Red Supergiant stars.

Both AGB and RGB stars exhibit a wide range of time-variable observables: luminosity, colour, colour index and the whole spectral energy distribution, as well as spectral features like line or band strengths, line shapes and radial velocities. The period itself may also be variable, as too can be the light-curve's shape, the chemical composition, the overall morphology, the diameter, the mass loss ... and many more features. Those observables and related derived quantities all vary on time-scales of a few days or even a few hours, up to 10^4 years. What a challenge to observational "time-domain" astronomy!

† "to which he has attached a short history of that new, wonderful star in the neck of Cetus, which is clearly visible during certain times of the year, but is afterwards completely invisible."

2. AGB Variables

2.1. *Pulsating atmospheres and mass loss*

The outer layers of very evolved stars on the AGB are strongly influenced by dynamical effects, and deviate significantly from those of less evolved red giants which on the whole show a hydrostatic configuration (Nowotny *et al.* 2010). Radial pulsations excited in layers below the photosphere lead to time-dependent variation of the atmospheric structure. On top of that, radiation pressure on dust particles which can form in the levitated cool layers cause an outflow of gas and dust. As the time-scale of pulsation differs from that of dust formation, the dynamic behaviour in the dust-forming region can differ from one object to another. Although it remains challenging to model the complex interplay of the physical processes present in the atmospheres of pulsating and mass-losing giants on the AGB, significant progress has been made in the last years (see Höfner 2009). Fig. 1 shows self-consistent dynamic model atmospheres simulating the dust-driven stellar wind occurring in the atmosphere of a C-type Mira. It starts with the initial hydrostatic model ($L_\star = 7000\,L_\odot$, $T_\star = 2600\,\mathrm{K}$, C/O $= 1.4$), and the dynamic effects are introduced by a variable inner boundary ($P = 490^{\mathrm{d}}$, $\Delta u_{\mathrm{p}} = 6\,\mathrm{km\,s}^{-1}$). As discussed in detail by Höfner *et al.* (2003) or Nowotny *et al.* (2010, 2011), it leads to a realistic description of the photometric variations ($\Delta m_{\mathrm{bol}} \approx 1^{\mathrm{mag}}$) and of the resulting mass loss ($\langle \dot{M} \rangle \approx 2.5 \cdot 10^{-6}\,M_\odot\mathrm{yr}^{-1}$, $\langle u \rangle \approx 7.5\,\mathrm{km\,s}^{-1}$).

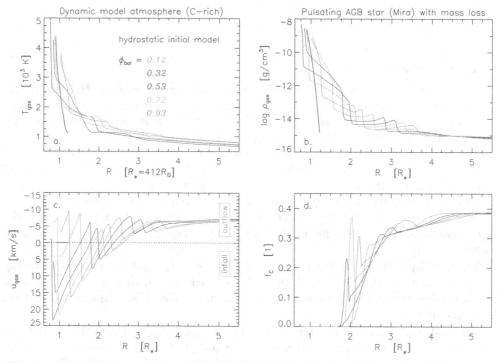

Figure 1. Radial structures of a model atmosphere resembling a typical Mira variable with dusty outflow. A few selected phases (ϕ_{bol}) during a pulsation cycle are compared with the corresponding hydrostatic model atmosphere (thick black). From Nowotny *et al.* (2010).

2.2. *Multi-colour light curves*

A fundamental approach to characterise a variable AGB star and to test modelling efforts as decribed above is the monitoring of the photometric variations throughout the

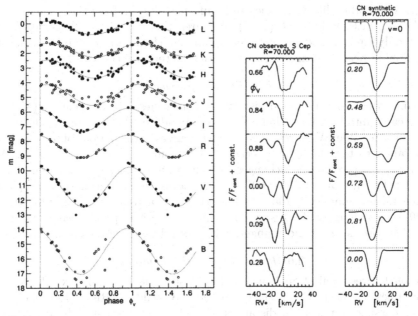

Figure 2. Left panel: Observed photometric variations of the C-type Mira RU Vir in various broad-band filters from the visual to the NIR (right-hand ordinate). Measurements from different periods (open/filled circles) are merged into a combined light cycle; each data point is then plotted twice. Dotted lines represent sinusoidal fits. From Nowotny *et al.* (2011). Two right hand panels: A comparison of observed (middle) and synthetic (right) line-profile variations. The plotted CN lines, which can be found near 2 μm in high-resolution spectra of long period variables, show characteristic behaviour throughout the light cycle, with line doubling around phases of maximum visual light. From Nowotny *et al.* (2010).

light-cycle from the visual to the NIR, thus covering the bulk of the spectral-energy distribution. Nowotny *et al.* (2011) compiled such multi-colour light curves exemplarily for the Mira RU Vir by adopting data from Eggen (1975; *BVRI*) and Whitelock *et al.* (2006; *JHKL*); the result is shown in Fig. 2.

2.3. *Velocities and line-profile variations*

Another interesting way to investigate dynamic effects in AGB stars is to study line-profile variations in high-resolution spectra of such objects. The complex velocity fields in AGB atmospheres (cf. Fig. 1c) strongly influence molecular line profiles (shapes, time-dependent shifts in wavelength, multiple components), see Fig. 2. Radial velocities derived from Doppler-shifted spectral lines provide information about velocities in the corresponding line-forming region. Time-series high-resolution spectroscopy (mainly in the NIR) thereby allows one to probe the atmospheric kinematics, such as the propagation of shock waves within the atmosphere (Hinkle *et al.* 1982; Nowotny *et al.* 2010).

2.4. *Pulsation, spatial resolution*

The large physical size and high absolute IR luminosity of AGB stars make them ideal targets for high-resolution spatial observations to resolve their stellar diameters. That ultimatively provides critical tests for current modelling efforts and our understanding of the stellar physics. As typical examples, Woodruff *et al.* (2008) were able to monitor the angular diameters of AGB variables at various wavelengths, probing different atmospheric layers and even dust-forming zones by means of near-infrared aperture masking using the Keck I Telescope. Paladini *et al.* (2009) estimated time- and wavelength-dependent interferometric observables like visibilities and uniform disk radii from their dynamical model atmospheres. Wittkowski *et al.* (2011) compared their VLTI-AMBER spectroscopy to self-excited dynamical model-atmosphere diameters (Ireland *et al.* 2011), demonstrating the potential of near- and mid-infrared interferometers like the VLTI.

2.5. *Changing periods*

For variables whose observational history can sometimes exceed a century (Mattei *et al.* 2002), research into long-term changes in period and possibly also in regularity or light-curve shape is feasible. Wood related observed changes in the periods of some well-studied Mira variables to the effects of a recent thermal pulse (Wood, 1975). Later, Wood & Zarro (1981) interpreted such changes as long-term luminosity evolution, though the explanation by Zijlstra *et al.* (2004) of a chaotic feedback of atmospheric opacity and pulsations takes a completely different line. However, a systematic study by Uttenthaler *et al.* (2011), searching for the dredge-up of the radio-active *s*-process element Technetium in stars that show such period changes, ruled out a link to recent thermal pulses.

2.6. *Period-Luminosity Relations*

Correlation between pulsational period, luminosity and colour index (PL or PLC) in Mira variables has been known since the 1960s (Osvalds & Risely 1961; Feast, 1963). Owing to missing or uncertain distances of field stars, the first tight PL relation was found for Miras in the LMC (Glass & Lloyd Evans 1981). Wood *et al.* (1999) then demonstrated multiple PL(K) relations from large MACHO and OGLE datasets. The distributions in these sequences have been related to (*inter alia*) radial modes, binarity, non-radial pulsation modes, convection, mass loss and chromospheres (Wood, 2010). The well-populated, tight sequences of the fundamental and first overtone radial-pulsation modes proved to be useful for distance measurements, both in the Milky Way field and in extragalactic systems.

2.7. *Extragalactic AGB stars*

In the tradition of the pioneering work by Glass & Lloyd Evans (1981) on LMC Miras, the next challenge was to work outside the local group. The most distant application of Mira PL-relations was made by Rejkuba (2004 and references therein) on Cen A halo long-period variables. They confirmed an intermediate-age population there, and compared the distances from a Mira PK-relation and the RGB tip at 4 Mpc. In fact they agree very well, to within 0.05 mag. There is a rich future for the up-coming E-ELT era!

Within the local group, Whitelock *et al.* (2009) and Menzies *et al.* (2011) have recently monitored the dwarf spheroidals in Fornax and Sculptor in the near infrared. Accurate distances could be derived for both galaxies, and also clues about the star-formation history. Lorenz *et al.* (2011) monitored in total more than 700 long-period variables for two years in the dwarf spheroidals NGC 147 and NGC 185, leading not only to improved distances for the systems but also to the classification of the variables into fundamental- and overtone-mode pulsators of defined atmospheric chemistry. The different star-formation history of the galaxies was related to the differing distribution of variability classes.

2.8. *Long-term changes in mass loss*

While it is well established that mass loss increases on average during AGB evolution (Habing 1996), there is also observational evidence (Olofsson *et al.* 1988) for more episodic processes. Interferometric mm-CO maps of so-called detached shells are intriguing examples of that (Olofsson *et al.* 2000). Similar structures are also evident from dust emisson (Kerschbaum *et al.* 2010; Decin *et al.* 2011) or scattered light (Mauron *et al.* 2000). The corresponding mass-loss modulations happen on time-scales of 10^2–10^4 years, i.e. much longer than pulsational ones and shorter than thermal inter-pulse times.

3. Asteroseismology and Exoplanet Missions

Somewhat similar to the realisation of valuable output from monitoring projects like OGLE or MACHO which were not originally designed for research into red variables, astroseismology and exoplanet space missions are now also revolutionizing this field by the unprecedented availability of high-cadence, high-precision photometric data sets.

Using observational data from CoRoT, Lebzelter (2011) investigated a sample of long-period variables for the small amplitude variations that are sometimes claimed on unusually short time-scales (hours or days instead of months or years) and found a quite low rate of such events: only 0.15 per star and year.

For RGB stars the KEPLER mission is proving to be extremely fruitful. Because of the large sample sizes, ensemble seismology of hundreds of objects is now possible (Huber *et al.* 2010), and also supports statistically meaningful conclusions. By analysis of the mode spacings, Bedding *et al.* (2011) succeeded in differentiating between H-shell- and He-core-burning objects for the first time. After long searches Gravity-Mode period spacings have also eventually been identified (Beck *et al.* 2011).

Whereas KEPLER focuses on relatively small fields and faint objects, the new mission BRITE-constellation (Kuschnig *et al.* 2009) to be launched in early 2012 will be devoted to objects brighter than about $V = 5$ mag, thus allowing easy follow-up observations with medium-sized telescopes and interferometers. The Canadian-Austrian-Polish cooperation will use 6 nanosats with 3-cm-aperture telescopes to make high-precision and high-cadence observations in two filter bands. Short-term phenomena that can be expected in late-type giants include flares, convection signatures, spots, acoustic-, mixed- and gravity modes, and dimmings related to mass loss.

4. "Wish list" for Variable Red Giants

The multitude of time-variable observables and the complex, interrelated physical phenomena in the field of red variables put special constraints on observational data. A few general desiderata are mentioned below.

- Multi-wavelength monitoring is often a key to successful comparisons with theoretical models.
- Do not neglect the red, and especially the near infrared. Simultaneous visual data are preferable for studying variable light-curve shapes and amplitudes in general. "Phased" material is less useful.
- In the field of asteroseismology both ground- and space-based surveys often monitor their fields for only relatively short periods. In order to increase the usefulness of such data for studies of red variables too, we urge that they monitor for longer than a few months—or return to the same field later.
- Photometry should be carefully cross-calibrated, and supplied with transformations, to enable the data to be combined with vintage material.
- Historic photographic plates constitute a valuable tool for probing long-term amd maybe even evolutionary-scale variations. Please make that information available in an astrometric- and photometrically calibrated form!
- Don't neglect bright objects! They are perfect for follow-up work at high spectral and spatial resolution.

Acknowledgements

The authors thank the *Austrian Science Fund (FWF)* for supporting this work under project number P23586-N16 and P21988-N16.

References

Beck, P. G., *et al.* 2011, *Science*, 332, 205

Bedding, T. R., *et al.* 2011, *Nature*, 471, 608

Decin, L., *et al.* 2011, *A&A*, 534, A1

Eggen, O. J., 1975, *ApJS*, 29,77

Feast, M. W., 1963, *MNRAS*, 125, 367

Glass, I. S. & Lloyd Evans, T., 1981, *Nature*, 291, 303

Habing, H. J., 1996, *A&A Rev.*, 7, 97

H. Habing and H. Olofsson (eds.) 2004, *Asymptotic Giant Branch Stars* (Berlin: Springer)

Hevelius, J., 1662, *Mercurius in Sole visus Gedani* (Danzig: Simon Reiniger), p. 146

Hinkle, K. H., *et al.* 1982, *ApJ*, 252, 697

Höfner, S., *et al.* 2003, *A&A*, 399, 589

Höfner, S., 2009, *ASPCS*, 414, 3

Huber, D., *et al.* 2010, *ApJ*, 723, 1607

Ireland, M. J., Scholz, M., & Wood, P. R., 2011, *MNRAS*, in press

Kerschbaum, F., *et al.* 2010, *A&A*, 518, L140

F. Kerschbaum, T. Lebzelter and R. F. Wing (eds.), 2011, *Why galaxies care about AGB stars II: Shining examples and common inhabitants* (San Francisco: ASPCS), 445

Kuschnig, R., Weiss, W. W., Moffat, A., & Kudelka, O., 2009, (San Francisco: ASPCS), 416, 587

Lebzelter, T., 2011, *A&A*, 530, A35

Lorenz, D., *et al.* 2011, *A&A*, 532, A78

Mattei, J. A., Menali, H. G., & Waagen, E. O. 2002, *AAVSO Monograph 15*, XIII + 14.

Mauron, N. & Huggins, P. J., 2000, *A&A*, 359, 707

Menzies, J. W., *et al.* 2011, *MNRAS*, 414, 3492

Nowotny, W., *et al.* 2010, *A&A*, 514, 35

Nowotny, W., *et al.* 2011, *A&A*, 529, A129

Olofsson, H., *et al.* 2000, *A&A*, 353, 583

Olofsson, H., Eriksson, K., & Gustafsson, B. 1988, *A&A*, 196, L1

Osvalds, V. & Ristey, A. M. 1961, *Publ. Leander McCormick Obs.*, 11, 147

Paladini, C., *et al.* 2009, *A&A*, 501, 1073

Rejkuba, M. 2004, *A&A*, 413, 903

Uttenthaler, S., *et al.* 2011, *A&A*, 531, A88

Whitelock, P., *et al.* 2006, *MNRAS*, 369, 751

Whitelock, P., *et al.* 2009, *MNRAS*, 394, 795

Wittkowski, M. *et al.* 2011, *The Messenger*, 145, 24

Wood, P. R. 1975, in: W.S. Fitch (ed.), *Multiple Periodic Variable Stars*, IAU Coll. 29 (Dordrecht: Reidel), p. 69

Wood, P. R. 2010, *Mem.S.A.It.*, 81, 883

Wood, P. R. & Zarro, D. M. 1981, *ApJ*, 247, 247

Wood, P.R., *et al.* 1999, in: T. Le Bertre, A. Lèbre and C. Waelkens (eds.), *Asymptotic Giant Branch Stars*, Proc. IAU 191 (San Francisco: ASPCS), p. 151

Woodruff, H. C., *et al.*, 2008, *ApJ*, 673, 418

Zijlstra, A. A., *et al.* 2004, *MNRAS*, 352, 325

New Horizons in Time-Domain Astronomy
Proceedings IAU Symposium No. 285, 2011
R.E.M. Griffin, R.J. Hanisch & R. Seaman, eds.

© International Astronomical Union 2012
doi:10.1017/S1743921312000403

Polarimetric Variability

Stephen B. Potter

South African Astronomical Observatory, Cape Town, South Africa
email: sbp@saao.ac.za

Invited Talk

Abstract. I present new observations of galactic and extra-galactic polarized variable sources, and demonstrate the science that one can obtain with the appropriate instrumentation.

1. Magnetic Cataclysmic Variables

Intermediate polars (IPs) consist of a red dwarf (known as the secondary or the donor star) that is filling its Roche lobe, and an accreting white dwarf (the primary; WD). They are sub-classes of the magnetic Cataclysmic Variables (mCVs) which in turn constitute about 20% of the known CV population. IPs show a large variety of observational properties which still need to be understood in terms of accretion and evolutionary state. In contrast to polars, the white dwarfs in IPs are not phase locked to the orbital period but instead spin at a different rate (asynchronous). A wide range of WD asynchronism seems to characterize this class, and has been discussed in terms of spin equilibrium driven by magnetic accretion (Norton & Wynn 2004) which can take place in a variety of ways ranging from magnetized accretion streams to extended accretion disks.

In a few systems the presence of a soft X-ray emission component, similar to that observed in the polars (Buckley 2000), raises the question of evolution: are these soft X-ray IPs the progenitors of the polars? Our understanding of the evolutionary relationships among mCVs and in particular of IPs is unfortunately still very poor, but the addition of new systems and the study of their properties in new ways has great potential to alleviate the problem.

The study of IPs not only advances our understanding of this class of object in its own right, but also provides an opportunity to explore a broad range of astrophysical phenomena, the most obvious being accretion. CVs have been pivotal in the development of accretion-disk theory, largely because they are nearby (and hence bright), they evolve on short time-scales (hours to weeks) and are therefore ideal for micro-arc-second imaging techniques such as Doppler tomography. The additional ingredient of a magnetic field in IPs adds a further dimension to the exploration of accretion.

IPs are also a source of gamma-ray, X-ray and cyclotron optical emission. The reason is that the accretion material is eventually magnetically channelled onto the surface of the WD, where a small shock region forms. The material in the shock becomes highly ionized, very dense and very hot (several 10s of KeV), leading to the emission of X-ray radiation (by Bremsstrahlung) and cyclotron radiation (optically polarized). The white dwarfs in intermediate polars typically rotate every 10–20 minutes, thus permitting an almost 180° probe of the shocks on short time-scales.

An additional recent surprise is the discovery of planets in CV systems (Potter *et al.* 2011, in press; Beuerman *et al.* 2011), which has certainly expanded the types of environments in which planet formation is thought to take place.

Fig. 1 demonstrates the importance of using the correct polarimetric instrumentation. It shows two sets of observations of the same target—the intermediate polar NY Lup.

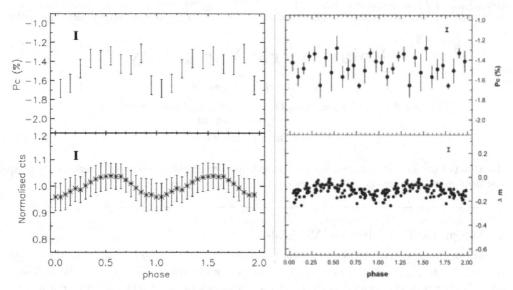

Figure 1. Photo-polarimetric observations of the IP NY Lup. The top and the bottom panels are the phase-spin-folded circular polarization and photometry, respectively. The left and right panels were made with the SAAO 1.9-m telescope and the VLT, respectively.

The upper and lower panels show spin-phase-folded circular polarization and photometry, respectively. Both data sets (left and right) show a detection of the spin period in the photometry; however, only the left data set shows a clear spin modulation in the circular polarization. Contrary to expectations, the superior data on the left were obtained with a 1.9-m telescope whereas those on the right were observed with the VLT. Moreover, the 1.9-m polarimeter (Potter *et al.* 2010) used photomultiplier detectors, which are several times lower in sensitivity compared to the CCD detector used on the VLT. The 1.9-m polarimeter also measured circular AND linear polarization simultaneously in two filters (not shown). The reason why the 1.9-m polarimeter does so well is because it is optimised to measure polarization variability. The exposure readout times of the photomultiplier tubes are very fast (sub-second), thus enabling polarization measurements to be made on a much shorter time-scale than the intrinsic variability of the polarization. The CCD readout times on the VLT were too slow, and led to the variability becoming smeared.

2. Blazars

Another source of polarimetric variability are blazars. Blazars, or BL Lacertae objects (BL Lacs), are radio-loud active galactic nuclei (AGNs) in which the relativistic jets are aligned close to the line-of-sight of the observer. They are characterised by intense and rapid variability across the electromagnetic spectrum; many of these objects demonstrate rapid and intense flaring episodes. For example, Aharonian *et al.* (2007) reported that during 2006 July PKS2155 experienced an outburst that corresponded to a Very High Energy (VHE, for $E > 200\,GeV$) flux increase of 35 times its quiescent level, some 20% Crab, i.e. 0.2 times the flux observed from the Crab Nebula (Abramowski *et al.* 2010). It has been shown that the radio and optical polarisation features of blazars are correlated (Gabuzda *et al.* 2006). The optical polarization can therefore be used as a tracer of the resolved radio state, but on time-scales relevant to VHE measurements. That implies

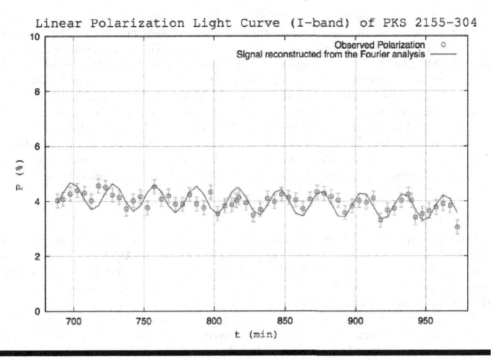

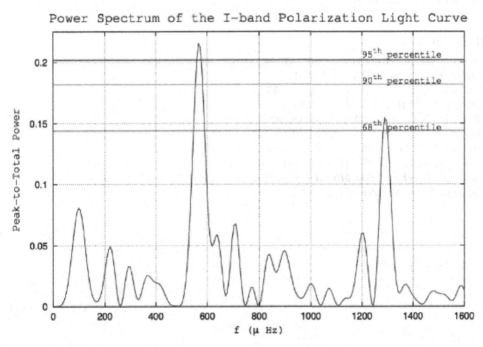

Figure 2. Variable linear polarization from the blazar PKS2155. The upper and lower panels show the linear polarization and corresponding Fourier analysis, respectively.

that we should be able to constrain the location of the emitting region in the jet as well as gaining information on the strength and orientation of the magnetic field of the jet; those are all important parameters for modelling blazar physics.

Fig. 2 shows new observations of PKS2155 taken with the polarimeter on the SAAO 1.9-m telescope. The observations were made during the 2009 flare event. The top panel shows the I-band linear polarization spanning $\sim$5 hours. What is immediately apparent is a possible quasi-periodic oscillation of $\sim$30 minutes as traced out by the solid curve. The lower plot shows the corresponding Fourier transform of the data, and indicates a strong peak at $\sim$30 minutes. Unfortunately the data do not continue long enough to confirm unambiguously that PKS2155 is a QPO, and are not some chance variability. However the time variability is intrinsic to PKS2155, and the time-scales can be used to put constraints on the sizes of emission regions, etc. These observations can also be combined with gamma-ray observations, particularly where various competing shocked-jet models (Holmes *et al.* 1984; Valtaoja *et al.* 1991) make different predictions.

3. Summary

With the correct polarimetric instrumentation, significant discoveries can be made with even a small telescope. Indeed, the polarimeter on the SAAO 1.9-m telescope out-performs the polarimeter on the VLT in high-time-domain polarimetry. In fact the VLT observations of the IP NY Lup did not detect the polarimetric spin modulation due to the spinning white dwarf.

The next step will be to obtain high-speed spectropolarimetry, and that will require high-speed CCD detectors on a 10-m-class telescope. The Robert Stobie Spectrograph on the Southern African Large Telescope has that capability, and is currently undergoing its commissioning phase.

References

Abramowski, A., *et al.* 2010, *A&A*, 520, 83
Aharonian, F., *et al.* 2007, *ApJ*, 664L, 71
Beuermann, K., *et al.* 2011, *A&A*, 526, 53
Buckley, D. A. H.. 2000, *New AR*, 44, 63
Gabuzda, D. C., *et al.* 2006, *MNRAS*, 369, 1596
Holmes, P. A., *et al.* 1984, *MNRAS*, 211, 497
Norton, A. J., Wynn, G. A., & Somerscales, R. V. 2004, *ApJ*, 614, 349
Potter, S. B., *et al.* 2010, *MNRAS*, 402, 1161
Potter, S. B., *et al.* 2011, *MNRAS*, 416, 2202
Valtaoja, E., *et al.* 1991, *AJ*, 101, 78

New Horizons in Time-Domain Astronomy
Proceedings IAU Symposium No. 285, 2011   © International Astronomical Union 2012
R.E.M. Griffin, R.J. Hanisch & R. Seaman, eds.   doi:10.1017/S1743921312000415

Gamma-Ray Waveband and Multi-Waveband Variability of Blazars

Stefano Ciprini[1,2]

[1]ASI Science Data Center, Frascati, Roma, Italy

[2]INAF Observatory of Rome, Monte Porzio Catone, Roma, Italy
email: `stefano.ciprini@asdc.asi.it` (for the Fermi LAT Collaboration)

Abstract. The Fermi Gamma-ray Space Telescope, as an all-sky survey and monitoring mission, is producing well-sampled gamma-ray light curves for dozens of blazars and other high-energy sources. We report highlights of gamma-ray variability properties, and outline multi-frequency observing campaigns that are targeted to new or known blazars which emit gamma rays.

Keywords. surveys, catalogues, gamma rays: observations, quasars: general, BL Lacertae objects: general, methods: statistical, radiation mechanisms: nonthermal

---

1. Mono- and Multi-Waveband Blazar Variability with FERMI

The Large Area Telescope (LAT), on board the FERMI Gamma-ray Space Telescope (Atwood *et al.* 2009), is a pair-conversion gamma-ray telescope, sensitive to photon energies from about 20 MeV up to >300 GeV, with a wide field of view (> 2.4 sr) and working in all-sky survey mode. The entire sky is observed every 2 orbits (~3 hours), representing a continuous monitoring of the variable and transient gamma-ray sky and producing regular daily- or weekly-sampled light curves for dozens of GeV sources. Multi-wavelength (MW) observing campaigns are limited only by difficulties in co-ordinating other telescopes. Irregular and aperiodic variability is found in blazars at all the timescales and at all the energies (MW variability). EGRET already showed that blazars have a high-energy component in their spectral energy distributions (SED) and are the largest class of variable gamma-ray sources, although it was limited by statistics, while FERMI is observing low gamma-ray brightness states too. Of the studied FERMI LAT blazars, 2/3 are variable and high states are less than 1/4 of the total light curve range (Abdo *et al.* 2010). Flat Spectrum Radio Quasars (FSRQs) and low energy peaked BL Lac objects (BL Lacs) show the largest relative variance (Fig. 1), while high-energy peaked BL Lacs display a lower variable but persistent emission. Sources like PKS 1510-08, PKS 1502+106 (Fig. 4), 3C 454.3 (Fig. 3; Ackermann et al. 2010), 3C 279 (Fig. 4), PKS 1830-211 (Fig. 2), 4C 21.35, 4C 38.41 (all FSRQs) and AO 0235+164, 3C 66A (both BL Lacs) are among the brightest, most variable, and isotropically luminous blazars seen by FERMI. For bright flares even intra-day light curves have been extracted (Figs. 2 and 3).

Discrete autocorrelation function (DACF) and structure function (SF, Figs. 1 and 3) analysis showed different patterns, autocorrelation times and power-law indices ($1/f^\alpha$ trends, where $f = 1/t$) implying different variability modes for each source (more flicker, $\alpha \simeq 1$ or more Brownian, $\alpha \geqslant 2$, dominated). 3C 454.3 is a fully Brownian gamma-ray source while other powerful blazars have values half-way between the two modes. Average power spectral density analysis (PSD) in the frequency domain over the blazar subclasses points out α slopes from 1.3 to 1.6 (Abdo *et al.* 2010; Ackermann *et al.* 2011). No evidence for persistent characteristic gamma-ray timescale(s) is found. Flare profiles are mostly symmetric (in part because of superimposing flares and large bin smoothing).

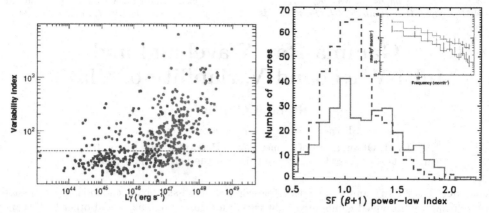

Figure 1. Results from gamma-ray light curves extracted in fixed 1-month bins over the first 2 years of FERMI LAT all-sky survey for the 886 blazars/AGNs of the clean source sample of the second FERMI LAT Source Catalog, 2FGL. Left panel: Variability index versus isotropic gamma-ray luminosity (red/grey: FSRQs; blue/black: BL Lac Objects). Dashed line represents the 99% confidence level for a source to be variable. Right panel: distribution of the temporal power spectral density (PSD) power-law indexes ($\alpha = \beta+1$) for the FSRQs (red/continuous line) and BL Lac objects (blue/dashed line) of the sample evaluated in time domain using first-order structure function (SF) analysis (blind power-law index β estimation using a maximum lag of 2/3 of the total light-curve range). Cumulative PSD for bright FSRQs (red/top line) and BL Lac objects sub-samples (blue/bottom line) showing similar slopes (inset). From Ackermann *et al.* (2011).

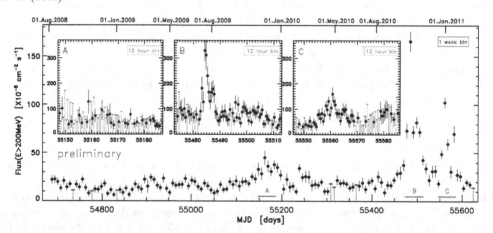

Figure 2. Main panel: 31.5-month (945-day) gamma-ray flux ($E > 200$ MeV) light curve in weekly time bins of blazar PKS 1830-211, from 2008 August 04, to 2011 March 7 (MJD 54682.65 to 55627.65) as an example of the FERMI LAT capabilities in high-energy temporal variability monitoring. Inset panels: 12-hour bin light curves detailing the period around the mild flare of October 2009 (A interval), detailing the period around the large outburst of October 2010 (B interval) and the secondary double flare of December 2010 and January 2011 (C interval).

The fractional variability during outburst appears similar to its longer-term mean in the few objects studied in detail.

FERMI-driven MW observing campaigns are shedding light on the PSD-SED plane (i.e., time-scale–energy parameter space). Broad-band MW studies mostly address cross-correlation and time-lag analysis, model time-resolved SEDs, search for orphan flares and spectral hysteresis and analyse gamma-ray *vs* synchrotron amplitude ratios and emission peaks, and study the radio–gamma-ray connection and source populations. Simultaneous

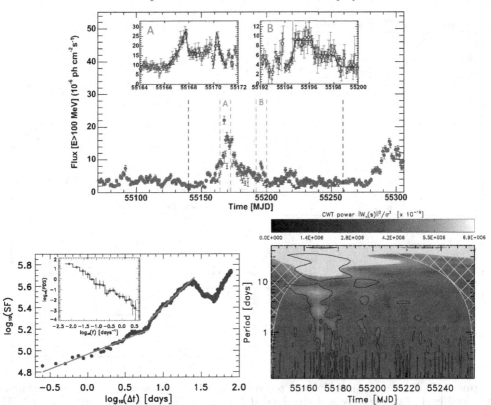

Figure 3. Top panel: daily-bin gamma-ray flux light curve of blazar 3C 454.3 (100 MeV - 200 GeV band, red/gray points, main panel) between 2009 Aug. 27 2010 Apr. 21 (MJD 5507055307). Dashed black lines mark period over which the SF, PSD and Wavelet analysis are conducted (using improved resolution with 3 hour bins). The light curve of the previous 2008 JulyAug. flare, shifted by 511 days, is also shown (violet/dark-grey tiny points). Insets show blow-ups of the two periods (A and B on the plot) when the largest relative flux increases took place (red/gray filled, blue/dark open, green/light-gray filled data points corresponding to daily, 6-hour, 3-hour bin fluxes, respectively). Bottom left panel: SF of the 3h-bin flux light curve for the period 2009 Nov. 5 - 2010 Mar. 4 (MJD 55140-55259, black dashed lines in the top panel) and corresponding PSD (inset). Bottom right panel: plane contour plot of the continuous Morlet wavelet transform power density for the same light curve (thick black contours: 90% confidence levels of true signal features against white/red noise background; cross-hatched region: cone of influence where spurious time-frequency edge effects are important). From Ackermann *et al.* (2010).

MW observations are crucial also for the identification of newly discovered gamma-ray sources.

In order to define and better constrain physical parameters, processes and emission components, to clarify the role of the central engine, jets and their interplay, the jet composition and structure in AGNs and blazars is necessary to collect more different and longer sequences of MW observations. Some clues are already emerging. (1) The knowledge of redshifts is crucial but ∼50% of BL Lacs have still unknown z. (2) Simple single-zone synchrotron self Compton (SSC) descriptions are vanishing. (3) Internal shock models with composite particle energy distributions work well. (4) Cross-correlation analysis with optical polarization provides important clues on jet physics. (5) The location of emission site can be both inside and outside the broad line region (BLR). (6) Magnetic fields are complex but can be highly ordered and jet-aligned during

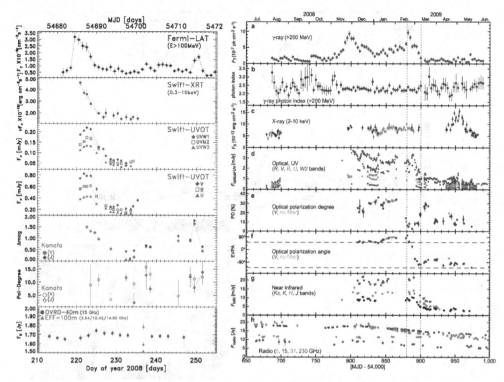

Figure 4. Radio, mm, optical (flux and polarization), UV, X-ray, gamma-ray fluxes light curves obtained by FERMI intensive and coordinated MW campaigns (PKS 1502+106, left, 3C 279, right). From Abdo et al. (2010b,c).

gamma-ray flares. (7) In some cases bright gamma-ray flares seems to occur after ejections of superluminal radio knots. Two examples of composite MW light curves obtained from FERMI blazar campaigns (PKS 1502+106 and 3C 279) are reported in Fig. 4 (Abdo *et al.* 2010b, Abdo *et al.* 2010c). In conclusion, FERMI LAT is demonstrating very good capabilities in the field of gamma-ray variability analysis and radio–gamma-ray connection and is showing an optimal synergy with the SWIFT mission.

Acknowledgements

The FERMI LAT Collaboration acknowledges support from a number of agencies and institutes for both development and the operation of the LAT as well as scientific data analysis. These include NASA and DOE in the United States, CEA/Irfu and IN2P3/CNRS in France, ASI and INFN in Italy, MEXT, KEK, and JAXA in Japan, and the K.A. Wallenberg Foundation, the Swedish Research Council and the National Space Board in Sweden. Additional support from INAF in Italy and CNES in France for science analysis during the operations phase is also gratefully acknowledged.

References

Abdo, A. A., *et al.* 2010a, *ApJ*, 722, 520
Abdo, A. A., *et al.* 2010b, *ApJ*, 710, 810
Abdo, A. A., *et al.* 2010c, *Nature*, 463, 919
Ackermann, M., *et al.* 2010, *ApJ*, 721, 1383
Ackermann, M., *et al.* 2011, *ApJ*, in press (arXiv:1108.1420)
Atwood, W.B., *et al.* 2009, *ApJ*, 697, 1071

New Horizons in Time-Domain Astronomy
Proceedings IAU Symposium No. 285, 2011
R.E.M. Griffin, R.J. Hanisch & R. Seaman, eds.

© International Astronomical Union 2012
doi:10.1017/S1743921312000427

Two Centuries of Observing R Coronæ Borealis

Geoffrey C. Clayton

Dept. of Physics & Astronomy, Louisiana State University, Baton Rouge, LA 70803, USA
email: gclayton@fenway.phys.lsu.edu

Abstract. R Coronæ Borealis was found to be variable in the year 1783, and was one of the first variable stars to be so identified. Its class, the R Coronæ Borealis (RCB) stars, are rare hydrogen-deficient carbon-rich supergiants. RCB stars undergo massive declines of up to 8 mag due to the formation of carbon dust at irregular intervals. The mechanism of dust formation around RCB stars is not well understood, but the dust is thought to form in or near the atmosphere of the star. Their rarity may stem from the fact that they are in an extremely rapid phase of the evolution, or are in an evolutionary phase that most stars do not undergo. Several evolutionary models have been suggested to account for the RCB stars, including a merger of two white dwarfs (WDs) or a final helium-shell flash (FF) in a PN central star. The large overabundance of ^{18}O found in most of the RCB stars favours the WD merger model, while the presence of Li in the atmospheres of five RCB stars favours the FF one. In particular, the measured isotopic abundances imply that many, if not most, RCB stars are produced by WD mergers, which may be the low-mass counterparts of the more massive mergers thought to produce type Ia supernovae. Understanding these enigmatic stars depends to a large extent on continuous monitoring to catch their irregular but rapid variations caused by dust formation, their variations due to stellar pulsations, and long-term changes that may occur over centuries. I will use observations of R Coronæ Borealis obtained over 200 years to demonstrate what kinds of monitoring are necessary for these and similar classes of variables.

Keywords. circumstellar matter, dust, evolution, surveys

---

1. Introduction

R Coronæ Borealis (R CrB) was one of the first variable stars identified. Its brightness variations have been monitored since its discovery over 200 years ago (Pigott & Englefield 1797)—see Fig. 1. Clayton *et al.* (2005, 2007) made the remarkable discovery that cool R Coronæ Borealis (RCB) stars with CO bands have $^{18}O/^{16}O$ ratios that are orders of magnitude higher than those seen for any other star. The RCB stars are a small group of hydrogen-deficient, carbon-rich supergiants. About 65 are known in the Galaxy and the Magellanic Clouds (Tisserand *et al.* 2008; Kraemer *et al.* 2005; Zaniewski *et al.* 2005; Alcock *et al.* 2001; Clayton 1996). Their defining characteristics are hydrogen deficiency and unusual variability—RCB stars undergo massive declines of up to 8 mag caused by the formation of carbon dust at irregular intervals. Two models have been proposed for the origin of an RCB star: the Double Degenerate (DD) and the final helium-shell flash (FF) models (Iben *et al.* 1996; Saio & Jeffery 2002). The former involves the merger of a CO- and a He-WD (Webbink 1984). In the latter, a star evolving from a planetary nebula (PN) central star expands to supergiant size by a FF (Fujimoto 1977; Renzini 1979). Three stars (Sakurai's Object, V605 Aql and FG Sge) have been observed to undergo FF outbursts that transformed them from hot evolved PN central stars into cool giants with spectral properties similar to RCB stars (Clayton *et al.* 2006; Asplund *et al.* 1998, 1999, 2000; Clayton & De Marco 1997; Gonzalez *et al.* 1998).

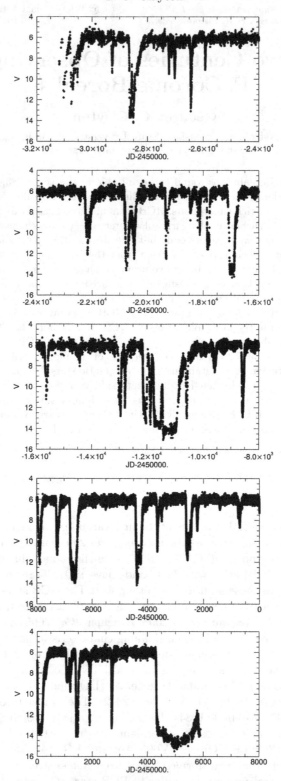

Figure 1. A 100-year light curve for R CrB. Each panel contains 8,000 days. The data plotted are visual AAVSO observations.

However, Sakurai's Object shows no evidence of ^{18}O (Geballe *et al.* 2002). Therefore, the FF stars on the one hand, and most of the RCB stars on the other, are likely to be stars with different origins. But the actual observations are somewhat contradictory. Indeed, five RCB stars, including R CrB itself, exhibit enhanced Li abundances, as does the FF star Sakurai's Object (Lambert 1986; Asplund *et al.* 2000). As shown by Herwig & Langer (2001), Li enhancements are consistent with the FF model. However, the production of ^{18}O requires temperatures large enough that would destroy completely any Li present. For that reason the simultaneous enrichment of Li and ^{18}O is not expected in the DD merger model. Unfortunately, all five of those stars are too hot to show CO, so their abundances of ^{18}O are unknown. The presence of Li in some RCB stars, and also in Sakurai's object, is extremely difficult to explain by a DD merger. Since ^{18}O strongly supports the DD merger/accretion model for most of the stars studied so far, the obvious conclusion is that there are (at least) two evolutionary channels for formation for the RCB stars, the DD perhaps being the dominant mechanism.

2. Understanding the RCB Stars

The presence or absence of circumstellar material and the morphology of this material provides a fossil record of previous evolutionary stages. Mass loss occurs in both models during the common-envelope or PN-ejection phases, but the observational manifestations will differ. In the DD model no fossil envelope should remain from the common-envelope phase when the two WDs finally merge to form an RCB star. About 10% of single stars will undergo a final-flash event (Iben *et al.* 1996), and about the same percentage of RCB stars (R CrB, RY Sgr, V CrA, and UW Cen) show evidence of resolved fossil dust shells in the far-IR (Walker 1994). At the present time, such a shell has been imaged in the visible around only two RCB stars: UW Cen and R CrB. The shells may be a fossil PN, now neutral, and detectable only in scattered light from dust in the shell. The sizes of the shells are consistent with a PN ejection several thousand years before the FF (Clayton *et al.* 1999, 2011). UW Cen and R CrB are two of the five RCB stars that show enriched Li, indicating that they may be the product of a FF rather than a DD merger. When in a deep decline, an RCB star can be imaged by using its own dust cloud as a natural coronograph: if it is surrounded by a neutral circumstellar shell, it will be detectable. By applying that technique we can start to get a better statistical sample of RCB stars in order to determine what fraction of them are the result of the FF rather than the DD process.

But RCB stars are very rare. Only about 55 have been discovered in the Galaxy so far (Clayton 1996; Alcock *et al.* 2001; Zaniewski *et al.* 2005; Tisserand *et al.* 2008). The rate of He- and CO-WD mergers is estimated to be about one per century, so if the typical lifetime as an RCB star were about 10^5 years then there would be $\sim 10^3$ in the Galaxy. That agrees with the extrapolation to the Galaxy from the number of RCB stars known in the LMC, where a better census can be made. So where are all the RCB stars? The RCB stars may be an 'old bulge' population and there may be ~ 250 RCB stars in the reddened 'exclusion' zone toward the bulge (Zaniewski *et al.* 2005).

Constraints on the spatial distribution and the formation rate of such stars are needed to understand their origins and to test them in the context of actual population-synthesis results. To do so, it is crucial to increase significantly the number of known RCBs. New sky surveys can be used to discover new RCB stars and thus increase the observed sample. Already the ASAS-3 survey has been used to find about 10 new RCB stars. Other surveys coming on line such as CRTS are also being used to look for new RCB stars.

It is hoped that the results of these surveys will help us distinguish between the FF and DD processes for the formation of RCB stars, and to understand better their population and lifetimes.

References

Alcock, C., *et al.* 2001, *ApJ*, 554, 298.
Asplund, M., Gustafsson, B., Kameswara Rao, N., & Lambert, D. L. 1998, *A&A*, 332, 651.
Asplund, M., *et al.* 1999, *A&A*, 343, 507.
Asplund, M., *et al.* 2000, *A&A*, 353, 287.
Clayton, G. C. 1996, PASP, 108, 225.
Clayton, G. C. & De Marco, O. 1997, *AJ*, 114, 2679.
Clayton, G. C., *et al.* 1999, *ApJ*, 517, L143.
Clayton, G. C., *et al.* 2005, *ApJ*, 623, L141.
Clayton, G. C., *et al.* 2006, *ApJ*, 646, L69.
Clayton, G. C., Geballe, T. R., Herwig, F., Fryer, C., & Asplund, M. 2007, *ApJ*, 662, 1220.
Clayton, G. C., *et al.* 2011, *ApJ*, 743, 44.
Fujimoto, M. Y. 1977, PASJ, 29, 331.
Geballe, T. R., Evans, A., Smalley, B., Tyne, V H., & Eyres, S. P. S. 2002, Ap&SS, 279, 39.
Gonzalez, G., *et al.* 1998, *ApJS*, 114, 133.
Herwig, F. & Langer, N. 2001, Nuclear Physics A, 688, 221.
Iben, I., Tutukov, A. V., & Yungelson, L. R. 1996, *ApJ*, 456, 750.
Kraemer, K. E., Sloan, G. C., Wood, P. R., Price, S. D., & Egan, M. P. 2005, *ApJ*, 631, L147.
Lambert, D. L. 1986, in: K. Hunger, D. Schönberner & N. Kameswara Rao (eds.), *Hydrogen Deficient Stars and Related Objects*, IAU Colloq. 87, *ASSL*, Vol. 128, p. 127
Pigott, E. & Englefield, H. C. 1797, *Phil. Trans. Royal Soc.*, 87, 33.
Renzini, A. 1979, in: B. E. Westerlund (ed.), *Stars and Star Systems*, ASSL Vol. 75, p. 155.
Saio, H. & Jeffery, C. S. 2002, *MNRAS*, 333, 121.
Tisserand, P., *et al.* 2008, *A&A*, 481, 673.
Walker, H., 1994, CCP7 Newsletter, 21, 40.
Webbink, R. F., 1984, *ApJ*, 277, 355.
Zaniewski, A., *et al.* 2005, *AJ*, 130, 2293.

New Horizons in Time-Domain Astronomy
Proceedings IAU Symposium No. 285, 2011
R.E.M. Griffin, R.J. Hanisch & R. Seaman, eds.

© International Astronomical Union 2012
doi:10.1017/S1743921312000439

On Rapid Interstellar Scintillation of Quasars: PKS 1257-326 Revisited

Hayley E. Bignall and Jeffrey A. Hodgson

International Centre for Radio Astronomy Research, Curtin Univer sity, WA 6845, Australia
email: H.Bignall@curtin.edu.au

Abstract. The line of sight towards the compact, radio loud quasar PKS 1257−326 passes through a patch of scattering plasma in the local Galactic ISM that causes large and rapid, intra-hour variations in the received flux density at centimetre wavelengths. This rapid interstellar scintillation (SS) has been occurring for at least 15 years, implying that the scattering "screen" is at least 100 AU in physical extent. Through observations of the ISS we have measured microarcsecond-scale "core shifts" in PKS 1257-326, corresponding to changing opacity during an intrinsic outburst. Recent analysis of VLA data of a sample of 128 quasars found 6 sources scintillating with a characteristic time-scale of < 2 hours, suggesting that nearby scattering screens in the ISM may have a covering fraction of a few percent. That is an important consideration for proposed surveys of the transient and variable radio sky.

Keywords. techniques: high angular resolution, ISM: structure, quasars: individual

---

1. Introduction

Radio sources of the order of 0.1 milli-arcseconds or smaller in angular size exhibit fluctuations in their measured flux densities at centimetre wavelengths caused by scattering in the Galactic ionised interstellar medium (Rickett, 1990). In the weak scattering régime, which occurs at frequencies above $\sim$4 GHz for most extragalactic lines of sight, the spatial scale of those fluctuations corresponds approximately to the Fresnel scale r_F at the distance of the scattering material, D; $r_F = \sqrt{D/k}$, where $k = 2\pi/\lambda$ is the wavenumber corresponding to observing wavelength λ. The corresponding characteristic time-scale of weak interstellar scintillation (ISS) ranges between minutes and days, depending on λ, D and v, the relative transverse velocity between source, screen and observer. Only the most compact sources scintillate; when the source's angular size, θ_S, significantly exceeds $\theta_F = r_F/D$, scintillation in weak scattering is suppressed owing to averaging over incoherently scintillating regions of the source. Below the transition frequency in the regime of strong scattering, compact quasars will exhibit refractive scintillation on longer time-scales, but are generally too large to exhibit the narrow-band, short-time-scale diffractive scintillation displayed by pulsars (Narayan, 1992). The recent MASIV VLA Survey showed that more than half of all compact, flat-spectrum radio quasars exhibit intraday variability due to ISS at 5 GHz with a duty cycle of at least 25%, and rms variations of typically 2–10% (Lovell et al. 2008). ISS can be used as a probe both of small-scale structure in the Galactic ISM and of source structure on microarcsecond (μas) scales. For further discussion of potential applications, see Koay et al., page 347.

In this paper we discuss the population of quasars which exhibit weak ISS on atypically short time-scales—minutes to hours—on account of nearby Galactic scattering screens. The closer the scattering screen, the smaller r_F and the shorter the characteristic scintillation time-scale for a given velocity, $t_c \sim r_F/v$. In addition, θ_F is large for small D, and as quasar angular sizes often exceed θ_F for $D \gtrsim 10$ pc their scintillation will tend to be dominated by nearby scattering plasma. In section 2.1 we present a new analysis of the μas-scale evolution of the long-lived rapidly scintillating quasar PKS 1257−326.

2. "Intra-Hour Variable" Quasars

During the past 15 years three quasars have been found to show large-amplitude ($\gtrsim 10\%$ rms) cm-wavelength variability over time-scales < 1 hour; they are PKS 0405−385—discovered with the Australia Telescope Compact Array (ATCA) by Kedziora-Chudczer et al. (1997), J1819+3845, discovered with the Westerbork Synthesis Radio Telescope by Dennett-Thorpe & de Bruyn (2000), and PKS 1257−326, discovered with the ATCA (Bignall et al. 2003; hereafter B03). That the rapid variability is entirely due to ISS of these sources has been shown by measurements of variability pattern arrival-time delays of typically minutes between widely separated telescopes (Jauncey et al. 2000; Dennett-Thorpe & de Bruyn, 2002; Bignall et al. 2006, hereafter B06), and by annual cycles in the characteristic variability time-scale (Dennett-Thorpe & de Bruyn, 2003; B03).

Annual cycles occur as the scintillation velocity (both speed and direction) changes with the Earth's orbit; in this case quasars can be considered fixed on the sky and the velocity of the scattering plasma is of order the same as the Earth's orbital velocity, ~ 30 km s^{-1}. Moreover, the scintillation patterns have been found to be highly anisotropic (Rickett et al. 2002; Dennett-Thorpe & de Bruyn 2003; B06; Walker et al. 2009). The scintillation time-scale thus increases at times of year when the relative transverse velocity between the Earth and the scattering plasma is small, or more specifically when the velocity component along the *minor* axis of the scintillation pattern is small. Measurements of scintillation time-scale and two-station pattern arrival-time delays at various times of the year can be used to determine the scattering screen velocity and the length scale, the anisotropy axial ratio and the position angle of the scintillation pattern. The pattern scale also constrains the distance to the scattering screen and the angular size of the source. In practice, in the case of a highly elongated scintillation pattern, the solution becomes degenerate and it may not be possible to constrain all parameters uniquely.

For PKS 0405−385 and J1819+3845, and also for a number of other intraday variable scintillating sources, the rapid scintillation has been found to be episodic (Kedziora-Chudczer, 2006; Cimò, 2008). Koay et al. (2011) argue that the episodic ISS of J1819+3845 is possibly related to the nearby scattering "screen" moving out of the line of sight, rather than to intrinsic expansion of the source or to fading of a compact scintillating component. The high anisotropy and intermittency of ISS is suggestive of highly magnetically-stressed structures, of which there is also evidence from pulsar scintillation. The origin of those highly localised scattering screens remains a mystery. Linsky et al. (2008) suggested that scattering may occur in turbulent regions of interaction between nearby partially-ionised warm interstellar clouds, and found cloud velocities consistent with screen velocities determined for background fast scintillators seen through the edges of the clouds. However, there is not yet sufficient evidence to confirm the association.

Recently Koay et al. (2011) searched for rapid scintillation in a sample of 128 flat-spectrum quasars and BL Lac objects observed over 11 days with the VLA, detecting 6 rapid scintillators with characteristic time-scales of less than two hours, although none with rms variations larger than 10%. Koay et al. showed that rapid *and* large-amplitude scintillation requires both relatively nearby screens (within ~ 250 pc for source sizes of 10μas, or within ~ 12 pc for larger source sizes of 200μas) and highly compact sources, i.e., with most of the observed source flux density contained in the scintillating component.

2.1. *PKS 1257−326 revisited: microarcsecond-scale source evolution*

Although PKS 1257−326 is an otherwise fairly typical flat-spectrum quasar at redshift 1.256, it is unique in that it has shown rapid scintillation in all observations between 1995 and 2011, with a repeating annual cycle in the characteristic time-scale. That implies a scattering structure at least ∼ 100 AU in extent, assuming a minimum velocity for the scattering screen consistent with the annual cycle and time-delay data presented in B06. In ATCA data from 2001 to early 2002, an offset was found between the light curves at 4.8 and 8.6 GHz, with the 8.6-GHz light curve consistently leading. The time delay between the two frequencies in those data showed an annual cycle over the first year of the monitoring programme (B03); in subsequent data, however, that offset was not observed. We relate the observed evolution of the frequency offset to long-term intrinsic variability in the inner jet of PKS 1257−326.

Assuming that the scintillation patterns at each frequency are similar and have the same degree of anisotropy, one can calculate the expected time offset by:

$$\Delta t = \frac{\mathbf{r} \cdot \mathbf{v} + (R^2 - 1)(\mathbf{r} \times \hat{\mathbf{S}})(\mathbf{v} \times \hat{\mathbf{S}})}{v^2 + (R^2 - 1)(\mathbf{v} \times \hat{\mathbf{S}})^2} \ , \tag{2.1}$$

where $\mathbf{r}$ is the displacement vector between the two frequency components of the scintillation pattern, $\mathbf{v}$ is the scintillation velocity (which is a function of the day of year), R is the axial ratio of anisotropy in the scintillation pattern, and $\hat{\mathbf{S}}$ is a unit vector in the direction of the major axis of the scintillation pattern. This is the same equation used by B06 to calculate the two-station time delay, but with the baseline replaced by the displacement between the scintillation patterns at the two frequencies. In the case of the frequency offset, the displacement vector $\mathbf{r}$ needs to be solved.

Although not uniquely constrained, a reasonable fit to the observed time delay between the 4.8 and 8.6 GHz light curves over the period from February 2001 to February 2002 can be obtained for a displacement vector of $\mathbf{r}(\alpha, \delta) = (-1, 2) \times 10^4$ km (Fig. 1). That assumes a scintillation velocity, pattern axial ratio and position angle corresponding to the preferred annual cycle model given in B06. If we adopt the screen distance $D \sim 10$ pc also determined by B06, the displacement corresponds to an offset between the centroids of the 4.8 and 8.6 GHz scintillating components of the source of ∼15 µas, < 0.1 pc at the redshift of PKS 1257−326, with the 4.8 GHz component being displaced to the northwest of the higher-frequency component. Such a displacement is readily explained as an optical-depth effect or "core shift", as has been measured in a number of quasars with VLBI astrometry (Kovalev *et al.* 2008). The large-scale jet of PKS 1257−326 observed at milli-arcsecond (with VLBI; Ojha *et al.* 2010) and arcsecond scales (with the VLA; B06) also extends in the same direction. The core shift deduced from the ISS patterns is consistent with synchrotron self-absorption at the base of the jet.

However, data obtained since mid-2002 are consistent with *no* offset between the two frequency components. By comparison to the long-term changes in mean total flux density and spectral index of PKS 1257−326 (Fig. 2) we see that the period until early 2002 where the frequency offset was observed corresponds to an outburst in total flux density, when the source showed an inverted spectrum between 4.8 and 8.6 GHz. At later times, when there was no offset detected between the two frequency components, the overall spectral index is close to zero. The outburst and core shift are readily explained as due to a region of enhanced emission (probably induced by shock acceleration of particles in the jet) that is initially optically thick at 4.8 GHz owing to synchrotron self-absorption, and later fades and becomes optically thinner. The observed relationship between the offset of the scintillation patterns at different frequencies and the intrinsic source evolution

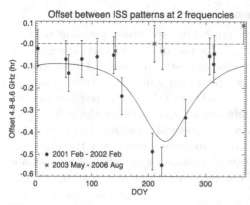
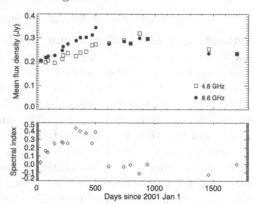

Figure 1. Offset measured by cross correlation of ATCA light-curves at 4.8 and
8.6 GHz. The solid line shows a model annual cycle corresponding to a core shift of
~15 μas.

Figure 2. Evolution of average flux density
and spectral index over a 5-year period. The
core shift was only observed when the source
had an inverted spectrum between 4.8 and 8.6
GHz.

provides strong evidence that the offset is indeed source-intrinsic, and not a refractive
effect of the intervening medium.

3. Summary

Variability at centimetre wavelengths due to ISS occurs over a range of time-scales.
That is an important consideration for proposed surveys for transient and variable radio
sources with future wide-field telescopes. Our results for quasar PKS 1257−326 demonstrate the potential of ISS observations at multiple frequencies for studying the physical
conditions in otherwise unresolved relativistic jets via the evolution of μas-scale structure.

References

Bignall, H. E., *et al.* 2006, *ApJ*, 652, 1050
Bignall, H. E., *et al.* 2003, *ApJ*, 585, 653
Cimò, G. 2008, in: The role of VLBI in the Golden Age for Radio Astronomy, *Proceedings of
    Science*, PoS (IX EVN Symposium) 046
Dennett-Thorpe, J. & de Bruyn, A. G. 2000, *ApJL*, 529, L65
Dennet-Thorpe, J. 2002, *Nature*, 415, 57
Dennett-Thorpe, J. 2003, *A&A*, 404, 113
Jauncey, D. L., *et al.* 2000, in: H. Hirabayashi, P. G. Edwards and D. W. Murphy, *Astrophysical
    Phenomena Revealed by Space VLBI* (Sagamihara: ISAS), p. 147
Kedziora-Chudczer, L. 2006, *MNRAS*, 369, 449
Kedziora-Chudczer, L., *et al.* 1997, *ApJL*, 490, L9
Koay, J. Y., *et al.* 2011, *A&A*, 534, L1
Kovalev, Y. Y., Lobanov, A. P., Pushkarev, A. B., & Zensus, J. A. 2008, *A&A*, 483, 759
Linsky, J. L., Rickett, B. J., & Redfield, S. 2008, *ApJ*, 675, 413
Lovell, J. E. J., *et al.* 2008, *ApJ*, 689, 108
Narayan, R. 1992, *Phil. Trans. R. Soc. Lond. A.*, 341, 151
Ojha, R., *et al.* 2010, *A&A*, 519, A45

New Horizons in Time-Domain Astronomy
Proceedings IAU Symposium No. 285, 2011
R.E.M. Griffin, R.J. Hanisch & R. Seaman, eds.

© International Astronomical Union 2012
doi:10.1017/S1743921312000440

Sonification of Astronomical Data

Wanda L. Diaz-Merced[1], Robert M. Candey[2], Nancy Brickhouse[3], Matthew Schneps[3], John C. Mannone[4], Stephen Brewster[1], and Katrien Kolenberg[3,5]

[1] Glasgow University, Scotland, UK.   email: wanda@dcs.gla.ac.uk

[2] NASA, Goddard Space Flight Center, Greenbelt. Maryland, USA

[3] Harvard-Smithsonian Center for Astrophysics, Cambridge, MA, USA

[4] Shirohisa Ikeda Project Gurabo, Puerto Rico

[5] Instituut voor Sterrenkunde, Leuven, Belgium

Abstract. This document presents JAVA-based software called XSONIFY that uses a sonification technique (the adaptation of sound to convey information) to promote discovery in astronomical data. The prototype is designed to analyze two-dimensional data, such as time-series data. We demonstrate the utility of the sonification technique with examples applied to X-ray astronomy and solar data. We have identified frequencies in the CHANDRA X-Ray observations of EX Hya, a cataclysmic variable of the intermediate polar type. In another example we study the impact of a major solar flare, with its associated coronal mass ejection (CME), on the solar wind plasma (in particular the solar wind between the Sun and the Earth), and the Earth's magnetosphere.

Keywords. Solar Wind, Sonification, xSonify, Variable Stars

1. Introduction

Sonification is a developing field. It integrates a wide variety of professional fields and broadens the interaction between users who are accessing information in a diversity of modalities (see http://www.icad.org, the International Community for Auditory Display). For scientific purposes, it employs the highly developed sense of hearing as an adjunct to data visualization in order to enhance current techniques for data analysis. Since the beginning of spacecraft experiments, it became evident that space plasmas were abundant with numerous kinds of nonlinear plasma waves. The new generation of spacecraft with advanced experimental techniques are forever engaged in an ongoing improvement in accuracy and time resolution to capture plasma, particle, radio, magnetic field, and X-ray data (to mention just a few) with increasing accuracy and time resolution. Developments in both the speed and storage capacity of computers have made it possible to perform numerical simulations with increasingly larger numbers of particles and more grid cells for a longer running time, and with higher phase-space dimensions.

A major problem limiting the utilization of visual displays is that data typically contain much more information than can be displayed effectively with currently available technologies. It is also important to consider the limitations imposed by the nature of the human eye. For example, even the best computer screens available today are limited to a range of spatial resolutions. That limitation affects the useful dynamic range of the display, reducing the amount of data scientists can study at any one time. Scientists currently work around such limitations by filtering the data so as to display only the information which they believe is important to the problem in hand. But since that involves making some guesses about the results they are searching for, many discoveries may be missed.

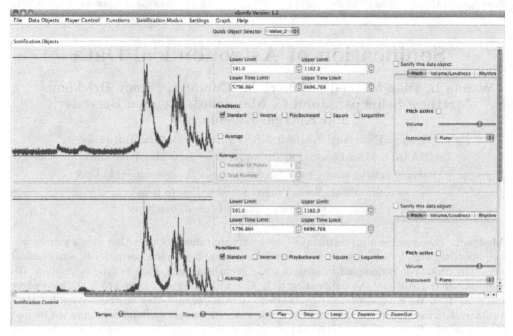

Figure 1. xSONIFY prototype

xSONIFY is a tool for analysing space physics data with an improved functional interphase and allowing a wide variety of file input formats. This tool will open up the SPDF space physics data collection (see http://www.spdf.gsfc.nasa.gov, NASA's Space Physics Data Facility) to a new community of researchers that are currently excluded from space physics research.

Sonification provides other benefits for researchers, including the ability to:

- Analyze complex or rapidly/temporally changing data
- Explore large datasets (particularly multi-dimensional datasets)
- Explore datasets in frequency dimensions rather than spatial ones
- Identify new phenomena which current display techniques miss
- Find correlations and patterns which are masked in visual displays
- Monitor data while doing something else (e.g., background event-finding).

For example, when searching for bow-shock and magnetopause crossings we expect distinctive signatures to be especially apparent above the background noise if the data are sonified. Those boundary signatures appear as characteristic changes in the whole spectrum, including the background noise; other emissions appear as tones at distinctive frequencies or time-frequency spectra, such as electron plasma oscillations, magnetic noise bursts in the magnetosheath, or whistler mode emissions marking regions of currents.

2. xSONIFY

xSONIFY sonifies 2-dimensional data from text files and sample files from the CDAWeb heliophysics holdings. It is based on JAVA 1.5 with JAVA Sound, MIDI, JAVASpeech, WebStart and Web Services technology. Currently it features three different sonification modes (Pitch, Loudness, Rhythm) with various controls (Play, Stop, Loop, Speed, Time point)—see Fig. 1. It has limited pre-processing of input data (Limits, Invert, Logarithm,

Averaging). We will expand it to handle all 2-dimensional plus multi-dimensional data, and add additional necessary features.

The software follows a modular data pipeline approach (Daudé and Nigay, 2003):

- Data Transformation: to appropriate geophysical units;
- Normalization: convert to abstract normalized view;
- Sonification Transformation: to abstract sound parameter space;
- Auditory Display Transformation: sonic rendering.

We have made the source code openly available at `http://xsonify.sourceforge.net` under the NASA Open Source Agreement (NOSA). As it becomes more stable and capable, we will advertise it widely and develop a community of active users. We have designed an iterative development procedure whereby we solicit requirements and feedback from volunteers (and other interested users) who are evaluating its usability; we then prioritize the requirements according to their importance, usefulness and difficulty of implementation, and produce the next version for further testing.

3. Examples using XSONIFY

The X-ray variable EX Hydrae. We sonified time-series of the EX Hya light-curve data in order to analyze the frequency content of the data. The registry space spanned (the relative high or low frequency of discrete or noise sound) can give information on instantaneous frequency changes. The notes are approached as the product of several frequencies (fundamentals, harmonics, etc.). Temporal fluctuation information is then portrayed as a simultaneously sounded cluster of pitches. When each data set was sonified we derived a pitch that changed according to phase, frequency and time variations. The different spans were grouped and then extracted from the data. Those correspond to spin and orbit parameters that have to be extracted from the data to be characterized. Data acquired at a sampling-rate resolution of 1 sec were converted from .FITS to .txt and imported into the sonification prototype. The data were heard in sets of 4,000, with each data set overlapping by ten minutes with the previous one. Two continuous segments thus overlapped by 10 minutes. As a reference to our ears and for the conversion from musical note to data frequency, we calibrated the frequencies in terms of the notes we recognized. A period search by extracting the harmonics using all the data then showed several statistically significant periods in the range between 250 and 800 sec (see Fig. 2).

Solar Wind: ACE, WIND, GOES. We attempted to use sonification techniques to analyze the impact of a major solar flare (X17 Halloween Storm) on the solar plasma, in particular, the solar wind between the Sun and the Earth, and the astrophysical cavity which is the Earth's magnetosphere. The focus was on sonification settings to analyze the X17 flare and its associated halo CME, initiated 2003 October 28. Particle flux and magnetic fluctuation data were extracted from a constellation of satellites: Advanced Composition Explorer (ACE), the WIND space probe, and the Geostationary Operational Environmental Satellite (GOES). The power spectrum was sonified, giving special emphasis to the harmonic content of the sound and restricting the sonification to a window span of 1 decade between 10–1000 mHz, detecting frequency-averaged level and descending tones. Those correspond to spectral indices, which in turn tell us something about the physical processes in the plasma. We approached instantaneous changes using sound and/or the notes produced by a timbre in terms of a sum of a number of distinct frequencies such as harmonics, fundamentals, partials, non-harmonics, etc. We listened to satellite time-series data to identify changes in the spectral index of the space plasmas.

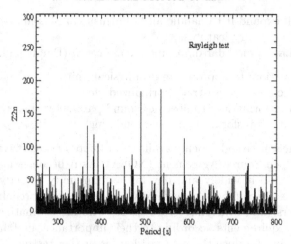

Figure 2. Rayleigh test periodogram, courtesy G. J. M. Luna

The sonification showed both expected and unexpected changes in the power spectra. The sounds were characterized using FFT algorithms.

The GOES satellite data preliminary spectral index −1.20 was not as negative as heard in the sonification. Sonification might indicate events for which a characterization of the descending tones have to be developed. Listening to turbulent data can be done meaningfully in frequency space, and it might also be possible to detect differences in phase velocity. Listening to power spectra may also be meaningful in the frequency space, by restricting the year-to-decade windows between 10–1000 mHz. The descending tones and frequency average levels will correspond to the spectral index.

Direct Mapping Sonification has so far been useful to detect unexpected changes in the data being analyzed (e.g., Kolmogorov power spectra −5/3 in SWIFT data). Sonification provides an inexpensive and accessible tool to examine the impact of a major solar flare on the solar wind, magnetosphere and ionosphere. It is necessary to develop a three-dimensional analysis of fast solar wind and slow solar wind using sound. We use the latter to extract vector components for further correlation.

4. Conclusions

Characterization of sounds using Fourier transformations provides a fair approximation to the data. Sonification might indicate local changes that are better characterized with non-stationary data assumptions. Documentation on how to listen to different space-physics data sets and on sonification techniques will be developed, to extend sonification as a reliable data analysis tool for use by the entire scientific community. Currently, perception experiments are being designed and carried out to demonstrate how sonification techniques may be used as an adjunct to astrophysical data visualization by the sighted.

Acknowledgements

This research is sponsored by the Smithsonian Women's Committee.

Reference

Daudé, S. & Nigay, L. 2003, *Proc. ICAD 2003*, p. 176

New Horizons in Time-Domain Astronomy
Proceedings IAU Symposium No. 285, 2011
R.E.M. Griffin, R.J. Hanisch & R. Seaman, eds.
© International Astronomical Union 2012
doi:10.1017/S1743921312000452

Probing Magnetic Mysteries with Stellar Flares

Rachel A. Osten

Space Telescope Science Institute, 3700 San Martin Drive, Baltimore, MD 21218, USA
email: `osten@stsci.edu`

Invited Talk

Abstract. Flares are a fact of life for stars in the cool half of the H-R diagram. The production of magnetic fields and the consequent dynamic interactions of field and plasma give rise to observational phenomena which span the electromagnetic spectrum. Stellar flares have an impact not only on the stellar atmosphere but also the stellar environment, which can include forming and already formed planets. This talk gave a brief review of our current state of understanding of stellar flares, highlighting some of the main unanswered questions by a panchromatic approach. It also emphasized commonalities between observations and analysis of stellar flares and other types of transient and variable sources.

Microscopy of the Interstellar Medium

Mark Walker

Manly Astrophysics, 3/22 Cliff St, Manly 2095, Australia
email: `Mark.Walker@manlyastrophysics.org`

Invited Talk

Abstract. Radio-wave scintillation arises from inhomogeneities in the ionised ISM. Only small columns of free-electrons are needed to introduce significant phase changes, and the relevant length scales (around the Fresnel scale) are tiny, making scintillation a sensitive technique for studying the ISM on small-scales. On these scales the ISM is surprisingly rich in features, being densely populated by compact structures which manifest high stresses. These structures appear to be an entirely new component of the Galaxy.

New Horizons in Time-Domain Astronomy
Proceedings IAU Symposium No. 285, 2011
R.E.M. Griffin, R.J. Hanisch & R. Seaman, eds.

© International Astronomical Union 2012
doi:10.1017/S1743921312000464

Towards a New Generation of Multi-Dimensional Stellar Models: Can Our Models Meet the Challenges?

I. Baraffe[1], M. Viallet[1] & R. Walder[2]

[1]Physics and Astronomy, University of Exeter, Exeter, EX4 4QL, UK

[2]CRAL, École Normale Supérieure, 69007 Lyon, France

email: Isabelle.Baraffe@ens-lyon.fr

Invited Talk

Summary. The talk described the first steps of development of a new multi-dimensional time-implicit code devoted to the study of hydrodynamical processes in stellar interiors. The main motivation stemmed from the fact that our physical understanding of stellar interiors and evolution still largely relies on one-dimensional calculations. The description of complex physical processes like time-dependent turbulent convection, rotation or MHD processes mostly relies on simplified, phenomenological approaches, with a predictive power hampered by the use of several free parameters. These approaches have now reached their limits in the understanding of stellar structure and evolution. The development of multi-dimensional hydrodynamical simulations becomes crucial to progress in the field of stellar physics and to meet the enormous observational efforts aimed at producing data of unprecedented quality (COROT, Kepler GAIA). The code we are developing solves the hydrodynamical equations in spherical geometry and is based on the finite volume method. The talk presented a global simulation of turbulent convective motions in a cold giant envelope, covering 80% in radius of the stellar structure. Our first developments show that the use of an implicit scheme applied to a stellar evolution context is perfectly thinkable.

Echo Mapping of AGNs

Keith Horne

SUPA Physics & Astronomy, St. Andrews, KY16 9SS, Soctland

email: kdh1@st-and.ac.uk

Summary. Time delays can be used to dissect on micro-arcsecond scales the structure of photo-ionised gas in active galactic nuclei. This talk discussed methods used for this type of indirect imaging, and presented some recent results.

Day 5:

Preparing for the Future

New Horizons in Time-Domain Astronomy
Proceedings IAU Symposium No. 285, 2011
R.E.M. Griffin, R.J. Hanisch & R. Seaman, eds.

© International Astronomical Union 2012
doi:10.1017/S1743921312000488

Exploring the Time Domain with Synoptic Sky Surveys

S. G. Djorgovski, A. A. Mahabal, A. J. Drake, M. J. Graham, C. Donalek and R. Williams

California Institute of Technology, Pasadena, CA 91125, USA
email: george@astro.caltech.edu

Invited Talk

Abstract. Synoptic sky surveys are becoming the largest data generators in astronomy, and they are opening a new research frontier that touches practically every field of astronomy. Opening the time domain to a systematic exploration will strengthen our understanding of a number of interesting known phenomena, and may lead to the discoveries of as yet unknown ones. We describe some lessons learned over the past decade, and offer some ideas that may guide strategic considerations in the planning and execution of future synoptic sky surveys.

1. Introduction: Exploring a New Domain

In the 1990s astronomy transitioned from a data poverty to an immense, exponentially-growing data richness. The main agents of change were large digital sky surveys that produced data-sets measuring from a few to a few tens of Terabytes, and they, in turn, were enabled by burgeoning information technology. The challenge of the effective scientific use of such data-sets was met by the advent of the Virtual Observatory (VO) concept. The data volume continues doubling on a scale of ~1–2 years, reflecting Moore's law that describes the growth of the technology that produces the data. There is also an accompanying growth of data complexity and data quality. We are now transitioning into the Petascale regime, and the main agents of change are the synoptic sky surveys that cover large areas of the sky repeatedly. Some of the current surveys include CRTS, PTF and PanSTARS in the optical and various SKA prototypes in radio, leading to the next generation of facilities (such as LSST, SKA and many others described at this Symposium) that will effectively operate in a time-domain mode, producing tens of TB daily.

Time-domain astronomy (TDA) opens a new discovery space, not just by the sheer growth of data rates and data volumes but also by opening the "time axis" (actually, many axes) of the observable parameter space (OPS). A distinction should be made between the OPS, which is limited by our technology and the physical limitations of measurements (things like the transparency of the Earth's atmosphere or the ISM, diffraction limit, quantum limits, etc.) and the physical parameter space, which is populated according to the laws of nature; the mapping of one onto the other is not trivial. This expresses the vision of a systematic exploration of the OPS first formulated by Zwicky (1957), who referred to it as the "morphological box". History has shown that every time technology enables us to open a new portion of the OPS we are likely to discover some new types of objects and phenomena (Harwit 1975). Specifically, exploration of the time domain ("monitoring sky for variability") was eloquently advocated by Paczynski (2000).

It is a rich territory to explore. Some phenomena, such as various cosmic explosions, accretion or relativistic behaviour, can be studied *only* in the time domain. As a whole,

TDA touches practically every field of astronomy, from the Solar System to cosmology and from stellar structure and evolution to extreme relativistic phenomena. Nor is it confined to the electromagnetic signals, as the neutrino and cosmic ray astronomy mature and gravitational-wave astronomy is born. This very richness makes the TDA a too diffuse concept, just as it makes little sense to talk about "spectroscopic astronomy" or "imaging astronomy". Rather, we can more meaningfully focus on the subjects of synoptic sky surveys or transient event discovery and characterization.

Recent discoveries of previously predicted phenomena such as supernova breakout shocks or tidal disruption events illustrate the scientific potential of TDA. It is reasonable to expect that a systematic exploration of the previously poorly covered parts of the OPS, in terms of the sensitivity, time cadences, area coverage, etc., may lead to a discovery of previously unknown phenomena.

TDA was also recognized as one of the most promising areas of the new, data-rich astronomy at the very onset of the VO concept (Djorgovski *et al.* 2001ab), and indeed it exercises every visualised VO functionality, and then some. As we argue below, a strong computational infrastructure is an essential enabling factor for the TDA.

2. Some Lessons Learned

The field is far too big to review adequately here. Our own experiences may be illustrative of the challenges involved, at least in the visible wavelength regime.

A search for highly variable and transient sources in the DPOSS plate overlaps (Mahabal *et al.* 2001; Granett *et al.* 2003) covered ~8,000 deg^2 with at least 6 exposures (2 in each of 3 filters) and time base-lines ranging from a few months to ~8 years. We found that, at those time base-lines, roughly half of the high-amplitude variable objects are Galactic stars (mainly CVs and flaring dM), and half are AGNs (mostly blazars). We also found that in a single snapshot there will be ~10^3 optical transients per sky down to ~20 mag, an estimate that has held well since then. It was clear that a variety of phenomena contribute to the population of optical transients (OTs), but that (near) real-time follow-up observations would be necessary to establish their nature.

The Palomar-Quest (PQ) survey (Djorgovski *et al.* 2008) and the concurrent NEAT project lasted ~5 years (ending in September 2008), with exploration of the time domain as the main science drivers. They resulted in discoveries of several hundred supernovæ (SNe), mostly in collaboration with the LBNL Nearby Supernova Factory, studies of AGN variability, and studies of the most variable sources in the sky (aside from SNe) which again turned out to be mainly CVs and beamed AGN (Bauer *et al.* 2009). For the last 2 years of the survey we processed the drift scan observations in real time, the PQ Event Factory, which in some cases led to follow-up spectroscopy within an hour of the initial OT detection. The scientific returns were limited mainly by the poor quality of the data, and by the availability of follow-up. PQ was succeeded at the same telescope by the PTF (Rau *et al.* 2009), which operates with a very similar model but with a much better camera and much more abundant follow-up resources.

Aside from the confirmation that an OT event stream will contain a broad variety of astrophysical phenomena, several key lessons emerged. First, that asteroids are the main contaminant, with ~10^2–10^3 asteroids for each astrophysical transient, so a joint data processing and analysis is necessary. Secondly, that adequate follow-up—and in spectroscopy in particular—is essential for scientific returns; this is still a critical issue, and it is getting worse. Thirdly, that rapid classification of transients is essential in order to distill the incoming event stream down to a manageable number of interesting events worthy of the expenditure of the limited follow-up resources. A part of this requirement is a reliable and robust elimination of various data artifacts: in a massive data stream, inevitably there will be many glitches, and even the most unlikely things will happen, and most of them can look like transient events to a data pipeline. And finally, that the cost of software development will dominate any current or future synoptic sky survey, accounting perhaps for ~80% of the total cost. One practical lesson was that the real-time processing demands must be accommodated in the overall system architecture, in addition to all that has been learned in the processing of single-epoch surveys.

We are currently conducting the Catalina Real-Time Transient Survey (CRTS; Drake *et al.* 2008, Djorgovski *et al.* 2011a, Mahabal *et al.* 2011). CRTS taps into a data stream used to search for NEO asteroids, thus both satisfying the need to separate asteroids from astrophysical OTs and illustrating yet again that the same data stream can feed many different scientific projects. CRTS has so far discovered ~1,000 SNe, including some novel or unusual types, a comparable number of CVs and dwarf novæ, variability-based IDs of previously unidentified *Fermi* gamma-ray sources, planets or other low-mass companions around white dwarfs, young stellar objects, and a plethora of variable stars and AGN (for example, Drake *et al.* 2010, 2011, 2012). CRTS imposes a very high detection threshold for OTs, and even this subset of the highest-amplitude events strains our follow-up capabilities. If we modify the pipeline to pick all statistically significant variables, the number of OTs would grow by at least an order of magnitude.

We are accumulating an unprecedented data-set of images and source catalogues (light curves) for $> 5 \times 10^8$ sources covering ~33,000 deg^2, spanning the time base-lines from 10 min to ~7 years and growing. This archival information is extremely useful for the interpretation of OTs, and it can enable a variety of archival TDA studies.

One lesson of CRTS is that a synoptic sky survey need not be photometric: its job is to discover transients, which can be done very efficiently in a single bandpass (or just an unfiltered CCD); the photometry is best done as a part of the follow-up. This relaxes many calibration and data quality demands faced by surveys that aim to be photometric. One should separate discovery of OTs from their characterization.

Another, iterated, lesson is that spectroscopic follow-up is already a key bottleneck, with only maybe ~10% of CRTS transients followed. The problem will get worse by orders of magnitude with the next generation of synoptic sky surveys. Thus, the need for effective automated classification of transient events is critical.

3. Cyber-Infrastructure for Time-Domain Astronomy

TDA is by its nature very data-intensive, requiring a strong cyber-infrastructure that includes data processing pipelines, archiving, automated event classification and distribution, assembly of relevant information from the new data and the archives, etc.

The ephemeral nature of transient events requires that they are electronically distributed (published) in real time, in order to maximize the chances of a necessary follow-up. To this effect, we developed VOEvent, a VO-compliant standard for the event information exchange. Our vision was to lay the foundations for the robotic telescope networks with feedback, that would discover and follow-up transients, involving a variety of computational and archival data resources, and to facilitate event publishing, brokering, and interpretation. The next step was to develop a concept of event portfolios, that would automatically accumulate the relevant information and make it both machine- and human-accessible, via the web services and various electronic subscription mechanisms. The current implementation is *SkyAlert* (Williams *et al.* 2009).

The challenge of an automated event classification and follow-up prioritization is still outstanding. All OTs look the same when discovered—a star-like object that has changed its brightness significantly relative to the comparison base-line—and yet, they represent a vast range of different physical phenomena, some of which are more interesting than the others. Nowadays, surveys generate tens to hundreds of OTs per night; LSST may find $\sim 10^5$ to 10^7 per night. Which ones are worthy of the expenditure of valuable and limited follow-up resources?

This problem entails some special challenges beyond traditional automated classification methods, which are usually done in some feature vector space with an abundance of homogeneous data. Here, the input information is generally sparse and heterogeneous, and often with a poor S/N; there are only a few initial measurements, and they differ from case to case with differing measurement errors; the contextual information is often essential and yet difficult to capture and incorporate in the classification process; many sources of noise, instrumental glitches, etc., can masquerade as transient events in the data stream; new, heterogeneous data arrive, and the classification must be iterated dynamically. The process must be automated, robust and reliable, with at most a minimal human intervention. Requiring high completeness (not missing any interesting events) and low contamination (a few false alarms), and the need to complete the classification process and make an optimal decision about expending valuable follow-up resources in (near) real time are substantial challenges that require some novel approaches (Donalek *et al.* 2008, Mahabal *et al.* 2008, Djorgovski *et al.* 2011b).

Most of the information about any given event initially, and often permanently, would be archival and/or contextual: spatial (what is around the event), temporal (what is its past light curve), and panchromatic (has it been detected at other wavelengths). Applying it may require a human (expert) judgment, and yet human involvement does not scale to the forthcoming event data streams. We are working on methods to harvest human pattern recognition skills and turn them into computable algorithms.

4. Concluding Comments

TDA—or simply astronomy with synoptic sky surveys—is intrinsically *astronomy of telescope-computational systems*. An optimal strategy may be to have dedicated survey telescopes, and surveys that are not overburdened by other requirements (e.g., multicolor photometry), and a hierarchy of follow-up facilities. For example, there may be a set of smaller, robotic telescopes providing multicolour photometry and helping to select the most promising events for spectroscopy. It would also make sense to coordinate surveys at different wavelengths to serve as a first-order mutual multi-wavelength follow-up.

There is an understandable trend to optimize a given survey's parameters (cadence, depth, etc.) for a given scientific goal such as SNe or NEO asteroids. That inevitably introduces selection biases against objects whose variability may not be captured well with that particular window function, thus diminishing the likelihood of truly novel discoveries. It would be good to have a broad spectrum of time base-lines that can capture a variety of phenomena, both known and as yet unknown. It would make sense if the competing surveys coordinate their sky coverage and cadences, and share the data.

Adequate and effective follow-up, especially optical spectroscopy, remains a key limiting factor. In a situation where there is a steady and abundant stream of events, the highly disruptive Target-of-Opportunity approach is not optimal; dedicated follow-up facilities are needed. The current-generation spectrographs on large telescopes tend to be optimized for highly multiplexed spectroscopy of faint objects, for example for studies of galaxy evolution. In contrast, the follow-up of transient events requires efficient single-object spectroscopy with relatively short exposures. Trying to repurpose existing equipment for a highly inefficient use makes little sense: telescopes and instruments dedicated for spectroscopic follow-ups of transient events should be designed accordingly.

All this has to be built on a strong cyber-infrastructure for data processing and archiving, event discovery, classification and publishing, etc. Automated, robust and reliable event classification is a key need for effective scientific returns and for optimal use of expensive facilities and resources. Given the importance of archival data for the early classification and interpretation of events, efficient, VO-type data services will be increasingly more important. Overall, a strong investment in astroinformatics, including facilities, software, and scientist training, is a major strategic need.

A transition from a data poverty regime to data overabundance will also change the astronomical sociology and operational modes: we are already in a situation where the producers of these massive data streams cannot fully exploit them in a timely manner. Thus, the focus of value shifts from the ownership of data to the ownership of expertise needed to make the discoveries. A key concept, promoted by the CRTS, is the completely *open data* philosophy: making the synoptic sky survey and event data streams available immediately to the world. While that trend was already apparent with the single-epoch surveys, it becomes critical with the synoptic sky surveys and the highly perishable transient events that they discover. As the data rates exceed the capabilities of any individual group to follow up effectively, it only makes sense to open them up, and thus engage a much broader segment of the astronomical community; in fact, it would be irresponsible to do otherwise. While the concept of a proprietary data period may still make sense for some types of targeted observations, it does not do so for the exponentially growing data streams of today or the future.

Finally, perhaps real-time astronomy with OTs is being overemphasized. There is a lot of excellent, non-time-critical science that can be done with the growing archives from synoptic sky surveys, such as a systematic search for variables of given types (like the RR Lyraes for mapping the Galactic structure), improved characterization of AGN variability as a constraint on theoretical models of accretion and beaming, etc.

Acknowledgments: This work was supported in part by the NASA grant 08-AISR08-0085, and the NSF grants AST-0407448, AST-0909182 and IIS-1118041. We thank numerous collaborators from the DPOSS, PQ, and CRTS surveys, and the VO and astroinformatics communities for the work and ideas that have shaped our own ideas and scientific results. We also thank the organizers for a most interesting Symposium.

References

Bauer, A., *et al.* 2009, *ApJ*, 705, 46

Djorgovski, S. G., *et al.* 2001a, *ASPCS*, 225, 52

Djorgovski, S. G., *et al.* 2001b, in *Mining the Sky*, eds. A. Banday *et al.*, *ESO Astroph. Symp.*, p. 305, Berlin: Springer-Verlag

Djorgovski, S. G., *et al.* 2008, *AN*, 329, 263

Djorgovski, S. G., *et al.* 2011a, in: eds. T. Mihara & N. Kawai (eds.), *The First Year of MAXI: Monitoring Variable X-ray Sources* (Tokyo: JAXA Special Publ.), in press

Djorgovski, S. G., *et al.* 2011b, in: A. Srivasatava, N. V. Chawla and A. S. Perera (eds.), *CIDU Conf. 2011*, (NASA), p. 174.

Donalek, C., *et al.* 2008 , *AIPC*, 1082, 252

Drake, A. J., *et al.* 2009, *ApJ*, 696, 870

Drake, A. J., *et al.* 2010, *ApJ*, 718, L127

Drake, A. J., *et al.* 2011, *ApJ*, 738, 125

Drake, A. J., *et al.* 2012, *ApJ*, submitted (arXiv/1009.3048)

Granett, B., *et al.* 2003, *BAAS*, 35, 1300

Harwit, M. 1975, *QJRAS*, 16, 378

Mahabal, A. A., *et al.* 2001, *BAAS*, 33, 1461

Mahabal, A. A., *et al.* 2008, *AN*, 329, 288

Mahabal, A. A., *et al.* 2011, *BASI*, 39, 387

Paczyński, B. 2000, *PASP*, 112, 1281
Rau, A. *et al.* 2009, *PASP*, 121, 1334
Tonry, J. 2011, *PASP*, 123, 58
Williams, R., *et al.* 2009, *ASPCS*, 411, 115
Zwicky, F. 1957, *Morphological Astronomy*, (Berlin: Springer-Verlag)

Appendix: A Figure of Merit for Synoptic Sky Surveys

It has become customary to compare surveys using the etendue, a product of the telescope collecting area A and the instrument field of view Ω as a figure of merit (FoM). However, $A\Omega$ simply characterizes the telescope and partly the instrument, and says nothing about the survey, such as its depth, coverage rate, cadence, etc. A more meaningful FoM is needed; see Tonry (2011) for a relevant discussion.

We propose the following indicator of a survey's discovery potential, a product of its spatio-temporal coverage rate, C, and the estimate of the depth, D, that may be reasonably expressed as proportional to the S/N ratio of the individual exposures. Thus:

$$C = R \times N_p \times f_{open} \quad \text{and} \quad D = [A \times t_{exp} \times \epsilon]^{1/2}/FWHM \sim S/N,$$

where R is the area coverage in deg^2/night (not counting repeated exposures), N_p is the number of passes per field in a given night, f_{open} is the fraction of open time averaged over the year, including weather losses, engineering time, deliberate closures, etc., A is the effective collecting area in m^2, t_{exp} is the average exposure time in sec, ϵ is the overall efficiency (throughput) of the instrument, and $FWHM$ is the typical seeing FWHM in arcsec. Clearly, all these parameters should be taken as typical or averaged over a year. Note that f_{open} and $FWHM$ characterize the site, A and ϵ, and partly R, characterize the telescope+instrument, and the remaining parameters reflect the chosen survey strategy.

CD represents a FoM for a discovery rate of events, so the nett discovery potential of a given survey would be CD multiplied by the number of years the survey operates. While this FoM accounts for most of the important survey parameters, it still does not capture factors such as the sky background and transparency, the total number of sources detected (which clearly depends strongly on the Galactic latitude), the cadence, bandpasses, angular resolution, etc.; nor does it account for operational parameters such as the data availability, the time delay between observations and the event publishing, etc. Nevertheless, we believe that CD is a much more relevant FoM than the traditional (and often mis-used) $A\Omega$, as far as a characterization of *surveys* is concerned.

The following table shows the estimated values of the relevant parameters and CD for the 3 components of the CRTS, and for several other current or future surveys. The assumed values of input parameters are based on our own experience or on published values, and may be consistently too optimistic. The values of CD are no better than $\sim 20\%$.

Survey	R	N_p	f_{open}	A	t_{exp}	ϵ	$FWHM$	C	D	CD
CRTS:CSS	1200	4	0.7	0.363	30	0.7	3	3360	0.92	3090
CRTS:MLS	200	4	0.7	1.767	30	0.7	3	560	2.03	1140
CRTS:SSS	800	4	0.7	0.196	20	0.7	3	2240	0.55	1240
CRTS total	2200	4	0.7	(2.326)		0.7	3	6160		5470
PTF	1000	2	0.7	1.131	60	0.7	2	1400	3.45	4820
SkyMapper	800	2	0.7	0.785	60	0.8	2	1120	3.07	3440
PS1	1000	4	0.7	2.54	30	0.8	1	2800	7.81	21860
LSST	5000	2	0.75	34.9	15	0.8	0.8	7500	25.6	192000

New Horizons in Time-Domain Astronomy
Proceedings IAU Symposium No. 285, 2011
R.E.M. Griffin, R.J. Hanisch & R. Seaman, eds.
© International Astronomical Union 2012
doi:10.1017/S174392131200049X

Pulsars, SKA and Time-Domain Studies in the Future

Michael Kramer[1,2]

[1] Max-Planck-Institut für Radioastronomie, 53121 Bonn, Germany
[2] Jodrell Bank Centre for Astrophysics, University of Manchester, Manchester, M13 9PL, UK
email: mkramer@mpifr.de

Invited Talk

Abstract. Pulsars are a classical example for time-domain astronomy. Using just the precisely measured arrival time of pulses from pulsars, we can study a wide range of physics, and in particular probe the validity of theories of gravity. Despite huge success, pulsar astrophysics will completely change with the Square Kilometre Array (SKA), lifting pulsar science into a new era of time-domain astronomy. I will review the applications using pulsars and the prospects for the SKA.

Keywords. (stars:) pulsars: general, gravitation, gravitational waves

1. Introduction

The Square Kilometre Array (SKA) will be the world's largest radio telescope, enabling transformational science. It will be a flexible, multi-purpose observatory that will be able to serve the whole astronomy community as *the* premier imaging and surveying instrument. Five Key Science Projects have been selected by the international community; those projects must drive essential parts of the design in order to ensure that their science goals can be achieved. Some of the most major advances in science can be expected for time-domain studies and pulsars in particular. The reason is the SKA's overwhelming sensitivity by virtue of its enormous collecting area, and is its defining feature. As the name suggests, the total collecting area for the SKA is planned to be close to one square kilometre, offering a target sensitivity of $2 \times 10^4 \, \mathrm{m^2 \, K^{-1}}$.

2. The Square Kilometre Array

The SKA is likely to consist of sparse Aperture Arrays (AAs) of tiled dipoles in the frequency range of 70 to 500 MHz, while above 500 MHz the following three implementations are being considered:

a) 3 000 15-m dishes with a single pixel feed, a sensitivity of 0.6 SKA units, $T_{\mathrm{sys}} = 30 \, \mathrm{K}$ and 70% efficiency covering the frequency range of 500 MHz to 10 GHz.

b) 2 000 15-m dishes with Focal Plane Arrays (FPAs) from 500 MHz to 1.5 GHz, a sensitivity of 0.35 SKA units, a Field-of-View (FoV) of 20 deg^2, $T_{\mathrm{sys}} = 35 \, \mathrm{K}$ and 70% efficiency and a single pixel feed from 1.5 to 10 GHz, with $T_{\mathrm{sys}} = 30 \, \mathrm{K}$.

c) A combination of dense AAs with a FoV of 250 deg^2, a sensitivity of 0.5 SKA units, covering the frequency range of 500 to 800 MHz and 2 400 15-m dishes with a single pixel feed covering the frequency range of 800 MHz to 10 GHz, a sensitivity of 0.5 SKA units, $T_{\mathrm{sys}} = 30 \, \mathrm{K}$ and 70% efficiency.

The signals of these receiving elements can be combined to form independent FoVs, resulting in a survey speed of 10,000 times that of existing telescopes.

The exact configuration of the SKA will be determined after the present global R & D phase is completed. However, most of the collecting area will be concentrated in a central core of the SKA: 20% of the collecting area will be located within a 1-km radius, and 50% within a 5-km radius. The remainder of the receiving elements will be distributed in stations (probably) along log-spiral arms with base-lines extending to 3,000 km in order to permit high-resolution imaging and high-precision astrometry too. Increased resolution will be achieved via intercontinental base-lines between Australia and Southern Africa.

The SKA will be built in three phases. Phase I: by 2010, about 0.1 SKA unit will be available in the core region covering the low- and mid-frequencies. Phase II: by 2022, the full array will be completed. Phase III: Beyond 2022, high-frequency capabilities for observations, potentially up to ~ 20 GHz, will be added.

3. Time-Domain Radio Astronomy

As demonstrated by other contributions in these proceedings (for instance, Keane, p. 342), the current phase space in time-domain astronomy is covered only sparsely by known sources. In the "coherent-emission" part of phase space, most of the sources represent the emission from neutron stars, and especially from pulsars. Indeed, the radio emission from pulsars shows a rich variety of intriguing properties with a wide range of observed time-scales. The observed phenomena, which cover an astounding 18 orders of magnitude of time, have been summarised by Stappers (p. 103). The range of physics that has been probed is similarly wide, extending from gravitational physics to extreme plasma physics, from solid state physics to studies of the Galactic structure, from stellar physics to binary evolution. The SKA will contribute enormously to all these areas. Its contribution to studies of gravitational physics by measuring pulsars orbiting black holes or by using them as cosmic gravitational-wave detectors has been identified as one of five SKA Key Science Projects (see Kramer *et al.* 2004). The questions to be addressed and answered include:

- What is the nature of gravity? Was Einstein right? Is gravity described by a tensor field or are there additional scalar fields, as it is sometimes proposed to explain Dark Energy?
- What are the properties of gravitational waves? Do gravitons have Spin 2? What is the mass of gravitons, and hence what is the propagation speed of gravitational waves?
- What happens in strong gravitational fields, in conditions of extreme curvature and near singularities? What are the properties of black holes? Do the "no-hair" and "cosmic censorships" theorems hold?

Answering these questions requires a survey for pulsars and the high-precision timings of a selected sample of them.

4. A Galactic Census for Pulsars

We currently know about 2,000 pulsars; they include 10 double neutron stars and one double pulsar, but no pulsar–black hole systems. Major continuing efforts to increase the numbers include the P-ALFA survey with the Arecibo telescope, the GBT-350 MHz survey or the High Time Resolution Universe (HTRU) Survey in the Northern hemisphere with the Effelsberg 100-m telescope, and in the Southern hemisphere with the Parkes telescope. Recently, deep searches of unidentified point sources seen by the

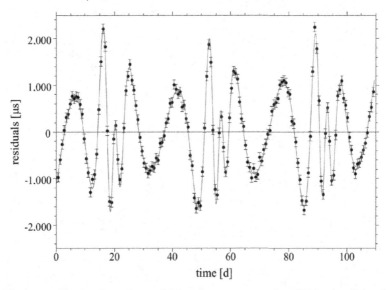

Figure 1. Timing residuals of a pulsar orbiting the super-massive BH in the Galactic Centre in a 0.1-yr orbit. Shown are three orbits with an eccentricity of $e = 0.4$ for an extreme Kerr BH. Even with a timing precision of only 100 μs, the characteristic periodic residuals caused by the BH's quadrupole moment are clearly visible (Liu *et al.* 2012).

FERMI gamma-ray space telescope have been tremendously successful. Soon, the Low Frequency Array (LOFAR) should identify many new pulsars in the solar vicinity (van Leeuwen & Stappers 2010), providing a local census. Still, all these very good results are only a prelude to what will be possible with the SKA.

Depending on the low end of the luminosity function and the collecting area that is deployable, one can find practically all active Galactic pulsars beaming towards Earth, providing a "Galactic Census" of pulsars (Cordes *et al.* 2004). We expect to find about 20,000 to 30,000 pulsars, including 1,000 to 3,000 millisecond pulsars and 100 relativistic binaries (Smits *et al.* 2009). This survey will be conducted at a frequency between 1–3 GHz, depending on Galactic latitude and frequency overlap of different antenna types. The outcome of the survey will be a sample of accurate millisecond pulsar clocks that we will time as a Pulsar Timing Array to study the nHz-gravitational wave sky. Furthermore, we should uncover the rare pulsar–black hole (BH) systems where a pulsar orbits a stellar BH in the Galactic plane or an intermediate BH in a globular cluster. Once the interesting sources have been identified, they will need to be timed with the full array to achieve the best possible precision. The timing therefore consists of two parts: a medium to low-precision timing programme to follow-up the 20,000 to 30,000 discovered sources to identify the "goodies", and a high-precision timing part of an estimated 100 to 300 sources in order to extract the science. For the latter sources, we can expect timing precisions to increase with the sensitivity of the SKA over the performance of current telescopes.

While the Galactic Census will deliver the bulk of the sources, a small piece of sky— the Galactic Centre (GC)—needs to be treated specially. The GC region is interesting not only because a large stellar density suggests the occurrence of BH-MSP systems, but also because we will discover pulsars orbiting the super-massive BH in the GC. Tracing the movement and rotation of pulsars in their orbits provides in principle an easier measurement of the BH spin and quadrupole moment since those quantities scale with the square and cube of the masses, respectively. Such a measurement would also extend

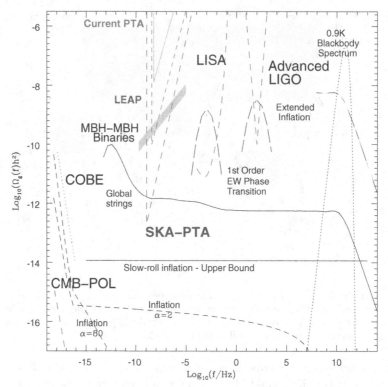

Figure 2. Summary of the potential cosmological sources of a stochastic gravitational-wave (GW) background, overlaid with bounds from COBE, current Pulsar Timing Array (PTA) experiments and the goals of CMB polarization experiments, LISA and Advanced LIGO. LEAP will improve the current best PTA limits by more than two orders of magnitude, enabling the detection of a GW background caused by the merger of massive black holes (MBHs) in early galaxy formation. The amplitude depends on the MBH mass function and merger rate, so uncertainty is indicated by the size of the shaded/yellow area. LEAP is the next logical step towards realizing a PTA with the SKA; the latter will improve the current sensitivity by about four orders of magnitude.

the mass range of BH properties studied, but the problem is that the interstellar medium in the GC causes severe interstellar scattering because of multi-path propagation. That can only be combated by observing at higher radio frequencies; frequencies between 10 and 15 GHz should be sufficient.

5. Black-Hole Properties

What makes a binary pulsar with a black-hole companion so interesting is that it has the potential to provide a superb new probe of relativistic gravity. As pointed out by Damour & Esposito-Farèse (1998), the discriminating power of this probe might supersede all its present and foreseeable competitors. The reason lies in the fact that such a system would be very sensitive to strong gravitational self-field effects, making it (for instance) an excellent probe for tensor-scalar theories. Wex & Kopeikin (1999) showed that, in principle, the measurement of classical and relativistic spin-orbit coupling in a pulsar-BH binary allows us to determine the spin and the quadrupole moment of the black hole. That would test the "cosmic censorship" conjecture and the "no-hair" theorem. While Wex & Kopeikin (1999) showed that, with current telescopes, such an experiment would be almost impossible to perform (with the possible exception of pulsars about the

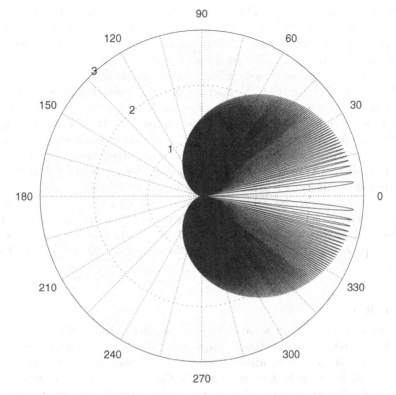

Figure 3. "Antenna pattern" of a single-pulsar timing response to a single monochromatic gravitational-wave source. For illustration purposes, the pulsar distance is chosen to be small with a value of 200 pc and the GW period is chosen as 5 years, in order to show the structure of the response pattern. The GW source is in the 0° angle position, the orbital plane inclination is 90° and the orbital plane coincides with the plane of the paper (Lee *et al.* 2011). The fine structure of this antenna pattern enables the precise localisation of gravitational-wave sources.

Galactic-Centre black hole), Kramer *et al.* (2004) pointed out that the SKA sensitivity should be sufficient. Indeed, this experiment benefits from the SKA sensitivity in multiple ways. On the one hand it provides the required timing precision, but on the other it also allows us to perform the Galactic Census which should eventually deliver a sample of pulsars with BH companions. As shown recently by Liu *et al.* (2012) , it should be "fairly easy" to measure the spin of the GC black hole with a precision of 10^{-4}–10^{-3}. Even for a pulsar with a timing precision of only 100 μs, characteristic periodic residuals would enable the measurement of the BH quadrupole moment to 1% or better. That would allow us to test the no-hair theorem with similar precision (see Fig. 1). Perhaps most importantly, by using the method described by Liu *et al.* it will be possible to determine whether the pulsar-SGR A* system is "clean", i.e. if the timing data are undisturbed by a nearby stellar population.

6. Gravitational-wave Astronomy

While pulsars already provide indirect evidence for the existence of gravitational waves, they can also be used to detect and study them directly using a "Pulsar Timing Array" (PTA). Such PTA experiments have a good chance of detecting GWs, but we are currently limited by the sensitivity of our telescopes. The Large European Array for Pulsars (LEAP), which combines the collecting power of Europe's largest telescopes (Kramer

& Stappers 2010), may get very close to detecting, or may even detect, a stochastic gravitational-wave background, possibly with the help of the International Pulsar Timing Array (IPTA) (see Hobbs *et al.* 2010), but in order to guarantee a detection—or even study the properties of gravitational waves—we need the overwhelming sensitivity of the SKA. Owing to the increased timing precision achievable with the SKA and the large number of suitable sources to be discovered, a PTA with the SKA would achieve a sensitivity that even exceeds that of LISA (Kramer & Stapppers 2010; see Fig. 2).

Different types of signals can be detected and studied with the SKA: stochastic, burst, and periodic signals. A stochastic gravitational-wave background should arise from a variety of sources. Cosmological sources include inflation, string cosmology, cosmic strings and phase transitions (Kramer *et al.* 2004). The expected correlation signal among the PTA pulsars can be measured with very high precision, so the polarisation and mass properties of gravitons can be probed and compared with the predictions of general relativity (Lee *et al.* 2008, 2010). For single GW sources, the superb astrometric precision that will be realised with the SKA (Smits *et al.* 2011) will enable us to pinpoint their locations to a precision that allows efficient electromagnetic follow-up (Lee *et al.* 2011; also see Fig. 3).

7. Conclusions

We have only touched upon a small amount of science topics that can and will be addressed by the SKA. Further topics range from studies of pulsar radio emission to attempts to understand the conditions for the formation of planets around pulsars, facilitated for the first time by having a Galactic Census of pulsars and hence a (nearly) complete sample. But already the SKA pathfinders will provide a great amount of exciting new discoveries, giving us a glimpse of what will be possible with the SKA. The global efforts will culminate in the construction of the SKA as the world's largest and most powerful telescope. As a versatile observatory it will revolutionise many areas of astrophysics, but in particular that of pulsars and fundamental physics.

References

Cordes, J. M., Kramer, M., Lazio, T. J. W., Stappers, B. W., Backer, D. C., & Johnston, S., 2004, *New Astr.*, 48, 1413
Damour, T. & Esposito-Farèse, G., 1998, *Phys. Rev. D*, 58 (042001), 1
Hobbs, G., *et al.* 2010, *Classical and Quantum Gravity*, 27(8), 084013
Kramer, M. & Stappers, B., 2010, in Proceedings of the ISKAF2010 Science Meeting, *POS*, 112, 34
Kramer, M., Backer, D. C., Cordes, J. M., Lazio, T. J. W.., Stappers, B. W., & Johnston, S., 2004, 48, 993
Lee, K. J., *et al.*, 2011, *MNRAS*, 414, 3251
Lee, K. J., Jenet, F. A., & Price, R. H., 2008, *ApJ*, 685, 1304
Liu, K., Wex, N., Kramer, M., Cordes, J. M., & Lazio, T. J. W.. 2012, *ApJ*, in press
Smits, R., Kramer, M., Stappers, B., Lorimer, D. R., Cordes, J. M., & Faulkner, A., 2009, *A&A*, 493, 1161
Smits, R., Tingay, T., Wex, N., Kramer, M., & Stappers, B. 2011, *A&A*, 528, 108
van Leeuwen, J. & Stappers, B. W., 2010, *A&A*, 509, A7
Wex, N. & Kopeikin, S., 1999, *ApJ*, 513, 388

New Horizons in Time-Domain Astronomy
Proceedings IAU Symposium No. 285, 2011
R.E.M. Griffin, R.J. Hanisch & R. Seaman, eds.

© International Astronomical Union 2012
doi:10.1017/S1743921312000506

From Hipparcos to Gaia

L. Eyer[1], P. Dubath[1,2], S. Saesen[1], D.W. Evans[3], L. Wyrzykowski[3], S. Hodgkin[3] and N. Mowlavi[1,2]

[1] Geneva Observatory, CH-1290 Sauverny, Switzerland
email: Laurent.Eyer@unige.ch

[2] ISDC, Department of Astronomy, University of Geneva, CH-1290 Versoix, Switzerland

[3] Institute of Astronomy, Cambridge University, Cambridge, UK

Invited Talk

Abstract. The measurement of the positions, distances, motions and luminosities of stars represents the foundations of modern astronomical knowledge. Launched at the end of the eighties, the ESA Hipparcos satellite was the first space mission dedicated to such measurements. Hipparcos improved position accuracies by a factor of 100 compared to typical ground-based results and provided astrometric and photometric multi-epoch observations of 118,000 stars over the entire sky. The impact of Hipparcos on astrophysics has been extremely valuable and diverse. Building on this important European success, the ESA Gaia cornerstone mission promises an even more impressive advance. Compared to Hipparcos, it will bring a gain of a factor 50 to 100 in position accuracy and of a factor of 10,000 in star number, collecting photometric, spectrophotometric and spectroscopic data for one billion celestial objects. During its 5-year flight, Gaia will measure objects repeatedly, up to a few hundred times, providing an unprecedented database to study the variability of all types of celestial objects. Gaia will bring outstanding contributions, directly or indirectly, to most fields of research in astrophysics, such as the study of our Galaxy and of its stellar constituents, and the search for planets outside the solar system.

Keywords. space vehicles, surveys, stars: variables: other, astrometry, techniques: photometric, techniques: spectroscopic

---

1. Introduction

Time domain plays a central role in the Hipparcos and Gaia missions. Indeed, to determine parallaxes and proper motions of stars, repeated observations of positions are required. The accuracy of ground-based astrometric measurements are limited because of atmospheric distortions, telescope deformations and difficulties related to the definition of the reference frame. Even if a set of distant stars is used, they may have streaming motions or residual parallaxes, leading to inaccuracies in parallax measurements.

In the realms of space, deformations caused by the presence of the atmosphere or gravity are avoided altogether. The issue of the reference frame is solved in a similar manner for both Hipparcos and Gaia. The basic measurements are wide angles between celestial targets. As the spacecraft is scanning the sky over an extended period of time, many such angles with many orientations are typically measured for a given target. The whole sphere is then reconstructed using some mathematical formulations, solving for the astrometric parameters of all targets globally. In that way, a full-sky positional frame of unprecedented accuracy can be self-calibrated (Lindegren & Bastian 2011).

Both Hipparcos and Gaia are projects of the European Space Agency (ESA). Other astrometric missions have been proposed, notably the USA's FAME and SIM, but were not in the end pursued. There is, however, a Japanese astrometric project, JASMINE (see http://www.jasmine-galaxy.org/index.html), the first step of which is a nano-satellite to be launched in 2013. JASMINE should achieve HIPPARCOS-quality accuracies. Although the principles of astrometric measurements are similar for both HIPPARCOS and GAIA, their scientific cases are different; the science case for HIPPARCOS was mostly stellar, whereas that for GAIA is more oriented towards the Galaxy, aiming to study its structure and formation history. However, GAIA will also have a very significant impact on asteroid studies, stellar astronomy and fundamental physics.

Both missions also provide very valuable data sets for variability studies. The cadence of the observations is imposed by optimizing the satellite's scanning routines for best astrometric performance. It turns out, however, that the resulting time-sampling is rather favourable for period determination (Eyer & Mignard 2005), since the one-day and one-year aliases (and related ones) which tend to plague ground-based observations are absent, and other aliases relating to spacecraft characteristic periods are of rather low amplitude. In addition, both missions are performing a homogeneous whole-sky survey with a single set of instruments and with fully standard data reduction procedures, thereby producing a very coherent data set. Though the results of HIPPARCOS were published in 1997, they are still unsurpassed in extent and quality. That is true, not only for astrometry but also for variability studies: HIPPARCOS is still the only whole-sky survey with a published systematic variability analysis.

2. HIPPARCOS

The HIPPARCOS mission collected observations for 118,000 stars down to magnitude $H_p = 12.4$, grouped into two main samples:

(*a*) A whole-sky survey sample of 52,000 stars brighter than $7.9\ V + 1.1\sin(b)$ magnitude for blue stars with $B - V < 0.8$ and brighter than $7.3\ V + 1.1\sin(b)$ for redder stars with $B - V \geqslant 0.8$ (*b* is the Galactic latitude).

(*b*) A set of 66,000 objects selected on the basis on their scientific interest.

The properties of these objects (position, proper motion, magnitude, colour, spectral and variability types, radial velocities and multiplicity) were compiled and published in the HIPPARCOS Input Catalogue (ESA 1992). There were also observational campaigns to collect complementary information during the mission. For example, the AAVSO (American Association of Variable Star Observers) measured the brightness of long-period variables in order to determine the correct exposure times for HIPPARCOS for those brighter than $H_p = 12.4$. In another campaign the H_p band was re-calibrated by measuring R, N and C stars simultaneously from the ground and from space.

The satellite was launched from Kourou (French Guiana) on an Ariane-4 rocket in 1989. The length of the mission was 3.3 years, during which an average of 110 measurements per object were collected. The final results were published in a 17-volume catalogue in 1997 (ESA 1997). The mission realized its original goal of obtaining astrometry at the milli-arcsecond level. In addition, the HIPPARCOS Star Mapper provided measurements which led to the Tycho (ESA 1997) and Tycho-2 (Høg *et al.* 2000) catalogues, which contain astrometric and photometric (B_T and V_T) measurements for 1 and 2.5 million stars, respectively. As expected, the Tycho epoch photometry in B_T and V_T has been more difficult to handle than the H_p-band photometry from the main mission, and has led to a relatively low number of variable stars (Piquard *et al.* 2001). Ten years after the publication of the HIPPARCOS Catalogue a new reduction of the astrometric data was undertaken (van Leeuwen 2007). Various systematic effects were reduced for bright stars, leading to errors more consistent with the photon noise limits. The project scientist of HIPPARCOS, Michael Perryman, wrote two books about the mission (Perryman 2009 and

Perryman 2010). The first summarizes and highlights the scientific harvest of HIPPARCOS during the 10 years following the publication of the catalogue. The second narrates the fantastic human, technological and scientific adventures of the HIPPARCOS mission.

2.1. *A Note on the Pleiades Distance Derived from* HIPPARCOS

Distances of open clusters have been based upon HIPPARCOS parallaxes. For the Pleiades, the distance of 118.3 ± 3.5 pc thus derived (van Leeuwen 1999) was in contradiction to the value of 132 ± 4 pc obtained by the customary main-sequence-fitting technique (Meynet *et al.* 1993). Other independent determinations (Pan *et al.* 2004; Zwahlen *et al.* 2004; Munari *et al.* 2004; Soderblom *et al.* 2005) favour the greater distance. Although, at the level of individual stars, the distances determined via the HIPPARCOS parallax measurements are acceptable within the individual error estimates, the discrepancy becomes more acute when averaging. The new HIPPARCOS reduction by van Leeuwen (2007) confirmed the smaller value of 120 ± 1.9 pc (van Leeuwen 2009). This is the kind of debate that GAIA will close. Unfortunately, for some individual stars such as Atlas, GAIA will not be able to provide a distance determination because its survey is limited to $V \sim 6$ mag.

3. GAIA

The GAIA programme anticipates astrometric, photometric, spectrophotometric and spectroscopic measurements for one billion celestial objects with magnitudes between $V \sim 6$–20. The mission is designed to last for 5 years, with a possible 1-year extension; the launch is currently planned for 2013. The number of measurements accumulated after 5 years varies between 40 and 250 per object, depending mostly on ecliptic latitude, and expects to average 70 (the estimate takes into account the dead-time estimation). Spectroscopic measurements with a resolution of 11,500 will be obtained for stars brighter than 16–17 mag. Recent performance estimates can be found on the GAIA Webpage†. It is important to note that the astrometric accuracies quoted were derived as errors in the parallax at the end of the mission.

It was realized early in the preparation of the mission that processing all the data will be a challenge of the highest order. People interested in certain aspects created Working Groups, which later evolved into Coordination Units (Mignard *et al.* 2008). There are currently 8 Coordination Units (CU); one of them, CU7, is responsible for all aspects relating to variability analyses of the data. The CU7 analyses start with calibrated data provided by other Coordination Units, but the variability and related time-series analyses have impacts on most Coordination Units. One additional Coordination Unit is currently being formed to develop the catalogue access interface for the scientific community. The GAIA Data Processing and Analysis Consortium (DPAC) now includes more than 500 scientists and software engineers. Our estimates of the number of variable sources that GAIA will detect range from 50 million to 150 million. The rather large range reveals limitations of our knowledge in the domain of variable phenomena, but it also reflects uncertainties in the ultimate precision reached once the satellite is in space. Some effort is currently being spent to improve these estimates using results from other space missions such as KEPLER and CoRoT. The diversity of variable phenomena is illustrated by the variability tree presented by Eyer & Mowlavi (2008). GAIA will detect most variability types from that tree.

† http://sci.esa.int/science-e/www/area/index.cfm?fareaid=26

3.1. *Variability Processing and Analysis*

CU7 is in charge of analyzing all variable objects that appear above a certain variability threshold. Specific criteria are set for detecting particular phenomena such as short-period variables, solar-like magnetic variables, planetary transits and periodic small-amplitude variables. Those characteristics are taken into account to lower the variability thresholds set for general purposes. That special variability detection can therefore select variable stars that would be considered constant based on the general variability criteria.

Once an object is considered a candidate for variability, its behaviour is characterized by certain parameters. Simple models such as a trend or periodicity model are tested to see if they can explain the variability. This characterization task produces a set of parameters ("attributes") that are used in the next step of the processing, which is the classification. The classification algorithms compute membership probabilities for each object on the basis of the specific values of the attributes. The intention is to use supervised, unsupervised and semi-supervised methods. Different supervised methods have been evaluated using HIPPARCOS data, and "random forest" proves to be one of the best algorithms in terms of overall performance (Dubath *et al.* 2011).

Once an object is classified, then if it belongs to a certain specific group of variables, further specific analysis is carried out. The different classes that will thus be treated in more detail currently include eclipsing binaries, solar-like variability (powered by magnetic fields), rotation-induced variable stars, RR Lyrae/Cepheid stars, long-period variables, main-sequence pulsators, compact oscillators, pre-main sequence oscillators, microlensing, cataclysmic variables, rapid phases of stellar evolution, and active galactic nuclei. In that step, some attributes will be refined and additional parameters computed, but another important activity is to evaluate the quality of the classification, since contamination and completeness are two fundamental criteria.

3.2. *Science Alerts*

DPAC is preparing a system for science alerts (Wyrzykowski & Hodgkin 2011). Science alerts are defined as events where "the science data would have little or no value without quick ground-based follow-up". Different types are foreseen from different instruments: astrometric alerts (including fast moving objects), photometric, and spectroscopic alerts. Examples of objects that could constitute photometric alerts include supernovæ (6,000, with 2,000 before they peak), microlensing events (1,000, with some having an astrometric signature), cataclysmic variables (novæ, dwarf novæ), eruptive stars (Be, RCrB stars, FU Ori) and possibly gamma-ray bursts, including orphan afterglows. The detection of these photometric alerts will be performed in near-real-time GAIA broad-filter photometry. The classification, however, will utilize all other data available immediately from GAIA, including preliminarily calibrated low-resolution spectra. This will assure the best possible recognition of an alert. For supernovæ, for example, it can provide a preliminary type and an estimate of the red-shift. GAIA will also provide alerts of Near Earth Objects and potentially hazardous asteroids, owing to its superb astrometry capability. One of the huge challenges for these science alerts is how to maximize completeness and avoid contamination from other uninteresting objects.

Since the data might sometimes accumulate on board the satellite before being transmitted to Earth, there are time constraints which limit in practice the window within which alerts can be treated. This time window is between a few hours and two days. The alerts will not start from the first day of operation, but only after a 3-month verification phase. The goal is to verify the quality of the products of the alert pipeline and—more precisely—the probabilities associated with the classification of these alerts. A group of ground-based telescopes is being set up to assist with the verification phase. Furthermore,

once the data are on public release it is agreed that scientific follow-up has to be organized in advance. Such structuring of the ground-based effort is being done now. With
the development of large ground-based surveys, the specificities of the GAIA alert system
should be compared to others such as PTF, CRTS, SKYMAPPER and Pan-STARRS, and
tuned to complement them as well as possible.

There has been a proposal to introduce a "Watch List", that is, a list of targets suggested by interested scientists. If one of those targets behaves in an unexpected manner,
an alert will be sent to the scientist(s) in question so that a follow-up can be organized.
That requires good coordination among groups with different scientific interests, but is
an effective way of enhancing the scientific return of GAIA for selected objects.

4. Conclusions

The HIPPARCOS mission has been a tremendous success. If the results of the GAIA
mission reach current expectations, there is no doubt at all that its multi-epoch survey
of astrometry, photometry, spectrophotometry and spectroscopy will have considerable
impacts on astrophysics. But we should recognize that GAIA is not the only large-scale
survey which is collecting data now, or is planned to do so: Pan-STARRS, LSST, OGLE,
Sky-Mapper, PTF and CRTS are not all. GAIA will nevertheless make a significant and
unique contribution because no other survey will reach the astrometric accuracy of GAIA.
In return GAIA will benefit from the other surveys, especially those having photometric
systems that extend into the ultraviolet or infrared.

A large public spectroscopic survey relating to GAIA was selected by ESO for the VLT;
it starts in 2012 January and is expected to have 30 nights per semester for four years,
with an additional fifth year subject to progress review.

Acknowledgements

We thank Michel Grenon, Joshua Bloom and Pierre North for valuable discussions.

References

Dubath, P., *et al.* 2011, *MNRAS*, 414, 2602
ESA. 1992, *The Hipparcos Input Catalogue* ESA SP-1136
ESA. 1997, *The Hipparcos and Tycho Catalogues*, ESA SP-1200
Eyer, L. & Mignard, F. 2005, *MNRAS*, 361, 1136
Eyer, L. & Mowlavi, N. 2008, *J. Phys. Conf Series*, 118, 012010
Høg, E., *et al.* 2000, *A&A*, 355, L27
Lindegren, L. & Bastian, U. 2011, *EAS Publications Series*, 45, 109
Meynet, G., Mermilliod, J.-C., & Maeder, A. 1993, *A&AS*, 98, 477
Mignard, F., *et al.* 2008, IAU Symposium, 248, 224
Munari, U., *et al.* 2004, *A&A*, 418, L31
Pan, X., Shao, M., & Kulkarni, S. R. 2004, *Nature*, 427, 326
Perryman, M. 2009, *Astronomical Applications of Astrometry: Ten Years of Exploitation of the* HIPPARCOS *Satellite Data* (Cambridge, UK: Cambridge University Press)
Perryman, M. 2010, *The Making of History's Greatest Star Map* (Berlin: Springer-Verlag)
Piquard, S., *et al.* 2001, *A&A*, 373, 576
Soderblom, D. R., *et al.* 2005, *AJ*, 129, 1616
van Leeuwen, F. 1999, *A&A*, 341, L71
van Leeuwen, F. 2007, *ASSL* (Dordrecht: Springer), Vol. 350
van Leeuwen, F. 2009, *A&A*, 497, 209
Wyrzykowski, L. & Hodgkin, S. 2011, arXiv:1112.0187
Zwahlen, N., 2004, *A&A*, 425, L45

New Horizons in Time-Domain Astronomy
Proceedings IAU Symposium No. 285, 2011
R.E.M. Griffin, R.J. Hanisch & R. Seaman, eds.
© International Astronomical Union 2012
doi:10.1017/S1743921312000518

The Future of the Time Domain with LSST

Lucianne M. Walkowicz

Department of Astrophysical Sciences, Princeton University, Princeton, NJ 08544, USA
email: lucianne@astro.princeton.edu

Invited Talk

Summary. In the coming decade LSST's combination of all-sky coverage, consistent long-term monitoring and flexible criteria for event identification will revolutionize studies of a wide variety of astrophysical phenomena. Time-domain science with LSST encompasses objects both familiar and exotic, from classical variables within our Galaxy to explosive cosmological events. Increased sample sizes of known-but-rare observational phenomena will quantify their distributions for the first time, thus challenging existing theories. Perhaps most excitingly, LSST will provide the opportunity to sample previously untouched regions of parameter space. LSST will generate 'alerts' within 60 seconds of detecting a new transient, permitting the community to follow up unusual events in greater detail. However, follow-up will remain a challenge as the volume of transients will easily saturate available spectroscopic resources. Characterization of events and access to appropriate ancillary data (e.g. from prior observations, either in the optical or in other passbands) will be of the utmost importance in prioritizing follow-up observations. The incredible scientific opportunities and unique challenges afforded by LSST demand organization, forethought and creativity from the astronomical community. To learn more about the telescope specifics and survey design, as well as obtaining a overview of the variety of the scientific investigations that LSST will enable, readers are encouraged to look at the LSST Science Book: http://www.lsst.org/lsst/scibook. Organizational details of the LSST science collaborations and management may be found at http://www.lsstcorp.org.

Optimal Strategies for Transient Surveys with Wide-Field Radio Telescopes

J.-P. Macquart, N. Clarke, P. Hall and T. Colegate

ICRAR, Curtin University, Bentley, WA 6102, Australia
email: jean-pierre.macquart@icrar.org

Summary. Fast-time-scale radio transients open new vistas on the physics of high brightness-temperature emission, extreme states of matter and the physics of strong gravitational fields. The next generation of wide-field, highly-sensitive large-N radio arrays such as LOFAR, ASKAP, the MWA and the SKA offer access to an enormous volume of new transients' parameter space. This talk described how we develop a formalism to investigate the optimal means of exploiting those facilities in terms of two metrics (event detection rate, and observed detection rate per cost of hardware and operations), and how we employ that framework to assess the relative merits of various antenna configurations and backend architectures. It discussed how we compare fly's-eye survey modes against incoherent and coherent combination of the interferometric signal, and concluded by examining, in the context of SKA-scale aperture arrays, backend and beamforming configurations that maximise scientific return for a given hardware cost.

New Horizons in Time-Domain Astronomy
Proceedings IAU Symposium No. 285, 2011
R.E.M. Griffin, R.J. Hanisch, & R. Seaman, eds.

© International Astronomical Union 2012
doi:10.1017/S174392131200052X

Next-Generation X-Ray Astronomy

Nicholas E. White

Sciences and Exploration Directorate, NASA Goddard Space Flight Center,
Greenbelt, MD 20771, USA
email: nicholas.e.white@nasa.gov

Invited Talk

Abstract. This review of future timing capabilities in X-ray astronomy includes missions in implementation (ASTRO-H, GEMS, SRG and ASTROSAT), those under study (currently NICER, ATHENA and LOFT), and new technologies that may be the seeds for future missions, such as lobster-eye optics. Those missions and technologies will offer exciting new capabilities that will take X-ray Astronomy into a new generation of achievements.

Keywords. Telescopes, X-rays: general, Instrumentation: detectors

1. Introduction

X-ray astronomy missions planned over the coming 5 years will have new capabilities that are the result of technology investments made over the past 20 years. In addition to the planned missions, both ESA and NASA are studying a number of concepts that will bring further advances. In this review I will highlight those missions individually, emphasizing how each new capability will bring a new view of the Universe.

2. Missions in Development

2.1. *NuSTAR: Nuclear Spectroscopic Telescope Array*

The NuSTAR mission will for the first time bring focusing X-ray optics to the 6-79 keV hard X-ray band, resulting in a factor of 10–100 increased sensitivity over previous instruments (Harrison *et al.* 2010). This band is important for a number of studies. The extragalactic X-ray sky is dominated by the X-ray background at 40 keV, so NuSTAR will be optimal for studying the evolution of massive black holes in active galactic nuclei through extra-galactic surveys. The band is also free from the effect of X-ray photo-electric absorption from the interstellar medium and other obscuring matter, which means that NuSTAR will be able to probe many obscured regions such as the population of compact objects and the nature of the massive black hole in the centre of the Milky Way. NuSTAR will image the Ti-44 isotopic line emission at 68 and 78 keV in young supernova remnants to study the birth of the elements and supernova dynamics. In combination with observatories at other wavelengths, NuSTAR will be optimal for probing particle acceleration in relativistic jets from active galactic nuclei.

The key NuSTAR technology is the use of depth-graded, multilayer-coated, grazing-incidence optics to increase X-ray reflectivity above 10 keV. The mission is part of NASA's small explorer (SMEX) program. The telescope requires a 10-metre focal length, which is achieved in the small launch volume of a Pegasus through on-orbit deployment of an extendable mast. The overall energy band of NuSTAR is 5–80 keV, with an angular resolution of ~50″ (Half Power Diameter). The field of view is 13′. Each focal plane consists of four CdZnTe pixel sensors with a resolution of 1.0 keV at 60 keV (FWHM) and a time resolution of 0.1 msec. The planned orbit of 550 km × 600 km with a 6-degree inclination avoids the South Atlantic Anomaly in order to ensure a low background. The current planned launch date is 2012 February.

2.2. ASTROSAT

ASTROSAT is a multi-wavelength space observatory of the Indian Space Research Organisation (Agrawal *et al.* 2006). It is to be launched in late 2012, and will offer a multiwavelength capability that allows simultaneous monitoring of targets from optical wavelengths to 100 KeV, with high timing precision. That will enable sky surveys in the hard X-ray and UV bands, broad-band spectroscopic studies of X-ray binaries, AGN, SNRs, clusters of galaxies and stellar coronæ, and studies of periodic and non-periodic variability of X-ray sources. The sky monitor will also be a trigger for Target Of Opportunity observations. The mission lifetime is planned to be least 5 years.

ASTROSAT will carry five instrument packages: (1) twin 40-cm Ultraviolet Imaging Telescopes (UVIT), (2) three Large Area Xenon Proportional Counters (LAXPC) covering medium X-rays from 3–80 keV with an effective area of 6,000 cm^2 at 10 keV, (3) a Soft X-ray Telescope (SXT) with conical foil mirrors and X-ray CCD detector, covering the energy range 0.3–8 keV (∼200 cm^2 at 1 keV), (4) a Cadmium-Zinc-Telluride codedmask imager (CZTI), covering hard X-rays from 10–150 keV, with ∼10° field of view and 1000 cm^2 effective area, and (5) a Scanning Sky Monitor (SSM) consisting of three one-dimensional position-sensitive proportional counters with coded masks on a rotating platform to scan the available sky once every six hours in order to locate transient X-ray sources.

2.3. *SRG: Spektrum Roentgen Gamma*

The Russian SRG mission will fly on a medium-class spacecraft platform (Navigator, Lavochkin Association, Russia). The launch from Bajkonur is currently planned for late 2013 using a Soyuz-2 rocket into an orbit around L2. There are two instruments: (1) eROSITA (extended Röntgen Survey with an Imaging Telescope Array) provided by Germany (Predehl *et al.* 2010) and (2) ART-XC (Astronomical Roentgen Telescope X-ray Concentrator) led by Russia (Pavlinsky *et al.* 2010). The primary eROSITA science goal is to detect 100,000 galaxy clusters up to redshift ∼1.3 in order to study the large scale structure in the Universe and test cosmological models, especially Dark Energy (Cappelluti *et al.* 2010). The results will provide complementary constraints on the Dark Energy parameters with a precision comparable to that of other Dark Energy experiments planned for later this decade. eROSITA will also detect 3 million active galactic nuclei (AGN) as well as many variable galactic objects (CVs, Novae, GRB afterglows, stellar flares, etc.).

The detection of 100,000 galaxy clusters drives the telescope design. The effective area of eROSITA is about twice that of one XMM-Newton telescope in the energy band below 2 keV, whereas it is three times less at higher energies, and is a consequence of the small f-ratio (focal length to aperture) of the eROSITA mirrors. The short focal length gives the larger field of view that is essential for the all-sky survey. The angular resolution averaged over the field of view is ∼28″, which is sufficient to distinguish extended clusters from point sources. SRG will scan the entire sky for four years (compared to ROSAT's 6 months), and will result in a final eROSITA sensitivity during this all-sky survey that is approximately 30 times deeper than ROSAT in the soft X-ray band. The 0.5–2 keV flux limit for galaxy clusters will be, on average, of the order of 3×10^{-14} erg cm^{-2} s^{-1}. For point sources detected in the all-sky survey the typical flux limit is 10^{-14} erg cm^{-2} s^{-1} and 3×10^{-13} erg cm^{-2} s^{-1} in the 0.5–2 keV and 2–10 keV energy bands, respectively.

The latter is ~100 times more sensitive than the HEAO-1 survey in the same 2–10-keV X-ray band.

2.4. GEMS: *Gravity and Extreme Magnetism SMEX*

GEMS will perform the first sensitive X-ray polarization survey of several classes of X-ray emitting sources characterized by strong gravitational or magnetic fields (Jahoda *et al.* 2010). These sources are expected to have intrinsic asymmetries with respect to a distant observer and are therefore likely to have emission with a net polarization. Polarization measurements are routinely made in other bands, providing important probes of astrophysical objects which are not available via imaging, spectroscopic, or timing observations. With this new capability GEMS promises to determine (a) how fast black holes spin, and where the energy is released in black-hole systems including disks, coronæ, and jets, (b) the location of the energy release in neutron-star systems powered by rotation, accretion, and the magnetar phenomenon and the physical mechanism responsible for magnetar emission, (c) whether the magnetic fields of shell supernova remnant shocks are tangled or aligned, and—if aligned—the direction, and (d) the field direction and regularity of pulsar-wind nebulæ.

As a result of photoionization in a proportional counter, a photoelectron is emitted preferentially in the direction of the absorbed photon's electric field. The distribution of emission angle gives an unambiguous determination of the polarization: micro-pattern gas detectors determine the photoelectron emission angle by forming images in the plane containing the ionization trail left by the photoelectron. With a pixel size (~100 microns) that is small compared to the path length (~1 mm) of the photoelectron, the emission angle can be determined with high accuracy (Costa *et al.* 2001). GEMS has two X-ray polarimeter instruments, each at the focus of an X-ray telescope with ~1ʹ.7 angular resolution. The mirrors will be deployed at the end of a coilable boom to reach their 4.5-m focal length. The two polarimeters will be mounted at different angles with respect to each other so that, if the detectors are identical, false modulations common to all detectors will cancel. The entire spacecraft will rotate around the science axis so that each detector angle is mapped with uniform exposure onto all sky angles. The nominal rotation period is 10 minutes, which is relatively rapid with respect to orbital changes and is not an integral multiple of the orbit period. Typical observations are greater than 10^5 sec and will thus include more than 100 spacecraft rotations. The spacecraft will be launched in 2014 into a circular orbit of 575 km altitude and 28°.5 inclination.

2.5. ASTRO-H

The joint JAXA/NASA ASTRO-H mission will be the sixth in a series of highly successful X-ray missions initiated by the Institute of Space and Astronautical Science (ISAS) in Japan (Takahashi *et al.* 2010). ASTRO-H will also be the next major observatory-class mission in X-ray astronomy. The launch date is currently planned for the summer of 2014. The science focus will be dedicated to investigating the physics of the high-energy universe by performing high-resolution, high-throughput spectroscopy with moderate angular resolution over a very wide energy range (from 0.3 keV–600 keV). ASTRO-H brings several new innovations that will represent a major advance over current capabilities, including the long-awaited X-ray micro-calorimeter array for high-resolution X-ray spectroscopy (~7 eV resolution) and hard X-ray imaging in the 5–80 keV band provided by multilayer coatings on grazing incidence optics. The mission will also carry an X-ray CCD camera as a focal plane detector for a soft X-ray telescope (0.4–12 keV) and a non-focusing soft gamma-ray detector (40–600 keV). Owing to limited space this review will concentrate on the next-generation capabilities of the X-ray micro-calorimeter.

The soft X-ray Spectrometer (SXS) is being developed by an international collaboration led by JAXA and NASA (Mitsuda *et al.* 2010). SXS is an integral-field spectrometer with 36 pixels and has an energy resolution of $\sim$7 eV between 0.3–12 keV. The array is cooled to 50 mK and makes precise measurements of the heat generated by the absorption of an X-ray photon. The array has a $3' \times 3'$ field of view. The effective area of SXS at 6 keV is 210 cm^2, with an angular resolution of $1'.7$, HPD (half-power diameter) with a goal of $\sim 1'$. The cooling system has redundancy that will protect the instrument against a single point failure.

The SXS will open a new era in X-ray spectroscopy. At E > 2 keV it will be both more sensitive and have higher resolution than current spectrometers. The Fe K emission region around 6 keV is particularly important, and will reveal conditions in plasmas with temperatures between 10^7 and 10^8 K, which are typical values for stellar accretion disks, SNRs, clusters of galaxies and many stellar coronæ. In cooler plasmas Si, S, and Fe fluorescence and recombination occurs when an X-ray source illuminates nearby neutral material. Fe emission lines provide powerful diagnostics of non-equilibrium ionization due to inner shell K-shell transitions from Fe XVII–XXIV. For example, ASTRO-H will observe clusters of galaxies—the largest bound structures in the Universe—to reveal the interplay between the thermal energy of the intra-cluster medium, the kinetic energy of sub-clusters from which clusters form, measure the non-thermal energy content and trace directly the dynamic evolution of clusters of galaxies.

3. Missions Under Study

3.1. NICER: *Neutron star Interior Composition ExploreR*
SEXTANT: *Station Explorer for X-ray Timing and Navigation*

NICER was selected in 2011 October for a 11-month phase A study by NASA as part of the Explorer program. The Principal Investigator is Keith Gendreau of the Goddard Space Flight Center. NICER offers order-of-magnitude improvements in time-coherent sensitivity and timing resolution beyond the capabilities of current X-ray observatories. The mission is optimized for addressing the following science goals: (1) to reveal the nature of matter at extreme densities through neutron star mass and radius measurements, (2) to establish the sites and mechanisms of radiation in their extreme magnetospheres, and (3) to measure definitively the stability of neutron stars as clocks, with implications for gravitational-wave detection and time-keeping.

Over the past year the NASA Office of the Chief Technologist has been studying how the capability provided by NICER can prove the concept of using pulsars as a deep-space navigation tool under the name Station Explorer for X-ray Timing and Navigation using X-ray timing (SEXTANT). SEXTANT is a technology demonstrator to validate space navigation using X-ray observations of milli-second pulsars. Pulsars provide a natural infrastructure for a GPS-like navigation solution that works throughout the Solar system. Navigation accuracies of 500 m in a day can be achieved, with 100 m in a few days for interplanetary spacecraft.

NICER/SEXTANT will achieve its goals by deploying a high-heritage X-ray timing instrument as a payload on the International Space Station (ISS), to be attached in the summer of 2016. The high collecting area is achieved with a collection of 56 X-ray concentrator/detector pairs that are simplified derivatives of the thin foil mirrors made by GSFC over the past 3 decades for ASCA and SUZAKU (Serlemitsos 2010). The telescopes together provide more than 2200 cm^2 of effective area with a $\sim 10' \times 10'$ field of view focusing onto Silicon Drift Detectors (SDD) from MIT Lincoln Labs. The SDD provide an energy resolution comparable to that of X-ray CCD detectors, with an absolute timing resolution better than 200 ns. The telescope array is pointed by using a 2-axis gimbal on a Zenith pointing Express Logistics Carrier (ELC) on board the ISS. The ISS offers nearly continuous contact with the instrument, and Target-Of-Opportunity observations will be possible within minutes.

3.2. ATHENA: *Advanced Telescope for High ENergy Astrophysics*

ATHENA is an observatory-class mission that will address key science challenges across astrophysics. It is designed to (a) map the innermost flows around black holes, measure their spins, and determine the equation of state of ultradense matter in the cores of neutron stars, (b) measure the energy flows giving rise to cosmic feedback, quantify the growth of supermassive black holes and the evolution of their obscuration over cosmic time and determine velocity and metallicity flows due to star-burst superwinds, (c) determine the evolution of the intracluster medium through temperature, metallicity and turbulent velocity changes with redshift, constrain dark energy as a function of redshift using clusters of galaxies and reveal the missing baryons at low redshift locked in the warm and hot intergalactic medium, and in addition it will (d) determine the physical conditions in hot plasmas covering a wide range of objects and phenomena, with profound impacts on astrophysics, from stars and planets, through supernovæ and the Galactic Centre.

Achieving those ambitious goals requires a major leap forward in high-energy observational capabilities. The X-ray optical system will utilise the innovative silicon pore optics technology pioneered in Europe to achieve the required 1 m^2 effective collecting area with $10''$ angular resolution (with a goal of $5''$). An assembly of two 12-m focal length telescopes will feed two instruments operating simultaneously. One of those is a next generation X-ray microcalorimeter spectrometer which provides integral field spectroscopy over a $2'$ field of view (using a 32×32 array) with 3-eV resolution. A Wide Field Imager is an active pixel sensor camera covering the full field of view given by the ATHENA flight mirror, providing wide-field survey capabilities and a high time-resolution capability and $\sim$100 eV energy resolution. If selected, ATHENA will be placed in orbit at L2, which provides uninterrupted viewing. The design assumes a 5-year mission lifetime, but has consumables for at least 10 years.

ATHENA is a simplified version of the International X-Ray Observatory (IXO) (Bookbinder 2010) that can be implemented by ESA alone while still addressing the highest priority goals of the IXO science case. JAXA and NASA participation is limited to instrument contributions. ESA will decide in 2012 whether to select ATHENA for a definition phase with a launch in $\sim$2022. In parallel to the ESA ATHENA study, NASA has made a call for ideas to implement the 2010 Decadal Survey priorities for X-ray astronomy in a more affordable cost profile to drive technology investments over the rest of the decade.

3.3. LOFT: *Large Observatory for X-ray Timing*

LOFT (Feroci *et al.* 2010) was one of four missions selected in 2011 February for a competitive study by ESA as part of its call for the third M-class mission to be launched in the 2023 time-frame. The key science goal is the use of high-time-resolution X-ray observations of compact objects to provide direct access to strong-field gravity, to the equation of state of ultradense matter, and to black-hole masses and spins. Those science goals require an order-of-magnitude increase in collecting area to 10 m^2 and an energy resolution of $\sim$260 eV over the 2–30 keV band. That extremely large collecting area will be achieved with an array of monolithic silicon drift detectors deployed in orbit like solar array panels, so that the detector array can fit within a launcher shroud. Since LOFT is observing the brightest galactic and extragalactic sources, an angular resolution of $\sim 1°$ is sufficient and can be provided by simple collimators.

4. Next Generation All Sky Monitors: Lobster Eye Optics

Angel (1979) proposed a wide field-of-view X-ray optic that mimics the way a lobster's eye works, and uses a curved array of square channels. Lobster-eye optics gives an order-of-magnitude or more improvement in point-source sensitivity over current all-sky monitors and offers a new capability for studying the variable X-ray sky in detail, including stellar capture events in galactic nuclei, super-flares from solar-type stars, gamma-ray bursts, supernova break-outs, thermonuclear bursts on accreting neutron stars, electromagnetic counterparts to gravitational-wave and cosmic neutrino sources, surveys of active and variable stars (dMe-dKes, Algols, RS CVns, CVs, etc.), magnetar outbursts, and AGN variability and blazar flares. There is also considerable scope for the discovery of unexpected high-energy time-variable phenomena. It is hoped that in the near future a mission will be selected to fly that technology and realize those science goals.

Lobster-eye optics work by reflecting incoming X-rays on two orthogonal walls into a central focus with arc-minute angular resolution. This gives an instantaneous grasp (collecting area field-of-view solid angle) that is much larger than Wolter-1 X-ray telescopes. Square pore (20-micron) glass micro-channel plate (MCP) arrays provide a practical means to implement a lobster-eye optic with fields of view of 10° or more. Test MCP arrays have been measured at Leicester University (G. Fraser, private communication) with a PSF of $\sim 2'$ (FWHM). So far there are no missions planned or being studied which make use of these optics for Astrophysics, but they are being incorporated in the Mercury Imaging X-ray Spectrometer (MIXS) on ESAs BepiColombo mission, due for launch in 2014, so is it hoped that an astrophysics application will follow soon.

5. Conclusions

The next 5 years in X-ray astronomy will bring exciting new missions that will open new vistas on the X-ray sky. For the 2015–2025 era several missions are being studied that build upon these new capabilities and will, it is hoped, bring other new technologies to bear. Those concepts will surely evolve over the coming years, but whatever missions finally emerge the new capabilities being developed promise spectacular advances.

References

Agrawal, P. C. 2006, *Adv. Space Res.* 38, 2989
Angel, J. R. P. 1979, *Ap.J.* 233, 364
Bookbinder, J. 2010, *SPIE Proceedings*, Vol. 7732, 77321B-1-12
Cappelluti, N., *et al.* 2010, *Mem. S.A.It.* 17, 159
Costa, E., *et al.* 2001, *Nature*, 41, 622
Feroci, M., *et al.* 2010, *SPIE Proceedings*, Vol. 7732, 77321V-1-14
Harrison, F. A., *et al.* 2010, *SPIE Proceedings*, Vol. 7732, 77320S-1-8
Jahoda 2010 2010, *SPIE Proceedings*, Vol. 7732, pp. 77320W-1-11
Mitsuda, K., *et al.* 2010, *SPIE Proceedings*, Vol. 7732. pp. 773211-1-10
Pavlinsky, M., *et al.* 2010, in: A. Comastri, M. Cappi and L. Angelini (eds.), *X-ray Astronomy 2009, in press*
Predehl, P., *et al.* 2010, *SPIE Proceedings*, Vol. 7732, pp. 77320U-1-10
Serlemitsos, P. J., *et al.* 2010, *SPIE Proceedings*, Vol. 7732. pp. 77320A-1-6
Takahashi, T., *et al.* 2010, *SPIE Proceedings*, Vol. 7732, pp. 77320Z-1-18

New Horizons in Time-Domain Astronomy
Proceedings IAU Symposium No. 285, 2011   © International Astronomical Union 2012
R.E.M. Griffin, R.J. Hanisch & R. Seaman, eds.   doi:10.1017/S1743921312000531

Technical and Observational Challenges for Future Time-Domain Surveys

Joshua S. Bloom

University of California, Berkeley, CA 94720, USA

email: jbloom@astro.berkeley.edu

Invited Talk

Abstract. By the end of the last decade, robotic telescopes were established as effective alternatives to the traditional role of astronomer in planning, conducting and reducing time-domain observations. By the end of this decade, machines will play a much more central role in the discovery and classification of time-domain events observed by such robots. While this abstraction of humans away from the real-time loop (and the nightly slog of the nominal scientific process) is inevitable, just how we will get there as a community is uncertain. I discuss the importance of machine learning in astronomy today, and project where we might consider heading in the future. I will also touch on the role of people and organisations in shaping and maximising the scientific returns of the coming data deluge.

Keywords. methods: data analysis—methods: statistical—techniques: photometric—surveys—stars: variables—sociology of astronomy—history and philosophy of astronomy

1. Introduction

Though the scientific interests in this conference were bewilderingly diverse, we were brought to Oxford with a common interest in understanding objects and events that *change* with time. To be sure, *all* astrophysical entities change with time—in brightness, in position on the sky, in physical size, in colour—but our perceptions of such changes have only just begun to broaden, as modern instrumentation techniques have matured. Indeed, we learned of the ambitions in the stellar variability community in studies at the micro-magnitude level (enabled so elegantly by Kepler and CoRoT), and at time-scales ranging from seconds to centuries. Explorations of (intrinsically) faint and fast (<week) transients at optical wavebands (enabled by projects like PTF) have revealed new classes of events (e.g. Kasliwal 2011). Likewise, the radio community has been in hot pursuit of new classes of variability at sub-second time-scales.

One of most exciting endeavours for time-domain astronomers (and, frankly, for most of the maverick-minded) is the discovery of the unknown. However, given the rising complexity and expense of new surveys, blind exploration as a singular goal is a dangerous impetus. As Tom Prince claimed over a particularly engrossing dinner, "If it's not worth doing, it's not worth doing well." In other words, the significant technical and observational challenges of pulling off a successful synoptic survey are only worth tackling when the science is compelling. That said, it is indeed tempting to look at the Kasliwal or Cordes phase-space plot in time-scale and peak luminosity and wonder what else might be lurking in the white-space. But there is, I claim, no serendipity without bread and butter science. And it is precisely the bread and butter science that provides the important technical challenges which, when met, will enable new (unexpected) discoveries.

Aside from presenting *Time* as the the unifying thematic approach, we were also struck by the similarity of challenges imposed by the sheer volume of the data now collected (and to be collected by the behemoth synoptic surveys envisaged over the next decade). The proverbial "data deluge" has indeed inundated the astronomical community. It is this enormity of high-quality digital data that forces us to address a sea-change in the way we conduct ourselves as scientists going forward. Yes, there are interesting technical hurdles in the acquisition, movement, management and access of those data, but the real paradigm change comes in the need for fundamentally different approaches to discovery and inference regarding those data. The eight-hundred million light curves updated almost daily by LSST cannot all be scrutinised by astronomers, nor by their spectrographs. Yet some small fraction, perhaps just a few rapidly-evolving events per night, will need the full involvement of the world's largest and most precious telescopes to extract the most science. In this context, the crucial question for time-domain science is this: "How do we do discovery, follow-up and inference when the data rates (and requisite time-scales) necessarily preclude human involvement?"

2. The Autonomous Data-Driven Workflow

In this modern data-driven workflow, where people are abstracted from the real-time loop, I distinguish the acts of *finding*, *discovery* and *classification* of astrophysical events. "Finding" might be considered the process of the extraction of candidate events from raw data into a more abstract (and compact) form. In the optical domain, an example would involve the reduction and subtraction of two frames, the identification of significant changes in the subtraction image, and the recording of metadata about each candidate into a database. "Discovery" would be the recognition that a candidate is indeed of some astrophysically varying source (and not an artefact) that might be of interest for further scrutiny. "Classification" would be the act of understanding and quantifying what that event is likely to resemble among the classes of known (and hypothesised) events. In that sense, Galileo was the first to *find* Neptune (Kowal & Drake 1980). But since he recorded it in his notebook as an uninteresting source (at least as compared to Jupiter)†, he is credited with neither its discovery nor classification. Just think how famous Galileo could have been!

3. Discovery

Each wavelength domain (and spectrum, for that matter) presents its own set of difficulties when trying to automate discovery. High-energy missions like FERMI are working at the Poisson noise limit, gravity-wave astronomers are working in the low signal-to-noise regime where template matching algorithms increase the number of effective statistical trials, and radio surveys contend with complex radio frequency interference (RFI) that can mimic the signal of interest. My group has focused its effort in automating discovery at optical wavebands, on candidates found in image differences. While image differencing is an improvement over catalogue-based discovery (such as minimising bias against discovery in crowded fields or around galaxies), the number of spurious ("bogus") candidates vastly outnumbers the *bona fide* ("real") astrophysical sources. In the Palomar Transient Factory (PTF) there about 1000 bogus for each single real candidate. We have developed a framework that uses a training set—based either on the aggregation of expert opinion (Bloom *et al.* 2011) or on retrospective samples with ground truth (e.g. from spectroscopic confirmation)—that allows us to identify rapidly and automatically the most promising real candidates during real-time runs of PTF. We use the labels to train a random forest (RF) classifier that allows us to identify astrophysical candidates reasonably (Bloom

---

† He did, however, note its apparent motion.

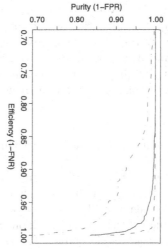

Figure 1. Machine-learned on-the-fly purity vs. efficiency in determining whether a new source from the PTF is a transient (i.e. an explosive event) as opposed to a variable star. The dashed (dashed-dotted) line shows this trade-off for sources within (outside) the SDSS footprint. The solid line shows the aggregate results for all PTF sources in the training sample.

et al. 2011). Just one month before this Symposium our real/bogus framework promoted a candidate event in M101 to the top of the scanning list of "local-universe" events, leading to the discovery of the very nearby Type Ia supernova SN 2011fe (Nugent *et al.* 2011). Future surveys, dealing with many more candidates per night than the 1.5 million found in PTF, will need to employ a similar approach to discovery.

4. Classification

Once a source makes it through the discovery hoop of being "astrophysical†", follow-up decisions must be made. Generating a probabilistic classification based upon all (but still limited) available data is used to inform follow-up decisions. One generic approach to classification is to map the available data, however heterogenous, into an m-dimensional real-number set; this is called "feature space", and once the data are coerced into this space we can apply existing machine-learned frameworks for classifying new sources. There are modern techniques (e.g. Stekhoven & Bühlmann 2011) for imputing missing values into feature space by predicting what those values would have been if they were available and not censored.

4.1. *On the Fly*

In the limit of only little data, context (the location of an event on the sky and what other sources it is physically associated with) becomes an important discriminator of class. Almost all of the initial classification in PTF, built upon another machine-learned trained set, relies on contextual information coerced into a set of well-defined features. Some of those data are available locally, within the same databases as those that house the candidates, but many of the data are distributed throughout the Web. The quality and information content in those ancillary datasets is critical for making accurate classification statements. This is illustrated in Fig. 1, where the lack of a rich set of SDSS information clearly diminishes the classification accuracy of new events.

† In PTF, so as to avoid asteroid discovery, we actually require two astrometrically coincident candidates to be of high real/bogus value as the criteria for discovery.

Since we rely heavily on Simbad, USNO-B1.0, NED and SDSS for contextual data, the main bottleneck in making rapid classification statements about a new event is in the speed and reliability of foreign Web-services. One clear lesson for future surveys is that having physically co-located and unrestricted access to information from other surveys and at other wavebands is crucial. That statement is somewhat contrary to the current push to standardised access protocols for remotely-managed data.

4.2. *In Retrospect*

As a survey continues to accumulate data on the same varying source, eventually the time-series data outweigh the importance placed on context data. For example, a source near the outskirts of a galaxy might rightfully be called a supernova or nova based on context, but if the source is found to be varying periodically with a period around 0.5 days with a certain peak-to-peak amplitude then we will eventually gain confidence in classifying the source as an RR Lyrae star. Much of the body of work on time-domain classification has focused on a *retrospective classification* of completed or nearly completed time-domain surveys (see Bloom & Richards 2011 for a review). On a set of thousands of HIPPARCOS and OGLE variable stars across 26 different classes, for instance, we found a classification error rate of about 22% using 53 periodic and aperiodic features (Richards *et al.* 2011b). Adding more features and dealing better with missing data reduces the error rate to ~15%. It is interesting that, when the taxonomy of variable stars is accounted for, the overall mis-classification rate across three broad classes (pulsating, eruptive and multi-star) plummets to about 5%. In other words, the machine-learned frameworks are better at distinguishing between grossly different physical processes even if the feature set was not specifically encoded to capture those physical processes.

The generation of features can be a non-negligible expense both in time (for Web-based features) and computationally (e.g. for periodogram analysis). However, what are even more costly are traditional fitting routines, such as those for eclipsing binaries, microlensing events and supernovæ. In that context, my suspicion is that the best retrospective classifiers will use a hybrid of generic classification tools on computed features and science-specific fitting routines on sources that are likely to belong to certain families of sources. Just how—from a perspective of classification, accuracy and expense—to architect optimally such a system is an open question.

4.3. *On People vs. Machine*

Despite (or perhaps because of) the enormity of the data, there is considerable interest in using coordinated, collaborative public input to aid classification. Such crowdsourcing† has been carried out in earnest with static-sky images, and is now being used in some time-domain applications (e.g. Smith *et al.* 2011). Though the outreach aspect of that effort may be incredibly important, it remains to be seen whether truly novel science will flow from astronomy crowdsourcing at a sustained level. One worry on the time-domain front is that, unlike software and hardware, expert opinions do not scale easily. To get increasingly refined classification statements on more and more data, for example, the number of those capable and willing to give opinions cannot keep pace with the data growth. Another concern I have is that, unlike algorithms, people do not behave in deterministic ways; I can re-run my code on all previous data and get back the same

---

† A note for future generations: "crowdsourcing," a blend of the words crowd and outsourcing, is a vernacular term used to describe the act of public collaboration in a project to create and annotate content. As of late 2011, the term had not yet been added to the Oxford-English Dictionary.

Classification of known SNe

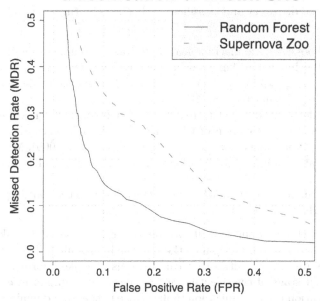

Figure 2. Supernova Zoo versus random-forest supernova discovery. We trained a classifier based on human labelling of PTF sources to predict which sources would turn out to be spectroscopically-typed SNe. By changing the SN discovery threshold from the random forest model, we trade a lower missed detection rate (MDF) for a higher false-positive rate (FPR). For a sample of 345 known PTF SNe, employing the RF score to select objects is *uniformly better*, in terms of MDF and FPR, than using the SN Zoo score. At 10% FPR, the RF criterion (threshold = 0.035) attains a 14.5% MDR compared to 34.2% MDR for SN Zoo (threshold = 0.4). From Richards *et al.* (2012).

result. That lack of determinism also means that the science-sacred concept of repeating and evolving experiments, notwithstanding the large "human cost", is lost in crowd-sourcing‡.

The John Henrys of this world have learned that we should never do with people what can be done as well (or nearly as well for a fraction of the cost) with machines. That is not to say that crowdsourcing might not have its useful role in time-domain science, but that great care must be placed on experimental design both from the perspective of "human subjects" and in the uniqueness of the expected result. One crucial role that humans will continue to play in enabling the automated real-time loop is that of *label experts*. Just as we did in the real/bogus exercise, trained humans can label a small subset of the data so that supervised learning algorithms can be performed on the data and applied in real time. At this Symposium we presented the results of machine-learned discovery/classification of PTF sources that are likely to be supernovæ. We used the Supernova Zoo (Smith *et al.* 2011) mark-up of tens of thousands of candidate events (collected through a specialised DB query at the end of each PTF night) to construct a classifier capable of discovering supernovæ efficiently. The results of that exercise are depicted in Fig. 2 showing that the classifier outperforms the human labelling.

‡ Peng (2011) provides a recent and useful discourse on repeatability in modern science.

5. Parting Thoughts

As the automated discovery and classification business matures, a number of interesting lines of inquiry will need to be studied. For example:

• How do we bootstrap the machine-learning process from one time-domain survey to the next, given the inherent differences in the ways that those surveys are conducted? Active learning (using expert opinions at optimised moments in the learning process) looks promising (e.g. Richards *et al.* 2011a).

• How do we detect and quantify real time-domain outliers—novel events that are not part of the established taxonomy? Clustering and semi-supervised learning seems an appropriate start (Protopapas *et al.* 2006; Rebbapragada *et al.* 2009). My view is that the way we get good at finding needles in a haystack is by getting really good at identifying hay.

• How can we imbue domain knowledge (and physics) into the learning process without having to use traditional domain-specific fitting routines?

No discovery or classifier engine will ever be perfect, in a sense of making statements precisely about the underlying origin of the observed variability. The best we can hope for are well-calibrated probabilities that can allow us to make informed decisions about moving to the next stage of the scientific process (see, e.g., Morgan *et al.* 2012). Viewed that way, classification is a maximisation tool: science in the time-domain will increasingly be conducted with major resource limitations—in the computational power available for discovery and classification, in follow-up telescope availability, in people's time and, ultimately, in capital cost.

Acknowledgements

I am grateful to the organisers and participants for the many enlightening discussions during the conference. I acknowledge support from a CDI grant (#0941742) from the National Science Foundation.

References

Bloom, J. S. & Richards, J. W. 2011, arxiv/1104.3142
Bloom, J. S., *et al.* 2011, ArXiv e-prints
Kasliwal, M. M. 2011, PhD thesis, California Institute of Technology
Kowal, C. T. & Drake, S. 1980, *Nature*, 287, 311
Morgan, A., *et al.* 2012, *PASP*, in press
Nugent, P.E., *et al.* 2011, *Nature*, in press
Peng, R. D. 2011, *Science*, 334, 1226
Protopapas, P., *et al.* 2006, *MNRAS*, 369, 677
Rebbapragada, U., Protopapas, P., Brodley, C. E., & Alcock, C. 2009, in: D.A. Bohlender, D. Durand and P. Dowler (eds.), Astronomical Data Analysis Software and Systems XVIII, ASP Conf. Ser. Vol. 411 (San Francisco: ASP), p. 264
Richards, J., *et al.* 2012, in prep
Richards, J. W., *et al.* 2011a, arxiv/1106.2832
Richards, J. W. 2011b, *ApJ*, 733, 10
Smith, A.M., *et al.* 2011, *MNRAS*, 412, 1309
Stekhoven, D. J.. & Bühlmann, P. 2011, arxiv/1105.0828

New Horizons in Time-Domain Astronomy
Proceedings IAU Symposium No. 285, 2011
R.E.M. Griffin, R.J. Hanisch & R. Seaman, eds.

© International Astronomical Union 2012
doi:10.1017/S1743921312000543

Summary: A Very Timely Conference

Rosemary F. G. Wyse

Physics & Astronomy Dept., Johns Hopkins University, Baltimore, MD 21218, USA

email: wyse@pha.jhu.edu

Invited Talk

1. Time Discovers Truth

The conference poster includes a very apt phrase that describes a primary motivation for this conference: Time discovers truth. This aphorism, attributed to Seneca, was certainly affirmed by the many exciting talks and discussions at this conference, in both formal and informal settings.

An incredible wealth of behaviour in astronomical objects of all scales and energies has been observed, manifesting a broad range of underlying physical processes. Phenomena that result in extremely small, time-dependent variations in optical photometry are now being revealed by dedicated surveys. For example, both external and internal processes can contribute to tiny variations in the light from stars: the detections of planetary transits of their host stars (transit depths of 0.01%) and stellar oscillations (the field of asteroseismology; ppm variations), are now routinely achievable for large samples of stars and stellar remnants, using both Earth- and space-based telescopes. Theories of planet formation and stellar structure and evolution are being tested as never before. Additionally, the public's imagination is being stimulated by the quest for Earth-like extra-solar planets.

Turning from these small masses to super-massive black holes, again precise measurements are giving credibility to estimates of mass. The differences in delay times, compared to the behaviour of the continuum light, for the variability of different spectral emission lines—probing different distances from the black hole—allows independent estimates to be made for the mass of the black hole, one for each emission line. Consistency of these estimates implies a robust result. Much recent effort in the development of theories of galaxy formation has been focussed on understanding the observed correlations between properties of the stellar content of a galaxy—such as the luminosity of the bulge component—and the mass of the super-massive black hole at its centre, and better quantification of these correlations will allow discrimination between theories.

In the radio regime, precision timing of pulsars is reaching even higher levels of sophistication, probing the pulsars themselves and the intervening interstellar medium—plus fundamental physics.

The physics of transient phenomena, as distinct from recurring variations, has also been under intense study. Prior to the advent of dedicated time-domain surveys, the distribution of optical luminous transients in the plane of characteristic time-scale of variability against peak luminosity was essentially bimodal, with low-luminosity novæ, varying on time-scales of around a day, well separated from the much more luminous supernovæ, which vary on tens of days. We have now identified populations that occupy loci between those two and even extend the occupied area to higher luminosities. Theories and theorists are validated and confounded. Radio transients have long been known to occupy many disparate regions of the time-luminosity plane but even here new types of variable objects are being discovered in hitherto empty loci (e.g., Rotating RAdio

Transients). Gamma-ray transients plausibly include emission (viewed at a favourable angle) after a star has been tidally disrupted upon getting too close to a super-massive black hole.

The truth revealed is that everything varies, in myriad ways. Even the gamma-ray emission from the Crab nebula, used to cross-calibrate telescopes, has now been detected to vary.

2. The Great Enemy of Truth is Very Often Myth

Several "unexpected" results, such as for the Crab nebula, reminded us to keep an open mind and to discard preconceived notions. As John F. Kennedy noted, "The great enemy of truth is very often myth, persistent, persuasive and unrealistic." Equally, "The unsought will go undetected" (Sophocles). Persistence and patience are required to reveal very long time-scale variability, while innovative application of (new) technology can enable the exploration of ever-faster variability. Time Allocation Committees should be adventurous (at least occasionally).

3. The Truth is Rarely Pure and Never Simple

What to make of this variability revealed by the time domain? Oscar Wilde's admonition that "The truth is rarely pure and never simple" is surely apt. As we were reminded by several speakers, simple detection of variability is not enough to claim a new discovery—understanding is required. For that, follow-up and/or simultaneous acquisition of data in other wavelength regions is critical. Indeed, many talks demonstrated the effectiveness of a multi-wavelength campaign, with added physical insight provided across the spectrum from gamma-ray to radio wavelengths. Surveys of the "static sky" are, in different ways, as important as surveys in the "time domain" for many fundamental questions. In that context, a medium-depth all-sky U-band imaging survey is badly needed—U-band data are crucial, for example, to enable reliable photometry-based estimates of the metallicity of stars like the Sun.

Imaging surveys with complementary capabilities should also be undertaken, such as a wide but shallow survey to augment the findings of a deep but narrow survey. A case study would be the issue raised at the conference as to the identification of possible hosts for the "orphan" supernovæ that have been detected in several wide-area imaging surveys on the lookout for transients. In order to rule out the existence of host low surface-brightness dwarf galaxies, deep-imaging follow-up with good sky-subtraction and flat-fielding would be needed. Different levels of follow-up are needed for objects of different perceived levels of scientific interest (which themselves change with time), and thus the design of the initial detection survey would ideally provide for automatic classification of newly identified variables, as input to their prioritisation for follow-up.

Indeed, ongoing and planned all-sky deep-imaging surveys need spectroscopic counterparts to aid much of the astrophysical interpretation. Photometric redshifts are not sufficient for much moderate- to high-redshift science. Spectra are required for the determination of line-of-sight motions of stars and nearby galaxies. Spectra with good enough signal-to-noise and resolution can be used to break degeneracies inherent in the analyses of solely photometric data, such as that between galactic star-formation rate and stellar mass function which is found when analysing the integrated light from galaxies. Spectral data allow AGN to be distinguished more readily from star-forming galaxies, and allow the progenitors and hosts of explosive events to be identified more robustly.

Spectroscopic surveys of stars provide estimates of important stellar parameters—metallicity, gravity and effective temperature—which in turn enable distances and ages to be estimated. By combining those with proper motions from time-domain imaging surveys, full 6-dimensional phase space can then be explored. The addition of chemical abundances provides yet more dimensions of space in which to look for structure that can be used to define stellar components of galaxies. "Near-field cosmology" exploits the conserved quantities in old, nearby stars to infer conditions long ago, when the stars formed. For example, the Initial Mass Function of long-dead massive stars can be constrained from the elemental abundances in the old, long-lived stars that they enriched. The ESA astrometric satellite GAIA will of course produce truly unprecedented data on positions and proper motions for stars throughout the Milky Way. Complementary spectroscopic data are already being acquired, as we heard.

The radio regime as yet lacks a comprehensive survey, but LOFAR, SKA and precursor surveys will provide that, and more, going a long way towards satisfying those who ask for "All the sky, all the time".

The high-energy regime is very active at the present. The unexpected detections enabled by dedicated all-sky surveys include M-dwarf stars emitting in the hard X-rays, relativistic outflows associated with a tidal disruption accretion event, and the highest redshift object (known at the present time)—a Gamma-ray Burster at a redshift of 9.3.

4. Survey Strategies

That surveys must maintain serendipity was a common refrain—the unknown and the unlikely happen. This is sometimes good! The Princes of Serendip succeeded through "accidents and sagacity" (according to Walpole), though it is hoped that astronomers depend more strongly on the latter while designing surveys and analysing data.

There was a general consensus that size alone is not sufficient as a measure of a survey's worth, no matter which algorithm is used to define "size". Clever ideas are also needed, in all aspects of survey design and execution. There must be careful selection of cadences and of the required precision of the data, in order to enable as much science as possible; maintaining flexibility is key to maximizing the discovery capabilities of a survey.

Saving the pre-pipeline data, or even raw data, if feasible, can also enable discoveries and analyses beyond the original science case for a survey or satellite mission (it is a truism that the actual scientific breakthroughs achieved by a given project are often far removed from the original science case). The tuning of a data reduction pipeline for the most efficient implementation of a particular science investigation can be viewed as indispensable, but can make other analyses effectively impossible.

The saving and archiving (after digitization) of "vintage" data can prove invaluable in identifying hosts/progenitors of transient events. Real-time stellar evolution can be observed using very long-term monitoring, but only provided that the older data are accessible. Significant efforts should be directed at data conservation.

There is undoubtedly a great challenge in the need to deal with the data deluge from large surveys, both coming and planned. Happily, astronomy is not alone in facing the deluge, and huge efforts are being made world-wide. Novel statistical techniques are being developed to aid such analyses. In this situation, as we heard, creative ideas beat data ownership.

5. Conclusions

There is clearly much that has been learned and understood from surveys of the variable sky, and much that remains to be learned and understood.

There were sentiments expressed that "the time domain" was such a large and growing field that this could well be the last all-encompassing conference on that topic. However, the lively discussions after the extremely stimulating talks argue strongly for the continuation of such a format. The more specialized workshops, held in the afternoons, provided a forum both for very rapid introductions to the most current topics in a given field and a means to discuss the most pressing issues. "Synergy" and "cross-fertilization" are much used and abused words, but were very apt when describing the outcome of the week's interactions.

However, "We are time's subjects, and time bids be gone" (Shakespeare)

I would like to thank the organisers, and especially Elizabeth Griffin and Bob Hanisch, for their vision, hard work and patience.

New Horizons in Time-Domain Astronomy
Proceedings IAU Symposium No. 285, 2011
R.E.M. Griffin, R.J. Hanisch & R. Seaman, eds.

© International Astronomical Union 2012
doi:10.1017/S1743921312000555

Workshop Reports

19 Workshops were held during IAU S285. 15 submitted reports of the discussions that took place, while for the remaining 4 we have reproduced the summaries that were available on our wiki prior to the Symposium. The reports are arranged in the following order:

New Horizons in Time-Domain Astronomy
Proceedings IAU Symposium No. 285, 2011
R.E.M. Griffin, R.J. Hanisch & R. Seaman, eds.
© International Astronomical Union 2012
doi:10.1017/S1743921312000567

The CoRoT and Kepler Revolution in Stellar Variability Studies

Pieter Degroote and Jonas Debosscher

Instituut voor Sterrenkunde, K.U. Leuven, BE-3001, Leuven, Belgium
email: `pieter.degroote@ster.kuleuven.be`

Abstract. Space-based observations of variable stars have revolutionized the field of variability studies. Dedicated satellites such as the CoRoT and KEPLER missions have duty cycles which are unachievable from the ground, and effectively solve many of the aliasing problems prevalent in ground-based observation campaigns. Moreover, the location above the Earth's atmosphere eliminates a major source of scatter prevalent in observations from the ground. These two major improvements in instrumentation have triggered significant increases in our knowledge of the stars, but in order to reap the full benefits they are also obliging the community to adopt more efficient techniques for handling, analysing and interpreting the vast amounts of new, high-precision data in an effective yet comprehensive manner. This workshop heard an outline of the history and development of asteroseismology, and descriptions of the two space missions (CoRoT and KEPLER) which have been foremost in accelerating those recent developments. Informal discussions on numerous points peppered the proceedings, and involved the whole audience at times. The conclusions which the workshop reached have been distilled into a list of seven recommendations (Section 5) for the asteroseismology community to study and absorb. In fact, while addressing activities (such as stellar classification or analysing and modelling light curves) that could be regarded as specific to the community in question, the recommendations include advice on matters such as improving communication, incorporating trans-disciplinary knowledge and involving the non-scientific public that are broad enough to serve as guidelines for the astrophysical community at large.

Keywords. methods: data analysis, methods: numerical, methods: statistical, techniques: photometric, surveys, stars: fundamental parameters, stars: oscillations, stars: variables: other

1. Introduction

Space-based asteroseismology started with the study of α UMa (Buzasi *et al.* 2000), which was observed with the star camera on board the WIRE satellite. The first satellite dedicated to asteroseismology (MOST, Walker *et al.* 2003b) was launched shortly afterwards, and was followed in 2006 with the launch of the larger CoRoT mission (Baglin *et al.* 2006). The latter mission was designed to have two concurrent observing strategies. Besides observing thousands of stars in its programme of exoplanet surveys, CoRoT was also the first satellite to acquire high-cadence photometric time-series of thousands of stars simultaneously. It was also the first satellite to observe targets, specifically selected for their seismic potential, continuously for several months, with a 32 s sampling. Then in 2009 the large NASA mission KEPLER (Koch *et al.* 2010) was launched. KEPLER is the first of its kind to be launched in an Earth-trailing orbit; the purpose was to minimize the influences of Earth-scattered light, the South Atlantic Anomaly, etc. KEPLER now raises the number of stars observed simultaneously to nearly 150,000. It also monitors the same field continuously for several years.

2. Two Revolutions

The CoRoT and KEPLER missions have revolutionized the study of stellar variability, for two reasons. In the first place they measured light curves for a vast number of stars, thereby enforcing the application and development of stellar classification and data mining tools. Secondly, almost all of the light curves for the individual stars have a duty cycle, precision and quality that is unachievable from the ground (Fig. 1 illustrates just a few examples of CoRoT light curves). The material can thus enable in-depth studies of particular targets, with or without acquiring additional data such as multi-colour photometry, spectroscopy, magnetic field measurements and the like.

Fig. 2 shows a comparison between KEPLER light curves and phase diagrams from a typical large-scale ground-based survey (Trans-Atlantic Exoplanet Survey, TrES; O'Donovan et al. 2009). The jump in data quality is evident. In the case of ground-based observations, one often needs to phase the data in order to see variability at a given frequency, making the biggest challenge to overcome the *detection* of variability. The sparse time-sampling which is most typical for ground based observations, and is due to night/day cycles, telescope allocation time, instrumental downtime, etc., severely influences the height of the side lobes in the window function, allowing for confusion in the frequency detection (i.e., aliasing), and thereby complicating the interpretation of the detections even more. In the case of space-based data, however, the *characterisation* of the variability is most often the problem, since the high precision and high duty cycle basically solve aliasing and improve detection limits significantly. A comparison between the phase diagrams from TrES and the full light curves from KEPLER makes it clear that phase diagrams hide part of the information present in the light curves. In CoRoT and KEPLER photometry, the shape of the light curves is in general immediately apparent, and can be tracked over the entire time-span of the observations, whereas the construction of a phase diagram in those cases would tend to *reduce* the amount of information present in the light curve. In particular, that opens up the possibility to detect frequency or amplitude changes, sometimes over short periods of time. Another example can be found in exoplanet research, where the detection of transit-timing variations is now possible.

An example in which phase diagrams can still be useful is stellar classification, where simple and few attributes are to be selected. In such cases there is often no need to embark upon the subtle variations from phase to phase. The frequency of the dominant variation, for example, can already carry enough information to distinguish one class of variable stars from another.

3. Two Approaches

The two revolutions which CoRoT and KEPLER have brought about (*viz.* the increase in the number of stars and the increase in precision), also gave us two approaches for treating the data: one can do case studies (for instance, those by Appourchaux *et al.* 2008 and Degroote et al. 2010), and one can do population studies, which usually involve classification algorithms or pipelines to extract useful parameters from the light curves (see, for example, Debosscher et al. 2009; Miglio *et al.* 2009). It is evident that both approaches are necessary to improve our knowledge of the stars in general and of stellar evolution in particular.

A common bias in case studies is that usually some information about the star is known (or gathered) in advance, such as its spectral type, and that the observed variability is interpreted in a matching framework. That partially solves the degeneracy that is often encountered and which is inherent to single-band light-curve analyses, i.e., that one morphological type can represent different types of underlying behaviour. Noteworthy examples of that effect include ellipsoidal deformation and mono-periodic pulsators, or spotted stars and (highly) multi-periodic oscillators. Unfortunately such an approach can lead to a false interpretation of data, and is thus not preparing us for the unexpected.

At least three methods of approach are suggested. One approach includes the treatment of the data under different assumptions about the underlying origin (perhaps together with a statistical model evaluation) or at least reporting alternative explanations. A second method, and perhaps a more objective approach, would be to refrain from physical interpretation as long as possible, but instead to describe the light curves as far as possible through purely observational parameters, and to approach observational data with an open mind. A third method is to gather more information, and it would be particularly useful to have multicolour photometric time series for the targets. Some degeneracy between effective temperature and reddening might exist, but at least in principle it can be accounted for. In that case, however, one should not rely too much on the Kepler and CoRoT input catalogues, where various fundamental parameters for the stars are available but were derived with certain assumptions that might not be valid for all stars—as discussed by Pinsonneault *et al.* (2011). Nevertheless, the inherent faintness of virtually all CoRoT and Kepler targets puts heavy constraints on the possibilities to acquire additional data.

Population studies of the CoRoT and Kepler fields are often the basis for further research, so any researcher can select those targets according to a few selection

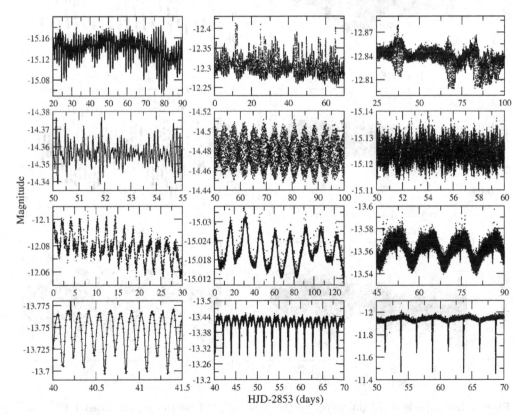

Figure 1. A sample of light curves measured by the CoRoT space mission, illustrating the enormous variety of stellar variability that can be observed with space-based instruments.

criteria. Most classifiers label their targets and put them in predefined classes (supervised classification), while failing to recognise the underlying continuous spectrum of characteristics or possible hybrid properties. An alternative approach might be to use a list of tags (for example, the star is red/blue, is a high/low amplitude pulsator, is a slow/rapid pulsator... instead of labelling it as—say—an RR Lyræ star) or even representing the tags by a continuous number to indicate *how* red or *how* blue, and so on. On the other hand, such an approach could imply wrongly that classification can be carried out in an unbiased frame, but in fact there is a general tendency to agree that a truly unbiased and blind classification is impossible and that one should perhaps refrain from such a pursuit. A possible approach to a fairly unbiased (but not all-blind) classification could be the use of the non-scientific community to classify light curves, in analogy with the Galactic Zoo programme. Indeed, proposals and projects are actually under way to set up such endeavours.

Many of the methods that are currently applied to analysing and interpreting variable stars in the CoRoT and KEPLER fields suffer from the bias arising from our prior knowledge. However, it is most difficult to account for the unexpected, so starting from what we already know is certainly the easiest way to start. We should be prepared to learn

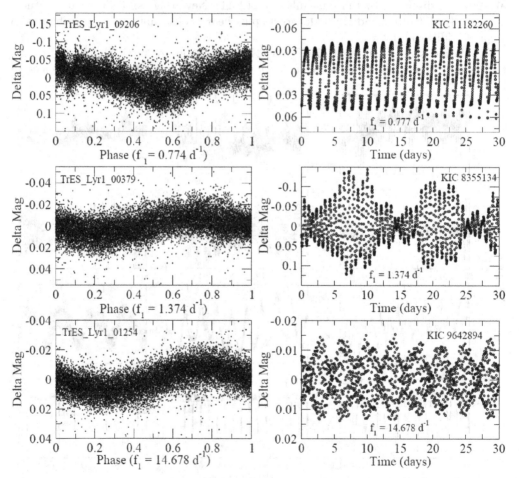

Figure 2. Comparison of light curves for the same targets from the ground-based TrES survey and the KEPLER space mission. The TrES light curves had to be phased according to the dominant detected frequency to show the variability.

more about the stars in the coming years, and incorporate that acquired knowledge in the further treatment of the CoRoT and KEPLER targets, ideally with a close involvement of theoreticians. We believe the knowledge and expertise, in contrast to the data quality, will not immediately make huge leaps (such as the sudden discovery of several new classes of variable stars). We also believe that this is even not desirable, because it would greatly increase the risk of false and hasty discoveries. The arrival of data with an unprecedented quality implies the use of new techniques, with which we are by definition not familiar. Instead of seeking the giant leaps, we should take small steps, and make sure we understand them before advancing further in the realm of the unknown and unexpected.

4. Long term Variability

Some stars have been observed photometrically for a century or more. That raises the question of whether it is possible to see real-time stellar evolution in some of them. It is the feeling of many, however, that even that time frame is still too short to witness the effects of evolution unambiguously, except in the case of the most rapidly evolving objects. The point is well illustrated by studies of Cepheid pulsators. From the O-C diagrams of these stars, it seems that the intrinsic phase-to-phase scatter of the periods is larger than the expected period change arising from evolutionary changes. On the other hand, some trends are noticeable in some Cepheid light curves—but the *sense* of the period change is often opposite to what is expected from evolutionary models. It is noted that that can be interpreted as evolutionary weather or noise, and reflects the fact that most of our models only reproduce long-term evolution, which for most stars implies longer than at least a century.

5. Recommendations

The Workshop discussed the above issues in some depth, and reached a variety of conclusions which we now list below as a number of recommendations:

(*a*) There are two areas where closer collaboration in the community could be beneficial. One is stellar classification, where the sharing of prior knowledge and newly acquired knowledge are critical to progress. The other is effective communication about the use of observational diagnostic tools and statistics in the treatment of light curves; the objective is to apply appropriate techniques for describing and analysing the exquisite light curves coming from CoRoT and KEPLER.

(*b*) Collaboration between groups can also be in the form of competition: the same data sets could be given to different groups for blind tests, in order to compare and evaluate different approaches.

(*c*) The inclusion of the non-scientific community is recommended, in particular for classification purposes. The non-scientific community could classify light curves visually (in a guided process), in an analogy to the Galaxy Zoo programme.

(*d*) The community is encouraged to communicate data and methods efficiently and uniformly. Various frameworks for that are already in place (e.g., the Virtual Observatory).

(*e*) We have entered a phase in variable-star research where input of new methods is particularly to be welcomed. We could benefit especially from looking at other scientific fields, such as geology, meteorology or computer sciences, which share an interest in time-series analysis and are (on some levels) more advanced than some of the practices that are common in the astronomical community.

(*f*) Observers are encouraged to separate theory from observation, and to try to parameterize light curves observationally. The optimal procedure is to use models that are independent from theory, but to use parameters which can be interpreted by theory (cf. the Gussian processes for modelling spots, instead of a spot model).

(*g*) The community is encouraged to be cautious, and to try to advance the field in small steps so as to acquire new knowledge in a reliable way. In particular one should be careful to draw firm conclusions from blind analyses, and to be fully aware of prior knowledge. Thus, instead of going for giant leaps we should take small steps, and should make sure we understand them before advancing further in the realm of the unknown and unexpected.

References

Baglin, A., Michel, E., & Auvergne, M., The COROT Team. 2006, in: *Proceedings of SOHO 18/GONG 2006/HELAS I, Beyond the spherical Sun*, (Sheffield: ESA Special Publication), p. 624.
Buzasi, D., *et al.* (2000), *ApJL*, 532, L133
Debosscher, J., *et al.* 2009, *A&A*, 506, 519
Degroote, P., *et al.* 2010, *Nature*, 464, 259
Koch, D. G., *et al.* 2010, *ApJL*, 713, L79
Miglio, A., *et al.* 2009, *A&A*, 503, L21
O'Donovan, F. T., *et al.* 2009, in: *NASA/IPAC/NExScI Star and Exoplanet Database, TrES Lyr1 Catalog*, p. 6
Pinsonneault, M. H., *et al.* 2011, *ArXiv e-prints.*
Walker, G., *et al.* 2003, *PASP*, 115, 1023

New Horizons in Time-Domain Astronomy
Proceedings IAU Symposium No. 285, 2011
R.E.M. Griffin, R.J. Hanisch & R. Seaman, eds.

© International Astronomical Union 2012
doi:10.1017/S1743921312000579

SWIFT: Opportunities, Capabilities and Data Handling

Rhaana Starling

Department of Physics and Astronomy, University of Leicester Leicester LE1 7RH, UK
email: rlcs1@star.le.ac.uk

Abstract. A focus session was held for those wanting to familiarise themselves with the SWIFT satellite and to consider its exploitation for specific scientific goals. An overview was presented, with questions throughout. Proposal preparation and and the automated science products from the X-ray Telescope were discussed. This account summarises the information given in the presentation and in the answers to the questions which were raised.

Keywords. space vehicles: instruments, methods: data analysis, X-rays: general, astronomical data bases: miscellaneous

1. Introduction and Aims

The SWIFT satellite (Gehrels *et al.* 2004) has proven a valuable tool for the exploration of the time domain. It is unique in its fast slew capability of typically 70–100 seconds, coupled with simultaneous coverage in gamma-rays, X-rays, UV and optical. All SWIFT data are immediately available in the public domain, and SWIFT responds promptly to "Target of Opportunity" requests in addition to carrying out a full Guest Investigator programme and Key Science Projects. To date, science with SWIFT has been extremely varied, ranging from studies of gamma-ray bursts, soft gamma repeaters and active galactic nuclei to X-ray binaries, supernovæ, cataclysmic variables, flare stars and comets.

The aims of this workshop were (a) to give potential SWIFT users an overview of the capabilities of each SWIFT instrument, illustrated with science highlights, (b) to outline the opportunities for observing with SWIFT, and (c) to answer any SWIFT-related questions. Discussions both on the ready-made science-grade products and on the core data analysis techniques and the data archives were invited.

2. Summary

The following topics were presented:

- SWIFT mission objectives
- Instrumentation
- Scientific capabilities: some highlights
- Observing with SWIFT: opportunities
- Access to the archive
- The UK SWIFT Science Data Centre
- Automated science-grade data products: images, astrometry, light curves, hardness ratios and spectra
- Data processsing, and where to find help

The primary goal of the NASA SWIFT mission, launched in 2004, was to uncover the origins of Gamma-Ray Bursts (GRB). SWIFT reacts autonomously to GRBs triggers. It has gone on to cover a broad spectrum of astrophysics, particularly areas which benefit most from SWIFTS' fast slewing capability and simultaneous broad-band coverage. Science highlights and general information about the mission can be found at http://swift.gsfc.nasa.gov/docs/swift/ (US site), http://swift.asdc.asi.it/ (Italian site) and http://www.swift.ac.uk/ (UK site).

SWIFT carries a wide-field (2 steradian) 15–350 keV gamma-ray telescope: the Burst Alert Telescope (BAT, Barthelmy et al. 2005), a 23×23 arcmin FOV 0.3–10 keV X-ray telescope (XRT, Burrows *et al.* 2005) and a UV/optical telescope with 7 lenticular filters and 2 grisms covering 170–650 nm (UVOT, Roming et al. 2005). More details can be read at http://www.swift.ac.uk/instruments.shtml.

Proposals are solicited for Guest Investigator (GI) and fill-in programmes yearly, while rapid observations can be requested at any time via a Target of Opportunity (ToO) request at http://www.swift.psu.edu/too.html. GI observations can be requested for Large Programmes, monitoring campaigns, observations coordinated with other facilities and ToOs for future events, and in addition US proposals may request funds to mine the SWIFT archives or to carry out theoretical projects.

SWIFT data are all public immediately, to facilitate scientific return. When the BAT triggers on a source, initial data products are immediately downlinked via the TDRSS network of satellites and available via the Gamma-Ray Burst Coordinates Network (GCN). Data recieved via ground-station passes are typically available 2 hours after downlink in a Quick-Look database. All data are subsequently archived.

The UK SWIFT Science Data Centre hosts a full SWIFT archive; it can be accessed at http://www.swift.ac.uk/swift_portal/archive.php. Science-grade X-ray data products are also available from that site. As explained in detail by Goad *et al.* (2007) and Evans et al. (2007, 2009), ready-made images, light curves, spectra, hardness ratios and astrometry have been created for GRBs (http://www.swift.ac.uk/results.shtml), while the same products can be created for any SWIFT-observed object of interest at http://www.swift.ac.uk/user_objects/.

Instructions for downloading and installing SWIFT software, and a walk-through description of the data reduction steps for all three SWIFT instruments, can be found in http://www.swift.ac.uk/proc_guide.shtml. One-day training sessions are available at the University of Leicester.

Participants raised a number of questions on the use of the UVOT, particularly for spectropscopic observations. An example of use of the UV grism is shown for GRB 081203A in Kuin *et al.* (2009). The grism calibrations are ongoing; users requesting grism observations are referred to the UVOT Software Guide on the SWIFT data analysis webpages.

Further questions should be directed to the Helpdesk at swift-help@star.le.ac.uk.

References

Barthelmy, S. D., *et al.* 2005, *SSRv*, 120, 143
Burrows, D. N., *et al.* 2005, *SSRv*, 120, 165
Evans, P. A., *et al.* 2007, *A&A*, 469, 379
Evans, P. A., *et al.* 2009, *MNRAS*, 397, 1177
Gehrels, N., *et al.* 2004, *ApJ*, 611, 1005
Goad, M. R., *et al.* 2007, *A&A*, 476, 1401
Kuin, N. P. M., *et al.* 2009, *MNRAS*, 395, L21
Roming, P. W. A., *et al.* 2005, *SSRv*, 120, 95

New Horizons in Time-Domain Astronomy
Proceedings IAU Symposium No. 285, 2011
R.E.M. Griffin, R.J. Hanisch & R. Seaman, eds.

© International Astronomical Union 2012
doi:10.1017/S1743921312000580

Optical & NIR Transient Surveys

Organized by Nicholas J. G. Cross[1] and S. G. Djorgovski[2]

[1] Scottish Universities' Physics Alliance (SUPA), Institute for Astronomy,
University of Edinburgh, Royal Observatory, Edinburgh, EH9 3HJ, Scotland
email: njc@roe.ac.uk

[2] California Institute of Technology, Pasadena, CA 91125, USA

Abstract. A workshop on *Optical & Near Infrared Transients* took place during the first afternoon of the Symposium. It ran for two sessions. The first was given over to talks about various current optical and near-infrared transient surveys, focussing on the VISTA surveys, the Catalina Real-Time Transient Survey, Pan-STARRS, GAIA, TAOS and TAOS2. The second session was a panel-led discussion about coordinating multi-wavelength surveys and associated follow-ups.

1. Introduction

The objective of the *Optical & Near Infrared Transient Surveys* Workshop was to address some of the key issues facing the current and forthcoming ground-based, optical and near-infrared synoptic sky surveys. The issues covered included key lessons learned from the current surveys, what the principal bottlenecks are (especially regarding follow-up observations), the scientific opportunities, the technological challenges, and how coordination between different surveys at all wavelengths might be improved.

The workshop was allocated two sessions, one of 90 minutes and one of 60 minutes. The first session set the scene and initiated discussions by arranging five short talks, each of about 15 minutes, which highlighted some of the current or imminent transient surveys, their scientific goals and the key outstanding challenges that they are encountering. The second session was a panel-led discussion that focused mainly on the issues of effective follow-up and coordination between different surveys, community engagement, and the cyber-infrastructure needs.

2. NIR Transient Surveys – Nicholas Cross

Near-infrared telescopes, instruments and surveys are very similar in their design to optical telescopes, instruments and surveys. However, current infra-red detectors are both smaller and significantly more expensive than their optical counterparts and the sky is brighter and more variable from the ground. Near-infrared transient surveys are therefore designed to probe areas where equivalent optical surveys are insufficient, such as finding planets around low mass stars (Sipőcz *et al.* 2011), looking for pulsating stars through the dust and nebulæ in the Galactic plane (Minniti *et al.* 2010), or studying Young Stellar Objects in star-forming clouds (Alves de Oliveira & Casali 2008), rather than more general transient searches that are better performed in the optical.

The two fastest current near-infrared survey instruments are the United Kingdom Infrared Telescope with the Wide Field CAMera (UKIRT-WFCAM; Casali et al. 2007) and the Visible and Infrared Survey Telescope for Astronomy (VISTA; Emerson *et al.* 2004). The data from both of these instruments are processed by the VISTA Data Flow System (VDFS), whose components are the Cambridge Astronomy Survey Unit (CASU) and the Wide Field Astronomy Unit (WFAU) in Edinburgh. CASU is responsible for daily data reduction and processing of individual data blocks and the astrometric and photometric calibration. WFAU is responsible for putting together data from different data blocks (deep stacks, multi-band catalogues, multi-epoch catalogues and variability statistics) as well as producing a queriable database (Hambly *et al.* 2008; Cross *et al.* 2009).

The most ambitious current NIR transient survey is the VISTA Variables in Via Lactea (VVV; Minniti *et al.* 2010). It covers 520 square degrees of the Galactic plane and bulge, and when complete will observe 1 billion objects over 100 epochs in the Ks band. The main science driver is to derive distances to different structures within the Milky Way using the period-luminosity relationships of RR Lyræ and Cepheid stars, so accurate post-calibration light curves are more important than real-time transient selection and follow-up. Most of the processing of light curves takes place during archive curation at least 6 weeks after the observations (Cross *et al.* 2009).

The main difficulties facing the VVV stem from the extreme source density of $> 10^6$ sources per deg^2, leading to difficulties in sky subtraction, deblending and occasional astrometric errors. The focal plane contains 16 2k×2k non-buttable Raytheon detectors that are laid out in a 4×4 pawprint. The design has 90% spacing in the X-direction and 45% spacing in the Y-direction, requiring half-detector overlaps to produce a 1.5 sq. deg. image tile. With highly variability both in the skies and in the PSFs within the duration of a data block significant processing is needed to remove the skies. The processing of a single tile is time consuming, requiring catalogue extraction on 6 pawprints as well as on the final tile.

The combination of dense fields and complex processing means that these fields are the ones that are most likely to need reprocessing. That adds large overheads to archive curation, especially when the processing is split between two data centres: data have to be re-transferred and re-ingested into the archive. However, using the VVV to deal with database tables containing $10^{10} - 10^{11}$ rows while also enabling users to have queries returned in sensible times is stretching the Microsoft SQL Server 2008 to its limits.

3. Catalina Real-Time Transient Survey – Andrew Drake

The Catalina Real-Time Transient Survey (CRTS; http://crts.caltech.edu) is a synoptic sky survey which uses data streams from 3 wide-field telescopes in Arizona and Australia. The data result from a NEO asteroid search led by Ed Beshore and Steve Larson of UAz LPL (http://www.lpl.arizona.edu/css). CRTS searches for highly variable and transient sources outside the Solar system. The collaboration illustrates how the same synoptic sky survey data stream can feed many different scientific projects in a highly efficient manner.

The survey covers the total area of $\sim 33,000$ deg^2, down to limiting magnitudes ~ 19–21 mag per exposure (depending on the telescope and the seeing), with time base-lines from 10 min to ~ 7 years (and growing). There are now typically ~ 300–400 exposures per pointing, and coadded images reach deeper than 23 mag. The basic goal of CRTS is a systematic exploration and characterization of the faint, variable sky. The survey has detected $\sim 4,000$ high-amplitude transients to date, including $> 1,000$ supernovæ, many of them unusual (in both 2009 and 2010, CRTS published more SNe than any other survey), nearly 1,000 CVs and dwarf novæ (the majority of them previously uncatalogued), hundreds of blazars and other AGN, highly variable and flare stars, etc.

CRTS has a completely "open data" philosophy: all transients are published electronically immediately, with no proprietary period at all, and all of the data (images, light curves) will be publicly available, thus benefiting the entire astronomical community. The events are published in real time using a variety of electronic mechanisms, as described on the survey website. Annotated events with links to data from other surveys and archives are also provided at the CRTS website, and at http://skyalert.org. About 10^8 light curves are now in process of being released through a web server.

In many ways, CRTS is a scientific and technological testbed and precursor for the grander synoptic sky surveys to come. More details and references are given in Drake *et al.* (2008), Djorgovski *et al.* (2011), Mahabal *et al.* (2011), and elsewhere in these Proceedings (page 306).

4. Pan-STARRS – Stephen Smartt

The survey is operating and producing transients. Details can be found at http://pan-starrs.ifa.hawaii.edu/public/home.html, and elsewhere in these Proceedings (page 71).

5. TAOS & TAOS2 – Matt Lehner

The Taiwanese-American Occultation Survey (TAOS) searches for occultations of stars by distant small bodies in the outer Solar system. Details can be found at http://taos.asiaa.sinica.edu.tw.

6. The GAIA Satellite as a Transient Survey – Simon Hodkin

GAIA is a European Space Agency (ESA) astrometry space mission, and a successor to the ESA HIPPARCOS mission. GAIA's main goal is to collect high-precision astrometric data (positions, parallaxes, and proper motions) for the brightest 1 billion objects in the sky. Those data, complemented with multi-band, multi-epoch photometric and spectroscopic data collected from the same observing platform, will allow astronomers to reconstruct the formation history, structure, and evolution of the Galaxy.

GAIA will observe the whole sky for 5 years with an average of 80 times per source, providing a unique opportunity for the discovery of large numbers of transient and anomalous events such as supernovæ, novæ and microlensing events, GRB afterglows, fallback supernovæ and other theoretical or unexpected phenomena. GAIA's focal plane is comprised of 106 CCDs with almost 1 billion pixels that are read out every 4.5 seconds. The Astrometric field G-band CCDs will sample the sky with 0.06×0.18-arcsecond pixels. The BP/RP spectrographs provide blue and red low-dispersion (R∼10–100) spectra for all GAIA sources. The Radial Velocity Spectrograph will provide velocities for stars with $V \leqslant 16$ to a precision $\leqslant 10$ km s^{-1}.

The Photometric Science Alerts team has been tasked with the early detection, classification and prompt release of anomalous sources in the GAIA data stream. Transient phenomena will be identified either via the discovery of new sources (new to GAIA), or through the detection of significant changes in magnitude and/or the spectrum. Precise astrometry, spatial morphology, photometry and spectroscopy will be combined with a history of the transient's environment to provide initial event classifications and associated probabilities. Ground- and space-based archival data will be cross-matched against each transient to refine its classification before publication of the event to the community at large via email, skyalert.org and other agreed mechanisms, within 24–48 hours of the observation. The aim is to publish all the data which have contributed to the discovery and classification of the event, including the transient's light curve, coordinates, morphology, thumbnail cutouts from the onboard SM and AF windows, BP/RP spectroscopy, and the cross-match results. Even though GAIA is primarily an astrometric mission, it provides real power for the early discovery and accurate classification of alerts, thanks to the unique combination of high sensitivity, high spatial resolution, and near-simultaneous low-dispersion spectroscopy.

7. Panel Discussion on Multi-wavelength Transient Programmes: Surveys and Follow-ups

The panel consisted of Peter Nugent, Neil Gehrels, Joseph Lazio, George Djorgovski and Nicholas Cross. Nugent, Gehrels and Lazio each gave a short introduction before questions were invited from the floor.

Discussion

PETER NUGENT (TO CROSS): Have you tried using parallel databases?

CROSS: Our tests so far show that Microsoft SQL should be sufficient for the VVV. We have not had the manpower to test other DBs like sciDB or MonetDB thoroughly enough, nor work out how we would have to change our curation software.

NUGENT: The expensive parts are the computing and follow-up. People try to get the most out of their own pipelines, but there is a lot of useful historical data ("historical follow-up").

LAZIO: There is a range of radio data from sub-millimeter to decametre. Radio wavelengths are important for astrometry in X-ray/gamma ray follow-ups and provide constraints on the physics.

GEHRELS: The good news is that gamma-ray transients get followed up, especially with telescopes with NIR spectroscopy for high-z events. The bad news is that everyone does the interesting GRBs and ignores the rest. For X-ray follow-ups, SWIFT started with 100 ToO requests per year and now receives 1000. It cannot handle any more. Also, it will not last forever; what will replace it?

HANISCH: We can now do a 1,000,000-row cross-match in a couple of seconds, but the difficulty is getting the data to the server. Real-time follow-ups are constrained by cross-matching to external data. We still need humans.

MUNDELL: The Liverpool GRB team have GRBs to 24 mag but no redshifts. The support amongst TACs for follow-ups of GRB detections is dropping just as the physics is getting better defined.

SCHMIDT: We have the wrong hardware for follow-ups. Multi-object spectrographs are not useful for following up a single bright object. Long-slit spectrographs are not useful for most transients either. We need dedicated hardware.

KULKARNI: The Palomar Transient Factory has follow-up telescopes too. A generous time allocation to collaboration investigators is not enough. Any idiot can discover a supernova, but following it up is more difficult and the follow-up telescope has to be bigger than the survey telescope.

MUNDELL: The Liverpool Telescope has spectroscopy and polarimity and a 10s change-over time between instruments.

DJORGOVSKI: Large surveys should be designed or organised to compliment each other by following each other up simultaneously.

LAZIO: There is also the need for robust archives since the radio often turns on much later.

DJORGOVSKI: We use layers of follow-up, gradually reducing the candidates until we have the ones that are interesting enough for spectra.

KULKARNI: This is the golden age of transient astronomy. It will be over by LSST.

DJORGOVSKI: I agree.

SCHMIDT: The main phase space of transients will be covered soon, perhaps in the next 12 months, and then we will move into a dour regime. But we cannot do the high-redshift universe well. We have the wrong hardware.

DJORGOVSKI: QSOs were spotted in the optical in the 1920s, but they were thought to be variable stars. They were only noticed as something special when they were matched to the radio.

SCHMIDT: Gravity waves will open up new studies.

PARTICIPANT: With GRBs detected by EXIST we will need NIR photometric follow-up from space to answer the dedicated science questions. This will not be possible in the present funding climate, but a 2-m to 3-m-class telescope in Antartica would cover half the sky for five months of the year.

RIDGWAY: What can we expect from LSST follow-up? No-one is doing the simulations to test this, they are too busy.

DJORGOVSKI: On the contrary, we have the opposite problem. Too many good people are wasting their time simulating data, instead of learning from the real data and event streams.

KULKARNI: We cannot extrapolate across 10 years. This is the decade of discovery: PTF, Pan-STARRS, SWIFT, HST, CHANDRA. Dark Energy Camera will also be around soon. LSST will be like KEPLER: people will use time-series analysis on the light curves rather than following up detections. It will be a physics machine, not a discovery machine: lots of chi-squared tests.

LAZIO: The precursors to the SKA will be interesting, and the metre and centimetre wavelengths. There will be new incoherent and coherent sources, and fast transients. We are still exploring the discovery space. What about high-speed optical surveys? There is TAOS, but it is not wide-field.

DJORGOVSKI: We have to think about the cadences as well as the wide field.

VESTRAND: We need to understand the whole system and have real-time knowledge of what is available. Sometimes it is like second graders playing soccer: everyone kicks at the ball. Everyone observes the same GRB at the same time and doesn't know that it is already being observed.

DJORGOVSKI: VOEvent was designed with problems like this in mind.

GEHRELS: Lots of small telescopes that are useful for follow-up are getting closed down.

KULKARNI: Lots of small telescopes need to be refurbished.

DJORGOVSKI: We need to use computers to observe archives as a historical follow-up.

DJORGOVSKI: This will be the first and last meeting dedicated to time-domain astronomy. The field will be too big for one meeting and there will be more specialised meetings.

8. Summary

The take-away messages from this workshop can be summarized as follows:

• There is not enough awareness in the community of all of the tools and facilities available, such as *SkyAlert*, or the various Virtual Observatory services, so the follow-up of transients is not as efficient as it could be.

• The main bottleneck in extracting science from existing synoptic surveys is the follow-up spectroscopy. The problem will get worse by orders of magnitude as we move towards the LSST. Dedicated spectroscopic facilities may be necessary. A lot of spectrographs developed for large telescopes are designed for many faint objects over a small FOV, whereas transients are observed one at a time.

• There should be more coordination between surveys at different wavelengths to maximise the scientific returns. The cadences and pointings need to be coordinated, sometimes with offsets as emission at some wavelengths (e.g. radio) can be delayed. Given such a variety of different technologies and science requirements, this is an extremely complex task.

References

Alves de Oliveira, C. & Casali, M. 2008, *A&A*, 485, 155

Casali, M., *et al.* 2007, *A&A*, 467, 777

Cross, N. J. G., *et al.* 2009, *MNRAS*, 399, 1730

Djorgovski, S. G. *et al.*, 2011, in: T. Mihara and N. Kawai (eds.), *The First Year of MAXI: Monitoring Variable X-ray Sources* (Tokyo: JAXA Special Publ.), in press

Drake, A. J., *et al.* 2009, *ApJ*, 696, 870

Emerson, J. P., *et al.* 2004, *The Messenger*, 117, 27

Hambly, N. C., *et al.* 2008, *MNRAS*, 384, 637

Mahabal, A. A., *et al.* 2011, *BASI*, 39, 387

Sipőcz, B., Kovács, G., Pinfield, D., & Hodgkin, S. 2011, in: F. Bouchy *et al.* (eds.), *Detection and Dynamics of Transiting Exoplanets*, EPJ Web of Conferences, 11, id.06003, 110, 600

Minniti, D., *et al.* 2010, *New Astronomy*, 15, 433

New Horizons in Time-Domain Astronomy
Proceedings IAU Symposium No. 285, 2011
R.E.M. Griffin, R.J. Hanisch & R. Seaman, eds.

© International Astronomical Union 2012
doi:10.1017/S1743921312000592

Gravitational Waves
and Time-Domain Astronomy

Joan Centrella[1], Samaya Nissanke[2] & Roy Williams[2]

[1]NASA Goddard Spaceflight Center, Greenbelt, MD 20771, USA
email: Joan.Centrella@nasa.gov

[2]California Institute of Technology, Pasadena, CA 91125, USA

Abstract. The gravitational-wave window onto the universe will open in roughly five years, when Advanced LIGO and Virgo achieve the first detections of high-frequency gravitational waves, most likely coming from compact binary mergers. Electromagnetic follow-up of these triggers, using radio, optical, and high energy telescopes, promises exciting opportunities in multi-messenger time-domain astronomy. In the decade, space-based observations of low-frequency gravitational waves from massive black hole mergers, and their electromagnetic counterparts, will open up further vistas for discovery. This two-part workshop featured brief presentations and stimulating discussions on the challenges and opportunities presented by gravitational-wave astronomy. Highlights from the workshop, with the emphasis on strategies for electromagnetic follow-up, are presented in this report.

Keywords. gravitational waves, compact binaries, time domain astronomy, multi-messenger astronomy

1. New Cosmic Messengers

Gravitational waves (GWs) are a new type of cosmic messenger, bringing direct information about the properties and dynamics of sources such as compact object mergers and stellar collapse. The observable GW spectrum spans over 18 orders of magnitude in frequency, ranging from phenomena generated in the earliest moments of the Universe to vibrations of stellar-mass black holes (BHs).

For time-domain astronomy, two GW frequency bands stand out as being especially promising. The high frequency band, ~ 1–10^4 Hz, will be opened by ground-based interferometric detectors starting around mid-decade. The strongest sources in that band are expected to be the mergers of stellar-mass compact objects, primarily BH binaries, neutron star (NS) binaries, and BH–NS binaries in the local universe out to several hundred Mpc. The low frequency band, $\sim 10^{-4}$–10^{-1} Hz, will be opened by space-based detectors in the 2020s. Merging massive black-hole (MBH) binaries, with masses in the range $\sim 10^3$–10^7 $M_\odot$ and detectable out to high redshifts ($z > 10$), are expected to be the strongest sources here and the ones of greatest interest to time-domain astronomy. Other low-frequency sources include inspirals of compact objects into central MBHs in galaxies and compact stellar binaries with periods of tens of minutes to hours†.

The GWs emitted by binaries typically evolve upwards in frequency with time. When the binary components are well separated and spiralling together because of GW emission, the waveform is a sinusoid increasing in frequency and amplitude, also called a chirp.

---

† The very low frequency band, $\sim 10^{-9}$–10^{-6} Hz, will be opened by pulsar timing arrays later this decade. The most likely sources in this region are binaries containing MBHs of $\sim 10^9 M_\odot$.

The final merger, in which the binary components coalesce, produces a burst of radiation; that is followed by a ringdown phase in which the merged remnant typically settles down to an equilibrium configuration. The time-scales for those phases depend primarily on the masses of the binary components. For compact remnants, the inspiral might be detectable for about a minute, while the merger and ringdown occur on time-scales of roughly a few to tens of milliseconds. The detailed nature of the merging objects, for example moduli of the neutron star crust, is revealed in those last few milliseconds. For MBH binaries, space-based detectors should be able to observe the inspiral for several months. The ensuing merger and ringdown then occur over minutes to hours.

The possibility of electromagnetic (EM) counterparts of these GW sources raises exciting prospects for multi- messenger astronomy in the time domain. An EM counterpart greatly boosts confidence in a GW detection (Kochanek 1993). EM counterparts may be a precursor (possibly associated with the binary inspiral), a flash (triggered during the merger and/or ringdown phases), or an afterglow. The timescales for these signals depend on emissions from gas in the vicinity of the binary and can vary widely depending on the type of EM emission produced. Afterglows in particular can be very long-lived.

The opportunities for GW–EM multi-messenger time-domain astronomy generated lively and fruitful discussions during a two-session workshop held during the Symposium. This report presents the highlights from that workshop.

2. High Frequency Gravitational Waves: Getting Ready for Detection

The GW window onto the universe is expected to open around the middle of this decade, when the the first detections of GW signals from compact binaries are made by ground-based interferometers with kilometre-scale arms. Those first detections will mark the culmination of years of development by hundreds of scientists world-wide, and will inaugurate a new era in GW astronomy. This Section presents a status report on those efforts, and highlights important questions and challenges for this category of multi-messenger time-domain astronomy.

2.1. *Status Update*

Currently there are three full-scale ground-based interferometric observatories: LIGO runs the 4-km observatories located in Hanford, Washington and Livingston, Louisiana, and the 3-km French-Italian Virgo detector located near Pisa, Italy. Both LIGO and Virgo were planned to be developed in stages. The initial detectors would be full-scale interferometers able to detect rare (nearby) events; the advanced detectors would be about a factor of 10 more sensitive, able to make multiple detections per year and to be true observational tools.

LIGO and Virgo have successfully reached their initial design sensitivities, completing several science data-taking runs in 2009–2010. According to current estimates of stellar-mass mergers, that many months of observation should have yielded a detection with probability < 2% (LSC 2010). Rather than waiting decades for a strong enough signal, LIGO and Virgo are undergoing upgrades to make joint detection rates at least yearly, perhaps weekly. Early science runs with Advanced LIGO/Virgo could start by mid-decade. As of September 2011, Virgo is still operating; the upgrade to Advanced Virgo is expected to follow the Advanced LIGO upgrade by about a year. Construction has also begun in Japan for the Large-Scale Cryogenic Gravitational-wave Telescope (LCGT; Kuroda 2010), an advanced detector that could start operating by the end of this decade. Another advanced detector may also be built in India.

The initial LIGO and Virgo detectors have been used to carry out science runs, deriving upper limits on sources within their observational reach, and to develop data-analysis and detection strategies. An EM follow-up programme and a blind-injection test were exercised during the 2009–2010 science runs, providing valuable experience for the advanced detector era (and of particular significance for this workshop). The follow-up programme incorporated a prompt search for EM counterparts triggered by GW transients (LSC 2011c). Candidate GW events and their possible sky locations were identified using a low-latency analysis pipeline. The most promising sky positions for EM imaging were selected using a catalogue of nearly galaxies and Milky Way globular clusters. Within ~30 minutes, that directional information was sent to partner telescopes around the globe and also to the SWIFT gamma-ray satellite (see Fig. 1). Nine such events were followed up by at least one telescope (LSC 2011c).

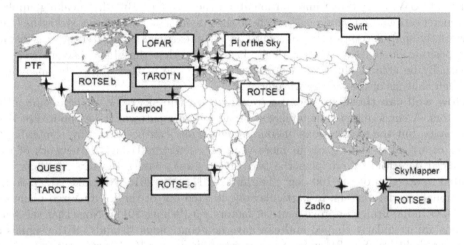

Figure 1. Telescopes that participated in the LIGO-Virgo EM follow-up exercise. Figure from LSC (2011c)

A blind injection test was also carried out during September 2010 (LSC 2011a). In this process—designed to provide a stress test for the full data analysis pipeline and science procedures—signals were secretly injected into the detector data stream by a small group of analysts. The parameters of those signals were sealed in an envelope, to be opened only after the entire collaboration had searched the data and carried out a full exercise of the processes, from potential detection to approval of a publication. In this case, a strong chirp signal was observed in the data shortly after injection. The data were vetted, sky positions were determined, and the GW trigger information was sent to several EM telescopes. As part of the process, LIGO and Virgo prepared a "data release"—a set of data from an injected signal, representing what could be released to the community when the first actual detections are made; see LSC (2011b).

2.2. *Expectations about GW Source Detection*

GWs are produced by the dynamical motion of massive objects in space/time. Since they couple very weakly with matter, GWs are ideal cosmic messengers for bringing information from dark, hidden regions within galaxies. This section highlights some key expectations gleaned from studies of GW source detection, focusing on compact binary sources.

How many compact binaries are expected to be observed by advanced ground-based GW detectors? Rates for detection by Advanced LIGO and Virgo are estimated using projected detector sensitivities, plus compact binary merger rates derived from either the observed sample of galactic binary pulsars or population synthesis results (LSC 2010). A network consisting of three advanced detectors is expected to detect between 0.4–400 NS–NS binaries per year (the most realistic estimate being about 40 per year), out to a distance of ~450 Mpc for optimally oriented sources (that is, face-on and located directly above the detector). For stellar BH binaries, in which each BH has a mass $\sim 10 M_\odot$, the rates range from 0.2–300 detections per year (most realistically ~20 per year), out to ~2000 Mpc. For NS–BH binaries, where the BH has a mass $\sim 10 M_\odot$, the rates range from 0.2–300 per year, the most realistic estimate being ~10 per year, out to ~900 Mpc. Averaging over the sky and the source orientation reduces all those distance ranges by a factor of about two (Finn 1993).

What properties of these sources can be measured by observing the GWs emitted? GWs carry direct information about their sources. Broadly speaking, applying parameter estimation techniques to analyze the gravitational waveforms yields the binary masses, the spins and the orbital elements, as well as extrinsic parameters such as distance and position on the sky (Cutler 1994). It is hoped that the global GW network will eventually be able to elucidate deep secrets of extreme matter, ripped by space itself, through analysis of the merger waveform.

How well can these sources be localized on the sky? GW detectors are all-sky monitors. A single interferometer has a broad antenna pattern. It has poor directional sensitivity, but the localizations of sources on the sky can be refined by comparing the times of arrival of the signals in more than one detector. Using the network of three advanced detectors (Virgo plus the two LIGO sites) enables the sky positions to be limited to a region of ~1–100 deg^2 (Fairhurst 2011, Wen 2010, Nissanke 2011). Adding additional interferometers, in particular one located out of the LIGO-Virgo plane such as LIGO India, brings improvements of factors ~3 (Schutz 2011). Note that the search regions are irregularly shaped and can have non-contiguous "islands" if the signals are near threshold (see Fig. 2), as is to be expected for many LIGO-Virgo sources.

What types of information can astronomers expect to receive from GW observatories? The advanced detector network will be able to supply the signal time of arrival and sky location of the source, along with an estimate of the false alarm rate (FAR). For merging compact binaries, the masses and spins of the components along with the inclination and luminosity distance will also be available. Other information may also be released: see LSC (2011b) and Fig. 2 for an example.

When will LIGO release to the public its rapid alerts for afterglow observers? Rapid release of triggers (~minutes) will begin under one of three criteria, according to the LIGO Data Management Plan (Anderson 2011): (a) after a GW detection has been confirmed and the collaboration agrees to begin the public programme, or (b) after a large volume of space-time has been searched by LIGO, or (c) if the detectors have been running for a long time. It is hoped that this release could be as early as 2016. It should be noted, however, that these criteria may be changed in the future and the release date brought forward.

2.3. *Observing EM Counterparts: Challenges and Opportunities*

Mergers of NS–NS and NS–BH binaries are expected to generate EM radiation in several wavelength bands. Coupling observations of EM radiation and GWs from those sources opens up exciting new avenues for exploration. Since many GW detections may be near threshold, the identification of an EM counterpart would provide additional confirmation of the event. Direct measurements of the binary properties through GW observations will allow testing and deeper understanding of the underlying astrophysical models which are

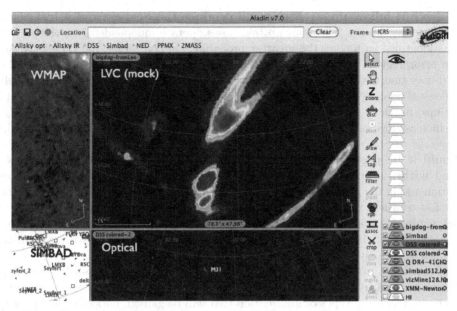

Figure 2. Non-contiguous search islands, illustrated with the Aladin "astronomical information system" (Fernique 2011). By converting the skymap to Healpix/FITS format, it can be compared directly with other astronmical images and catalogues. The panel 'LVC (mock)' corresponds to the skymap released as (LSC 2011b), with at least seven islands visible. Other panels show the same sky with other maps for comparison; left: WMAP data, lower-left: SIMBAD graduated catalog, and right: the Optical sky (DSS), where the galaxy M31 can be seen.

currently inferred only from EM radiation. This section captures exciting and challenging issues in this arena.

Which astrophysical phenomena generate promising EM counterparts to NS–NS and NS–BH mergers? These mergers are expected to produce collimated jets, observed near the axis as short-duration gamma-ray bursts (SGRBs). Afterglows from interactions of a jet with gas around the burst can be observed in the optical (near the axis, on time-scales of hours to days) and radio (isotropically, over weeks to years). A "kilonova" may also be produced from radioactive decay of heavy elements synthesized in the ejecta, yielding weak optical afterglows lasting several days (Li 1998). Note that SGRBs are rare within the distance observable by Advanced LIGO and Virgo. Thus the isotropic emissions, particularly those from kilonovæ, are likely to be the most promising observable EM counterparts; see Metzger (2011) for details.

What information will astronomers want from the GW observatories? Most astronomers will first want the time and sky location of the event in order to plan the follow-up. Information about the binary components and orbits will also be valuable for understanding the underlying astrophysics. A smaller group of scientists may also want the gravitational-wave time-series information. A means for distributing that information is currently being worked out.

Which astronomical instruments will be especially useful in searching for EM counterparts? Satellites such as SWIFT and FERMI are needed to find GRBs, which are expected to be coincident with GW detections but only if the axis of the inspiral system points to the Earth ("beaming"). Other gamma or X-ray satellite observatories, current and planned, include MAXI, SVOM, and AstroSat. Rapid follow-ups by ground-based optical telescopes with wide fields of view (such as PTF-2, Pan-STARRS and LSST) and radio telescopes (LOFAR, ASKAP, EVLA, etc.) will search for the afterglows.

Which EM follow-up strategies will produce the best astrophysical results?
In the simplest scheme, all follow-up telescopes operate independently, pointing at the
regions identified as having the greatest probability of containing the target. Since the
source error regions on the sky from GW events will be relatively large irregularly-
shaped ones that may have non-contiguous islands, many EM counterparts, particularly
afterglows of relatively short duration, may be missed by this approach. Coordinated EM
follow-ups, in which many telescopes operate co-operatively to cover the source region,
could increase dramatically the odds of imaging successfully an EM counterpart (Singer
2011).

**Would it be possible to observe the immediate optical "flash" that is ex-
pected within seconds of the merger time?** There are all-sky optical monitors in
operation and in the planning. The *Pi Of The Sky* observatory (Sokolski 2011) can locate
a very fast transient to within seconds, down to magnitudes fainter than 12. The planned
global monitoring system from Los Alamos and LCOGT (Wren *et al.* 2010) will have the
advantage of a much higher duty cycle, because of its distributed nature and multiple
telescopes at each site.

How early a notice of a GW trigger is useful, and possible? Since early
notification increases not only the odds of imaging successfully the counterpart but also
raises the amount and types of information that will be gained, notices should be sent as
early as possible. In the recent follow-up program carried out by LIGO and Virgo, notices
were sent out within 30 minutes of detection (LSC 2011c), and efforts are continuing to
reduce that to 1 minute. Detecting a GW signal during a binary inspiral (before the
merger) and releasing the information before the burst occurs is even more interesting—
and challenging; Cannon (2011) provides an early study.

Which triggers should be followed up? How low a significance is tolerable?
Strong, nearby GW sources should have a low false-alarm rate (FAR) and produce more
robust data, including more accurate values for the binary parameters and sky location.
However, signals near threshold might be more common. The answers to those questions
will depend on the actual detection rate.

**What more information about the local universe is needed in order ro
prepare for the data from Advanced LIGO and Virgo?** Two items in particular are
needed: a complete publicly-available catalogue of nearby galaxies (White 2011, Kulkarni
2009), and an inventory of known transients within reach of Advanced LIGO and Virgo.
For faint transients, it may be that the lack of such a catalogue, and the large numbers
of sources, can make it very difficult to identify a GW afterglow even if there is a wide,
deep telescope available. By using spatial coincidences with a complete galaxy catalogue
for those events within 200 Mpc, the number of false-positve transients in a typical GW
error box of $\sim$10 deg^2 is reduced by three orders of magnitude (Kulkarni 2009).

3. Low-Frequency Gravitational Waves: Looking to the Future

The low-frequency window contains a wealth of astrophysical sources. Because of noise
from fluctuating gravity gradients in the Earth at frequencies below a few Hz, low-
frequency GWs can only be observed with space-based detectors. The most highly developed proposal is the Laser Interferometer Space Antenna (LISA), which consists of three
satellites orbiting the Earth in a triangular configuration with arm lengths $\sim 5 \times 10^6$ km
(Jennrich 2009). Detectors such as LISA will observe coalescing MBH binaries inspiralling
over a period of several months, followed by the final merger and ringdown, as well as
inspirals of compact objects into central MBHs in galaxies and compact stellar binaries
with periods of minutes to hours.

What is the current status of low frequency GW detectors? Owing to budgetary problems, ESA and NASA terminated their partnership to develop LISA in the spring of 2011, and both agencies are now looking at lower-cost concepts. On the ESA side, studies are underway for the New Gravitational-wave Observatory (NGO, also known informally as LISA-lite, EuLISA, and eLISA), which is similar to LISA but with shorter arm lengths of $\sim 1 \times 10^6$ km. In the US, NASA is also examining concepts for a Space Gravitational-wave Observatory (SGO).

What can be learned from observing low-frequency GWs from MBH mergers? These space-based interferometers can observe MBH mergers (Sesana 2011) over a period of several months at relatively high signal-to-noise, allowing precision measurements of the binary properties, plus sky localization to < 100 deg^2. The expected merger time can be predicted and broadcast weeks or months in advance, providing excellent opportunities for EM follow-ups. The rates for MBH mergers are expected to be at least several per year, with the actual values depending on the instrument sensitivity.

What are the prospects for EM counterparts of MBH mergers? MBH mergers are astrophysically rich systems, with a variety of possible EM signals as precursors, flashes and afterglows (Schnittman 2010). Since MBH mergers are considered central to our understanding of galaxy and MBH assembly history and demography as well as galaxy-MBH co-evolution, the astrophysical payoffs will be significant (Komossa 2003).

4. Summary

The GW window onto the universe will open this decade, when Advanced LIGO and Virgo make the first detections of high-frequency GW signals that are expected to come from merging compact binaries. Strategies for EM follow-ups of the GW triggers are being designed and tested to search for radio, optical and high-energy counterparts. Searches for coincident GW and high-energy neutrinos will be made in co-ordination with the IceCube project (Bartos 2011). Co-ordinated searches, coupled with complete catalogues of galaxies and transients in the local universe, are needed in order to maximize the science from these multi-messenger studies. In the next decade, space-based interferometers will open the low-frequency window with observations of merging MBH binaries. These observations, and their EM counterparts, will provide important information on the evolution of structure and MBHs over cosmic time.

Acknowledgements

It is a pleasure to thank Eric Chassande-Mottin, Ed Daw, Stephen Fairhurst, Jonathan Gair, Mansi Kasliwal, Stefanie Komossa, Brian Metzger, Larry Price, Jeremy Schnittman, Alberto Sesana, and Leo Singer, who delivered excellent short presentations at the GW workshops that stimulated interesting and important discussions.

References

LSC 2010: Abadie, J., *et al.* 2010, LIGO Scientific Collaboration and Virgo Collaboration, *Class. Quant. Grav.*, 27, 173001
LSC 2009: Abbott, B. P., *et al.* 2009, LIGO Scientific Collaboration and Virgo Collaboration, *Reports on Progress in Physics*, 72, 076901
LSC 2011c: Abbott, B. P., *et al.* 2011c, LIGO Scientific Collaboration and Virgo Collaboration, http://arxiv.org/abs/1109.3498
Anderson, S & Williams, R.D. *LIGO Data Management Plan*
https://dcc.ligo.org/cgi-bin/DocDB/RetrieveFile?docid=9967
Apostolatos, T., Kennefick, D., Ori, A., & Poisson, E. 1993, *Phys.Rev.*, D47, 5376

Bartos, I., Finley, C., Corsi, A., & Marka. S. 2011, http://arxiv.org/abs/1108.3001

Cannon, K., *et al.* 2011, submitted

Cutler, C. & Flanagan, E. 1994, *Phys. Rev.*, D49:2658

Fairhurst, S. 2011, *Class. Quantum Grav.*, 28, 105021

Fernique, P., Boch, T., *et al.* 2011, http://aladin.u-strasbg.fr/

Finn, L. S. & Chernoff, D. F. 1993, Phys. Rev. D, 47:2198

Harry, G. M., *et al.* 2010, *Class.Quant.Grav.*, 27:084006, 2010.

Jennrich, O. 2009, *LISA technology and instrumentation. Class. Quant. Grav.*, 26:153001, 2009.

Kochanek, C. S., and Piran, T. 199 ApJ, 417, L17

Komossa, S. 2003, in: J. M. Centralla (ed.), *The Astrophysics of Gravitational Wave Sources*, AIP Conf. No. 686, (Berlin: Springer), p. 161

Kulkarni, S., & Kasliwal, M.M. Proc. RIKEN Symp., Astrophysics with All-Sky X-Ray Observations, 312, arXiv:0903.0218

Kuroda, K., The LCGT Collaboration. 2010, *Class. Quant. Grav.*, 27, 84004

Li, L.-X. & Paczyński, B. 1998, *ApJ*, 507, L59

LSC 2011a: LIGO Scientific Collaboration and Virgo Collaboration, 2011a, http://www.ligo.org/news/blind-injection.php

LSC 2011b: LIGO Scientific Collaboration and Virgo Collaboration. 2011b, *Mock data release GW100916* Colloq. *'Big Dog Data'* http://www.ligo.org/science/GW100916

Metzger, B. D. & Berger, E. 2011, http://arxiv.org/abs/1108.6056

Samaya, M., Nissanke, J. L., Sievers, N. D., & Holz, D. E. 2011, *ApJ*, 739, 99

Schnittman. J. D. 2011, *Class. Quant. Grav.*, 28, 94021

Schutz, B. F. 2011, *Class. Quant. Grav.*, 28, 125023

Sesana, A., Gair, J., Berti, E., & Volonteri, M. 2010, *Phys. Rev.*, D83, 044036, 2011.

Singer, L., Price, L., & Speranza. A., https://dcc.ligo.org/cgi-bin/DocDB/ShowDocument?docid=G1100983

Sokolowski, M. http://grb.fuw.edu.pl/

Wen, L. & Chen, Y., *Phys. Rev.*, D81:082001, 2010.

White, D. J., Daw, E. J., & Dhillon, V. S. 2011, *Class. Quant. Grav.*, 28, 85016

Wren, J., Vestrand, W. T., Wozniak, P., & and Davis, H. 2010, *SPIE Proc.*, 7737, 773723

New Horizons in Time-Domain Astronomy
Proceedings IAU Symposium No. 285, 2011
R.E.M. Griffin, R.J. Hanisch & R. Seaman, eds.

© International Astronomical Union 2012
doi:10.1017/S1743921312000609

The Future of X-Ray Time-Domain Surveys

Daryl Haggard[1] and Gregory R. Sivakoff[2]

[1] Center for Interdisciplinary Exploration and Research in Astrophysics (CIERA),
Department of Physics and Astronomy, Northwestern University, Evanston, IL USA
email: dhaggard@northwestern.edu

[2] Department of Physics, University of Alberta, Edmonton, AB, Canada

Abstract. Modern X-ray observatories yield unique insight into the astrophysical time domain. Each X-ray photon can be assigned an arrival time, an energy and a sky position, yielding sensitive, energy-dependent light curves and enabling time-resolved spectra down to millisecond time-scales. Combining those with multiple views of the same patch of sky (e.g., in the CHANDRA and XMM-NEWTON deep fields) so as to extend variability studies over longer baselines, the spectral timing capacity of X-ray observatories then stretch over 10 orders of magnitude at spatial resolutions of arcseconds, and 13 orders of magnitude at spatial resolutions of a degree. A wealth of high-energy time-domain data already exists, and indicates variability on timescales ranging from microseconds to years in a wide variety of objects, including numerous classes of AGN, high-energy phenomena at the Galactic centre, Galactic and extra-Galactic X-ray binaries, supernovæ, gamma-ray bursts, stellar flares, tidal disruption flares, and as-yet unknown X-ray variables. This workshop explored the potential of strategic X-ray surveys to probe a broad range of astrophysical sources and phenomena. Here we present the highlights, with an emphasis on the science topics and mission designs that will drive future discovery in the X-ray time domain.

Keywords. accretion, accretion disks, stars: flare, (stars:) novæ, cataclysmic variables, (stars:) supernovæ: general, stars: winds, outflows, galaxies: active, galaxies: nuclei, X-rays: general

---

1. X-ray Astronomy's Broad Reach

X-ray data span an enormous dynamic range within astrophysical régimes. In the coming decades X-ray observatories, in concert with instruments across the electromagnetic spectrum, will systematically tackle the exciting "time domain". They will have enough power to reveal the progenitors to gamma-ray burst (GRBs), to probe the physics behind supernova (SN) shock breakout, to identify and characterize tidal disruption events, to constrain models of the accretion physics in X-ray binaries (XRBs) and active galactic nuclei (AGN), and to determine the rates and driving mechanisms behind stellar flares and their impact on space weather—to name a few examples. X-ray detectors are unique in that every photon is time tagged, energy tagged, and assigned an accurate sky position. X-ray observations also cover time-scales from sub-millisecond to ~40–50 years, span orders of magnitude in spatial resolution, and achieve a decade in energy coverage with decent energy resolution. The result is sensitive, energy-dependent light curves and time-resolved spectroscopy for every target. A wealth of high-energy time-domain data already exist, from which variability on time-scales ranging from microseconds to years has already been identified in a wide variety of objects.

In this workshop we discussed the missions that would be optimal for discovering and characterizing X-ray transients and variables. We were motivated in part by the enormous interest that has been expressed by the astronomical community (and evidenced by this very Symposium) in optical and (more recently) radio transients. We felt that the X-ray community is not, at present, making a compelling case for the power of X-ray observatories which are optimized for time-domain studies (with a few notable exceptions such as RXTE and SWIFT). We hope that the guiding questions outlined below, together with those generated by workshop attendees, will bring into focus the kinds of efforts needed to lobby most effectively for those missions, archives, cadences and science objectives in order to ensure that X-ray astronomy is well resourced in the future and thus able to contribute substantially to the exploration of transient phenomena.

The time domain is already expanding rapidly. To optimize these many transient domain studies we must connect the targets and the science at multiple wavelengths. Competition can be inimical to progress; although some might fear lest the optical and radio communities absorb resources away from the X-ray transient community, there was general consensus that the deepest insights into physics, and hence the highest science impact, result from coordinated, multiwavelength observations.

This overview of the workshop does *not* explore the full scope of science accessible in the X-ray domain, nor advocate any particular mission. Both the science and the technology are rapidly evolving, and attempts to place the entirety of X-ray astronomy under a single umbrella may be a questionable exercise—as explained in Martin Elvis' response to NASA's recent call for "Concepts for the Next NASA X-ray Astronomy Mission"†. Instead, we hope to prompt the astronomical community into thinking about the central role which X-rays have played and still can and should play, in our exploration of astronomy's time domain.

2. Guiding Questions

We asked workshop attendees to discuss these guiding questions:

(*a*) In recent years optical and radio transient science have increasingly gained attention among the general astronomical community. At the same time, X-ray transient surveys seem to be ceding ground, both financially and scientifically. What are the most compelling science cases for current and future X-ray transient studies? What efforts does the X-ray transient community need to undertake to lobby most effectively for the importance of X-ray transient studies (past and present) to the general astronomical community?

(*b*) The Rossi X-ray Timing Explorer (RXTE) has been a tremendous boon for studies of X-ray transients. However, it will cease operation at the end of this year. While some of its scientific capacities can be shifted to current instruments like SWIFT and MAXI, other capacities are unique to RXTE among currently flown instruments. What steps do we need to take to transition from the era of RXTE to the era without it? What important lessons have we learned from RXTE? How will new planned or soon-to-be-launched instruments support X-ray transient surveys? What inventive ways can we develop to utilize new instruments that may not have been designed originally for X-ray transient studies?

(*c*) The scientific output of X-ray transient surveys can be greatly increased through multi-wavelength observations. How do we best coordinate multi-wavelength observations, especially for X-ray transient surveys? Do we need to develop an X-ray Transient Network, or are existing infrastructures like the Gamma-ray Circular Network and the Astronomers Telegram sufficient? What cadences are needed to achieve various science

† NASA has recently solicited the community to suggest new X-ray mission concepts for advancing the goals of the Physics of the Cosmos (POC) programme (NASA RFI NNH11ZDA018L). These submissions are public, and are available on the POC webpage: http://pcos.gsfc.nasa.gov/studies/x-ray-mission-rfis.php.

priorities at different wavelengths? Are there opportunities for "citizen science" with X-ray transient surveys?

3. Workshop Highlights

The workshop was structured as a pure discussion—there were no formal science talks. Some of the most active discussions that took place are outlined below. A 1.5-hour audio recording of the workshop, together with a written transcript, are available at http://faculty.wcas.northwestern.edu/~dha724/xray_transients_2011/.

3.1. *X-ray Transients and Variables*

Our first discussion was of transients (unanticipated [dis]appearance or flaring) as opposed to variables (periodic or repeated fluctuations). Are X-ray studies more likely to uncover "variables" than "transients"? The majority opinion was that most X-ray variables were initially identified as transients (as is indeed the case with optical/radio transients), and that in most cases the distinction is driven by the detection limits of individual surveys. For example, on very deep optical data (to $\sim 28^{\text{th}}$ magnitude) one may begin to see progenitors of Type Ia supernovæ (which are themselves are probably variable) in addition to novæ, X-ray binaries and the like.

3.2. *The Science Case(s) for X-ray*

It is essential to state the most compelling science cases for current and future X-ray transient studies—to identify what is unique about the X-ray domain and why it should be compelling to fund an X-ray mission rather than a UV or IR one. *Strong gravity* and *accretion physics* are both areas to which the X-ray time domain brings a unique view. The most interesting individual science cases for X-ray time-domain studies included:

- Gamma-ray bursts (black-hole birth, cosmological probes)
- Supernova shock break-out
- Tidal disruption events
- X-ray variability of AGN and XRBs
- Giant hard X-ray flares (from flare stars and blazars)
- Impact of stellar flares on space weather/planetary habitability
- Variability in SgrA*
- Accreting millisecond pulsars
- Coherent pulsations and QPOs in neutron stars
- Galactic black-hole and neutron-star populations

In addition, other X-ray variables, not yet recognized as such, might supply the most compelling physical insights, though it is difficult if not impossible to base an X-ray mission on only an anticipated benefit. Many of the phenomena cited were originally discovered in the X-ray domain (though most remain only poorly characterized). However it was felt that, in coming years, the impetus will most likely come not from X-ray missions but from optical or radio telescopes, reflecting an enthusiasm for the "new" (LOFAR, ATA, PTF and potentially LSST) as opposed to the "old" or established (RXTE All-Sky Monitor, SWIFT, XMM-NEWTON and CHANDRA surveys). If we make out that the X-ray sky is a known entity, then the potential for discovery is perceived to be greater at less known wavelengths, making the latter seem more exciting. An X-ray transient mission therefore needs some goal like testing general relatively to bolster its case, i.e., something that can only be done through X-ray science.

X-ray variability in AGN and XRBs probes the physics of the inner accretion disk. These, in particular, test strong gravity. The same is true for tidal disruption events. The structure of the variability and its time-scale may assist in distinguishing between radiatively efficient and inefficient accretion flows and the mechanisms responsible for launching jets and winds. Sensitivity to very rapid variations (coherent pulsations, QPOs, fractional variability) is critical for understanding local XRB sources, and may shed light on more distant sources by analogy.

From the multiwavelength perspective, radio quenching and radio flaring have been seen in X-ray binaries within days. Hence, having missions that have the capacity to observe an XRB daily after an outburst has proved critical. The difficulty now is coordinating efficiently with other observatories; in the radio (for example) coordinations with EVLA have improved with the introduction of dynamic scheduling, but are still of the order of a few hours.

For AGN the relevant time-scale is weeks to years. As pointed out, AGN go into a deep low state and stay there for days or weeks; that is when the X-ray spectral complexity is most pronounced, and when distinguishing between different inner disk models is most effective. Thus, for AGN it might be the dips in their light curves, not the flares, which prove more interesting. Monitoring tidal disruption flares is also most effective on time-scales of weeks, but time-scales of minutes have not yet been explored for blazars. At GeV energies we are limited by statistics, but there is sub-day variability, and presumably it is the X-ray non-thermal component that is varying. As described by Kulkarni (p. 55), the time-scale for the X-ray shock break-out from supernovæ is hours or less.

In stellar coronal variability, both sensitivity and wavelength coverage are important. Greater sensitivity allows one to look for flares from stars at greater distances or for weaker flares from stars less distant, but we require the multi-wavelength context (soft X-rays, hard X-rays, UV/optical) to facilitate a full interpretation. X-ray emission from stars provides information about the coronal material and the coronal dynamics, and which cannot be obtained from other wavelengths; X-rays show how the tenuous coronal plasma is reacting to magnetic reconnection. There are many aspects regarding stellar flares that are of outstanding interest. What drives the extreme energy release? How do

Prospects for X-ray Time Domain Surveys

Discovery & Monitoring	Rapid Response	High Time Resolution
• All Sky Monitor	• Rapid slew (< hr)	• Sub-ms timing
• MAXI/AstroSAT, Lobster-eye technology, LOFT	• Swift, AstroSAT, LOFT, a suite of Swifts!?	• AstroSAT, Athena, LOFT?
• Science drivers: GRBs, SNe shock breakout, accretion physics, tidal disruptions	• Science drivers: GRBs, stellar flares/space weather, transient response	• Science drivers: Strong gravity, neutron star physics, XRB physics, QPOs
• ~Daily Cadence	• High Availability	• High Sensitivity

Figure 1.

flares affect the stellar environment (both in the context of young stars where planets are forming in a disk, and for older stars where planets have already formed)? How might flares affect habitability? Flares need to be understood in the context of larger magnetic processes, and dynamo processes. To study stellar flares in detail requires high time-resolution, e.g., responses within minutes, because most of the energy in the initial flare is released in the first few minutes in the so-called "impulsive" phase, when one expects to see hard X-rays and radio emission; later the flare transitions to the "gradual" phase when thermal X-rays and the UV/optical responses begin to dominate. Observations at different time-scales thus probe different physics. Statistics of stellar flares can usefully be derived on all times-scales: minutes, hours and days.

Another extension of stellar flare science involves space weather and the impacts upon the Earth. Studying the solar corona might in principle teach a great deal about flares on just one class of star but would teach little about its past and projected future behaviour; broadening the sample to many different types of star suggests how the Sun behaved in the past, and how it might behave in the future. The inverse of the argument is to regard observations of stars as proxies for modelling how the Sun's influence on space weather might evolve. Strength might then be given to potential new missions by opening them to other scientific communities (and their resources), though careful crafting of the science case would be imperative.

Clearly, the cadences required (minutes to years) depend crucially on the class of sources being explored. Different science goals are best accomplished with different technologies (see Fig. 1). Very fast transients represent territory that is largely unexplored, while at the other end of the scale all-sky X-ray monitoring programmes have mission lifetimes that are poorly matched to the long (rest-frame) variability time-scales of AGN. At present there is too much reliance upon serendipity; the 1999 flare of Sgr V4641, for instance, or the recently reported outburst in the Arches cluster could easily have been missed.

3.3. *Multi-wavelength Coordination*

At several junctures the workshop discussed practices for coordinating multi-wavelength observations and sending alerts to the community. For example, is there a need to develop an X-ray Transient Network, or are existing infrastructures like The Gamma-ray Circular Network and the Astronomer's Telegram (ATel) sufficient? (see p. 221). One existing problem is a degree of confusion in nomenclature. If different groups use different names to identify the same sources, it results in complications and leads to duplicate follow-up observations. This seems to be particularly true when the Galactic Centre is up and is being observed by INTEGRAL. A "transient wiki" could keep track of everything that is currently active. An increase in the number of joint proposals allowed (e.g., NASA+ESO or ESA+NOAO/Australian facilities) might also be important.

In general, ATels and other alert services seem to be serving the community well. Moreover, inside ATel there is now the AtelStream, which is a scheme for unifying announcements. There is also no doubt that the situation regarding joint proposals has improved tremendously over the past decade, but the need for a continued push for time-share agreements and joint proposal opportunities, especially for projects which require strictly simultaneous data, is strongly supported. In practice, it currently requires a significant commitment of time to coordinate a multi-wavelength campaign, possibly because of identifiable structural issues: very few observatories are set up for multi-wavelength collaborations. One successful example is the excellent inter-agreement between SAO/CHANDRA and NRAO/EVLA, within which it is quite straightforward to obtain simultaneous X-ray and radio observations. Two modes are involved: the "discovery mode" for the transients, requiring rapid slew and other time-critical follow-ups, and the "follow-up" mode when multiple instruments need to bear down on the same target. The latter mode requires either robotic streams or actual structural changes to the way in which time is granted and/or scheduled.

Multi-wavelength follow-up of X-ray targets can also suffer from a mismatch in timing resolution. For example, X-ray data are time-tagged, and events can be resolved easily at the millisecond level, but that information is of little help when trying to coordinate those data with an IR observation, where integrations run for minutes or longer, and the outcome is a comparison of two completely different time domains. One solution might be to use large-format optical/IR photon-counting detectors which automatically incorporate time tagging, discussed by O'Brien (p. 385) and Welsh (p. 99). Absolute timing stamps can also be incorporated. However, very high time-resolution detectors generate enormous quantities of data and huge files.

3.4. *Optimizing Existing and Future X-ray Missions for Time-Domain Science*

The workshop discussed the following past, present, and future X-ray missions in detail and how they might accomplish the science goals outlined above:

Planned/Proposed: NuSTAR, AstroSAT, ASTRO-H, eROSITA, GEMS, SVOM, Athena, LOFT, WFXT, JANUS, Lobster, Smart-X, also earlier footnote
Active (+Archival): CHANDRA, XMM-NEWTON, *Suzaku*, SWIFT, INTEGRAL, MAXI
Archival: RXTE, ROSAT, *Einstein*

The recording and transcript include descriptions of individual missions; see also White (p. 159). The instruments which are now current are also providing extensive archives that will be particularly useful for time-domain studies involving longer baselines.

The relative merits of an X-ray all-sky monitor, rapid slew missions and missions or instruments optimized for high time-resolution came in for considerable discussion. There had been broad support at a HEAD meeting 11 years ago for an X-ray all-sky monitor, but as the demand for sensitivity increased the payload grew, and soon it faced much stiffer competition as a stand-alone mission—and lost. The landscape may be different now owing to the rapid growth of, and huge investment in, ground-based programmes like the Palomar Transit Factory (PTF) and Pan-STARRS, and the Large Synoptic Survey Telescope (LSST) promised in the next decade. When the radio equivalent is also added, the demand for X-ray all-sky sensitivity at least an order of magnitude better than present values will surely increase. Among the missions that might fill that niche are Janus and Lobster-eye detectors, described in White (p. 159).

In discussing the lessons learned from RXTE, the workshop recognised that flexibility in responding to target of opportunity requests (ToO) is critical for X-ray timing studies, that the discipline needs a capability to observe how the timing properties themselves change in time (they are sharper probes than changes in the energy spectra), and that some of the work done by RXTE—in particular the all-sky monitoring—can be done in the optical and infrared from the ground because most X-ray binaries (except the highly extincted ones) show enhanced optical and infrared emission during outburst. SWIFT has been fantastic in its rapid response to ToOs and can take over nicely from RXTE in certain régimes, but lacks the effective area for RXTE's timing work, specifically for the study of pulsations and QPOs. It is possible that the Indian mission AstroSAT (the launch is planned in 2012) will recover many more of RXTE's capabilities, and may even improve on them through its increased sensitivity. The AstroSAT data will be proprietary, but the possiblity for real-time release of transients remains open. The data archive will be housed at the Inter-University Centre for Astronomy and Astrophysics (IUCAA), but the plans for access to the data are unclear.

Data access and availability of funding, particularly for serendipitous and archival studies, influence which science and which missions gain traction in the astronomical community. One drawback of mission designs like that of SWIFT is its lack of funding for scientists pursuing ToOs. A similar problem affects the many X-ray mission archives such as ROSAT, CHANDRA, XMM-NEWTON, etc., that could be used for transient and variability science, as well as the utilization of multi-wavelength archives like GALEX and SDSS. Unfortunately, since most archival research is funded through soft money, competition for that funding influences the type of science that gets done since proposals need to be tailored to the preferences of the funding agencies. One possible funding programme is NASA's *Research Opportunities in Space and Earth Sciences* programme, which funds research connected with NASA missions, including FERMI, CHANDRAand GALEX. Indeed, radio astronomers studying compact objects could access NASA monies to do the radio follow up. In the EVLA's model for data sharing, the so-called RSRO time (Resident Shared Risk Observing), an observing team could be awarded pre-commissioning time in return for at least one team expert taking "in residence" status at the facility, but such a programme is unlikely to be workable within a space-based context. However, pipeline and software development was proposed as one area where data exchange might be feasible.

Future advances are likely to require yet higher time resolution and higher energies. The proposed Large Observatory For X-ray Timing (LOFT; possible launch ∼2020) is a high-sensitivity time-domain mission that could be the sort of instrument required; one possible science driver could be the spectral timing of black holes and AGN. An alternative might be a vast improvement in "gamma-ray burst" type capabilities, such as an instrument with the solid angle of BAT but 10 times more sensitivity, better source localization, and with an IR telescope; it would open up a huge phase space which has never been probed before. Such a mission would specifically support a range of science projects, from SN shock break-outs and tidal disruptions to moderate redshift, gamma-ray bursts at $z > 9$ and searches for the periods of ultra-luminous X-ray sources. Meanwhile, the Wide Field X-ray Telescope, a proposed medium-class NASA mission, could be a powerful instrument for transient detections in the distant universe and could consider targeting the LSST Deep-Drilling fields repeatedly during its lifetime.

3.5. *Opportunities for Citizen Science*

As a final topic for discourse, the workshop explored possibilities to involve citizen science in X-ray transient studies. So far there has been little involvement by non-specialists in X-ray or high-energy programmes. However, citizen science is rapidly becoming recognised as a way of getting interesting science done and—more importantly—of engaging the public and also achieving certain tasks that need to be carried out in order to justify the investment in support of science.

It was felt that the amateur community who normally worked in the optical domain would be enthusiastic about following up X-ray transients. Help could be enlisted through a message to the American Association of Variable Star Observers (AAVSO) or similar organization, seeking observers willing to follow a 14, 15 or 16th magnitude object. Even though it is unlikely that the faintest targets could thus be followed, the benefit to X-ray science would be the adaptation of abundant capabilities across multiple wavelengths. Indeed, many of the AAVSO data are of exceptionally good quality, and amateur observers collectively have the advantage of wide longitudinal coverage, something which is not possible for many professional astronomers.

It should be recognized, however, that searches which make extensive use of existing data can be computationally intensive—for instance, if one tried to find every possible transient in the INTEGRAL or BAT archives, or looked for transients in the CHANDRA and XMM-NEWTON deep fields on every possible timescale.

Such archival searches are often very RAM intensive and may not be adaptable to software that runs on unused cycles in the same way that (say) SETI@home or Einstein@home can be run. Another suggestion was to coordinate amateurs to monitor dense regions of the sky in some systematic way in order to observe new X-ray binaries in outburst. Many of those systems rise quickly to $\sim 16^{\text{th}}$ magnitude, and nowadays that is within reach for a large number of amateurs.

4. Summary

The X-ray time domain uniquely probes strong gravity, accretion physics, supernova shock break-out and stellar flares. Specific tests of the first two involve the inner accretion disks of X-ray binaries and AGN. Changes in the X-ray variability and in its time-scale probe the structure of accreting degenerate systems, and increased sensitivity to rapid variations enables studies of XRB pulsations, QPOs, and fractional variability. The spectral timing of black holes and AGN may also reveal the structure, and deep low states in AGN may give a particularly clean glimpse into their spectral complexity. Shock break-out has been an exciting topic that has featured throughout the Symposium, and the race is on for the first observations of an X-ray shock break-out. X-ray emission from stellar flares probe coronal material and its dynamics. Most of the energy from the initial flare appears in the X-ray and radio, usually within a few short minutes. Stellar and solar flares are critical to our understanding of space weather, and may have a profound impact on the habitability of planets.

Acknowledgements

We thank the 30 or so scientists who attended this workshop and contributed to our discussion, and in particular we thank Phil Charles, Stephane Corbel, Boris Gaensicke, Stefanie Komassa, Shri Kulkarni, Ashish Mahabal, Roberto Mignani, Rachel Osten, Danny Steeghs, Tom Vestrand, Barry Welsh, Peter Williams, and Patrick Woudt for generating lively dialogue. We also appreciate input from Niel Brandt, Craig Heinke, Tom Maccarone, and Richard Mushotzky, who commented on our guiding questions though they were not able to attend.

New Horizons of Time-Domain Astronomy
Proceedings IAU Symposium No. 285, 2011
R.E.M. Griffin, R.J. Hanisch & R. Seaman, eds.

© International Astronomical Union 2012
doi:10.1017/S1743921312000610

Gravitational Microlensing

Ł. Wyrzykowski[1,2], M. Moniez[3], K. Horne[4] and R. Street[5]

[1]Institute of Astronomy The Observatories, Cambridge, CB3 0HA, UK

[2]Warsaw University Astronomical Observatory, 00-478 Warszawa, Poland
email: wyrzykow@ast.cam.ac.uk, lw@astrouw.edu.pl

[3]Laboratoire de l'Accélérateur Linéaire, IN2P3-CNRS, Université de Paris-Sud,
91898 Orsay Cedex, France

[4]SUPA Physics & Astronomy, University of St Andrews, KY16 9NS, Scotland

[5]LCOGT, 6740 Cortona Drive, Suite 102, Goleta, CA 93117, USA

Abstract. Gravitational microlensing is a well established and unique field of time-domain astrophysics. For two decades microlensing surveys have been regularly observing millions of stars to detect elusive events that follow a characteristic Paczyński lightcurve. This workshop reviewed the current state of the field, and covered the major topics related to microlensing: searches for extrasolar planets, and studies of dark matter. There were also discussions of issues relating to the organisation of follow-up observations for microlensing, as well as serendipitous scientific outcomes resulting from extensive microlensing data.

Keywords. gravitational microlensing—planets—dark matter—variable stars

1. Extrasolar Planets and the Future of Microlensing (KH)

Microlensing happens when a compact object passes near the line of sight to a distant star, causing bending of the light rays and thus magnifying and forming multiple images of the distant source. On the scale of the Galaxy it is observed as a temporary brightening of a star, the light curve following a unique shape known as the "Paczyński lightcurve". If a lensing star harbours a planet, the curve may exhibit a very short blip, a flash or dip, whose duration is of the order of hours to days, depending on the mass ratio of the planet and the star. A suitably high observing cadence is necessary to detect planets in that way.

Thus far, combining the efforts of two wide-angle surveys (OGLE, MOA) and a number of follow-up networks (PLANET, μFUN, RoboNet, MiNDSTeP), some 13 planets have been found, mostly in large (0.5–5 AU) orbits, and with masses ranging from a few Earths to a few Jupiters. Selected highlights among these cool planet detections are:

OGLE-2005-BLG-071 : a cool Super-Jupiter of about 3 M_J at 2 or 4 AU, found as a 3-day anomaly at high magnification (Udalski *et al.* 2005),

OGLE-2005-BLG-390 : a small rock/ice planet of about 6 $M_\oplus$ at 2.9 AU, found as a 12-hour anomaly at low magnification (Beaulieu *et al.* 2006),

OGLE-2006-BLG-109 : a half-scale Solar System analogue hosting two planets with mass ratios and orbit sizes similar to Jupiter and Saturn, found from a long complex anomaly (Gaudi *et al.* 2008),

MOA-2009-BLG-266 : a cool Neptune (10 Earth masses), found as a 2-day anomaly at intermediate magnification (Muraki *et al.* 2011).

Microlensing is currently the only available technique that can detect small cool planets beyond the snow line (see Fig. 1), a population complementary to the small hot planets found by the KEPLER mission. Moreover, with upgraded surveys (MOA-2, OGLE-IV)

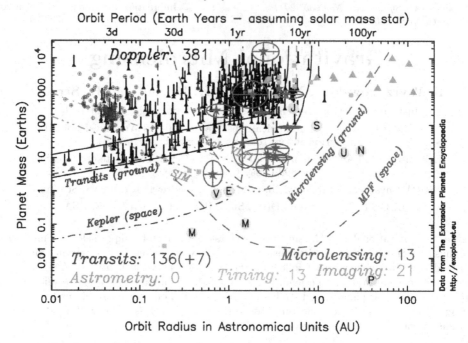

Figure 1. Exoplanets discovery space for different techniques. Different points show planets detected so far: small triangles with error bars for Doppler, circles with crosses for transit method, triangles for imaging, squares for timing and stars with error bars for microlensing. Letters in circles denote planets from the Solar System. A colour version of this figure is available on the Symposium's Website.

and consequent improvement in time-sampling of the light curves, microlensing is also sensitive to free-floating planets. The duration of an event towards the Galactic Centre provides an approximate lens mass, $M \sim M_J \, (t_E/\mathrm{day})^2$. The MOA survey finds an excess of short-timescale ($\sim$1d) events relative to predictions for mass functions for stars and brown dwarfs, interpreted as a population of unbound or distant planets, suggesting roughly 2 Jupiter masses in unbound planets per solar mass of stars (Sumi *et al.* 2011).

The Future of Microlensing. The field of microlensing has been developing for over a decade. To detect cool planets, the current strategy relies on wide-field survey telescopes to detect the events, and on follow-up networks to find and characterise anomalies. With recent MOA and OGLE upgrades, plus a new wide-field camera on the Wise Observatory 1-m telescope in Israel, the wide-field surveys may soon be delivering high-cadence quasi-continuous coverage of many events in parallel. The planned 1.6-m Korean Microlensing Telescope Network will have similar capabilities. Also promising is the deployment of 1-m robotic telescope networks with Lucky Imaging cameras (LCOGT, SONG) aiming to deliver sub-arcsecond imaging routinely in the crowded Bulge fields. These systematic ground-based surveys should deliver cool planet population statistics down to or below $M_\oplus$. In space, the European GAIA mission will soon be observing microlensing events, and the Dark Energy missions EUCLID and WFIRST may also undertake microlensing surveys capable of finding hundreds of cool Earth-mass planets.

2. Robotic Follow-Up (RS)

Responding to microlensing alerts is the most demanding of observing campaigns. The events never repeat, but last ≤few months, distributed throughout a ~6 month season and must be monitored continuously for many days around the peak. Objects in the large (~1500) target list can change in importance in a matter of minutes, so new data must be reduced and re-analysed immediately in order that priorities can be reassessed. Robotic telescope facilities, distributed in longitude, offer a highly efficient way to monitor ongoing events and to respond very quickly to alerts without the overheads of a large team of manual observers. Moreover, a completely deterministic prioritisation process is desirable so that the yield of planets detected and characterised in this way can be more easily applied to understand the underlying population of planets beyond the snowline. Of course, the challenges of this approach lie in the availability of robotic telescopes over several sites, and in developing reliable software capable of executing such a dynamic observing campaign.

Las Cumbres Observatory Global Telescope Network (LCOGT). LCOGT is building a global network of robotic telescopes at six sites in both hemispheres. Each site will host a cluster of telescopes, typically 2 1-m ones plus 3 0.4-m ones. To date, 2-m telescopes are already in operation: at Faulkes Telescopes North (FTN; Hawai'i) and South (FTS; Siding Spring, Australia). The complete network will have about 12 1-m telescopes and some 2 dozen 0.4-m ones.

All telescopes of a given class will have a homogenous complement of instrumentation, including imagers (in operation), Lucky Imaging Cameras (being commissioned at the present time) and spectrographs (under development—see p. 408). Rather than allocate a given observation to a specific telescope, programmes will be coordinated across the network, allowing telescopes at other sites to take over in the event of poor weather at one location. Unlike the SONG and MiNDStEP networks, LCOGT is not dedicated to microlensing. In partnership with SUPA/St. Andrews, LCOGT's southern ring is being deployed first, with site construction underway at the Cerro Tololo Interamerican Observatory, the South African Astronomical Observatory and McDonald Observatory, while the Siding Spring site is being expanded. The aim is to have 1 1-m telescope and 2 0.4-m ones operational at CTIO and SAAO by mid-2012; the rest of the network is to be deployed between 2012–2014.

Robotic Control Systems. The RoboNet Project exploits the LCOGT network, plus the Liverpool Telescope on La Palma, to obtain photometric follow-up of high-priority microlensing events in a completely robotic way. The first step is to prioritise algorithmically the list of events active at any given time according to their sensitivity to planets. The system implements this algorithm, WebPLOP (Horne *et al.* 2009), and incorporates a publically-accessible portal at `robonet.lcogt.net`. It subscribes to the ARTEMiS (Dominik, *et al.* 2010) system for anomaly alerts, but can also receive overriding orders from operators. Web-PLOP is queried robotically about every 30 min by the Observation Control software, which receives a list of current targets, and requests observations automatically from the telescope network. It also handles the returning data, preparing them for reduction. ObsControl provides a Web-based interface too, so that team members around the world can request additional observations, and issue target-of-opportunity overrides as needed. In the final stage the data are received by the data reduction pipeline, which uses the DanDIA (Bramich, 2008) software package to perform difference image analyses. The pipeline is fully automated, including target identification, making the photometry available publicly via the Website. The software also has an on-line portal to allow team members to interact with the data reductions.

Results and Future Developments. The robotic system has successfully obtained good coverage of almost all the lensing planets discovered in recent years, including MOA-2010-BLG-0266 (Muraki *et al.* 2011), MOA-2009-BLG-0319 (Miyake *et al.* 2011) and MOA-2009-BL-0387 (Batista *et al.* 2011). Additional planetary lensing events are still being analysed.

When increased telescope resources become available in the near future, it is anticipated that more complete coverage will be provided in future seasons. The Lucky Imaging cameras that are currently being commissioned at FTN and FTS will provide high spatial resolution imaging that will resolve blends in the dense star-fields of the Galactic Bulge. The plan is to install them on all the telescopes in the 1-m network.

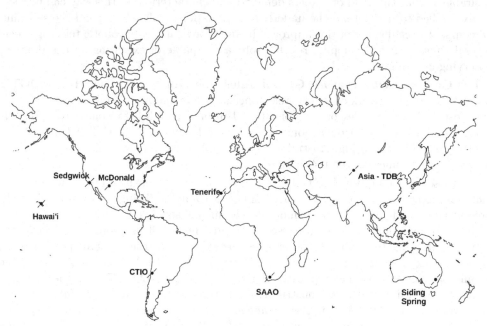

Figure 2. Map of LCOGT sites, including existing sites in Hawai'i and Australia, sites under construction, and possible future sites under consideration.

3. Dark Matter: Probing Galactic Structure with Microlensing (MM)

Microlensing is a powerful tool for probing the Milky-Way structure. Searches for microlensing towards the Magellanic Clouds (LMC, SMC) were originally intended to measure the optical depths through the Galactic halo in order to quantify dark matter in the form of massive compact objects. The searches towards the Galactic plane (Galactic centre and spiral arms) also enable measurements of the optical depth that is due to ordinary stars in the Galactic disk and bulge.

The microlensing optical depth up to a given source distance D_S is defined as the instantaneous probability for the line of sight of a target source to intercept a deflector's Einstein disk, that corresponds to a magnification $A > 1.34$. This probability is found to be independent of the deflectors' mass function:

$$\tau(D_S) = \frac{4\pi G D_S^2}{c^2} \int_0^1 x(1-x)\rho(x)\mathrm{d}x\,, \tag{3.1}$$

where $\rho(x)$ is the mass density of deflectors located at a distance xD_S. The *mean*

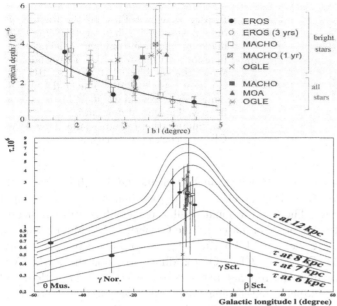

Figure 3. *Upper:* The optical depth through the Galactic plane as a function of the absolute Galactic latitude $|b|$. The line shows the fit (3.3). *Lower:* The optical depth at $< b >= -2.5°$ as a function of Galactic longitude; MACHO: open circles, EROS: filled circles, OGLE: crosses. The lines show the predicted optical depths as a function of latitude at 6, 7 (thick line), 8, 9, 10, 11 and 12 kpc [Rahal 2009], The measured optical depths around $l = 0°$ are compatible with the expected value at 8.5 kpc.

optical depth towards a given population of sources defined by the distance distribution $dn_S(D_S)/dD_S$ is defined as:

$$< \tau >= \int_0^\infty \frac{dn_S(D_S)}{dD_S} \tau(D_S) D_S^2 \, dD_S \ / \int_0^\infty \frac{dn_S(D_S)}{dD_S} D_S^2 \, dD_S. \qquad (3.2)$$

The estimate of $< \tau >$ is obtained from the distribution of the characteristic times t_E of the detected events, and needs the knowledge of the detection efficiency as a function of t_E, the most delicate aspect of this measurement. The rest of this review focusses on the results from the microlensing surveys towards targets with resolved stars, EROS, MACHO, OGLE and MOA.

Results towards the Galactic plane. More than 4000 events have been detected towards the Galactic bulge (but only a fraction have been used to estimate optical depths under controlled efficiency), and 27 towards the Galactic Spiral Arms (22 used for optical depth determination). The optical depth measurements are summarized in Fig. 3. The variation with the latitude deduced from the largest sample (EROS, Hamadache *et al.* 2006) is fitted well by:

$$\tau/10^{-6} = (1.62 \pm 0.23) \exp[-a(|b| - 3°)], \text{ with } a = (0.43 \pm 0.16) \text{ deg}^{-1}. \qquad (3.3)$$

This fit agrees with the results of MACHO (Popowski *et al.* 2005) and OGLE (Sumi *et al.* 2006) and with the Galactic models of Evans & Belokurov (2002) and Bissantz *et al.* (1997). There is also a satisfactory agreement between the measured optical depths towards the Galactic spiral arms and the model expectations, and there is no indication for a population of hidden compact objects in the disk.

Interpreting the Magellanic cloud surveys: halo versus local structures. The main result from the LMC/SMC surveys is that compact objects of mass within a $[10^{-7}, 10] \times m_\odot$ interval are not a major component of the hidden Galactic mass (Fig. 4).

The considerable differences between the EROS, MACHO and OGLE data sets may explain the apparent MACHO vs EROS/OGLE discrepancies: MACHO used fainter stars

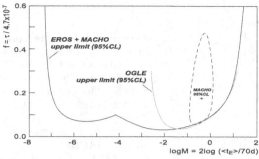

Figure 4. Constraints on the fraction f of the standard spherical Galactic halo made of massive compact objects as a function of their mass M. The solid line shows the combined EROS+MACHO upper limit [Moniez 2010]. The OGLE upper limit [Wyrzykowski *et al.* 2011] is shown in grey. The closed domain is the 95% CL contour for the f value claimed by MACHO (Alcock *et al.* 2000).

in denser fields (1.2×10^7 stars over $14 \deg^2$) than EROS and OGLE (0.7×10^7 stars over $90 \deg^2$). The faint source stars also made the efficiency estimates more complicated owing to larger blending effects. The observational results at their face value indicates that the hypothesis of an optical depth dominated by the Galactic halo—almost uniform through all the monitored LMC fields and 1.4 times larger towards SMC—is wrong, because it cannot explain the EROS–MACHO–OGLE differences nor the LMC–SMC differences. That said, the apparent discrepancy between the surveys can be understood by considering local structures inducing self-lensing; such structures (populations of foreground lenses as well as of background sources (Alcock *et al.* 2000) may be responsible for the variability of the optical depth with the monitored zones.

Conclusion and perspectives.

The hypothesis that compact objects make a substantial contribution to a standard halo is now clearly excluded. Observations towards both the SMC and the spiral arms find that compact objects belonging to a flattened halo or a thick disk are also scarce. Whether there is a small dark-matter component in the form of compact objects is still an open question that will be addressed by the infrared VVV project (using VISTA facility) and the LSST project, possibly with contributions from space missions like GAIA.

4. Serendipitous Science with Microlensing Data (ŁW)

This section contains a brief, subjective review of recent highlights of serendipitous science from the vast time-domain data collected by OGLE, a long-term microlensing survey which was initiated in 1992 by Bohdan Paczyński (Princeton) and Andrzej Udalski (Warsaw). Since 1996 OGLE has used a dedicated 1.3-m telescope at Las Campanas Observatory, Chile (Udalski *et al.* 2008). In 2010 March it entered into its fourth phase, when the detector was upgraded to a 32-chip mosaic CCD camera covering 1.4 $\deg^2$.

Photometric maps. Since microlensing events are predominantly being found in dense sky areas, OGLE regularly observes hundreds of square degrees towards the Galactic Bulge, Magellanic Clouds and Galactic Disk. Superb quality observations in I and V give rise to detailed colour-magnitude diagrams for the observed fields (Udalski *et al.* 2008; Szymański *et al.* 2010; Szymański *et al.* 2011). All the photometric data are made available through OGLE's Webpage†. They are useful in tasks such as mapping interstellar extinction, studying stellar populations, or studying the Bulge and Magellanic Cloud structures using stars in the Red Clump and the Tip of the Red Giant Branch.

† http://ogle.astrouw.edu.pl

Variable stars. A natural by-product of long-term photometric monitoring is numerous discoveries of variable stars. The catalogue based only on OGLE-III data from 2001–2009 already consists of 13 parts, each devoted to a different type of variable. The number of catalogued variables exceeds all known catalogues to date, e.g., 24,906 RR Lyræ stars found in the LMC alone (Soszyński *et al.* 2009) and 16,836 in the Bulge (Soszyński *et al.* 2011). Vast numbers of variable stars enable detailed statistical studies of those populations, but also reveal rare examples of peculiar objects, including RR Lyraes or Cepheids in eclipsing systems (Pietrzynski *et al.* 2010), or R CrB-type stars (Soszyński *et al.* 2009).

Pulsating variables and period-luminosity relation. The long time-base-line of observations enables the detection of periodic variable stars with periods ranging from hours to years. The well-known period-luminosity relation for classical Cepheids was extended by OGLE variables in all directions, with δ Scuti and RR Lyræ stars at the short-period end and OSARGS, Miras and LSPs at the long-period end (see Soszyński *et al.* 2007; Poleski *et al.* 2010). The large statistical sample of all pulsators provided by OGLE is supporting detailed theoretical studies of those stars, and is also being used for measurements of distance.

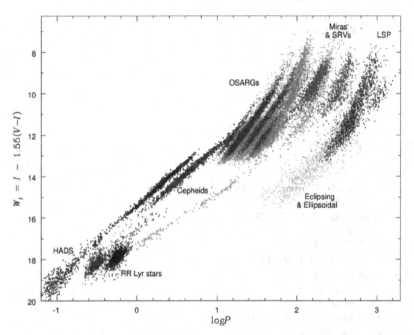

Figure 5. OGLE's variable stars period-luminosity relation. A colour version of this figure is available in the on-line edition.

Transiting planets. OGLE was among the first to conduct a massive search for transiting planets. Thanks to the milli-mag precision of its photometry, OGLE found a few dozen candidates (e.g. Udalski *et al.* 2004), among which about a dozen were confirmed by radial-velocity observations.

Miscellaneous. Owing to the long time-span between observations of the same regions of the sky, it is possible to measure the proper motions of the fastest stars (Soszyński *et al.* 2002), of asteroids and of Kuiper Belt Objects. Continuing observations are also revealing numerous supernovæ behind the Magellanic Clouds (Udalski 2003), as well as classical and dwarf novæ. Among other highlights is the recent capture of a nova explosion following an apparent merger of the components of a contact binary (Tylenda *et al.* 2011).

Acknowledgements

KH and RS acknowledge Qatar Foundation support through QNRF Grant NPRP-09-476-1-78. LW and the OGLE project acknowledge support of European Community's FP7 ERC Grant no. 246678.

References

Alcock, C., *et al.*, 2000, *ApJ*, 542, 281
Alcock, C., *et al.*, 2000, *ApJ*, 552, 582
Batista, V., *et al.* 2011, *ApJ*, 529, A102.
Beaulieu, J.-P., *et al.* 2006, *Nature*, 439, 437
Belokurov, V., Evans, N. W., & Le Du, Y. 2003, *MNRAS*, 341, 1373
Bissantz, N., Englmaier, P., Binney, J., & Gerhard, 0 .E., 1997, *MNRAS*, 289, 651
Bramich, D. 2008, *MNRAS*, 386, 77.
Dominik, M., *et al.* 2010, *AN*, 331, 671.
Evans, N. W. & Belokurov, V., 2002, *ApJ*, 567, L119
Gaudi, B. S., *et al.* 2008, *Science*, 319, 927
Hamadache, C., *et al.*, 2006, *A&A*, 454, 185
Horne, K., *et al.* 2009, *MNRAS*, 396, 2087.
Miyake, N., *et al.* 2011, *ApJ*, 728, 120.
Moniez, M., 2010, *GRG*, 42, 2047
Muraki, Y., *et al.* 2011, *ApJ*, 741, 22
Pietrzynski, G. *et al.* 2010, *Nature*, 468, 542
Poleski, R,. *et al.* 2010, *AcA*, 60, 1
Popowski, P., *et al.*, 2005, *ApJ*, 631, 879
Rahal, Y. R., 2009, *A&A*, 500, 1027
Sumi, T., *et al.* (OGLE collaboration), 2006, *ApJ*, 636, 240
Sumi, T., *et al.* 2011, *Nature*, 473, 349
Soszyński, I., *et al.* 2002, *AcA*, 52, 143
Soszyński, I., *et al.* 2007, *AcA*, 57, 201
Soszyński, I., *et al.* 2009, *AcA*, 59, 1
Soszyński, I., *et al.* 2009, *AcA*, 59, 335
Soszyński, I., *et al.* 2011, *AcA*, 61, 1
Szymański, M. K., *et al.* 2010, *AcA*, 60, 295
Szymański, M. K., *et al.* 2011, *AcA*, 61, 83
Tylenda, R., *et al.* 2011, *A&A*, 528, A114
Udalski, A. 2003, *AcA*, 53, 291
Udalski, A. *et al.* 2004, *AcA*, 54, 313
Udalski, A., *et al.* 2005, *ApJ*, 628, L109
Udalski, A. *et al.* 2008, *AcA*, 58, 69
Udalski, A. *et al.* 2008, *AcA*, 58, 329
Wyrzykowski, L. *et al.*, 2011, *MNRAS*, 416, 2949

New Horizons in Time-Domain Astronomy
Proceedings IAU Symposium No. 285, 2011
R.E.M. Griffin, R.J. Hanisch, & R. Seaman, eds.

© International Astronomical Union 2012
doi:10.1017/S1743921312000622

Light Echoes

Organized by Howard E. Bond[1], Misty C. Bentz[2], Geoffrey C. Clayton[3], and Armin Rest[1]

[1]Space Telescope Science Institute, 3700 San Martin Dr., Baltimore, MD 21218 USA
email: bond@stsci.edu

[2]Dept. of Physics & Astronomy, Georgia State University, Atlanta, GA 30303 USA

[3]Dept. of Physics & Astronomy, Louisiana State University, Baton Rouge, LA 70803 USA

Abstract. The first "light echo"—scattered light from a stellar outburst arriving at the Earth months or years after the direct light from the event—was detected more than 100 years ago, around Nova Persei 1901. Renewed interest in light echoes has come from the spectacular echo around V838 Monocerotis, and from discoveries of light echoes from historical and prehistorical supernovæ in the Milky Way and Large Magellanic Cloud as well as from the 19th-century Great Eruption of η Carinae. A related technique is reverberation mapping of active galactic nuclei. This report of a workshop on Light Echoes gives an introduction to light echoes, and summarizes presentations on discoveries of light echoes from historical and prehistorical events, light and shadow echoes around R CrB stars, and reverberation mapping.

Keywords. scattering, methods: data analysis, techniques: polarimetric, novae, cataclysmic variables, reflection nebulae, galaxies: active, galaxies: nuclei

1. Introduction to the Workshop

The phenomenon of a light echo occurs when light from a transient event, such as a supernova or other eruptive stellar outburst, scatters off nearby dust and reaches the Earth at later times than the light from the event itself. The first light echo was seen over a century ago, following the outburst of Nova Persei 1901. The apparent superluminal motion of these echoes excited considerable interest among astronomers at the time, but the subject then sank into relative obscurity. However, light echoes have recently received renewed interest because of the spectacular echo around V838 Monocerotis (which produced iconic *Hubble Space Telescope* images), and the interstellar and circumstellar echoes around SN 1987A.

Recently, there have been remarkable discoveries of light echoes from ancient supernovæ and other luminous transients in the Magellanic Clouds and Milky Way, allowing spectroscopic observations with modern equipment of events that occurred centuries ago. In extragalactic astronomy, "reverberation mapping" is a closely related technique used to study the structure of active galactic nuclei.

A 90-minute Workshop on the topic of "light echoes" was held during the Symposium. There were four semi-formal presentations, as described below, along with question-and-answer sessions and general discussion.

2. Light Echoes (H.E. Bond)

There are three requirements to produce an observable light echo:
(1) there must be an illuminating star that varies rapidly in brightness;
(2) the star must be intrinsically luminous (typically a nova or supernova); and
(3) there must be circumstellar or interstellar dust in the vicinity of the variable star.

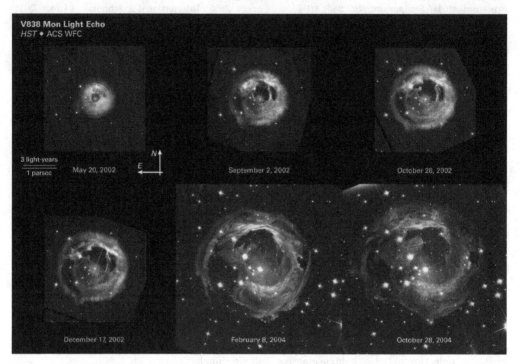

Figure 1. *Hubble Space Telescope* images showing the evolution and apparent superluminal expansion of the light echoes around V838 Mon. The images were obtained with the Advanced Camera for Surveys, and are all at the same angular scale.

It is extremely rare for all three requirements to be satisfied simultaneously, and thus light echoes have historically been detected only rarely.

The geometry of a light echo is simple (Bond *et al.* 2003, and references therein): at a time t after the outburst, the illuminated dust lies on the paraboloid given by $z = x^2/2ct - ct/2$, where x is the projected distance from the star in the plane of the sky, z is the distance from this plane along the line of sight toward the Earth, and c is the speed of light. This geometry explains why it is typically possible for an echo to appear to expand at many times the speed of light, when there is illuminated material well in front of the star.

The eruption of the previously unknown variable star V838 Monocerotis in 2002 produced the most spectacular light echo of modern times (see Fig. 1 for examples of *HST* images). The star itself belongs to a newly recognized class of "intermediate-luminosity red transients" or "luminous red novæ," whose maximum luminosities lie between those of classical novæ and supernovæ (see Bond *et al.* 2009). They generally evolve to extremely cool temperatures as their outbursts proceed. V838 Mon itself evolved from a K spectrum, through type M, and eventually was classified as the only known L-type supergiant.

Because of the superluminal expansion, direct images of light echoes are incapable of providing a geometric distance to the illuminating star. However, it is possible to use polarimetric imaging to identify highly linearly polarized locations within the echo, at which the scattering angle is close to $90°$; in this case, the above equation, with z set to zero, shows that the x distance is simply ct, and thus the angular radius yields a geometric distance. This novel method of astronomical distance determination was proposed by Sparks (1994), and applied with success to the V838 Mon echo by Sparks *et al.* (2008), based on polarimetric imaging with the *HST* Advanced Camera.

In the case of a periodic luminous variable star surrounded by a reflection nebula, there will be a train of nested light echoes propagating away from the star. The luminous Galactic Cepheid variable RS Puppis (with a 41.4-day period) provides a spectacular example of these nested echoes (Kervella *et al.* 2008). We are currently analyzing *HST* polarimetric imaging of the RS Pup nebula in order to determine a geometric distance to this important Cepheid (see Bond & Sparks 2009 for an outline of the method).

3. Light Echoes from Historical and Prehistorical Events (A. Rest)

In the last few years, the technique of difference-imaging of fields adjacent to known luminous transients has been exploited to make a number of discoveries of light echoes.

Among the most spectacular results have been the detection of a light echo from the 19th-century Great Eruption of η Carinae, allowing us to obtain the first spectroscopic observations of the star during this cataclysmic event (Rest *et al.* 2012). We have also discovered echoes from, and obtained spectra of, such historical events as the Cas A and Tycho supernovæ (Rest *et al.* 2008a, 2011). Moreover, we have detected light echoes from previously unknown supernovæ in the Large Magellanic Cloud, most of which can be associated with known young SN remnants in the LMC (Rest *et al.* 2008b).

One exciting application of the use of light echoes, permitted by the spectroscopic observations of supernovæ, is the possibility of seeing the event from several different viewing angles. As an example, this has allowed the discovery of anisotropic ejection in the case of the Cas A supernova (Rest *et al.* 2011).

4. Light and Shadow Echoes around R CrB stars (G. Clayton)

R Coronae Borealis (RCB) stars are luminous, hydrogen-deficient, carbon-rich supergiants (Clayton 1996; also p. 125). Two leading evolutionary models for producing these objects are a merger of a double white dwarf, and a final helium shell flash in a (single) star at the top of the white-dwarf cooling track, following ejection of a planetary nebula.

We obtained *HST*/ACS images of the reflection nebula surrounding the RCB star UW Cen in 2004 (see Clayton 2005). This nebula, first detected in 1990, is the only known reflection nebula around an RCB star. It has changed its appearance significantly since discovery.

At the estimated distance of UW Cen, the reflection nebula is approximately 0.6 light years in radius, so the nebula cannot have physically altered in such a short time. Instead, the morphology of the nebula appears to change as different parts are illuminated by light from the central star, modulated by shifting thick dust clouds near its surface. These dust clouds form and dissipate at irregular intervals, causing the well-known declines in the RCB stars. In that way, the central star acts like a lighthouse shining through holes in the dust clouds and illuminating different portions of the nebula.

When new dust forms, a "shadow echo" can be seen to move out from the star through the nebula. The existence of this nebula provides clues to the evolutionary history of RCB stars, specifically providing a possible link between them and planetary nebulæ; this would support the final helium shell flash model.

5. Reverberation Mapping of Active Galactic Nuclei (M.C. Bentz)

Active galactic nuclei (AGNs) are some of the most energetic objects in the universe, but are difficult to study in detail because of their large distances. The compact inner region of photoionized gas in an AGN only spans micro-arcseconds in angular size, and is spatially unresolvable for the forseeable future. Reverberation mapping (Blandford & McKee 1982; Peterson 1993) has therefore become the most useful tool for probing AGN structure because it can take advantage of their flux variability and, in effect, substitute time resolution for spatial resolution.

The basic technique involves monitoring an AGN spectroscopically over a period of several weeks to months or years. The continuum flux in an AGN is produced near the central black hole (BH) (probably in the surrounding accretion disk) and has been observed to vary on time-scales of hours to days. The gas in the photoionized broad-emission-line region (BLR) responds to the continuum variations with a time delay τ, the magnitude of which varies directly with the luminosity state of the AGN.

The average time delay between continuum fluctuations and the response of a broad emission line gives a measurement of the average size of the emission-line region. Combining τ with the velocity-broadened width of the emission line through the virial theorem provides a direct gravitational probe of the mass of the central supermassive BH. To date, some 50 BH masses have been determined in this way (Peterson *et al.* 2004; Bentz *et al.* 2009) and provide the basis for *all* BH mass determinations in AGNs at cosmological distances.

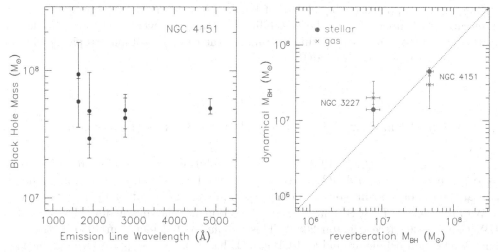

Figure 2. *Left:* Black-hole masses for NGC 4151 derived from reverberation mapping of several optical (Bentz *et al.* 2006) and UV (Metzroth *et al.* 2006) broad emission lines, all of which show a general consistency. *Right:* Reverberation masses for NGC 3227 and NGC 4151, compared to the masses derived from stellar dynamical modelling (circles) and gas-dynamical modelling (crosses) of spatially-resolved observations of the centres of these galaxies. The dotted line is the line of equality.

Several independent lines of argument provide evidence that reverberation measurements yield reliable measurements of the central BH mass. The BLR has been observed to exhibit ionization stratification, in which the more highly ionized emission lines (such as He II and C IV) respond to continuum variations with shorter time delays than the Balmer lines (Peterson 2001; Bentz *et al.* 2010a). The widths of these more highly ionized emission lines are broader and follow the expected $V \propto \tau^{1/2}$ relationship for gas undergoing gravitationally dominated motion, resulting in a constant BH mass determination from multiple emission lines (Peterson & Wandel 2000; Kollatschny 2003; Bentz *et al.* 2010a; see Fig. 2). Furthermore, although reverberation masses currently rely on a scaling factor that is derived by assuming that all galaxies follow the same $M_{BH} - \sigma_\star$ relationship (Onken *et al.* 2004; Woo *et al.* 2010), it has been possible to determine the BH masses of two AGNs independently through dynamical modelling.

The results of both gas-dynamical and stellar-dynamical modelling from spatially resolved observations of the centers of NGC 4151 and NGC 3227 give dynamical masses that are consistent with the masses derived from reverberation mapping (Davies *et al.* 2006; Onken *et al.* 2007; Hicks & Malkan 2008; see Fig. 2). A search is currently underway to identify additional candidates for such studies.

Reverberation mapping has the potential to map out directly the response of the gas in the BLR as a function of time delay and velocity, thereby providing direct observational constraints on the geometry and kinematics of the BLR gas. Recent spectroscopic monitoring campaigns have begun to achieve the data quality necessary to carry out a full decomposition of the line response (Bentz et al. 2010b), and a substantial amount of effort is currently being invested in carrying these analyses further and in obtaining additional high-quality monitoring datasets.

As our observational strategy and analysis methods continue to improve, reverberation mapping has to potential to provide, among other things, a solid foundation for the determination of BH masses in galaxies at cosmological distances. The observed tight correlations between host-galaxy properties and central BH masses imply that BHs strongly influence their host galaxies. By understanding the masses of those BHs, not only in nearby galaxies but also at cosmological distances, we can hope to unravel the mysteries of galaxy assembly and evolution across cosmic time.

6. Summing Up

Light echoes and reverberation maps are providing exciting new capabilities and discoveries, including the remarkable ability to obtain spectroscopic observations with modern instrumentation of events that occurred decades and even centuries ago. Full exploitation of these techniques requires careful scheduling of synoptic observations and the capability of quick reaction to targets of opportunity.

References

Bentz, M. C., *et al.* 2006, *ApJ*, 651, 775
Bentz, M. C., *et al.* 2009, *ApJ*, 705, 199
Bentz, M. C., *et al.* 2010a, *ApJ*, 716, 993
Bentz, M. C., *et al.* 2010b, *ApJ*, 720, L46
Blandford, R. D. & McKee, C. F. 1982, *ApJ*, 255, 419
Bond, H. E., *et al.* 2003, *Nature*, 422, 405
Bond, H. E., *et al.* 2009, *ApJ*, 695, L154
Bond, H. E. & Sparks, W. B. 2009, *A&A*, 495, 371
Clayton, G. C. 1996, *PASP*, 108, 225
Clayton, G. C. 2005, *Planetary Nebulae as Astronomical Tools*, 804, 180
Davies, R.I., *et al.* 2006, *ApJ*, 646, 754
Hicks, E. K. S. & Malkan, M. A. 2008, *ApJS*, 174, 31
Kervella, P., *et al.* 2008, *A&A*, 480, 167
Kollatschny, W. 2003, *A&A*, 407, 461
Metzroth, K. G., *et al.* 2006, *ApJ*, 647, 901

Onken, C. A., *et al.* 2004, *ApJ*, 615, 645
Onken, C. A., *et al.* 2007, *ApJ*, 670, 105
Peterson, B. M. & Wandel, A. 2000, *ApJ*, 613, 682
Peterson, B. M. 1993, *PASP*, 105, 247
Peterson, B.M., *et al.* 2004, *ApJ*, 613, 682
Rest, A., *et al.* 2008a, *ApJ*, 681, L81
Rest, A., *et al.* 2008b, *ApJ*, 680, 1137
Rest, A., *et al.* 2011, *ApJ*, 732, 3
Rest, A., *et al.* 2012, *Nature*, in press
Sparks, W. B. 1994, *ApJ*, 433, 29
Sparks, W. B., *et al.* 2008, *AJ*, 135, 605
Woo, J. H., *et al.* 2010, *ApJ*, 716, 269

New Horizons in Time-Domain Astronomy
Proceedings IAU Symposium No. 285, 2011
R.E.M. Griffin, R.J. Hanisch, & R. Seaman, eds.

Using the VO to Study the Time Domain

Rob Seaman[1], Roy Williams[2], Matthew Graham[2], and Tara Murphy[3]

[1]National Optical Astronomical Observatory, Tucson AZ USA
email: seaman@noao.edu

[2]California Institute of Technology, Pasadena CA USA

[3]University of Sydney, Australia

Abstract. Just as the astronomical "Time Domain" is a catch-phrase for a diverse group of different science objectives involving time-varying phenomena in all astrophysical régimes from the solar system to cosmological scales, so the "Virtual Observatory" is a complex set of community-wide activities from archives to astroinformatics. This workshop touched on some aspects of adapting and developing those semantic and network technologies in order to address transient and time-domain research challenges. It discussed the VOEvent format for representing alerts and reports on celestial transient events, the SkyAlert and ATELstream facilities for distributing these alerts, and the IVOA time-series protocol and time-series tools provided by the VAO. Those tools and infrastructure are available today to address the real-world needs of astronomers.

1. Introduction

Virtual Observatory activities relating to time-domain astronomy have focused on the VOEvent celestial transient alert protocol since 2005.† VOEvent is a message format for conveying reports of time-varying astronomical observations (Seaman *et al.* 2011). The notion of VOEvent is to permit the several different pre-existing astronomical telegram standards and transient alert protocols to be expressed in a common form that will permit interoperability, while enabling modern virtual technologies to be used to construct autonomous workflows. The goal is to close the observational loop from the discovery of transient phenomena, for instance by large synoptic surveys, to their follow-up by robotic or human-mediated instrumentation and telescopes.

VOEvent is necessary but not sufficient. A common message format requires an interoperable transport infrastructure layer. Three of these—SkyAlert, ATELstream and the VAO Transient Facility—are discussed below. Others such as the VOEvent-enabled Gamma-ray Bursts Coordinates Network (GCN) and the IAU CBAT service (Central Bureau of Astronomical Telegrams) have also been adapted to VOEvent compliance. VOEvent, like any living standard, must also evolve to meet the needs of the community, so the recent VOEvent v2.0 Recommendation (standard) of the International Virtual Observatory Alliance (IVOA, an activity of Commission 5 of the IAU) is also discussed below.

A key aspect of time-domain astronomy is the collection and interpretation of time-series data sets. This is related to VOEvent, but is also a key area for IVOA-compliant data archives. Recent work on time-series tools, performed by the U.S. Virtual Astronomical Observatory (VAO), was demonstrated during the workshop.

† IVOA VOEvent Working Group Wiki: http://voevent.org

2. Skyalert

Skyalert‡ is a clearing-house and repository of information about astronomical transients, each described by a collection of VOEvent packets that may have multiple authors. The components of Skyalert are:

- A Web-based event broker, allowing subscription so that information about transients can be delivered to users and their telescopes immediately upon receipt.
- A Web-based authoring system, so that authenticated users can inject events directly from automated discovery pipelines, or fill in Web forms, that may be delivered rapidly to others.
- An event repository, storing all events that come through the broker, and allowing bulk queries and drill-down.
- A *click or code* paradigm that allows people Web-based access and machines Web-service access.
- A way to see recent and past transients: as tables, multi-layered Web pages, or with popular astronomical software.
- A development platform for building real-time decision rules about transients, and for mining the repository.
- Open-source software to allow local implementations as well as the Web-based application.

The crucial standard that enables interoperable exchange of events is called VOEvent, now a Recommendation of the International Virtual Observatory Alliance. Reading that standard, as described above, is a good basis for understanding more of this document. Skyalert installation shows on the front page all the recent events (last 200) ingested into the system, as clickable dots in a semi-log timescale, with the present moment at the right, and older events further top the left. Clicking on any of the dots brings up the portfolio for that event. Also available from the front page is a collection of Atom feeds of recent events (both system feeds and your custom feeds).

2.1. *Event Portfolios*

Each transient will have a collection of data that we call a data portfolio: a collection of numbers, links, images, opinions, search results, etc. A portfolio is defined through a citation mechanism inherent in the VOEvent packet, where one event can cite another. Thus, an event with no citation becomes its own portfolio, but an event with a citation to another joins the portfolio to which the other belongs.

As noted above, the portfolio detail page can be accessed by clicking on the dot for a recent event. One can also select a specific event stream via *Browse Event Streams*, choose the required table of portfolios, and then narrow the search to a specific event. A third route to get to a specific event is by selecting a feed (one's own or a system feed), and then choosing a portfolio at that point. There are three representations that are available for each event of a portfolio:

- Overview: created by running the event data through the overview template.
- Params: a table of parameters and their values, plus representation of any Tables in the event.
- XML: showing the actual XML that was loaded into Skyalert

‡ http://skyalert.org

2.2. *Alerts*

An alert is a means of determining whether a portfolio is "interesting" in some way, and what to do if it is. From each "rule" (see below) is automatically generated a feed of interesting portfolios. It may be that the action which results from an interesting portfolio can cause another event to be loaded to the same portfolio, and that might in turn cause another rule to be satisfied, and another action to be taken.

Each alert has a collection of streams; for the rule to operate on a portfolio, it must have an event drawn from each of its needed streams. The simplest rules, however, need only events from one stream; for example, an event from a stream called *apple* might be interesting if it is bright; if there is a Param named *magnitude* then the trigger expression might be:

```
apple[magnitude] < 18
```

The trigger expression is interpreted by Python, within a sandbox environment that allows only math. functions. Thus, a trigger expression like `os.system(rm *)` will fail because the `os` module cannot be imported into the sandbox environment.

A rule runs on a collection of events (i.e. a portfolio). The simplest rule considers only one stream. Rules can be created only by a user who is registered and logged in. To build a rule, the user first clicks on *my feed and alerts*, then on *for a new alert*, and then selects an event stream (but should not click the advanced option). A name for the alert is entered, to start to make the trigger expression. The simplest expression is *True*, meaning that all of the events are interesting. The user must explicitly save the rule before the latter can do anything; the expression is checked for syntax before being saved, so that only syntactically correct trigger expressions get into the database. The alert-editor screen also has a button to show all the past events that would have satisfied the trigger. More complex rules can use multiple event streams to make a joint decision on the portfolio; suppose (for example) that an event from the *apple* survey has a follow-up from the `fruitObserver` catalogue, and that bright apple events are required which are also bright in the `fruitObserver` stream:

```
apple[magnitude] < 18 and fruitObserver[gMag] < 18
```

This a joint criterion on two different event streams, authored by different people. In one stream the author chose to use the Param called *magnitude*, and in the other stream the author chose to use the name *gMag*. Because of the underlying VOEvent model, Skyalert is able to integrate information from multiple authors.

2.3. *Layering facilities on VO-compliant protocols: VOEvent2*

The IVOA VOEvent standard was first defined, and a prototype created, in 2005, and reached official Recommendation status in 2006. It has been adopted successfully by many astronomical time-domain projects since then. It was recognized from the start that more advanced features would be needed in order to grapple with the challenging time-domain projects looming in the near future. The recently-adopted VOEvent2 standard embraces such features.

2.4. *VO compliant protocols: Distributed Transient Facility*

The time-domain community wants robust and reliable tools to enable the production of, and subscription to, community-endorsed event notification packets (VOEvents). The proposed Distributed Transient Facility (DTF) is being designed to be the premier broker-ing service for the community, not only collecting and disseminating observations about time-critical astronomical transients but also supporting annotations and the application of intelligent machine learning to those observations. Two types of activity associated with the facility can therefore be distinguished: core infrastructure, and user services. The prior art in both areas were reviewed by the workshop, and planned capabilities of the DTF were described. In particular, it focused on scalability and quality-of-service issues required by the next generation of sky surveys, such as LSST and SKA.

3. ATELstream

Bob Rutledge of the Astronomer's Telegram described to the workshop the new ATEL-stream† facility that provides a UNIX socket-based XML-driven messaging service for celestial transient event notices. Much interest was exhibited in ensuring that ATEL-stream and VOEvent remain interoperable and that the community meld together the two technologies for the benefit of all.

4. VOEvent2

This workshop was only the latest in a long line of workshops, meetings and sessions over the past half-dozen years concerned with using Virtual Observatory standards and protocols for studying the time domain. All have been tied to VOEvent, but also come more generally under the banner of *Hot-wiring the Transient Universe*‡. Copies of the same-named book (Williams, Emery Bunn & Seaman 2010) were made available for distribution.

The system architecture of the IVOA consists of numerous protocols, data models and services. The VOEvent Recommendation is one of the many diverse international standards of the IVOA. VOEvent refers to, and relies on, several of those, and other standards may in turn depend on VOEvent. VOEvent is also engaged in numerous external projects; several are currently distributing messages in VOEvent format, others are connected to major future surveys, and still others represent existing projects with a stake in enhancing their interoperability.

Planning is underway for a third-generation transport infrastructure. The diverse prototype technologies of the original VOEventNet were followed by the deployment of the operational SkyAlert system discussed above. Future celestial transient event transport infrastructure within the Virtual Observatory is anticipated to develop from the VAO transient facility project described below. In the mean time, numerous efforts will continue at working with other transient alert technologies and projects to ensure interoperability both across the community and for all types of time-varying celestial phenomena. Building such an infrastructure is an exercise in creative bootstrapping.

The IVOA VOEvent standard became a Recommendation of the IVOA in 2006. Evolution of the format was planned, and a major milestone was reached in 2011 with the acceptance of the VOEvent v2.0 update to the standard. Its major features include:

• The definition of VOEvent *Streams* in support of ongoing work to enhance the registration of VOEvent resources in the Virtual Observatory.

• VOEvent has always been transport-neutral. In recognition of the rapidly changing landscape of transport technologies, the explicit description of transport options has been removed from the standard.

• The VOEvent ⟨*Param*⟩ element has been generalized.

† http://blogs.astronomerstelegram.org/atelstream/
‡ http://hotwireduniverse.org/

- A facility was added to embed general-purpose tabular data structures within VO-Event packets.
- Time series will be supported as tables with *utypes* referencing an IVOA time-series data model.
- The VOEvent ⟨*Reference*⟩ element has been generalized.

In addition to changes to the VOEvent format, the v2.0 efforts focused on creating a more robust XML schema and on enhanced libraries and Web access compatible with the format and schema. The fundamental nature of the VOEvent format has remained the same between versions. VOEvent exists to support the engineering of empirical workflows (Seaman 2008) and with key elements:

- ⟨*Who*⟩ – author's provenance
- ⟨*What*⟩ – empirical measurements
- ⟨*WhereWhen*⟩ – targeting in spacetime
- ⟨*How*⟩ – instrumental signature
- ⟨*Why*⟩ – scientific characterization
- ⟨*Citations*⟩ – building threads of transient-response follow-up

5. VAO Time-Series Tools

Systems to disseminate event notifications are a major component of the VO's infrastructure for supporting time-domain science. Equally as important, however, are the tools and services that enable and facilitate the discovery and analysis of collections of time-series data. In view of the increasing number of new time-domain surveys now in progress or being planned, providing a framework to interconnect the data in distributed archives and appropriate services can only aid both the discovery of previously unknown phenomena and improve our understanding of already known ones. Through a number of activities, the VAO aims to create such interoperability, allowing astronomers to locate the data they want and then effortlessly connect the data providers to different types of available tools, such as a periodogram service or a time-series modeller, as part of a workflow or scripted analysis session.

As an illustration of such a system, one might consider an astronomer who is using variable stars, say RR Lyræ types, to study Galactic structures such as spiral arms, stellar streams or the like. A major contaminant in that kind of analysis can be eclipsing binaries, and the usual light curves of both classes can be difficult to distinguish. However, the binaries can easily be filtered out using phased light curves. The astronomer could therefore create a pure data set for the analysis by identifying suitable data through the VAO Time Series Archive Interconnectivity Portal and sending them to the periodogram service at the NASA Exoplanet Archive. The VAO infrastructure will handle the data transfer for the astronomer and the result that is returned is a phased light curve (see Fig. 1).

A pathfinder for this kind of collaboration is being developed by the VAO, initially connecting the Harvard Time Series Center, the NASA Exoplanet Archive and the Catalina Real-Time Transient Survey. As more data sets and tools become available, they will be integrated seamlessly. Provision for bulk activities, such as the large-scale characterization of time series, is also being considered.

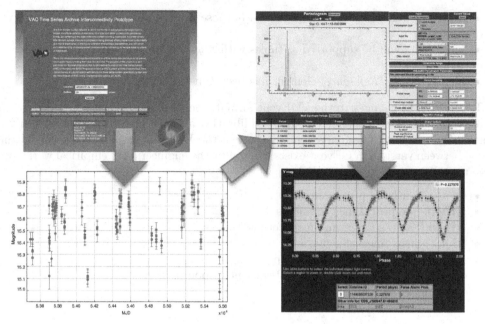

Figure 1. Time-series data workflow facilitated by the VAO.

6. Looking Ahead

During the workshop a wide-ranging discussion followed (and often interrupted) the presentations. There appeared to be a strong consensus that not only would the time domain increase in importance for astronomy in the future, but that to take advantage of its full potential virtual and semantic technologies would be critical. The often lively discussion addressed future directions for the virtual observatory time-domain facilities as a whole.

References

Seaman, R. 2008, *Astron. Nachr.*, 329, 237.
Seaman, R., *et al.* 2011, *IVOA Recommendation*, arXiv:1110.0523
R. Williams, S. Emery Bunn, & R. Seaman (eds), in: *Hot-wiring the Transient Universe* 2010,
   http://www.lulu.com/product/paperback/hotwiring-the-transient-universe/13205496

New Horizons in Time-Domain Astronomy
Proceedings IAU Symposium No. 285, 2011
R.E.M. Griffin, R.J. Hanisch & R. Seaman, eds.

Astrotomography

Keith Horne[1], Raymundo Baptista[2], Misty C. Bentz[3], and Danny Steeghs[4]

[1]SUPA Physics & Astronomy, University of St Andrews, St Andrews KY16 9SS, Scotland
email: kdh1(at)st-and.ac.uk

[2]Departmento de Fisica, Universidade Federal de Santa Catarina, Campus Trinidade,
88040-900 Florianopolis, SC, Brazil

[3]Department of Physics and Astronomy, Georgia State University, Atlanta, GA, 30303 USA

[4]Department of Physics, University of Warwick, Coventry CV4 7AL, UK

Abstract. Astrotomography refers to a suite of indirect imaging techniques that achieve micro-arcsecond angular resolution by measuring projections obtained from time-resolved spectroscopic observations. The projections arise from Doppler shifts, eclipses or time delays, combined with rotation of the star or binary system being imaged. At our workshop we reviewed and discussed state-of-the-art methods for imaging the surfaces and magnetic fields of rapidly rotating stars, the accretion flows in compact binary star systems and the broad emission-line regions in active galactic nuclei.

Keywords. accretion, accretion disks, black hole physics, methods: data analysis, techniques: high angular resolution, stars: binaries: eclipsing, galaxies: Seyfert

1. Astrotomography

Tomography reconstructs an image from measurements of its projections. It is widely used in geophysics to probe rock layers beneath the Earth's surface, and in medicine, where scanners project X-rays through a patient's body in various directions to obtain the projections needed to compute detailed diagnostic images of a patient's internal organs.

The Radon transform (Radon 1917) first demonstrated the equivalence of an image and its projections. Bracewell & Riddle (1967) pioneered the first practical applications, reconstructing maps of the radio sky using projections secured from lunar occultations or from fan beams of a linear antenna arrays. Modern interferometry—recovering images of the sky from measurements of projections of the image onto sine and cosine patterns with various wavelengths and position angles corresponding to the projected base-lines of pairs of telescopes—is closely related to tomography, where the projections are onto a basis of delta functions rather than sine and cosine functions.

In Doppler imaging of spotted stars (Vogt & Penrod 1983), Doppler shifts effectively project the visible hemisphere of a rotating star onto the observable radial-velocity profile of the star's spectral lines. The Least Squares Deconvolution (LSD) method (Donati & Collier Cameron 1997) collects information from thousands of weak spectral lines to form a mean velocity profile with a very high signal-to-noise ratio. Each starspot appears as a little bump in the line profile, rotating into view on the blue-shifted limb, crossing the face of the star and disappearing on the red-shifted limb. The spot's longitude determines the rotational phase at which the bump crosses the centre of the line profile; its latitude determines the range of phases and velocities spanned by the spot as it crosses.

In Zeeman-Doppler Imaging (Donati *et al.* 1997), high-resolution circular spectropolarimetry detects the polarised Zeeman components of rotating stellar magnetic fields.

The vertical, toroidal and poloidal field components have distinct polarisation signatures that march across the Doppler profile. From observations at a variety of rotation phases, the magnetic field vector is determined at each position on the surface of the rotating star.

Astrotomography methods for mapping accretion flows onto compact objects in binary systems and in active galactic nuclei were reviewed at our workshop by three expert practitioners, as reported below. In the discussion that followed we concluded that eclipse and Doppler mapping of accretion in binary systems is highly advanced observationally, with as yet untapped potential to validate models of accretion-disk atmospheres and chromospheres. In contrast, the echo mapping of accretion flows onto black holes in AGNs is currently limited by difficulties in obtaining suitable datasets, though there are prospects for improvement.

2. Eclipse Mapping of Accretion Disks

The light collected from Cataclysmic Variables (CVs) is the combination of emission from several distinct light sources, including the mass-donor star, the white dwarf (WD) at disk centre, the bright spot (BS) where infalling gas hits the outer disk rim, hotter inner regions plus the outer cooler regions of the accretion disk, and possibly vertically extended emission from a disk wind. Thus, interpreting and modelling the observed light from a CV is a degenerate problem that is usually plagued by the ambiguity associated with composite spectra. Nevertheless, for high-inclination systems ($i \geqslant 70^o$) the mass-donor star progressively covers different regions of the accretion disk, the WD and BS once per orbit, leading to eclipses. The shape of the eclipse has information about the surface brightness distribution of the occulted light sources. The maximum entropy eclipse mapping method (Horne 1985) is an inversion technique that assembles the information contained in the shape of the eclipse into a map of the accretion-disk surface brightness distribution plus the flux contribution of any uneclipsed source (e.g., the mass-donor star itself or a vertically-extended disk wind). It allows one to resolve structures spatially at angular scales of micro-arcseconds—well beyond the capabilities of present direct-imaging techniques.

2.1. *Principles and performance*

The mapping surface, know as the eclipse map, is defined as a grid of intensities centred on the WD. It can be either a flat grid in the orbital plane (standard eclipse mapping) or a conical surface with an outer cylinder (to account for a disk opening angle and for emission from the outer disk rim, 3-D eclipse mapping). The eclipse geometry is specified by the binary mass ratio q ($=M_2/M_1$, where M_2 and M_1 are the masses of the donor star and the WD, respectively) and the inclination i (see Horne 1985). Given the eclipse geometry, a model eclipse light curve can be calculated for any assumed brightness distribution in the eclipse map. A computer code then iteratively adjusts the intensities in the map (treated as independent parameters) to find the brightness distribution of the model light-curve which fits the eclipse lightcurve data within the uncertainties. The quality of the fit is checked by consistency statistics, usually the reduced χ^2. Because the one-dimensional light-curve data cannot fully constrain a two-dimensional map, additional freedom remains to optimize some map property. A maximum entropy procedure (e.g., Skilling & Bryan 1984) is used to select, among all possible solutions, the one that maximizes the entropy of the eclipse map with respect to a smooth default map, usually chosen as an axi-symmetric average of the eclipse map itself (see Baptista 2001).

The quality of an eclipse map depends on the phase resolution and the signal-to-noise ratio (SNR) of the input light curve. Poor phase resolution degrades the spatial resolution of the eclipse map, while low SNR limits the ability to recover faint brightness sources. Good-quality maps can be obtained for $\Delta\phi < 0.005\,P_{orb}$ and $SNR > 20$, while acceptable results can still be obtained for $\Delta\phi \leqslant 0.01\,P_{orb}$ or $SNR \geqslant 8$. Examples of performance under extreme conditions are discussed by Baptista (2001).

2.2. *A range of applications*

Early applications of the technique were instrumental in showing that accretion disks in outbursting dwarf novæ (Horne & Cook 1985) and in nova-like variables (Rutten *et al.* 1992) closely follow the expected $T \propto R^{-3/4}$ dependence of temperature with radius for a steady-state opaque disk, and in inferring disk mass accretion rates. Eclipse mapping of time-resolved spectrophotometric data delivers the spectrum of the accretion disk at any position on its surface. For example, spectral mapping analysis of the nova-like variable UX UMa (Rutten *et al.* 1993; Baptista *et al.* 1998) shows that its inner accretion disk is characterized by a blue continuum filled with absorption bands and lines which cross over to emission with increasing disk radius, while at the same time the continuum emission becomes progressively fainter and redder, reflecting the radial temperature gradient. Physical conditions in the disk photosphere might be inferred by fitting those spectra properly with disk atmosphere models.

Time-resolved eclipse mapping can be used to trace the evolution of the disk surface-brightness distribution through a dwarf nova outburst cycle, and it allows critical tests of existing dwarf-nova outburst models. For example, time-lapse analysis of the dwarf nova EX Dra (Baptista & Catalán 2001) suggests that its outbursts are the response of a viscous disk to a burst of enhanced mass transfer from the donor star, in support of the mass-transfer instability model (Bath 1975). Eclipse mapping is also a valuable tool for revealing the existence of complex structures in accretion disks, such as tidally-induced spiral shocks in the dwarf nova IP Peg in outburst (Baptista *et al.* 2000) and the changing illumination pattern of disk regions produced by the fast spinning magnetosphere of the WD in the intermediate polar DQ Her (Saito & Baptista 2009).

Flickering mapping is the most recent frontier tackled with eclipse-mapping techniques. By combining a large ensemble of light curves one can measure the flickering amplitude as a function of orbital phase, and produce a map of the flickering sources from its eclipse shape (Baptista & Bortoletto 2004). Flickering mapping of the dwarf nova HT Cas in quiescence reveals a disk-related flickering component strongly concentrated near disk centre (Fig. 1).

Assuming that the disk flickering is caused by fluctuations in the energy dissipation rate induced by MHD turbulence in the disk's atmosphere (Geertsema & Achterberg 1992), one finds that the disk viscosity parameter is large in the inner disk and decreases with radius as $\alpha \propto R^{-2}$ (Baptista *et al.* 2011).

3. Doppler Tomography of Accretion Flows

Rapid optical variability and the presence of strong and broad emission lines are two key signatures of accretion onto compact objects in close binaries. Strong emission lines are produced by the hot accreting gas, and it was soon recognized that the line-profile shape is determined by bulk motion of the gas. Most commonly, the flow takes the form of an extended accretion disk reflecting the orbital angular momentum of the transferred gas. The observed line profile can thus be considered as a projection of the velocity distribution of the emitting gas in terms of its radial-velocity component. Time-series of

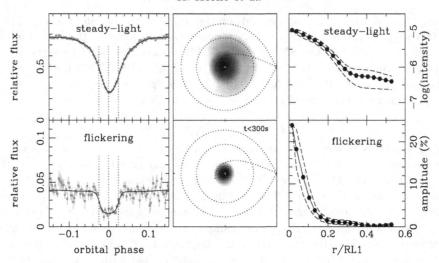

Figure 1. Flickering mapping of the dwarf nova HT Cas in quiescence. Left panels show data light curves (dots) and model light curves (solid lines). Middle panels display the corresponding eclipse maps in log intensity scale, while the right panels show the radial run of the steady-light intensity (top) and the relative amplitude of the flickering (bottom).

such profiles slice that emitting gas distribution as a function of the binary phase while the binary components orbit their common centre of mass. This presents a classic case of image reconstruction via a series of projections, which led to the development of Doppler tomography as a tool to image the emission-line distribution in compact binaries (Marsh & Horne 1988).

Although some explicit modelling of emission-line profiles can be done, asymmetries and complexities in the accretion flow make a model-driven reconstruction very difficult. However, thanks to the large dynamical velocities involved, detailed line-transfer effects that may set the local line-profile shape are overwhelmed by the bulk Doppler shifts. A model-independent reconstruction such as offered by Doppler tomography then presents a powerful and versatile tool. It has traditionally been employed for studying accreting white dwarfs in cataclysmic variables (Morales-Rueda 2004), but is applicable to a wide variety of systems including accreting neutron stars and black holes in X-ray binaries (Bassa *et al.* 2009), double degenerates (Roelofs *et al.* 2006) and Algols (Richards 2004). This variety also covers a range of flow geometries, from the more traditional accretion-disk dominated flow to streams (Kafka *et al.* 2010), propellers (Welsh *et al.* 1998) and magnetically controlled flows (Schwope *et al.* 2004).

The original version employed a regularised fitting procedure driven by maximum entropy constraints analogous to eclipse mapping (see above). Filtered back-projections offer a more straightforward implementation, but with the improvement of computational power the entropy-regularised implementation is the more popular choice. It gives some notable advantages, including having an explicit goodness of fit statistic driving the reconstructions, dealing with data imperfections and gaps, handling line blends, and being easily extendible (Marsh 2001; Steeghs 2003). The required input data are the emission-line flux resolved in both velocity and time. The spectral resolution effectively sets the velocity resolution of the reconstructed tomogram in the radial direction, while the time/phase resolution sets the azimuthal resolution. When planning Doppler tomography observations, a sensible compromise between spectral resolution and exposure time needs to be made that still offers an adequate S/N.

The central concept in Doppler tomography is the realisation that an emission-line source location can be characterised not just by its location in spatial coordinates but also in terms of its velocity vectors. The vector encodes both the size and orientation of the source velocity, and is sufficient to establish the projected radial-velocity components of such a source as a function of time. The observed line profile at a given binary phase is then a simple summation of all emission line sources as specified by their individual velocity vectors. The image we then seek to reconstruct is an intensity map in velocity coordinates. By working in velocity space, the problem is a straight de-projection from the data. We do not need to assume the velocity field of the flow that would be necessary to convert between spatial and velocity coordinates. The downside of this approach is that the reconstructed images are in a somewhat unfamiliar coordinate frame. However, it is recommended that one maps models to the velocity frame for direct comparison with tomograms in velocity space, as opposed to reconstructing the data in spatial coordinates (Steeghs & Stehle 1999).

Doppler tomography is now a widely-used tool that has supported a large volume of published studies which employ Doppler tomograms in a variety of binaries. When S/N is limited, one is often constrained to using only the strongest lines; however, it is increasingly clear that the Balmer lines are by no means the best lines to map despite their relatively large strengths. For example, van Spaandonk *et al.* (2010) highlight the sensitivity of the Ca II triplet both in terms of providing sharper images of the accretion disk and also in revealing emission components from faint donor stars.

Beyond mapping the accretion-flow dynamics, Doppler tomography can also be used to constrain fundamental binary parameters. Simple dynamical models such as the trajectory of the in-falling ballistic stream, or the solid-body rotation of the mass donor star, are often sufficient to constrain the binary mass ratio and the orbital velocities of the binary components directly. Such models can then be compared with the observed locations of stream-disk interactions or donor star emission components (Steeghs & Casares 2002; see also Fig. 2).

Larger telescopes and more efficient detectors and spectrographs now permit Doppler tomography of rather faint objects. At the same time, the application of higher S/N and spectral resolution to systems previously studied offers a step change in tomogram quality. It is unfortunate that we are often not in a position to model the observed tomograms in great detail. Emission-line formation processes in these accreting systems are complex, and sufficiently realistic models that predict emission-line profiles at the level of detail offered by many data sets are not really available. That also prevents us from exploiting quantitatively, beyond broad trends, the information contained in mapping multiple emission lines within the same system.

Ample opportunities remain to develop the technique further. They include considering full 3-D velocities, exploiting the direct constraints on the spatial location of a source during eclipses by combining Doppler tomography and eclipse mapping, and reconstructing physical-parameter maps as opposed to single emission-line distribution maps.

4. Reverberation Mapping of AGN Broad-Line Regions

Active galactic nuclei (AGNs) are some of the most energetic objects in the universe. The emission from photoionized gas near a supermassive black hole can rival that of all the stars in the host galaxy, yet it is emitted from a region comparable in size to our Solar System (1000 AU, or $\sim$0.01 pc). Although such regions are spatially unresolvable in even the most nearby AGNs, studies of the gas in the broad-line region (BLR: the region of photoionized gas from which the broad emission lines in AGN spectra are

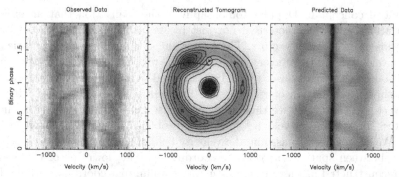

Figure 2. Doppler tomogram showing the distribution of He I emission in the double degenerate binary GP Com, compared to the observed input data and the reconstructed data. We see the near-circular accretion disk with a bright spot in the location where the in-falling gas stream hits the outer disk. In GP Com the accreting white dwarf is also an emission-line source visible near the center of mass. By fitting the locations of such features one can constrain binary parameters directly.

emitted) are critical to our understanding of galaxy and black-hole growth and co-evolution throughout cosmic time.

Black-hole masses for active galaxies are most often determined through the technique of reverberation mapping (Blandford & McKee 1982; Peterson 1993). The rarity of AGNs means that they are generally found at distances that are prohibitive for dynamical mod-elling (which is limited by spatial resolution); however, reverberation mapping replaces spatial resolution with time resolution. By monitoring the AGN's continuum flux (which is produced near the black hole, probably in the accretion disk) and the time-delayed response of the broad emission lines to changes in that continuum flux, it is possible to de-duce the average radius of the BLR. Combining that radius with the velocity-broadened width of the emission line through the virial theorem provides a measure of the black-hole mass, though with a scaling factor that depends on the geometry and kinematics of the BLR gas. To date, some 50 black-hole masses have been determined in this way (Peterson *et al.* 2004; Bentz *et al.* 2009) and provide the basis for *all* black-hole mass determinations in AGNs at cosmological distances.

The limiting factor in the accuracy of these black-hole masses is the scaling factor, f. The current practice is to assume that the $M_{\mathrm{BH}} - \sigma_\star$ relationship for quiescent galaxies with dynamical black-hole masses is the same as the $M_{\mathrm{BH}} - \sigma_\star$ relationship for active galaxies with reverberation masses. The scaling factor is then taken to be the average multiplicative factor required to bring the two relationships into agreement, and has been found to be $\langle f \rangle \sim 5$ (Onken *et al.* 2004; Woo *et al.* 2010). However, reverberation mapping has the potential to map directly the response of the gas in the BLR as a function of time delay and velocity, thereby providing stringent constraints on the geometry and kinematics of the BLR gas. With such constraints, f could be determined on an individual basis rather than an average basis, and the black-hole mass would be constrained directly without relying on the sample of objects with masses determined by dynamical modelling.

The challenge has been to acquire data of the quality necessary to control observational systematics and recover the full response of the broad emission lines. However, recent spectroscopic monitoring campaigns have been achieving high sampling cadences and high S/N spectra over long durations, and are now allowing us to reach the level of observational accuracy that is necessary to see differences in the mean response as a function of velocity across the line profile (Bentz *et al.* 2009; Denney *et al.* 2009). A

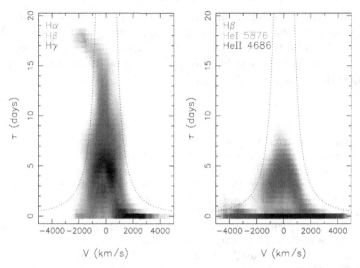

Figure 3. Optical broad emission-line response as a function of time delay and velocity for the
$z = 0.02$ AGN Arp 151. From Bentz *et al.* (2010).

substantial effort is currently being invested in mapping the full BLR gas response, and
is meeting with some success.

Fig. 3 shows the recovered response of five broad emission lines in the spectrum of
Arp 151 as a function of velocity and time delay (from Bentz *et al.* 2010). For comparison,
Fig. 4 shows the expected response for three different kinematic models of an AGN BLR.
The higher ionization lines in Arp 151 are seen to vary on much shorter timescales than
the lower ionization lines, as has been observed for many AGNs, indicating ionization
stratification in the BLR. The Balmer-line response is somewhat consistent with emission
arising from a flattened-disk geometry; however, the response is distinctly asymmetric,
with a strong response at short time delays in the red-shifted wing of the emission lines
and with much longer time delays in the blue wing. This asymmetry in the line response
may arise from a strong infall component in the BLR of Arp 151, or it may arise from a
disk asymmetry such as a warp or hotspot.

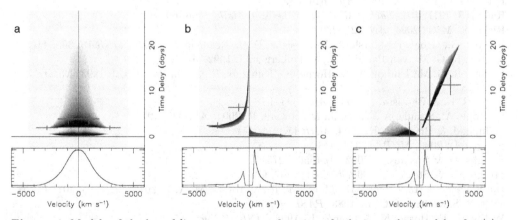

Figure 4. Models of the broad line response as a function of velocity and time delay for (a) a
Keplerian disk, (b) an infalling BLR and (c) a constantly-accelerated outflow. From Bentz *et al.*
(2009).

It is vital that our observations and analysis continue to improve, because reverberation mapping has the potential to answer fundamental questions about the fuelling of AGNs, to provide a solid basis upon which to determine black-hole masses at cosmological distances, and to increase our knowledge of the growth and co-evolution of galaxies and black holes throughout cosmic history.

References

Baptista, R., *et al.* 1998, *MNRAS*, 298, 1079

Baptista, R., Harlaftis, E. T., & Steeghs, D. 2000, *MNRAS*, 314, 727

Baptista R. 2001, in: H.M.J.Boffin, D.Steeghs & J.Cuypers (eds.), *Astrotomography: Indirect Imaging Methods in Observational Astronomy* (Berlin: Springer-Verlag), p. 307

Baptista, R. & Catalán, M. S. 2001, *MNRAS*, 324, 599

Baptista, R. & Bortoletto, A. 2004, *AJ*, 128, 411

Baptista, R., *et al.* 2011, Proceedings of *Physics of Accreting Compact Binaries*, (Universal Academic Press Inc.), in press. (arXiv 1105.1382)

Bassa, C. G., Jonker, P. G., & Steeghs, D. 2009, *MNRAS*, 399, 2055

Bath, G. T. 1975, *MNRAS*, 171, 311

Bentz, M. C., *et al.* 2009, *ApJ*, 705, 419

Bentz, M. C., *et al.* 2010, *ApJ*, 720, L46

Blandford, R. D. & McKee, C. F. 1982, *ApJ*, 255, 419

Bracewell, R. N. & Riddle, A. C. 1967, *ApJ*, 150, 427

Denney, K. D., *et al.* 2009, *ApJ*, 704, L80

Donati, J.-F. & Collier Cameron, A. 1997, *MNRAS*, 291, 1

Donati, J.-F., Semel, M., Carter, B. D., Rees, D. E., & Collier Cameron, A. 1997, *MNRAS*, 291, 658

Geertsema, G. T. & Achterberg, A. 1992, *A&A*, 255, 427

Horne, K. 1985, *MNRAS*, 213, 129

Horne, K. & Cook, M. C. 1985, *MNRAS*, 214, 307

Kafka S., Tappert C., Honeycutt, R. K. 2010, *MNRAS*, 403, 755

Marsh, T. R. 2001, in: H.M.J.Boffin, D.Steeghs & J.Cuypers (eds.), *Astrotomography: Indirect Imaging Methods in Observational Astronomy* (Berlin: Springer-Verlag), p. 1

Marsh, T. R. & Horne, K. 1988, *MNRAS*, 235, 269

Moralez-Rueda L. 2004, *AN*, 325, 193

Onken, C. A., *et al.* 2004, *ApJ*, 615, 645

Peterson, B. M. 1993, *PASP*, 104, 247

Peterson, B. M., *et al.* 2004, *ApJ*, 613, 682

Radon, J. 1917, *Ber. Sächs. Akad. Wiss. Leipzig, Math. Phys*, 69, 262

Richards, M. T. 2004, *AN*, 325, 229

Roelofs, G., Groot, P., Marsh, T. R., Steeghs, D., & Nelemans, G. 2006, *MNRAS*, 365, 1109

Rutten, R. G. M., van Paradijs, J., & Tinbergen, J. 1992, *A&A*, 260, 213

Rutten, R. G. M., Dhillon, V. S., Horne, K., Kuulkers, E., & van Paradijs, J. 1993, *Nature*, 362, 518

Saito, R. K. & Baptista, R. 2009, *ApJ*, 693, L16

Schwope, A., Staude, A., Vogel, J., & Schwarz, R. 2004, *AN*, 325, 197

Skilling, J. & Bryan, R. K. 1984, *MNRAS*, 211, 111

Steeghs, D. 2003, *MNRAS*, 344, 448

Steeghs, D. & Casares, J. 2002, *ApJ*, 568, 273

Steeghs, D. & Stehle, R. 1999, *MNRAS*, 307, 99

van Spaandonk, L., Steeghs, D., Marsh, T. R., & Torres, M. A. P. 2010, *MNRAS*, 401, 1857

Vogt, S. S. & Penrod, G. D. 1983, *PASP*, 95, 565

Welsh, W. F., Horne, K., & Gomer, R. 1998, *MNRAS*, 298, 285

Woo, J. H., *et al.* 2010, *ApJ*, 716, 269

New Horizons in Time-Domain Astronomy
Proceedings IAU Symposium No. 285, 2011
R.E.M. Griffin, R.J. Hanisch & R. Seaman, eds.

Small and Robotic Telescopes in the Era of Massive Time-Domain Surveys

M. F. Bode[1] and W. T. Vestrand[2]

[1] Astrophysics Research Institute, Liverpool JMU, Birkenhead, CH41 1LD, UK
email: mfb@astro.livjm.ac.uk

[2] Center for Space Science and Exploration, Los Alamos National Laboratory,
NM 87545, USA

Abstract. We have entered an era in time-domain astronomy in which the detected rate of explosive transients and important ephemeral states in persistent objects threatens to overwhelm the world's supply of traditional follow-up telescopes. As new, comprehensive time-domain surveys become operational and wide-field multi-messenger observatories come on-line, that problem will become more acute. The goal of this workshop was to foster discussion about how autonomous robotic telescopes and small-aperture conventional telescopes can be employed in the most effective ways to help deal with the coming deluge of scientifically interesting follow-up opportunities. Discussion topics included the role of event brokers, automated event triage, the establishment of cooperative global telescope networks, and real-time coordination of observations at geographically diverse sites. It therefore included brief overviews of the current diverse landscape of telescopes and their interactions, and also considered planned and potential new facilities and operating models.

1. Robotic Telescopes

Iain Steele (Liverpool JMU) gave a brief presentation on "Robotic Telescopes". A robotic telescope is one that can operate autonomously, that is, without night-time supervision. It is important to remember that small and robotic are not synonymous. There are many smaller telescopes that are not robotic, and several larger telescopes that are. It is generally assumed that such facilities are relevant only to the optical/NIR, and that was certainly the focus of this session. It would be interesting to consider what contribution could be made at other wavelengths (especially the radio) by such facilities. Can we rely just on facilities such as LOFAR and the SKA precursors to monitor and follow-up the time-domain radio sky?

For the optical/NIR, Hessman provides a very useful webpage† where he lists a whole raft of robotic-telescope projects. At the time of writing the list contained 122 projects, of which around 17% are general-purpose facilities and the remainder are dedicated to one or two science goals such as surveys for exoplanets or Gamma-Ray Burst follow-ups. Those in the latter class are usually run as "experiments", which can result in considerable cost savings compared to a common-user robot.

Another advantage of experiments can be the optimisation of telescope and instrument design for the science goal. As an example, consider the question of "small and quick" versus "large and slow" for the follow-up of transient sources. For Gamma-Ray Bursts, the photon rate in the afterglow phase scales as t^{-1}. Experience shows that the typical slew time of a telescope to a source scales roughly as the mirror diameter, D. The photon rate will therefore be proportional to D^{-1}, and since detection area scales as D^2

---

† http://www.uni-sw.gwdg.de/~hessman/MONET/links.html

the number of photons detected will be roughly proportional to D. *In this particular case,* the advantage of aperture therefore wins over that of speed. Similar calculations can be made for other experiments too, and show (for example) that for survey work the advantage of larger sky area versus depth can favour smaller facilities.

An important question that we should address relates to the balance of instrumentation available on robotic facilities. In general robotic telescopes are equipped with CCD imagers, and very few have NIR cameras or spectrographs and even fewer have polarimetric facilities. Is this balance the correct one in an era when LSST and other large sky surveys will provide regular photometric monitoring? Whatever instrumentation is used, however, it is very important that it be supported by some sort of data-reduction pipeline which can reduce the data to a level sufficient for reliable use in planning and triggering further observations in a timely fashion. This latter point was returned to later by Vestrand (Section 4).

Some of the major "feeder" facilities of the future are not common-user ones, and preparations need to be made well ahead of time to ensure that the necessary follow-up agreements are in place.

2. The AAVSO Perspective

Arne Henden (AAVSO) emphasised the role of the amateur community in this endeavour, and spoke particularly about the various networks that now exist. The AAVSO has established a global network of 19 robotic telescopes, from 60-mm to 80-cm aperture. The telescopes are heterogenous, but the software is not. All the images are pipeline-processed at AAVSO HQ, and images are transferred to Amazon Cloud for users to access. Work is under way to build a master scheduler based on VOEvent.

The challenges faced by the network include continuity and longevity of private sites, internet access at remote locations, data archiving and access to that archive, maintenance, problems typical of a "volunteer" situation, time allocation, and the presence of non-imaging instruments (spectrographs) at two of the larger telescopes which add to the potential science return but also add to the complexity. Several of those issues are of course common to professional networks too (see Section 4).

3. Follow-up of GAIA Alerts

Łukasz Wyrzykowski (Cambridge) focussed on GAIA, ESA's milestone mission which is to be launched in 2013 June. The mission's main goal is to map the Galaxy with ultra-precise astrometry and photometry of a billion stars, but owing to the repeated nature of the observations GAIA will also be very sensitive to anomalous and transient events. An alert will normally be issued during an interval of 2–24 hours after the detection of such an event. The depth of the survey is about $V = 20$ mag. The sampling is sparse, yielding on average 80 observations per object over the 5-year mission. Anomaly detection will therefore be based on as few as 1 or 2 GAIA data points, which will not be enough to characterise fully the detected phenomena or to exploit maximally the science involved. GAIA therefore requires an extensive ground-based follow-up capability for the verification and classification of the alerts that it reports. With hundreds to thousands of alerts expected every day, the situation calls for a network of robotised telescopes sited all around the globe.

The community is proposing to organise a formal consortium of telescopes and observers under the auspices of the European OPTICON network†, primarily with GAIA alerts in mind but also more generally for following up all sorts of time-critical cosmic events. The network is currently being assembled in the wake of two GAIA Science Alerts Workshops‡ held in Cambridge, UK. The network will be comprised of small (< 2-m) telescopes, mostly robotised, which are able to provide both multi-band imaging and low-to high-dispersion spectroscopy. The observational outputs will need to be standardised. There will also need to be clear rules on data policy, publishing, etc.

4. Factors to Consider for Effective Robotic Telescope Networks

Ashish Mahabal (Caltech) gave a brief presentation in which he emphasised that following up transient sources is complex for various scientific and sociological reasons. Different telescopes can observe only a subset of possible transients because (1) their aperture limits it, (2) the available instruments determine it, (3) the interests of the people running the telescope interfere with it (potentially)—plus a host of other reasons like weather, RA/Dec coverage, and (most importantly) the state of preparedness (robotic versus automated versus semi-automated versus manual).

The problem is going to get far worse as more transients are found, a majority of which (it will turn out after some observations) are of a relatively common nature (i.e. we already know a lot about their classes). It will therefore be critical to make early classification with minimal data in order to determine which transients are in fact most worthy of following up. One helpful procedure will be to obtain priors of different telescope/instrument combinations so that one might decide how likely it will be for an observation from a given instrument set-up to disambiguate the possible classes to which the transient might belong.

Another crucial factor is prompt feedback from the observatories. It is not enough for the observatories just to receive events: without feedback, further classification and informed follow-up will not be possible. Those stages also need a degree of automated pipeline processing.

Some work on these issues is already underway at Caltech, and in collaboration with the GAIA consortium, in particular the GAIA Alerts group at Cambridge.

Tom Vestrand (LANL) began by stating some of the assumed characteristics that went into developing an effective network of robotic telescopes for real-time follow-up of unpredictable events. They included real-time inter-communication between telescopes, heterogeneous instrumentation and general capabilities, and free exchange of information and knowledge between systems. It was felt that in the future there could ultimately be more robotic instruments than astronomers, with each telescope playing a dynamic role.

Currently however, the way we handle time-domain events is very much like the "second-grade soccer" problem—exciting, enthusiastic, but disorganised. An effective network would be "self-aware"—knowing what tasks each telescope is doing now, what the state of health of each system is in terms of instrument configuration, general capability, etc., what are the sky conditions at the telescope's location and whether it can observe effectively at the target location.

Locally, each telescope's "decision engine" would have to answer questions such as whether it has the sensitivity to make the required observation, what is the relative importance of the observation compared to what it is currently doing, how well it matches

---

† http://www.astro-opticon.org/
‡ http://www.ast.cam.ac.uk/ioa/wikis/gsawgwiki/index.php/Workshop2011:main

the telescope's "uniqueness" parameters, what the other network nodes are doing, and what is the highest-impact observation that the telescope can make at that moment.

In terms of what needs to be done, it is important to all follow-up projects, particularly with such a prospective deluge of alerts likely to be seen in the coming years, that event classification be as automated and reliable as possible. Similarly, event brokers need to become fully developed. Real-time feedback is also going to be critical, in the sense of knowing what is being found at each instant and then determining what the next step should be. Finally, "trust models" for all the sources of information being fed into and among the network components need to be developed.

To summarise: we are entering an era where a global ecosystem of robotic telescopes could make unpredicted measurements and discoveries in time-domain astronomy. Most of the technical elements are available or under development. The biggest problem is likely to be the sociology of humans.

5. Summary and Conclusions

- It is evident that in the era of massive time-domain surveys there is an increasingly important role for telescopes of modest aperture, but the required reaction speeds demand that they be robots.

- For effective and efficient responses for single telescopes, but particularly for networks, we need to embrace a "human out of the loop" philosophy—this includes almost every facet of automated data reduction.

- Amateurs can make important contributions in this regime with their relatively small telescopes but increasingly high levels of sophistication.

- Alert characterisation is crucial. For example, in the coming era, what *not* to observe will be a central theme.

- For robotic telescope networks, a full understanding of the roles of individual instruments is needed, plus the development of "trust models" and real-time "state of health" detection.

- The balance of sites, apertures and types of instrumentation needs to be considered carefully as it is likely to be sub-optimal for future scientific demands.

- The network needs to have efficient ways to interrogate its components in real time, and dynamic ways to determine "what next".

- Similarly, there is a great deal of work to be done on the integration of telescope ecosystems with well-defined roles for individual instruments.

- In some instances agreements will need to be struck with upcoming feeder facilities well ahead of their operational debuts, but overall, optimising the scientific return in the future will depend on developing focused science experiments.

New Horizons in Time-Domain Astronomy
Proceedings IAU Symposium No. 285, 2011
R.E.M. Griffin, R.J. Hanisch & R. Seaman, eds.

© International Astronomical Union 2012
doi:10.1017/S174392131200066X

Binarity and Stellar Evolution

R. E. M. Griffin[1] & Slavek Rucinski[2]

[1] Herzberg Institute of Astrophysics, Victoria, BC, V9E 2E7, Canada
email: elizabeth.griffin@nrc.gc.ca

[2] University of Toronto, George Street, Toronto, ON, Canada

Abstract. Models of stellar evolution constitute an extremely powerful, and for the most part apparently very successful, tool for understanding the progression of a star through its lifetime as a fairly compact entity of incandescent gas. That success has led to stellar evolution theory becoming a crutch when an observer is faced with objects whose provenance or current state are in some way puzzling, but how safe a crutch? The validity of the theory is best checked by examining binary systems whose component parameters have been determined with high precision, but it can be (and needs to be) honed through the many challenges which non-conformist single stars and triple systems also present. Unfortunately the lever of observational parameters to constrain or challenge stellar evolution theory is not as powerful as it could be, because not all determinations of stellar parameters for the same systems agree to within the precisions claimed by their respective authors. What are the sources of bias—the data, the instrument or the techniques? The workshop was invited to discuss particularly challenging cases, and to attempt to identify how and where progress might be pursued.

Keywords. stars: fundamental parameters stars:binaries stars:evolution (stars:) binaries: general (stars:) binaries: spectroscopic (stars:) binaries: visual stars: variables: other

1. Background

The theory of stellar evolution does rather well to model and explain the changes which a star undergoes (a) from its first emergence as a celestial entity and early journey as a relative youngster on to the main sequence (of HR-Diagram language), (b) during its extended life as a hydrogen-burning star, (c) as it digs deeper for new energy sources and in so doing becomes a cooler but inflated red giant, (d) during relatively rapid vacillations as new energy sources are tapped but mass loss impoverishes it, until (e) finally it undergoes some form of explosive change, or more meekly throws in the towel, and dwindles to a state of near-permanent insignificance. In qualitative terms the sequences that are predicted seem to be borne out by observation in remarkable detail; frequently the time-scale is the only substantial feature that cannot be checked against observation.

However, when we need to compare a model prediction in actual quantitative terms we find that there can be a great shortcoming on the side of the observations, or their interpretation: how can we pin down the temperature or colour indices if we cannot also pin down the degree of reddening? Can we be sure that the assumptions underlying the derivation of surface gravity from a stellar spectrum do not bias the answers? Does overall metallicity, or an abundance-ratio anomaly, influence how we derive 'best' stellar parameters from observation? How well can we determine the mass of a single star?

In real life, stars are anything but uniform or homogeneous in their nature, behaviour and constitution. Many exhibit evidence of energy expenditure that is noticeably over and above what is required simply to exist. A-type stars pulsate, M-type stars flare, B-type stars show emission, cool giants have flickering chromospheres, etc. Does stellar-evolution theory deal in a predictive manner with those cases? If not, is it because evolution models need to be in 3-D in order to avoid smearing out local ripples, or are such effects not in fact constructive—or destructive— enough to notice on the grand scale of evolution?

Binary stars are critically important here, particularly double-lined ones, and especially if the components have quite different temperatures—as in composite-spectrum binaries. The workshop discussed some prime examples, all relatively bright stars, so most of their basic physical parameters should surely be known by now with rather high precision. But that does not seem to be the case; there are examples in the literature of disagreements in the published values for some parameters which reach as much as 10 times the σ values claimed by the respective authors.

The participants at the workshop appeared to be principally observers, but the prime representative of the theory school (Peter Eggleton) held his own remarkably well despite questions and challenges from numerous of his "opposition".

2. The Case of o Leo

An improvement in theoretical modelling was well demonstrated in the *ad hoc* re-calculation of atmospheric diffusion as the cause of "metallicity" in Am and Fm stars. Since early days in the development of Michaud's diffusion theory (Michaud 1973; Michaud *et al.* 1976) it had been stated fairly categorically that diffusion could not become established below an effective temperature of about 6300 K (Vauclair & Vauclair 1982). The successful unravelling of the 4th-magnitude composite-spectrum binary o Leo, in which both components proved to have "metallic" properties and the cooler one had a T_{eff} (6100 $\pm$ 200 K) that was slightly *cooler* than the stated limit for diffusion (Griffin 2002), gave theorists a new observational fact to work with, resulting in a small modification to the mass of atmosphere involved and which had been assumed as an initial condition.

3. The Case of Capella

Although Capella is such a bright binary, its orbital parameters and the physical parameters of its component stars are not as well determined as individual authors claim. Whether or not the orbit solution includes all the historic radial velocities from the literature seems to make a difference (Branham 2008) of as much as 17% percent to the derived mass ratio. Both components are G-type giants, one somewhat cooler than the other, and the hotter one is still rotating rather rapidly—as is often typical of a "Hertzsprung-gap" giant. Measurements of the radial velocity of the secondary in the presence of a very similar but sharper-lined companion may well be affected by systematic errors which bear directly upon the derived mass ratio. Since the masses of the two giants are rather similar, uncertainties in their mass ratio become translated into serious uncertainties concerning the evolutionary status of the cooler star—in particular, whether it is on its first or later ascent of the red-giant branch.

To be specific, Weber & Strassmeier (2011) obtain a value for the hotter component's radial-velocity amplitude of 26.840 $\pm$ 0.024 km s^{-1}, while Torres *et al.* (2009) obtained 26.260 $\pm$ 0.087 km s^{-1}. Those differ by either 7 or 24σ. Until such parameters are constrained better by the observers it is difficult to apply very stringent tests to the models from which the evolutionary tracks are calculated. If this particular binary gives rise to such controversies it is fair to ask whether it is more useful in the long run to observe with high precision a small number of bright but possibly unrepresentative stars, or to rely on huge surveys yielding statistical information for faint stars only but yielding little as to how near (or how far) the statistical norm deviates from reality.

4. The Case of V 1309 Sco

Difficulties in accommodating the mass ratios and apparent evolutionary status for a number of well-studied binary systems have prompted Eggleton & Kiseleva (1996) to attribute the mis-match to a merger; such a merger might have taken place in either the current primary or the secondary component. However, until the startling behaviour of the contact binary V 1309 Sco (Nova Sco) over the past 10 years was revealed (Tylenda et al. 2011) the existence of such mergers was only postulated (though believed to be very likely nonetheless). Two outbursts in our Galaxy, V838 Mon (Munari *et al.* 2002; Bond *et al.* 2003) and V4332 Sgr (Martini *et al.* 1999) have been attributed plausibly to mergers, but the behaviour of Nova Sco which lifted it from the realms of a possible to an almost certain merger event was the rapid change of orbital period before the outburst and the disappearance of its eclipses—and indeed its orbit—afterwards; both of those are in keeping with a merger of the system's two components into one. Nevertheless, although V 1309 Sco is currently a red star (and is presumably undergoing evolution towards the giant branch right now), it does not seem to be very well agreed how cool it was before the merger. Observers are doubtless keen to monitor similar systems in the hope of finding more such events actually in progress.

5. Other Binary Systems

Good-quality parameters for detached binary systems have been compiled and published by Andersen (1991), and more recently by Torres, Andersen & Giménez (2010). However, almost all the stars listed there are dwarfs, not giants, and so do not present such an exacting challenge to stellar-evolution theory. Roger Griffin's excellent series of orbits for binary stars is also limited in that context inasmuch as almost all are for single-lined systems. To obtain a full set of stellar parameters requires in addition knowledge of a system's inclination, which is normally derived from an astrometric orbit—or can be assumed accurately enough if a system eclipses. The hope is that surveys like OGLE or MACHO will be able to return useful results for such purposes; recent results are certainly encouraging in that respect.

6. Stellar Metallicities

For many of the main-sequence binaries with well-determined physical parameters, the metallicity remains totally unknown, particularly for binaries with periods less than about 1 day. Why should that be? The metallicity is troublesome to determine when the orbital period is shorter than a few days because the accompanying rotational broadening of the lines adds serious problems of line blending. Since the location of a theoretical evolutionary track depends critically upon metallicity, the lack of that constraint obviously introduces a substantial uncertainty into the fitting of tracks to observation.

Probably the most important impact of metallicity on theoretical modelling comes from the work of Asplund *et al.* (2000, 2005), who found that the metallicity of the Sun has to be substantially revised (lowered) relative to what was the standard assumption of previous decades. This is a result of using 3-D modelling of the convective surface rather than the usual simplistic 1-D models. It has had the effect of making it much harder to match theoretically the observed and very detailed spectrum of helioseismic oscillations; agreement was very satisfactory prior to 2000. For the present (Asplund *et al.* 2010), no resolution of this conflict is in sight.

7. Conclusions

The current state of affairs offers much to encourage hope for better things to come. Several 3-D stellar-structure codes are now in operation (Asplund et al. 2000; Eggleton et al. 2006; Meakin & Arnett 2006; Stancliffe et al. 2011), while massive spectroscopic surveys promise to give better ideas as to the relative distributions of different stellar types and luminosities. GAIA should be a fresh source of precise physical parameters for a large number of stars.

There is probably no substitute for more and better (= more precise) data with which to test and address ongoing matters associated with theories of stellar evolution adequately and fully. However, it was not clear to the workshop whether a "broad sweep" approach (to coin a phrase from Bernard Pagel) would necessarily and incontrovertibly prove more valuable than concentrating on understanding a few bright systems better (the approach of "ultimate refinement"). Almost certainly the two approaches are complementary and not alternatives, but need to be well lubricated by improved communication. The light curves which huge new surveys for optical transients are now producing will not be beneficial in this context without some very careful calibration of the population of target objects, if such calibration is even possible. The fact that refined studies of bright stars have shown up so many oddities and oddball cases is rather depressing, inasmuch as the assumption of an "average" or "typical" stellar type or characteristic may be so gross that any theory which aspires to model such cases is likely to be rather impotent at understanding and predicting the wider range of varieties and variability which actually populate the heavens.

Acknowledgements

Our thanks and admiration go to Peter Eggleton for weathering the storm of challenges so adroitly.

References

Andersen, J. A. 1991, A&ARv, 3, 91
Asplund, M., Grevesse, N., Sauval, A. J., & Scott, P. 2009, ARA&A, 47, 481
Asplund, M., Grevesse, N., & Sauval, A. J. 2005, ASPC, 336, 25
Asplund, M., Nordlund, Å. A., Trampedach, R., & Stein, R. F. 2000, A&A, 359, 743
Bond, H.E., et al. 2003, Nature, 422, 405
Eggleton, P. P., Dearborn, D. S. P., & Lattanzio, J. C. 2006, Sci, 314, 1580
Eggleton, P. P. & Kiseleva, L. G. 1996, in: R.A.M.J. Wijers & M.B. Davies (eds.), Evolutionary Processes in Binary Stars, NATO ASI Series C, 477 (Dordrecht: Kluwer), p. 345
Griffin, R. E. 2002, AJ, 123, 988
Martini, P., Wagner, R. M., Tomaney, A., et al. 1999, AJ, 118, 1034
Michaud, G., 1973, ApL, 15, 143
Michaud, G., Charland, Y., Vauclair, S., & Vauclair, G. 1976, ApJ, 210, 447
Munari, U., Henden, A., Kiyota, S. et al. 2002, A&A, 389, L51
Stancliffe, R. J., Dearborn, D. S. P., Lattanzio, J. C., Heap, S. A., & Campbell, S. W. 2011, ApJ, 742, 121
Torres, G., Andersen, J. A., & Giménez, A., 2010, A&ARv, 18, 67
Torres, G., Claret, A., & Young, P. A. 2009, ApJ 700, 1349
Tylenda, R., et al. 2011, A&A, 528, 114
Vauclair, S. & Vauclair, G., 1982, ARA&A, 20, 37
Weber, M. & Strassmeier, K. G., 2011, A&A, 531, 89

New Horizons in Time-Domain Astronomy
Proceedings IAU Symposium No. 285, 2011
R.E.M. Griffin, R.J. Hanisch & R. Seaman, eds.

© International Astronomical Union 2012
doi:10.1017/S1743921312000671

Historical Time-Domain: Data Archives, Processing, and Distribution

Jonathan E. Grindlay[1] and R. Elizabeth Griffin[2]

[1] Harvard Observatory & Center for Astrophysics, Cambridge, MA 02138, USA
email: jgrindlay@cfa.harvard.edu

[2] NRC Herzberg Institute of Astrophysics, Victoria, BC V9E 2E7, Canada

Abstract. The workshop on Historical Time-Domain Astronomy (TDA) was attended by a near-capacity gathering of ~30 people. From information provided in turn by those present, an up-to-date overview was created of available plate archives, progress in their digitization, the extent of actual processing of those data, and plans for data distribution. Several recommendations were made for prioritising the processing and distribution of historical TDA data.

Keywords. astronomical data bases: catalogs, surveys; STARS: variables; galaxies: active

1. Introduction and Workshop Goals

Historical Time-Domain Astronomy (TDA) data are defined here as significant photographic-plate collections containing exposures of both images and spectra. The importance of those data, not only for current surveys (such as PTF, Pan-STARRS and CRTS) but also for future ones (particularly LSST), was discussed by Grindlay, who emphasized that measurements of the long-term variability of known objects and of historical transients provide long-term comparisons and context—particularly for rare objects or events—for current and future TDA surveys.

The first objective of the Workshop was to gather information about the health and status of astronomy's major plate collections.

The second objective was to ascertain the status of, or plans for, digitizing those collections in order to render the observations fully and freely available and useable. It is only by digitizing the images and spectra that those historic observations are transformed into electronic formats and thus made useful for—and useable by—modern astrophysics investigations and TDA surveys. However, digitization by itself is not adequate; converting the bits is a specialized endeavour and requires specific software that (a) incorporates features which make use of modern all-sky photometric catalogues for both global and local calibration, (b) deals with the point spread function (psf) of photographic rather than CCD images, and (c) involves laboratory spectra for converting stellar spectra into linearized output with the spectrum referenced to an object anchored on the world coordinate system. Plate digitization therefore needs to be accompanied by those additional stages in order to enable accurate scientific analysis, light curves, etc.

The third objective of the Workshop was to survey the state of software for photometric analysis of digitized plate data.

Since these data and analysis tools are only useful if made available in publications and on websites for community access, the fourth objective of the Workshop was to learn about plans and respective time-scales for hosting datasets and analysis tools.

2. Principal Historical TDA Plate Archives and Digitization Status

Seven archived plate collections and their digitization status (objectives 1 and 2) were summarized by their representatives at the Workshop. The archives thus described represented a significant fraction of the world total, which is estimated at 2 million direct-image plates and a further million spectrograms—all in need of specialist digitization for maximum utility in current astrophysics. The respective status of the high-speed scanner in Brussels and of the Wide-Field Plate Database were also discussed.

2.1. *Harvard Observatory, Cambridge, USA*

Grindlay outlined the *DASCH* project (http://hea-www.harvard.edu/DASCH/) to digitize the Harvard College Observatory plate collection and make the photometric data available (Grindlay et al. 2009; see p. 29). The total number of direct-image plates is ~450,000, covering the full sky (with approximately uniform coverage) from ~1890–1990 except for the "Menzel Gap" between ~1954–1970. The custom-designed *DASCH* scanner (Simcoe *et al.* 2006) and its operating software and calibration systems have been further optimized over its ~5-year development, during which ~19,500 plates have been scanned from 5 selected fields. A semi-automated plate-cleaning machine is in the final stages of development; it will enable cleaning the glass back-side of each plate faster than the 80 or so seconds that it takes for an operating sequence of loading–scanning– unloading a standard-size plate of $30\,\text{cm} \times 25\,\text{cm}$ (these are scanned two at a time). The "A" series plates are the deepest and have higher resolution ($59''.57\,\text{mm}^{-1}$) and are both standard-size (~5700 plates) and larger in size ($43\,\text{cm} \times 36\,\text{cm}$; ~20,000 plates). The larger plates are scanned singly and require an additional ~80 secs per plate to scan. Full "production scanning" and processing (~400 standard plates/day) can begin when the two cleaning machines are ready (anticipated for 2012 July). Scanning of the nearly half-million plates together with associated photometric reduction, and the derivation of light curves for all resolved objects, will require ~4 years. The ~5,000 spectrogram plates in the plate collection will be digitized after the direct plates have been completed.

2.2. *DAO/HIA, Victoria, Canada*

Griffin summarized the archive of stellar spectrograms at the Dominion Astrophysical Observatory (DAO). The total of ~110,000 spectra is the combined output from two telescopes: >93,000 from the 1.8-m telescope (inaugurated in 1918) and operated primarily at Cassegrain, and >16,000 from the 1.2-m telescope, built with a dedicated coudé in 1962. Since its inception the DAO has pursued the determination of spectroscopic orbits, often for objects with periods of months or years, and while most of the spectrograms exposed for that purpose have probably been measured in the respect of just one variable (the radial velocity), other aspects of the spectra and variable phenomena will have received qualitative assessment only and for the great majority parameters such as temperature, luminosity class and metallicity have never been investigated. Eclipsing variables also feature in the archive, and in particular the chromospheric eclipses exhibited by binaries of the ζ Aurigæ type. Digitization of both sub-sets of the archive has recently commenced using an upgraded in-house PDS scanner, and semi-automated reduction pipelines are now in place for data acquisition, extraction and reduction. Priority is being given to variable objects such as novæ, long-period variables and peculiar stars, but reasonable requests for scanning are also accepted. The fully-reduced data are being mounted on the CADC's public Website (http://www.cadc.hia.nrc.gc.ca/cadc/). The current tally of spectra thus available is approaching 1000 mark, and is steadily rising. The logbooks are also being keyed-in manually. When funds become available, a second PDS will be upgraded and operated in parallel.

2.3. *Sofia, Bulgaria*

As Tsvetkov explained, while astronomers at Sofia are responsible for the Wide-Field Plate DataBase project, Sofia itself is a repository for the Rozen Observatory (Bulgaria) collection of ∼9000 plates, plus minor sets from Byurakan Observatory in Armenia, and ESO (La Silla). The combined collection includes direct images from Schmidt telescopes, thereby providing coverage of both the northern and the southern sky; in time they span approximately 1970–1995. Scanning has been started, using Epson scanners.

2.4. *Sonnenberg Observatory, Sonneberg, Germany*

Tsvetkov described how P. Kroll has digitized (with commercial scanners) ∼215,000 of the total of nearly ∼300,000 plates. Those plates (of the northern sky) cover the period from 1923–1995, and were exposed in various "sky patrol" programmes which were pursued at Sonneberg for decades. Data from Kroll's digital images are being stored on CDRom as compressed TIFF files.

2.5. *Norman Lockyer Observatory, Devon, UK*

Goulev summarized the history and status of the collection of ∼6000 photographic spectra now at Norman Lockyer Observatory in SW England. The Solar Physics Observatory, London, had accumulated "numerous" spectra of bright stars on 8 × 3 cm and 16 × 8 cm plates from 1880 to 1912 (when it closed). At that point the plates, *inter alia*, were sent to Cambridge University Observatory (and were largely neglected), but Sir Norman Lockyer, Director of the London site from 1885, continued his work at what was then known as "Hill Observatory" in Devon. Now run largely as an educational facility by local volunteers, the NLO plans to generate a workable database from its own collections of observations, which include a significant number of novæ spectra, in particular of Nova Per (1901). Preservation efforts are under way, and some of the series of plates are already inventoried. When the Cambridge Observatories recently disbanded its own plate store, the collection sent there from London in 1912 was donated to the NLO. Part of any funds raised on the occasion of NLO's Centenary (February 2012) will contribute towards digitizing the scientific observations.

2.6. *Asiago Observatory, Italy*

Barbieri described the collection of ∼16,000 plates at the AO, and a current programme for digitization with an Epson scanner. The time required to digitize one plate electronically at 1600 × 1200 dpi, including cleaning it first, is ∼20–30 min. The digitization programme has been guided by science topic, with an emphasis on AGN variability which has yielded evidence for the binary supermassive black hole in BL Lac itself. The logbooks for the plates have also been digitized, and are on-line.

2.7. *Ukraine Observatory, Kiev, Ukraine*

Pakuliak outlined progress towards putting on-line the photographic observations, totalling some 300,000 direct plates, exposed between 1895–1996 at six Ukrainian Observatories: MAO/NAS, Mykolaiv, Crimean, Kiev, Lviv and Odessa. Each observatory is conducting complementary scientific investigations—from comets to quasars—with its archived material. Resources are limited, but nearly 75% of the plates have been digitized at low resolution using commercial flatbed scanners, and the plan is to do higher-resolution scans based on science priority. Two scans have been made per plate, one rotated through 90°, to quantify and thence minimize instrumental effects.

2.8. *Royal Observatory of Belgium, Brussels*

The ROB was unfortunately not represented at the Workshop. A purpose-built high-speed scanner, code-named *DAMIAN*, has now started scanning, though not solely astronomical material since it is a shared instrument for aerial-photographic films. It has for some time been planned that it serve the broad community by digitizing Europe's astronomical plate archives, but it is still early days. There is also the hope that it will eventually lead to full and public release of data so that, when combined with *DASCH* , astronomy's historical coverage will be enhanced to its maximum potential.

2.9. *Wide-Field Plate DataBase*

A very considerable effort towards realizing the Workshop's first two objectives has been made by the team who have been working very productively on the Wide-Field Plate DataBase (WFPDB), which has its headquarters in Sofia (Bulgaria) under the very capable direction of Milcho Tsvetkov and Katya Tsvetkova. The project started as an IAU initiative under what was then Commission 9, but following the disbandment of IAU Commission 9 *per se* in 2006 it has continued as an independent entity. The development of the Wide-Field Plate Database (http://www.skyarchive.org; Tsvetkov 1992), summarised by Tsvetkov, has been accumulating meta-data for $> 2.4 \times 10^6$ wide-field plates in various parts of the world but chiefly in Europe. Details of that project are to be found on p. 417. A synopsis of the current status of the Wide Field Plate Archive (WFPA), which can be accessed at http://www.wfpa.bas.bg/catalogue.html, brought the Workshop up to date, and kindled the hope that such comprehensive inventorying be seen through to completion. As Tsvetkov demonstrated, the main Catalogue is a summary table of the characteristics and coverage of 475 astronomical plate collections; another on-line table, the Wide-Field Plate Indexes, contains the parameters of over half a million plates from 133 archives, and can serve up quick-scan (preview) images of some of those; about 250,000 in total have been scanned to date.

3. Photometric Analyses and Data Distribution Plans

The provision of accurate and precise photometry from direct plates, and of spectrophotometry from spectrograms, is essential to reap the full science reward from historic photographic plates. While procedures for coping with the non-linear response of a photographic emulsion to incident light have been further developed since first evaluated by Hurter & Driffield (1890), one can only approach photometric calibration asymptotically. For direct (imaging) plates, the calibration is best done (in practice, can only be done) by using modern full-sky photometric catalogues to derive spatially- and magnitude-dependent fits to the digital images, as described for *DASCH* in Section 3.1.

For photometric calibration of spectrograms, unless full information in the form of auxiliary exposures to calibration devices is available, approximations are inevitable, particularly where exposures are rather heavy or rather light. The faint end of an exposure also needs careful handling on account of reciprocity failure: an emulsion will fail to record low-level light below a certain threshold, depending on the emulsion type and its state of hypersensitization, if any. Coupled with the provision of calibration marks, such as dots, strips or wedge-readings, is the need for physical measurements that render these informative. Even then, those physical measurements may not provide complete information. If (as not infrequently happened) the source illuminating the raster of holes or strips did not illuminate them evenly, an iteration to correct for that non-uniform illumination is required, though not many observers apparently knew of the need. The matter is non-trivial, since the shape of an emulsion's "characteristic curve" is rarely represented by enough points to define it unambiguously if some are incorrect through the effect mentioned above.

3.1. *DASCH*

Photometric analyses developed by Laycock *et al.* (2010) and Tang *et al.* (2012) yield rms uncertainties of ~0.10 mag over the full range of ~100 years of data from the 9 or so series of plates that contribute to a typical light curve, despite differences in plate scale, image quality and systematic effects. The basic approach employs SExtractor as the object detection and isophotal photometry engine for instrumental magnitude determination. Calibration curves are first derived in annular bins to account for vignetting, with instrumental magnitudes fitted against Hubble Guide Star Catalog (GSC2.3) magnitudes (B) for an initial photometric solution. This initial calibration is followed by local corrections to remove spatially-dependent plate effects (usually caused by the emulsion) or sky-related effects (atmospheric extinction, clouds, even trees near the horizon!). The GSC2.3 catalogue is not ideal since its photometric precision is only ±0.2mag and it is predominantly in a single band (photographic B). Fortunately, the all-sky APASS CCD survey (www.aavso.org/apass) described by p. 95 has Johnson B and V as well as Sloan g', r', i', which will improve significantly both precision and, particularly, colour corrections for *DASCH* photometry.

The *DASCH* Pipeline (Los *et al.* 2010) and database software run on a high-speed computer cluster and RAID disk system that can process (overnight) the full Pipeline for about 400 plates scanned per day, in production and processing mode, to populate a MySQL database with photometric values and errors for each of the resolved stellar images (typically ~100,000 on a standard plate; more on A plates). Light curves are generated very rapidly for any object by extracting from the database the magnitudes thus determined from all plates, or only those with magnitude measures meeting a set of user-selected criteria. Variability measures and tests of their validity can then be derived readily. Additional variability analysis tools are being developed, and will be made available when the full database becomes public. To complete each "tile" of sky, all adjacent tiles need to be scanned first to ensure complete coverage; the tile in question can then be released (see p. 29). A "demonstration tile", 6° centred on the open cluster M44, is available on the *DASCH* Website; light curves and digital images (thumbnails) centred on a given object can be displayed or downloaded for each point in the light curve, and light curves can be readily plotted to either eliminate or include points that have passed a wide variety of Pipeline "filters" (e.g., to remove blended images, plate defects, etc.).

The full *DASCH* output database of ~450,000 plate images and derived magnitudes for each resolved object (total ~1Pb!) will be made available on disk for world access as it is completed incrementally. The present plans are to digitize the northern sky at Galactic latitudes $|b| > 10°$ first, which enables comparisons first with existing surveys such as SDSS, PTF, Pan-STARRS1 (PS1) and the Catalina Real Time Survey (CRTS). That stage could be completed by mid-2013. Given the difference in limiting magnitudes (<14–18 for *DASCH* against <20–22 for the modern surveys), such comparisons will be mainly for context or extreme transients. The southern sky above/below the Galactic plane will be next, with expected completion in early 2015, followed by the Galactic Plane by mid-2016—or well before LSST. The reason for doing the Galactic Plane and Bulge last is to allow time to develop crowded-field photometry analysis further by invoking point-spread function (psf) and image subtraction techniques in order to improve the present isophotal photometry used for SExtractor. Experiments have recently been undertaken to optimise the use of PSFEx+SExtractor for magnitude- and position-dependent fitting of the plate psf.

3.2. *Asiago*

The photometry from scanned images, as presented here by Barbieri, has achieved photometry results with derived uncertainties in the range ∼0.1–0.15 mag (Johnson *et al.* 2005). The Asiago plates are primarily from just two telescope series (see the WFPA website referenced above) and are therefore more homogenous than the much larger number of plate series that contribute to the light-curve data of the *DASCH* output. The psf-fitting routine described in Johnson & Winn (2004) will be investigated for crowded-field photometry for *DASCH*. Unfortunately the limited resources for the Asiago project may preclude full photometric processing, data archiving and distribution.

3.3. *Ukraine*

The photometric quality of the digital scans of the several plate collections from the six Ukraine observatories is being investigated. However, lack of adequate resources does curtail full-scale processing and data distribution.

4. Prioritization and Future Plans: Message from the Workshop

(1) The benefits of Historical TDA depend critically on the "100-year" dimensions of astronomy's archived resources. This unique window drives the efforts referred to above and opens new discovery space. However, most of the afore-mentioned projects are seriously short of resources, mostly funding. It seems likely that full processing and public release of historical TDA data will become a reality for the Harvard-/DASCH data, but not for other projects. Compared to major current and future surveys, the requirements echoed by this Workshop are small. Investment of even part of the necessary resources will take our science to uncharted depths (e.g., see Tang *et al.*, p. 447).

(2) The Harvard archive suffers from the "Menzel gap" of ∼15 years (caused by HCO Director Menzel cutting the Harvard plate programme from 1954–1965, but which was only recovered completely by ∼1970), and priority needs to be given to all data which will help to fill that gap.

Acknowledgements

We gratefully acknowledge support for the *DASCH* programme by the HCO, the NSF (grants AST0407380 and AST0909073), and from the *Cornel and Cynthia K. Sarosdy Fund for DASCH*. We are also grateful to the HIA (NRC, Canada) for temporary funding to continue and enhance voluteer efforts in the plate-scanning department.

References

Grindlay, J. E., *et al.* 2009, *ASPC*, 410, 101
Hurter F. & Driffield, V. C. 1890, *J. Soc. Chem. Ind.*, 9, 455
Johnson, J. A. & Winn, J. 2004, *AJ*, 127, 2344
Johnson, J. A., *et al.* 2005, *AJ*, 129, 1978
Laycock, S., *et al.* 2010, *AJ*, 140, 1062
Los, E., Grindlay, J., Tang, S., Servillat, M., & Laycock, S. 2010, *ASPC*, 442, 269
Simcoe, R., *et al.* 2006, *Proc. SPIE*, 6312, 17
Tang, S., *et al.* 2012, *ApJ*, in preparation
Tsvetkov, M. 1992, IAU WGWRI Newsletter, 2, 51

New Horizons in Time-Domain Astronomy
Proceedings IAU Symposium No. 285, 2011
R.E.M. Griffin, R.J. Hanisch & R. Seaman, eds.

© International Astronomical Union 2012
doi:10.1017/S1743921312000683

Data Management, Infrastructure and Archiving for Time-Domain Astronomy

David Schade

Canadian Astronomy Data Centre, NRC Canada, Victoria, BC, V9E 2E7, Canada
email: David.Schade@nrc-cnrc.gc.ca

Abstract. The workshop on Data Management issues for Time-Domain Astronomy was conceived as a forward-looking discussion of the primary issues that need to be addressed for science in the time domain. The very broad diversity of the science areas presented in the main Symposium made it clear that most of the general issues for astronomy data management—for example, large data volumes, the need for timely processing and network performance—would be pertinent in the time domain. In addition, there might be other tight time constraints on data processing when the output was required to trigger rapid follow-up observations, while science based on very long time-baselines might require careful consideration of long-term data preservation and availability issues. But broadly speaking, data management challenges in the time domain are not at variance to any significant degree with those for astronomy or data-intensive research in general. The workshop framed and debated a number of questions: What is the biggest challenge faced by future projects? How do grid and cloud computing figure in data management plans? Is the Virtual Observatory important to future projects? How are the issues of data life cycle being addressed?

Keywords. Data analysis, astronomical databases, surveys, standards, sociology of astronomy

---

1. Introduction

The purpose of this workshop was to invite both a general scientific audience and a few representatives of some of the major astronomy projects of the near future to discuss what the future will look like for astronomy data management. The present-day environment features fast networks, cloud computing, cheap mass storage, and the Virtual Observatory. Governments have funded major computing infrastructure to support research, and the private sector can deliver on-demand scalability for a price. Major data centres have grown in sophistication and power. This rapidly-evolving technology landscape offers a range of options to observational astronomers. How have they reacted? What choices are being made for projects of all sizes that are near enough that data management plans have been developed in some level of detail?

There were people present who identified themselves as representatives of some of the major coming projects: LSST, GAIA, Pan-STARRS and others. There was representation from the international Virtual Observatory (VO) community and from several data centres. This report was based on an audio recording of the workshop; it simply summarises the main areas of discussion, the most salient points raised, and the general consensus on an issue, if one presented itself. Some of the statements about individual projects may not be consistent with the "official" project position since they may not have been made by project representatives.

2. What is the Biggest Infrastructure or Data Management Challenge?

The first area of discussion focussed on identifying the challenge which this group considered the biggest data management or infrastructure problem that they faced. Was it processing or storage capacity? Was it I/O challenges or networking—either internally or between the data collection and the users? Was it the cost of infrastructure?

Some projects, including GAIA, reported that the most serious obstacle had not yet been clearly identified. One smaller project reported that some of the obstacles were rooted in funding rather than technical problems. It is a question of both money and human resources. Does each project need to build its own data centre to house its collection, or can shared facilities be used to reduce costs? It was reported by an LSST team member that one of the biggest challenges is to establish the bandwidth to get the data from Chile to the United States. Large investments were planned to guarantee that the necessary bandwidth would be available, and will involve private priority access to network capacity—a challenge that is far more onerous than obtaining the processing capacity. There was some discussion of whether it was better to do major processing near the data in Chile, with the goal of reducing the quantity that needs to flow through long-distance fibres. It was stated that the baseline plan was to use major computing power (e.g., NCSA) that would be available only in the USA.

It was pointed out that there are at least two distinct areas where network capacity might be a challenge. The first is moving data from the observatory where they are produced to the processing and storage facility. Then a second problem might be the effective transfer of (possibly) Terabyte-sized datasets from the storage facility into the hands of other end users. It was possible, if one had extra processing power, to use that processing power to reduce storage requirements through compression. Of course, many major data centres and projects use compression as a routine part of their data management plans.

A participant familiar with VISTA made the point that they were starting to find that the catalogues which they produce are larger than the image data from which they were derived. This surprising situation can easily occur if many parameters are measured from image data for a single object. VISTA was experiencing a number of bottlenecks in data transfer between Garching, Cambridge and Edinburgh, and they were forced to use shipment of hard drives part of the time. Some participants experienced in the shipment of hard drives remarked that they found that such shipment was a very unsatisfactory means of data transfer, and was neither cheaper nor faster than the internet in the long run. Others pointed out that internet transfer might be best from some locations, for example La Palma to the United Kingdom, but that international transfers from countries like South Africa were simply not feasible using the internet. We clearly live in a world that is not as well connected as we like or need. In fact, it is normally the case that network performance is substantially below what might be expected. Some participants had experienced research internet connectivity that was nominally 1 gigabit per second between two locations but which in fact performed at only a small percentage of that figure.

In summary, moving data was frequently identified as a more serious challenge than data storage, and storage was seen as a greater challenge than access to sufficient processing power.

3. Grid and Cloud Processing

In Canada, the USA and Europe, governments have made major investments in computational infrastructure whose purpose is to support research. Facilities such as Compute Canada, TeraGrid, and now the Extreme Science and Engineering Discovery Environment in the USA and the European Grid Initiative are operated and managed to solve large-scale computing problems, including ones similar to those faced by astronomy and by its major survey projects in particular. In addition, there now exists a number of commercial clouds, such as Amazon Elastic Compute Cloud and Microsoft Windows Azure, whose ability to deliver on-demand processing and storage services might be attractive to large projects or to small teams as components of their data management systems.

Are any of the major projects trying to exploit the situation by using either government-funded or commercial facilities rather than building their own project-specific infrastructure? It was remarked by several participants that the commercial clouds are too expensive. Fully-costed estimates of storage, networking (I/O) and processing requirements indicate clearly that this is not a cost-effective way to approach the major problems of observational astronomy's data management. There were no opinions contrary to this viewpoint. A few participants mentioned that their projects had used commercial clouds in an experimental mode to evaluate performance and cost, but no one reported that those clouds were now in their long-term production data systems.

What about the research or academic clouds that are funded by governments as shared facilities? Are those resources being exploited? The history of high-performance computing had developed without data-intensive computing as part of its mandate—a situation that is only now beginning to change. According to one speaker, LSST has used Tera-Grid for some of the data challenges and to do preliminary evaluations of their processing problems. However, the components of the official LSST data management plan does not include any public facilities, nor are there plans to utilize major shared facilities in the future, either public or commercial. It was considered that a custom-designed processing and storage system for LSST was the only way to make a system that was efficient enough to handle the scale of the problem.

The Canadian community is involved in a project that is a collaboration of universities and national astronomy data management infrastructure. Its specific aim is to bring the national high-performance research computing (HPC) infrastructure (Compute Canada) into service for observational astronomy research. Like other HPC organizations Compute Canada has not been deeply involved in data-intensive research. A great deal of progress has been made, and a petabyte of storage has been allocated as well as access to significant computing power. This is the first project where production-level cloud computing has been done on Compute Canada systems. Both Compute Canada and the user community is slowly, and somewhat painfully, getting used to operating in a new way. The vast publicly-funded research infrastructure that is available to the research community makes this approach potentially extremely important and valuable, but astronomy teams do not yet easily find ways to interface to this system.

One participant described the experience of users of the EVLA, where the data rates may produce several terabytes of data for a small-to-medium sized programme, and could take weeks to process. Where should the processing be done? It is not easy to *ftp* the datasets back to one's home institution for processing, so is the solution to use some grid local to the observatory or the US TeraGrid? It may be important to realize that NRAO itself does not have sufficient facilities locally to store and process that quantity of data, leaving astronomers to scrounge for disk space and processing capacity.

To summarize this part of the discussion, both major and even moderate-sized, projects feel that they need to create custom-designed data management infrastructure for their projects. Although national research computing infrastructures may be able to provide sufficient processing power they have other shortcomings that prevent them from being used: they are not, in general, configured for data-intensive research; massive storage is not tightly coupled to fast internal networks; problems of efficient I/O persist, and the geographical layout of major shared facilities is, in general, not appropriate. Furthermore, there are issues of long-term commitment of resources to the projects.

None of the major projects has plans to use shared cloud or grid computing facilities, be they are academic or commercial.

4. Virtual Observatory and Time-Domain Astronomy

Most, if not all, major projects include in their data management plan statements such as: "Our data services will be compliant with the Virtual Observatory". But what do statements like that really mean? Do they mean that the application of the products of the VO will help their projects achieve their science goals? Do they mean that the projects will help the astronomy community reach the goal espoused by the VO, namely that all major sources of data be capable of a scientifically useful level of interoperability?

A number of the participants at the workshop are involved with the International Virtual Observatory Alliance (IVOA), which coordinates the work of the national VO groups. The IVOA considers it essential that major new projects participate in the development of VO standards and that they apply those standards in their own data systems. If not, it cuts the legs off the VO and defeats the main thrust of VO work, which is to create a unified data management system for astronomy, at least from the user's perspective. The research users are one potential beneficiary of the work of the IVOA. But what do survey teams or major projects gain from the use of IVOA protocols?

Is there really substantial buy-in to the use of VO standards? What do we mean by "substantial buy-in"? One answer is that organizing a project's observational metadata in a VO-standard way as close to the moment of observation as possible would deliver the maximum benefit to the projects' data management plan, in the sense that downstream from that point any VO standard could be incorporated into the design of that plan. A good example might be the Hubble Legacy Archive, where the Simple Image Access Protocol (SIAP—an IVOA standard) was built into the core of the system, so its data products were ready instantly for VO access. That is the way we would like to go but clearly there may be other reasons, such as efficiency, for structuring data in a specific way that differs from the IVOA standards. The IVOA does not provide a complete, end-to-end set of protocols that can be used as the basis for a project data management system.

Major projects would be motivated to adopt IVOA standards as parts of their systems if it were clear to them that it would lead to increased efficiency or effectiveness or reduced cost. Is it possible to make that case? There are examples, like the Hubble Legacy Archive, that demonstrate value for users. There is the example of the Canadian Astronomy Data Centre (CADC), where IVOA standards have been used in a variety of roles in the internal design of their data management system. In that example it was clear that using those standards saved effort in designing some parts of the system, although implementation was not always perfectly straightforward. It was also reported that an agreement was being developed between LSST and the VAO (Virtual Astrophysical Observatory, the VO organization of the USA) to define what those two organizations can do for one another to ensure VO access to LSST data products.

Are VO standards easy to implement? It was remarked that a presentation to a LIGO (Laser Interferometer Gravitational-Wave Observatory) meeting elicited the question: "Where can I find the software to implement VO standards?" It was agreed that there exists some software but not really enough to implement major pieces of the overall VO set of protocols. This is an area that clearly needs improvement if projects are to be able to derive benefits from the VO.

According to one participant a search of the IVOA documentation failed to turn up a clear explanation as to how one should proceed with applying IVOA standards and protocols. The VO documentation was desribed as "opaque" and difficult to understand. Several other participants echoed and endorsed that view. It was remarked that the IVOA had been in existence for a decade and had, over that time, successfully developed its own culture and jargon; it was referred to as a "maze of standards and documents" that was difficult for astronomers to navigate. There was a perception that there are no clear directions from the VO to the data providers on how to structure their data and systems. These points constitute serious criticism of the accessibility of IVOA products.

Proponents of the IVOA and of the VO in general suggested that it should be an IVOA priority to provide data managers with server-side and client-side libraries of software that support implementation of standards and protocols. In fact, there do exist some repositories of such software (e.g., `code.google.com/p/opencadc/`). A more complete set of tested and documented software would be a great step forward for the IVOA.

The adoption of standard data models was discussed. It was argued that the most fundamental benefit that could be delivered by the IVOA would be a clear and usable data model that met the needs of the vast majority of projects. It was pointed out that standard VO data models are not intended to cover every last eventuality and treat every piece of metadata, since specific observatories and instruments will always employ some set of unique characteristics. But there is a large subset of the attributes of an observation that are common across most datasets (an obvious one being the pointing on the sky), and where there are common attributes then we should adopt a common language to describe them. If we could develop a data model that standardized an extensive set of common attributes, then most observatories could adopt them as their starting point for their metadata systems.

In summary, there was a feeling that the VO has not made its products as accessible as they need to be if they are to be adopted by the developers of project data management systems. The main goal of research projects is to achieve their science goals; anything that assists might be adopted, but additional effort cannot be expended on the VO unless it is an assistance rather than an annoyance. *Those immediate considerations of reaching project science goals will always trump downstream considerations such as preserving data for the benefit of the community.*

5. Long-Term Preservation and Data Life Cycle

Many established data centres are philosophically bound to the idea that astronomical data have long-term value and therefore need to be preserved, but in reality they have no mandate or funding from external parties to support them in the pursuit of long-term data curation. The strong commitment that NASA has made in the past toward good data management and long-term archiving now appears to be suffering serious erosion. There was a worry that NASA, as an organization, was backing off from its understanding of the value of well-calibrated and well-archived data products. As a research community, we need to be vigilant in order to keep data preservation on the agenda.

In the past decade or two the cost of maintaining archives of astronomical data from earlier generations of instruments has been negligible because each new generation has produced orders-of-magnitude more data than had the previous generation. In consequence, we have not had to face seriously one of the important questions related to data life cycle: when are we justified in discarding data? Remarks were made that every survey that has ever been done has required re-processing at least several times, so losing raw data would have been scientifically catastrophic. However, things seem to be changing in that the data volumes produced by some projects greatly exceed storage and network capacity. Hard decisions must therefore be made at some point in the foreseeable future.

It is anticipated that facilities like the SKA will drive us toward different models for our data management systems, and different approaches to data life cycle. Careful cost-benefit analyses, driven by science, will need to be done. None of the projects represented at this workshop expressed the intention of discarding data at some point in the data life cycle. It was widely agreed that, so far, little has changed in our ability to process data optimally the first time. Data therefore need to be kept.

Keeping data over a long term implies not only that the data exist but that they be accessible; otherwise their value is minimal. It implies continued investment in archiving far into the future. What about archiving analysis and processing software? This has been discussed for many years but has rarely been implemented.

Most of the projects represented at this workshop were focussed clearly on achieving their science goals. Considerations of data life cycle and long-term preservation are, justifiably, secondary and will be considered seriously only at a later time.

6. Conclusions

- Most projects intend to develop their own custom-designed infrastructure to support their data management systems.
- VO is not delivering major benefits to the data management plans for most projects.
- Commercial clouds and government-funded academic research clouds are not part of the planned infrastructure for any of the projects represented at this workshop. Commercial clouds are too expensive and academic clouds are generally still not designed with data intensive research as a high priority.
- Processing power is not a major challenge.
- Storage can be costly but is not a major technical challenge.
- The most challenging activity is moving data between the components of a data management system, both internally (local networks and I/O) and externally (long internet transfers from observatory to data management site, and from thence to users).

In a nutshell, the people attending the workshop and the projects which they were representing are committed to the principle of building project-specific data management systems from the ground up. Alternatives such as exploring the use of grid or cloud computing, the VO, or using expertise and facilities in data centres and existing projects and sharing facilities are not (yet) being considered.

Acknowledgements

I would like to thank all the participants at this workshop for a lively and interesting discussion, and particularly JJ Kavelaars for helping conduct the workshop.

New Horizons in Time-Domain Astronomy
Proceedings IAU Symposium No. 285, 2011
R.E.M. Griffin, R.J. Hanisch & R. Seaman, eds.
© International Astronomical Union 2012
doi:10.1017/S1743921312000695

Amateur Community and "Citizen Science"

Arne A. Henden

AAVSO, 49 Bay State Road, Cambridge, MA 02138, USA
email: `arne@aavso.org`

Abstract. Citizen Science is the act of collecting or analyzing data by enlisting the help of volunteers who may have no specific scientific training. The workshop discussed how "Citizen Science" fits into time-domain astronomy, what the roles of such volunteers might be, and how amateur astronomers can help in the new era of surveys.

Keywords. surveys, techniques: photometric, stars: variables: other

1. Introduction

Amateur astronomers have been involved in every aspect of astronomy for centuries; in fact, the concept of a professional astronomer is only about a century old. Most early researchers had other careers (for example, Johannes Kepler was a mathematics teacher at a seminary; Isaac Newton was Lucasian Professor of Mathematics at Cambridge, UK) and only made astronomical observations as a hobby.

The concept of using a team of amateur astronomers as volunteers to help professional astronomers in their research is relatively recent. Harvard College, for example, enlisted the aid of amateurs to monitor the variable stars that they were discovering as part of the photographic all-sky survey which was being conducted at the turn of the last century. That group of volunteers formed the core of the American Association of Variable Star Observers (AAVSO), founded in 1911.

Other professions, such as ornithology, have also used amateur researchers for decades. Cornell is currently conducting a programme called "eBird", launched in 2002, where amateur bird-watchers document and enter on-line the date, time and location of their sightings of various bird species. The information is then used to gain a better understanding of patterns of bird occurrence and the environmental and human factors that influence them. The Christmas Bird Count, run in the USA by the Audubon Society (`http://birds.audubon.org/christmas-bird-count#`), has been in progress since 1900, pre-dating even the AAVSO. The advent of consumer electronic devices such as "smart phones" has greatly enhanced the ability of volunteers to provide useful scientific data, e.g., accurate time and location of bird identifications, and the possibility of taking pictures or recording song.

The Citizen Science workshop concentrated on the following questions:

- What is the difference between crowd-sourcing and true citizen science, and their roles in TDA?
- How should amateur contributions be properly acknowledged?
- How can we put professionals in touch with amateur(s) appropriate for their needs?
- How can we use the diverse professional skills (such as database expertise or website design) of the amateurs within scientific projects?
- What is the role of amateurs in the new age of surveys?

This report summarises the lively discussion which ensued as the group of ~30 work-shop participants tackled those questions in turn.

2. Crowd-Sourcing or Citizen Science?

The largest crowd-sourcing experiment currently in process (Raddick, et al. 2010) is Zooniverse (http://www.zooniverse.org). The first "zoo" was *Galaxy Zoo*, which displayed SDSS galaxies to users and asked them basic questions about shape and structure. This task required no knowledge of what a galaxy was; it just used humans to do the image classification. It was launched in 2007 and, when completed, had 10 million galaxies classified by 30 individual researchers each, with a total of about 83,000 participants. The Zooniverse itself reports about 500K volunteers divided among the dozen or so active projects.

Crowd-sourcing can also be applied profitably to time-domain astronomy. An example is the *Planet Hunters Zoo*, an experiment to find extra-solar planets by their transit signatures in the light curves generated through KEPLER public release datasets. The analyst is required to look at graphs (harder and less interesting!), pick out rare dips in brightness, and then see if there is a periodicity to multiple dips. The voluteers have discovered about 30 candidates that have been missed by the KEPLER team itself. The training and background is far more extensive than for *Galaxy Zoo*, and falls more into the amateur data-mining category.

One could imagine a similar team of amateur researchers looking at the time-domain transients found by LSST, or helping train the neural network for PTF that discriminates against image artifacts.

Results from *Citizen Sky* (http://www.citizensky.org) also sparked discussion. *Citizen Sky* was a true Citizen Science project in which amateurs and interested general public were trained in how to make brightness estimates of the $3^{\rm rd}$-magnitude star ϵ Aurigae, an eclipsing binary with a 27-year period and which happened to start its 2-year-long eclipse during IYA 2009. The volunteers were encouraged to learn about variable stars, create teams to work on research topics, and learn how to write up their results in papers for submission to a scientific journal.

Variable star astronomy is an obvious topic for citizen science, as amateurs can make brightness estimates with very crude equipment such as binoculars and the human eye, or with sophisticated robotic telescopes and CCD cameras. All such observations have their value. For example, participants noted the advantage of large numbers of observers dispersed in longitude when determining when an outbursting object starts to increase in brightness, alerting professionals quickly for detailed monitoring. Very little equipment is needed for such discovery, yet the amateur then has at least a small chance of becoming very famous. Several researchers using satellites may get alerted by amateurs when an object goes into outburst, so that they can schedule Target of Opportunity programs. Sometimes it is the opposite: objects that might go into outburst need to be monitored just prior to a satellite observation in order to know certain parameters when the object is in quiescence; this was the case for the recent V455 And observations for HST.

Where amateurs really become indispensable is when time-series observations must be made. Certainly cataclysmic variables are obvious candidates, since in outburst one may detect modulations from the accretion disk (such as super-humps), or can detect eclipses more readily than when the system is in quiescence. Time-series observations are also important in many related fields, such as the confirmation of exo-planet discoveries. Many variable stars have multiple periods, such as the Blazhko RR Lyr variables, where many nights of time-series data are required in order to understand fully their behavior. This is nearly impossible to do through prevailing policies of Telescope Allocation Committees for most major professional telescopes. For bright objects, even DSLR cameras can be used for high-precision results, one good example being the light curve of U Sco derived by Loughney (2010) and shown in Fig. 1.

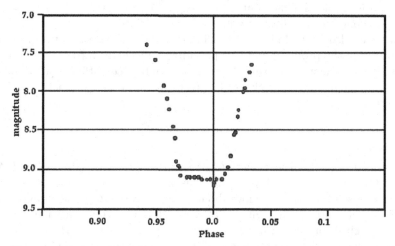

Figure 1. U Sco eclipse light curve, obtained with a digital SLR camera

Amateur participation can also be important when geographical location is an essential ingredient to success, as is the case (for example) for asteroid occultations. During such events a narrow shadow track, with width roughly equivalent to the asteroid diameter, crosses land masses, and in order to get sufficient chords of the event to model the asteroid shape, the event needs to be observed by many stations. Amateurs have even created portable systems (Degenhardt 2006) to monitor events of that kind, setting them up perpendicular to the predicted path so as to obtain as many chords as possible. That kind of activity is impossible with fixed-location professional observatories, neither is the volunteer effort required to set up such portable stations usually possible for professional researchers.

One concern raised by participants was the potential loss of data. Since contributions from amateurs are becoming more valuable, the observations need to be archived carefully. The AAVSO does a good job of archiving individual photometric measures of variable stars (the database contains more than 21 million variable-star observations, mostly by amateurs), but CCD images are not archived by the AAVSO and must be done individually by the amateurs. There are potentially important observations on many of those those images but they are effectively lost unless either the images or at least the photometry of all stars on them are stored in a publicly accessible manner. An example of the need for that kind of access is the recent supernova in M101, where valuable precursor photometry could in principle be obtainable if data archives were set up by the large number of amateurs who record deep-sky images of such objects on a routine basis.

3. Acknowledging Amateurs

Suppose you have worked with a team of amateur astronomers, some of whom have contributed data for your project, and others have done a literature search or have mined existing data catalogues for useful pertinent information. What criteria are you going to use for deciding which participants get recognition as co-authors, which will get an acknowledgement in the body of the paper, and which will be ignored?

The usual model for inclusion is the one established by the Center for Backyard Astrophysics (Patterson 2001), which generally requires that an observer must contribute one percent of the total data used for a paper in order to be considered for co-authorship. Most of the workshop participants felt that amateurs should be acknowledged in a fashion similar to that for professional co-authors, or might merit even stronger acknowledgement since they are contributing their volunteer time and effort.

Amateurs can also be included in the "broader impact" section of a grant proposal, thereby bolstering the proposal and also providing a mechanism to obtain funding, especially for new instrumentation. Amateurs participate beause of they enjoy the process, so the more encouragement that they can be given, the more likely that they will continue to contribute.

4. Facilitation

The American Astronomical Society (AAS) had a "Pro-Am" (Professional–Amateur) working group for a number of years, but disbanded it a couple of years ago on the grounds that most professionals know about amateur capability and individually find a method of locating and working with amateurs with appropriate equipment and skills to be useful for a given project. Since there are European Websites which have databases to link people who wish to share rides to various destinations, could the same basic structure be used to list the capabilities of amateurs so that professionals could make contact more easily with ones that meshed well with their projects? In fact, the AAS Pro-Am working group tried that at one stage but found that the Web design needed a dedicated volunteer, while a long-term coordinator was essential in order to keep entries up to date. With the right volunteers, however, such a facilitation Web-site might yet be workable.

Facilitation works well when it comes to reaching the general public. IYA 2009 demonstrated the (outstanding!) ability of the amateur community to band together and create outreach activities, far more extensively and successfully than those created by the professional community. IYA 2009 is a great example of a successful outreach programme. Amateurs are better connected with the public, so if they can be a part of a research project then they can bring that project to the public, whether through a talk at a local astronomy club or participation at a star party.

5. Use of Talents

Except for crowd-sourcing, the range of amateur expertise is extremely wide. For high-precision, high-accuracy photometry, for example, only the top amateurs can come close to the professional requirements of transforming their data, using proper error analyses, understanding their equipment, and so on. Bad habits are easy to acquire and hard to eliminate. Would it help to create a standardized system, whether of equipment, software or observational techniques, so that more amateurs could provide truly high-quality data?

One concern is the lack of quality software and textbooks for spectroscopy. Very few amateurs are currently spectroscopists, and since there is a lack of adequate training, even they may not be providing spectra of the professional quality of which their equipment could be capable. This is truly a growth area for amateur astronomers, so the professional community needs to devote time and effort into fostering it. However, some amateurs are indeed providing quality spectra, examples being the high-precision radial-velocity work on Polaris by a German amateur (Bücke 2006), the supernova spectroscopy from the French amateurs led by Christian Buil (Desnoux & Buil 2005), and the high-precision

Figure 2. Bench-mounted radial velocity spectrograph

exoplanet radial velocities from the Arizona Spectroshift project (Kaye, *et al.* 2005). An example of a bench-mounted spectrograph for precision radial velocities designed by Kaye is shown in Fig. 2.

It is universally agreed that amateurs have the right attitudes for observing, and that they offer an obvious advantage in nearly limitless telescope time. Their instruments often have a bigger field of view than professional telescopes do. Many have dogged determination, and will follow objects such as WZ Sge systems that might go into outburst once every few decades, or look for eclipses in long-period binary stars where only one eclipse has been seen to date and may nor recur until next century. They also hugely enjoy their contacts with the professional community. Their main limitation is a financial one, since amateurs are self-funded.

It is unfortunate that the word "amateur" does not gain much respect in today's world, even though its basic meaning is simply someone who loves what he or she is doing. Should amateur astronomers be re-branded as "semi-pros"? (that is, professional in quality but not having the status of a professional). "Citizen Scientist" might also be a suitable term, though is somewhat more formal.

It is important to reflect that most amateurs are professionals in other fields, and can bring that expertise to their scientific work, whether as database designers, colour-imaging experts, physicists, statisticians, or the like. Chemists often understand laboratory spectroscopy far better than the average professional astronomer does. Where possible, all that expertise should be tapped and applied to support a research project.

The amateur community has also helped professional astronomy in another way, by generating the incentive for vendors to create high-quality, low-cost equipment and software. Many of the vendors of astronomical telescopes and instrumentation are in fact themselves amateur astronomers. Professional astronomers no longer have to design and build their own equipment, even for some cutting-edge research projects, if the equivalent can be supplied commercially. For example, the MEarth project (Charbonneau *et al.* 2008) to search for transiting exoplanets around M-dwarf stars, uses commercially available RCOS telescopes, Software Bisque mounts, and Apogee cameras.

6. Surveys

There are many surveys underway like the All-Sky Automated Survey, the Palomar Transient Factory, the Catalina Real-Time Survey, SuperWASP, etc., and several—such as the Large Synoptic Survey Telescope (LSST)—in stages of advanced planning. Each covers a large area of the sky, and discovering many new transient objects on a nightly basis. Their archived images offer excellent opportunities to monitor for known variable stars, and have been mined by a number of researchers for new variables or exotic objects. In fact, most of the currently-known variable stars have come from the ASAS, OGLE, MACHO, Kepler and similar surveys, and the LSST team has forecast that it will detect 100K or more transient objects every night.

While on the surface such surveys might appear to be the death knell for amateur astronomy, in reality the case is exactly the opposite. Surveys produce data on large numbers of interesting objects, far too numerous to be followed by existing professional observatories. The amateur community can be relied upon to help in those cases, much as they were in the early 1900s during the last great era of surveys (the photographic plate sky patrols).

Citizen Scientists will also be needed in many of the software tasks relating to surveys and the generation of transient lists. They can examine the transients that are automatically detected by the pipeline software, and classify them. Such classification can then be used for training the software to do a better job on future transients that are detected. The software will also identify a large set of objects for which no classification can be found; those are actually the ones with the greatest potential of being interesting, but require much manual inspection to eliminate any garbage.

7. Summary

Time-domain astronomy is an ideal arena in which Citizen Scientists can participate. Many follow-up requirements dictate the use of multiple telescopes worldwide, and the amateur community is ready for just such involvement. More training is essential, however, both to increase the number of observers who can provide quality data and to improve the quality of those data.

The workshop was unanimous in its support of the work of amateur astronomers.

References

Bücke, R. 2006, *VdS Journal für Astronomie*
Charbonneau, D., *et al.* 2008, *BAAS*, 21, 4402
Degenhardt, S. 2006, *Occ. Newslet.*, 13, n.4
Desnoux, V. & Buil, C. 2005, *Proc. Soc. Astro. Sci.*, 24, 129
Kaye, T., Vanaverbeke, S., & Innis, J. 2005, *JBAA*, 116, 78
Loughney, D. 2010, *JBAA*, 120, 157
Patterson, J. 2001, *Int. Am-Pro PEP Comm.*, no.84, 4
Raddick, M. J., *et al.* 2010, *Astr. Ed. Rev.*, 9, 103

New Horizons in Time-Domain Astronomy
Proceedings IAU Symposium No. 285, 2011
R.E.M. Griffin, R.J. Hanisch & R. Seaman, eds.

© International Astronomical Union 2012
doi:10.1017/S1743921312000701

Workshop on Stellar Tidal Disruption

Organized by Glennys R. Farrar

Center for Cosmology and Particle Physics, New York University, USA
email: gf25@nyu.edu

Abstract. The past year has seen major advances in the observational status of Stellar Tidal Disruption, with the discovery of two strong optical candidates in archived SDSS data and the real-time X-ray detection of Swift J1644+57, plus rapid radio and optical follow-up establishing it as a probable Tidal Disruption Flare (TDF) in "blazar mode". These observations motivated a workshop devoted to discussion of such events and of the theory of their emission and flare rate. Observational contributions included a presentation of Swift J2058+05 (a possible second example of a TDF in blazar mode), reports on the late-time evolution and X-ray variability of the two SWIFT events, and a proposal that additional candidates may be evidenced by spectral signatures in SDSS. Theory presentations included models of radio emission, theory of light curves and the proposal that GRB101225A may be the Galactic tidal disruption of a neutron star, an interpretation of Swift J1644+57 as due to the disruption of a white dwarf instead of main-sequence star, calculation of the dependence of the TDF rate on the spin of the black hole, and analysis of the SDSS events, fitting their SEDs to profiles of thoretical emission from accretion disks and showing that their luminosity and rate are consistent with the proposal that TDEs can be responsible for UHECR acceleration.

Keywords. Tidal Disruption, Black Hole, Flare, AGN, SNe

1. Overview

This workshop, organized by G.R. Farrar and attended by about 30 people, occupied two afternoon sessions. The first focussed on observations, with contributions by S. van Velzen, A. Levan, A. Zauderer, B. Cenko, and S. Komossa, plus predictions for radio emission by B. Metzger and S. van Velzen. The second session was more theoretical, beginning with theories of emission (G. Lodato, E. Rossi and T. Piran) and continuing with rate and modelling (M. Kesden and G. Farrar). There was very lively discussion throughout. This summary emphasizes material that was not presented elsewhere in the Symposium.

2. Observations of Probable Tidal Disruption Flares

2.1. Tidal Disruption Events in SDSS Stripe 82

Van Velzen described the results of van Velzen *et al.* (2011), in which two probable tidal disruption events (TDE) were identified in archived SDSS "Stripe 82" data. Stripe 82 is a 300-deg^2 region in which 2.6 million galaxies were observed on average 70 times each, over eight years. The identification pipeline minimized contamination from variable AGNs by excluding hosts in a QSO colour locus, and from SNe by requiring flares to be nuclear; Fig. 1 (left) shows the fit of the observed separations to nuclear+stellar flares. Requiring 3 detections in u, g and r, and also that $d < 0''.2$, leaves 42 nuclear flares out of 342 in the original sample. All but two of these nuclear flares are removed by eliminating spectroscopic AGN and requiring no variability in the off-seasons. The colours and cooling of these two candidate tidal disruption flares (TDFs) are shown in

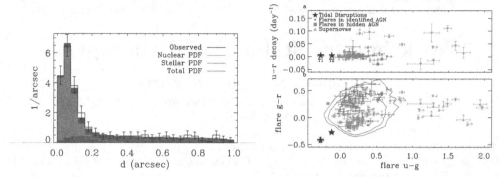

Figure 1. Left: Distribution of flare offsets (van Velzen *et al.* 2011a). Right: Comparison of candidate TDE flares to other flares in the analysis.

Fig. 1 (right), where it is seen that they are far removed from other AGNs or SNe in Stripe 82. The relative flux increase in the TDF candidates is much larger than has been observed for any other AGNs in the sample, with estimated probabilities of 10^{-7} and 10^{-5} that they are sampled from the AGN flare distribution. The probability of finding such low off-season variability for AGNs is estimated to be 10^{-6} and 10^{-5}, and the radio emission is < 1 mJy for both. Nor are they explained by being SNe; it is particularly telling that both are detected in the UV more than 2 years after the flare—see Fig. 2.

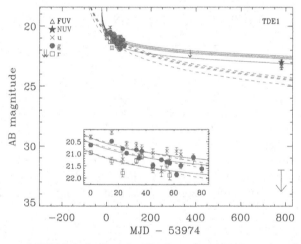

Figure 2. Light curve of TDE1, including the UV detection 800 days after the optical detection. The largest possible UV flux for a SN origin for this flare is shown by the large downward-pointing arrow.

The flaring-state spectrum of TDE2 is unlike any known SN, and the likelihood of it being a Type IIn SN, so close to the nucleus, is $< 0.08\%$. It cannot be ruled out that they are a new kind of "nuclear" core-collapse SN, but that would require a factor 1000 suppression of such events outside the nuclear region. Van Velzen *et al.* (2011) conclude that they are most likely TDFs, from black holes of mass $\sim 10^{7-8}\,M_\odot$. Extrapolation from the observed rate and sensitivity suggests that about 10/yr are expected in CRTS, QUEST and Pan-STARRS-1 Medium-Deep Field, and ~ 4000/yr in LSST. This presentation drew many comments and questions, especially regarding the possibility of a new type of nuclear SN.

2.2. *Swift J1644+57 in the Optical and X-ray*

Levan reviewed the key optical and X-ray observations of Swift J1644+57; details of those observations can be found at http://adsabs.harvard.edu/abs/2011Sci...333..199L, —/2011Sci...333..203B and —/arXiv1107.5307, and there was further extensive discussion of that event throughout the Workshop. Levan also reported that, since the publication of that paper, the source had continued a steady overall decline in X-ray emission, but recently more rapid X-ray variability seemed to have become apparent. The IR source continues to fade, while presumably becoming progressively more host-dominated; the IR fading lowers the black-hole (BH) mass estimate (based on the host stellar mass) by a factor of two. There is a low but marginally significant IR linear polarization ($7.4 \pm 3.5\%$) about 20 days after the outburst. How can we find more of these events? Where is the best place to search—radio, hard X-ray, soft X-ray?—and can we identify them easily? Are there smoking guns for their origin for which we should be looking? Can the examples that have been found so far still yield new physics from late observations?

2.3. *Swift J1644+57 in the Radio*

Zauderer described the EVLA discovery of radio emission from Swift J1644+57, highlighting that early localization (within a day of the trigger from SWIFT) was consistent with the centre of a normal galaxy at $z = 0.354$. Radio observations from 1–345 GHz obtained in the first month after the burst with many facilities (EVLA, CARMA, SMA, AMI Large Array, VLBA, OVRO 40-m) support the conclusion that Swift J1644+57 is a TDE: (a) The positional alignment between the optical image of the galaxy and the VLBA centroid, $0''.11 \pm 0''.18$, or 0.5 ± 0.9 kpc, is consistent with a nuclear origin. (b) SMA and CARMA verify a rising spectrum: a ν^2 behavior all the way to 345 GHz at early times, not seen in long GRBs. (c) The data can be fitted by synchrotron self-absorption, and the derived parameters show the source expanding relativistically—consistent with the Mar 25–28 range for the TDE (Zauderer *et al.* 2011). Follow-up is continuing (see Berger *et al.* 2012). She told a fascinating story of the difficulty of having a fast-time response and the serendipity involved in this case, and underlined the importance of knowing clearly, for future events, what signatures to look for at the different wavelengths, since response time is critical.

2.4. *Swift J2058+05*

Cenko discussed a second SWIFT event, J2058+05, which also has characteristics of a TDF (Cenko *et al.* 2011). It was discovered by the SWIFT BAT as a hard X-ray transient in 4-day all-sky co-added data. Follow-up XRT observations revealed a bright X-ray source, with a faint optical counterpart from GROND and a radio counterpart detected by the EVLA. The host galaxy is at $z = 1.19$, giving $L_X \sim 2 \times 10^{47}$ erg s^{-1} for the flare. There is no sign of AGN activity in the spectrum, and the SED is much bluer than for a blazar. Fig. 3 (upper plot) compares the optical and X-ray luminosities of Swift J2058+05 and Swift J1644+57 to other extreme objects. The table in the lower half of Fig. 3 summarizes the key properties of Swift J2058+05 and Swift J1644+57. HST and Keck observations imply $M_{BH} < 10^7 M_\odot$. Cenko pressed for answers to several questions: What are the fundamental characteristics that define this class of sources? With incomplete information, how can we distinguish blazar-mode TDFs from blazars? How can we measure collimation? What fraction exhibits relativistic jets, and why?

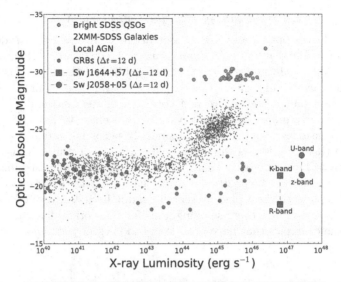

	Sw J1644+57	Sw J2058+05
L_x peak	1.00E+48	3.00E+47
α_x	-1.7 (?)	-2.2
Δt_{var}	80	< 1E4
E_x(iso)	2.00E+53	3.00E+53
β_{opt}	-1.0 (?)	2.0
M_V (transient)	> -18	-21.5
β_{rad}	1.3	0
νL_ν(rad)	1.00E+41	1.00E+42
M_{BH}	< 8e6	< 1e7
Γ	1.3	> 3

Figure 3. Upper: Comparison of optical and X-ray luminosities of Swift J2058+05 and Swift J1644+57 to those of other extreme objects. Lower: Properties of Swift J2058+05 and Swift J1644+57.

2.5. *Tidal Disruption Events from SDSS Spectra*

Komossa reported finding about 5 examples of TDE in SDSS spectra, in which she argues that high-energy ionization lines (of Fe, He) show a light echo from a UV flare. Details will be given in a paper in preparation.

2.6. *Predictions for Radio Emission from TDEs*

Two complementary models of radio emission from tidal disruption events were put forward. Metzger reported a model (Metzger *et al.* 2011) for the radio afterglows of TDEs, assuming a two-stage jet whose late-time behaviour follows the Blanford-McKee model. The radio emission is expected to peak ~1 year after a TDE. The radio is beamed, but much less than the X-ray, for Sw1644, according to observations. van Velzen reported a model (van Velzen *et al.* 2011b) of radio emission based on the jet–disk symbiosis relationship and adopoting an accretion rate from the SDSS TDE flares, but said that the model does not give good representation of the observations of Sw1644.

3. Theory and Modelling of Tidal Disruption Flares

3.1. *Modelling the Light Curves of TDEs*

Lodato reviewed the theory of emission from TDEs. The classic $t^{-5/3}$, where t is the time since pericentre passage, describes only the late-time fallback rate; at early times, $t \leqslant 0.2$ years, and the density profile and compressibility of the star produces a range of behaviour. Furthermore, although the bolometric luminosity from the disk (thermal emission) decreases as $t^{-5/3}$, a single-wavelength light curve does not exhibit the same scaling. At long wavelengths and early times, the disk thermal emission is in the Rayleigh-Jeans tail and $L \sim t^{-5/12}$. Attempts have been made (Cannizzo et al. 2011) to model Sw 1644, for which the X-ray emission is dominated by the jet so it is expected to scale with the fallback rate. The early light curve and event rate supports the argument that this was a deeply plunging event with a very short fall-back time of 1–2 days. However, the later light curve shows a much slower decay, with a fall-back time of ∼20 days that is best fitted with a model for low stellar mass, but it is not clear just how well the model fits. Moreover, GRB101225A has a light curve which can be understood as a Galactic TDE in which a minor object falls onto a neutron star (Campana et al. 2011).

3.2. *Tidal Disruption Model: the Wind*

Rossi discussed super-Eddington mass loss via winds, which obscures the disk emission at early times. The peak is first in the optical region, and then moves to higher frequencies. It is critically important to observe the wind, because super-Eddington accretion is not well understood; observations of the wind, and then of the disk, will therefore give tighter constraints on the physical parameters of the system. Many assumptions are needed for the modelling, so better constraints from observation would be valuable (for instance, to determine that a disk does form in the first place.) The wind should only be important for $M < 10^7 M_\odot$, and will be more prominent at optical than UV frequencies. It should show broad absorption lines in the UV (C IV, Lα, O VI)—broad and blue-shifted—owing to matter in the wind absorbing photons that are coming from deep in the wind/disk photosphere. The apparent properties of the jet depend strongly on observation angle.

3.3. *Model for J1644+57*

Piran concurred with the general view that J1644+57 is a TDF, but has argued (Krolik & Piran 2011) that the complex time-structure of its light curve, with multiple time-scales and intensity jumps, constitutes evidence that the disrupted star was a white dwarf rather than a main-sequence star. Specifically, Piran & Krolik argue for the existence of 3 sub-flares, each lasting about 1000–2000 seconds. Within each there is strong variability on time-scales of 100 seconds: about 3×10^4 seconds between sub-flares, and minima intensities between sub-flares which are suppressed by a factor of 600 compared to the maxima: see Fig. 4 (upper). The interpretation proposed by Krolik & Piran (2011) is illustrated in Fig. 4 (right): a WD experiences a partial disruption and enters an orbit with period of few 10^4 sec. Each time the WD returns to peri-centre, more material is stripped and a new accretion disk forms. The flare duration of ∼1000–2000 secs is the length of time that the accretion disk takes to drain out. The 100-sec time-scale is interpreted as the characteristic time of the accretion disk. They postulate that the above has a duration of ∼2×10^5 sec followed by a gradual decay. The tidal disruption radius for a WD is smaller than for a main-sequence star, owing to its compactness. If the maximum mass of a BH which can disrupt a main-sequence star without swallowing it whole is $10^8 M_\odot$, the corresponding limit for the disruption of a WD is ∼$3 \times 10^5 M_\odot$, which seems difficult to accommodate given the observed host galaxy. This proposal generated considerable debate, and triggered objections on several grounds, especially the required BH mass. [Taking into account the results discussed in next contribution from M. Kesden in fact loosens the BH mass constraint.]

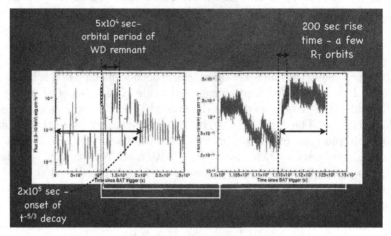

A white Dwarf disrupted by a 5 10^5 M$_\odot$ black hole - a model for Swift 1644

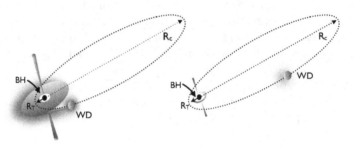

Figure 4. Upper: Sw J1644 light curves. Lower: Piran-Krolik picture of Sw J1644.

3.4. *Tidal Disruption of a Star by a Spinning Black Hole*

Kesden reported recent calculations which show that spinning black holes extend considerably the range of Black Hole masses which produce observable TDFs. Making that change, but otherwise following the rate calculation by Wang & Merritt (2004), leads to the conclusion that: (a) General Relativity is important for $M_{BH} > 10^7 \, M_\odot$, (b) the maximum BH mass that can produce an observable TDE increases from 10^8 to $10^9 \, M_\odot$, and (c) including the reduction in rate due to the existence of direct-capture orbits (applicable even for lower-mass BHs but apparently not previously taken into account) reduces the predicted TDE rate by a factor 2/3 for $10^7 \, M_\odot$, or 1/10 for $10^8 M_\odot$ (see Fig. 5). The upper limit to the BH mass for the tidal disruption of a white dwarf would increase correspondingly, improving the consistency of the Piran-Krolik model of Sw J1644+57 with the properties of the host galaxy.

3.5. *SDSS TDFs: Rate, Light Curves and UHECR acceleration*

Farrar outlined phenomenological aspects of the SDSS tidal disruption events. The TDE rate inferred from the two observed SDSS TDFs is $10^{-5.5\pm0.5}$ per galaxy per year, with

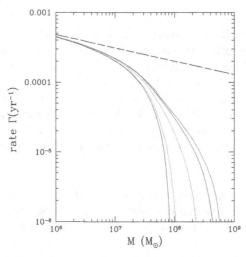

Figure 5. Tidal disruption rates for real galaxies, for a range of BH masses.

maximum sensitivity of the SDSS observations being to black holes with masses of $10^{7.5-8} M_{\odot}$. The SEDs of both events are fitted well by an accretion-disk model (Strubbe & Quataert 2009; Lodato & Rossi 2011), as shown in Fig. 6. Those fits imply that, as observed, $L_{\rm bol} > 10^{45} \, {\rm erg \, s^{-1}}$ for both flares. The flares thus satisfy a critical requirement of the proposal by Farrar & Gruzinov (2009) that TDEs can be the sources of ultra-high-energy cosmic rays (UHECR), namely that the bolometric power of a flare that accelerates UHECRs must be at least 10^{45} $\rm erg \, s^{-1}$. Combining the rate of TDEs with their potential UHECR power shows that TDEs appear to fit the bill perfectly as accelerators of UHECRs.

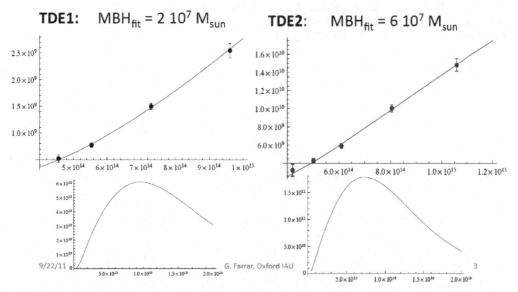

Figure 6. Fitting an accretion-disk model to the SEDs of the two TDEs from SDSS.

4. Summary

The radio, optical and X-ray evidence points strongly to the interpretation that Swift J1644+57 is a tidal disruption flare seen in "blazar" mode as opposed to "normal accretion" mode as for the SDSS TDFs (Bloom *et al.* 2011; Zauderer 2011; Levan *et al.* 2011). A similar, unusual flare in June, Sw J2058+05, may be a second example (Cenko *et al.* 2011). Complementing these two dramatic events, in which we are most likely seeing emission from a recently-formed jet, are two optically-detected flares found in SDSS archived data, whose SEDs and lightcurves are well described by emission from an accretion disk and which would be very difficult to explain as anything but tidal disruption events.

Acknowledgements

I am is grateful to A. Zauderer and S. van Velzen for taking notes, and to all the participants who made the Workshop a great success. I also thank the Editors for their helpful suggestions with this contribution, and the co-Chairs for organizing an excellent and stimulating meeting.

References

Berger, E., *et al.* 2011, *ApJ*, in press.
Bloom, J. S., *et al.* 2011. *Science*, 333, 203
Campana, S., *et al.* 2011, *Nature*, 480, 69
Cannizzo, J. K., Troja, E., & Lodato, G., 2011, *ApJ*, 742, 32
Cenko, S. B., *et al.* 2011, *ArXiv e-prints*
Farrar, G. R. & Gruzinov, A. 2009, *ApJ*, 693, 329
Krolik, J. H. & Piran, T., 2011. *ApJ*, 743, 134
Levan, A. J., *et al.* 2011. *Science*, 333, 199
Lodato, G. & Rossi, E. M. 2011, *MNRAS*, 410, 359
Metzgerm B. D., Giannios, D., & Mimica, P. 2011, *ArXiv e-prints*
Strubbe, L. E. & Quataert, E. 2009, *MNRAS*, 400, 2070
van Velzen, S., *et al.* 2011, *ApJ*, 741, 73
van Velzen, S., Körding, E., & Falcke, H. 2011, *MNRAS*, 417, L51
Wang, J. & Merritt, D., 2004. *ApJ*, 600, 149
Zauderer, B. A. 2011, *Nature*, 476, 425

New Horizons in Time-Domain Astronomy
Proceedings IAU Symposium No. 285, 2011
R.E.M. Griffin, R.J. Hanisch & R. Seaman, eds.
© International Astronomical Union 2012
doi:10.1017/S1743921312000713

Workshop on
Faint and Fast Transients

Organized by Mansi Kasliwal[1,2] and Lars Bildsten[3]

[1] Observatories of the Carnegie Institution for Science, Pasadena, CA 91101, USA
[2] Dept. of Astrophysical Sciences, Princeton University, Princeton, NJ 08544, USA
email: mansi@astro.caltech.edu
[3] Kavli Institute for Theoretical Physics, Univ. of California, Santa Barbara, CA 93106, USA

Summary. In the past five years systematic searches, serendipitous discoveries and archival searches have yielded over a dozen transients that are brighter than novæ but fainter than supernovæ. The observed properties of these "gap" transients tend to place them in distinct classes, some being about 100 times brighter than novæ and durations of nearly 100 days (e.g., M85 OT, PTF10fqs), while others (such as SN2002bj and PTF10bhp) nearly reach supernovæ-luminosities but fade in five days. The state of theoretical understanding varies substantially across the class of objects, and is ripe for progress.

The workshop commenced with a brief summary of the observational discoveries and theoretical models of transients in the gap. Most of the workshop was devoted to a round-table discussion of the observed objects, possible theoretical models for them, and ways forward (both observationally and theoretically) for developing a coherent understanding of these particular explosions and their place in stellar evolution.

The round-table session considered in turn (a) calcium-rich Halo transients, (b) Type .Ia explosions, (c) NS-NS and NS-WD binaries, (d) low-velocity transients, and (e) intermediate-luminosity red transients. It then concluded with a general discussion.

New Horizons in Time-Domain Astronomy
Proceedings IAU Symposium No. 285, 2011
R.E.M. Griffin, R.J. Hanisch & R. Seaman, eds.

© International Astronomical Union 2012
doi:10.1017/S1743921312000725

Workshop on Extreme Physics

Organizers: Carole Mundell[1] & Mark Sullivan[2]

[1] Astrophysics Research Institute, Liverpool John Moores Univ., Birkenhead, CH41 1LD, UK
email: `cgm@astro.livjm.ac.uk`

[2] Department of Physics, University of Oxford, Oxford, OX1 3RH, UK

Summary. Never before has there been such a wealth of versatile ground- and space-based facilities with which to detect variable emission across the electromagnetic spectrum and beyond, to non-EM signals such as neutrinos and gravitational waves, to probe the most extreme phenomena in the Universe. The variable sky is already providing a wealth of new and surprising observations of phenomena such as GRBs, SNe and AGN that are pushing current theories beyond the state of the art. Multi-messenger follow-up will soon become *de rigeur*, and upcoming radio and optical all-sky transient surveys will revolutionise the study of the transient Universe. In addition to the technical and data challenges presented by such surveys, a major new challenge will be the interpretation of the wealth of available data and the identification of the underlying physics of new classes of variable (and potentially exotic) objects. Theoretical predictions will be vital for interpreting these future transient discoveries.

The goal of this workshop was to bring together theorists and observers in order to identify unexplored synergies across three main research areas of extreme physics: gamma-ray bursts, supernovæ and, more generically, relativistic jets. It aimed to discuss key outstanding questions in these rapidly moving fields, such as the composition and acceleration of GRB and AGN jets, GRB progenitors and central engines, the origin of the wide range of observed variability time-scales in GRB prompt and after-glow light curves and related cosmological applications, the physics of the newly-discovered ultra-luminous SN-like optical transients—as well as to speculate on what we might hope to discover with future technology.

The workshop absorbed two 90-minute sessions, selecting 3 main science topics (Relativistic Jets, GRBs and SNe) which it organised as structured discussions driven by a series of short but provocative questions. The final session featured a panel-led debate but with full audience participation.

New Horizons in Time-Domain Astronomy
Proceedings IAU Symposium No. 285, 2011
R.E.M. Griffin, R.J. Hanisch & R. Seaman, eds.

© International Astronomical Union 2012
doi:10.1017/S1743921312000737

Workshop on Algorithms for Time-Series Analysis

Organizer: Pavlos Protopapas

Harvard-Smithsonian Center for Astrophysics, Cambridge, MA 02138, USA
email: `pprotopapas@cfa.harvard.edu`

Summary. This Workshop covered the four major subjects listed below in two 90-minute sessions. Each talk or tutorial allowed questions, and concluded with a discussion.

Classification: Automatic classification using machine-learning methods is becoming a standard in surveys that generate large datasets. Ashish Mahabal (Caltech) reviewed various methods, and presented examples of several applications.

Time-Series Modelling: Suzanne Aigrain (Oxford University) discussed autoregressive models and multivariate approaches such as Gaussian Processes.

Meta-classification/mixture of expert models: Karim Pichara (Pontificia Universidad Católica, Chile) described the substantial promise which machine-learning classification methods are now showing in automatic classification, and discussed how the various methods can be combined together.

Event Detection: Pavlos Protopapas (Harvard) addressed methods of fast identification of events with low signal-to-noise ratios, enlarging on the characterization and statistical issues of low signal-to-noise ratios and rare events.

New Horizons in Time-Domain Astronomy
Proceedings IAU Symposium No. 285, 2011
R.E.M. Griffin, R.J. Hanisch & R. Seaman, eds.

© International Astronomical Union 2012
doi:10.1017/S1743921312000749

Workshop on Radio Transients

Organized by Steve Croft[1] & Bryan Gaensler[2]

[1] Department of Astronomy, University of California, Berkeley, CA 94720, USA)
email: `scroft@astro.berkeley.edu`

[2] Sydney Institute for Astronomy, The University of Sydney, NSW 2006, Australia

Summary. We are entering a new era in the study of variable and transient radio sources. This workshop discussed the instruments and the strategies employed to study those sources, how they are identified and classified, how results from different surveys can be compared, and how radio observations tie in with those at other wavelengths. The emphasis was on learning what common ground there is between the plethora of on-going projects, how methods and code can be shared, and how best practices regarding survey strategy could be adopted.

The workshop featured the four topics below. Each topic commenced with a fairly brief introductory talk, which then developed into discussion. By way of preparation, participants had been invited to upload and discuss one slide per topic to a wiki ahead of the workshop.

1. *Telescopes, instrumentation and survey strategy.* New radio facilities and on-going projects (including upgrades) are both studying the variability of the radio sky, and searching for transients. The discussion first centred on the status of those facilities, and on projects with a time-domain focus, both ongoing and planned, before turning to factors driving choices of instrumentation, such as phased array versus single pixel feeds, the field of view, spatial and time resolution, frequency and bandwidth, depth, area, and cadence of the surveys.

2. *Detection, pipelines, and classification.* The workshop debated (a) the factors that influence decisions to study variability in the (u,v) plane, in images, or in catalogues, (b) whether, and how much, pipeline code could potentially be shared between one project and another, and which software packages are best for different approaches, (c) how data are stored and later accessed, and (d) how transients and variables are defined and classified.

3. *Statistics, interpretation, and synthesis.* It then discussed how (i) the choice of facility and strategy and (ii) detection and classification schemes influence what is seen (in terms of types of object and rates) by different surveys, (iii) how results from different surveys could be compared, and (iv) how what we know from existing surveys drives choices (i) and (ii), particularly as regards finding new classes of object.

4. *Multiwavelength approaches.* The workshop concluded by discussing what information is needed from wavelengths other than radio in order to classify transients and variables adequately and predict their rates as a function of topics (1), (2) and (3). It asked what the constraints are on responding to, and issuing triggers for, follow-up observations, and how that might feed back into considerations for designing our telescopes and surveys.

New Horizons in Time-Domain Astronomy
Proceedings IAU Symposium No. 285, 2011
R.E.M. Griffin, R.J. Hanisch & R. Seaman, eds.

Poster Papers

111 posters were displayed at IAU S285. They were divided arbitrarily into two groups, the changeover taking place between Tuesday and Wednesday.

Poster presenters were given the option of converting their posters into short papers. 55 so chose, and this Section contains those contributions, in alphabetical order of the lead author.

Summaries of the other 56 posters commence on page 429.

New Horizons in Time-Domain Astronomy
Proceedings IAU Symposium No. 285, 2011
R.E.M. Griffin, R.J. Hanisch & R. Seaman, eds.

Cepheids in Galactic Open Clusters: An All-sky Census

Richard I. Anderson, Laurent Eyer and Nami Mowlavi

Observatoire de Genève, CH-1290 Versoix, Swizterland
email: richard.anderson@unige.ch

Abstract. We perform an all-sky search for classical (type I) Cepheids that are members of Galactic Open Clusters. Our approach is multi-dimensional, using all available spatial and kinematic parameters. The quantification of errors is crucial for this analysis, so care is taken to find adequate and realistic representations of parameter uncertainties. The data employed in the calculation are taken from published catalogues and the literature, supplemented by specific radial-velocity observations. Our work in progress is outlined here, and issues related to the inhomogeneity of cluster radii in the literature are discussed in some detail.

Keywords. (stars: variables:) Cepheids, (Galaxy:) open clusters and associations: general, stars: evolution, stars: distances, astrometry, stars: kinematics, catalogues, astronomical data bases: miscellaneous, (cosmology:) distance scale

1. Introduction

Cluster Cepheids (CCs) have been the object of attention for a long time, owing in part to the ability to calibrate the zero-point of the well-established period-luminosity relationship (PLR); see, e.g., Sandage & Tammann (2006), Turner (2010). A solid determination of cluster membership for PLR calibrators is obviously of the utmost importance. Investigations of CC membership therefore usually take into account multiple membership constraints including position, distance and radial velocity or proper motion, as well as considerations of evolutionary status (age). Nevertheless, the membership of some CCs is disputed, for example SZ Tau in (or not) NGC 1647 (Baumgardt *et al.* 2000). Furthermore, studies of individual Cepheid-Cluster pairs suffer from data inhomogeneity, resulting in a lack of inter-comparability. In this paper we outline our work in progress that aims to enlarge the number of known Cluster Cepheids significantly; it is based on a self-consistent analysis of the literature and newly-observed data. We aim to avoid a direct dependence on stellar models by not considering evolutionary status as a membership constraint.

2. Overview

2.1. Compilation of Literature Data

We consider Open Clusters mentioned in Kharchenko *et al.* (2005) [hereafter K05] and Bukowiecki *et al.* (2011) [hereafter B11]. Those two catalogues were chosen because of their comparable definitions of tidal radii (see Section 3) in order to ensure comparability of our results. The compilation of cluster data and parameters by Dias *et al.* (2002) [from hereon DAML] was of great help for supplementing average proper motions or radial velocities. A list of classical Cepheids was compiled using the GCVS (Samus *et al.* 2011) and VSX (Watson 2006) catalogues. Additional information was added from many other sources through cross-matching, for example from the Fernie *et al.* (1995) database of Cepheids.

Selected high-precision radial-velocity observations were conducted at the Euler (La Silla, Chile) and the Mercator (La Palma, Canary Islands) telescopes to supplement this otherwise purely literature-based study.

2.2. *Computation of Membership Probabilities*

After making a cone search to identify CC candidates, we calculated "membership probabilities", that is, the probabilities of the null hypotheses (membership) being true for the candidates. The main assumptions made are Gaussian uncertainties, independent measurements, spherical symmetry for the clusters, accurate representation of parameters. In other words, starting from a proximity argument to investigate cluster membership, we filter out cases that are inconsistent with membership. Good cases remain, and will have to be checked individually.

The computation compares position (on-sky separation weighted by cluster radius), distance, proper motions in each direction, and median radial velocities. If x is the vector of differences between Cluster and Cepheid parameters, and Σ the sum of the covariance matrices, then we can calculate the unitless parameter c according to Robichon *et al.* (1999) and Baumgardt *et al.* (2000):

$$c = x^{\mathrm{T}} \cdot \Sigma^{-1} \cdot x \quad , \tag{2.1}$$

where $c \sim \chi_N^2$ and N denotes the degrees of freedom. We can then calculate the p-value associated with c and thus obtain our "membership probability".

3. Open Cluster Radii

In this project, we combine very inhomogeneous data from multiple sources. That leads to various problems that cannot all be covered here in full; here we discuss solely some issues relating to cluster radii, and refer to our article (currently in preparation) for an in-depth discussion concerning all parameters.

The most extensive catalogue of Open Cluster parameters is probably the [DAML] catalogue. It is in part based on WEBDA (available at `www.univie.ac.at/webda/`), is repeatedly updated, and is subject to quality controls. It lists diameters for 99% of all clusters.

However, multiple definitions of "cluster diameters" exist in the literature, such as "apparent diameters" (from visual inspection), "half-light radii", "core radii", "tidal radii" and "limiting radii" (based on star counts or density profiles). Some of these may be similarly defined, yet no single and precise definition of the quantity "cluster radius" exists. As a consequence, the compilation of diameters in [DAML] cannot adhere to a single definition. That creates a problem for our analysis, since we depend on well-defined cluster radii in two ways: 1) the initial selection of Cepheids in nearby clusters is obtained through a cone search around the cluster position, within an area defined by twice the limiting cluster radius, and 2) the membership probability calculated in Eq. 2.1 depends on the radius, since we use the radius in an ad-hoc definition of a positional membership probability distribution centred around the cluster's coordinates. We thus choose to limit our list of clusters to the combined list of [K05] and [B11], whose cluster radii are similarly, but not equally, defined ("coronal" *vs.* "limiting" radii). Unfortunately, that results in the exclusion of around 840 clusters listed in [DAML] in favour of ensuring comparability of our results.

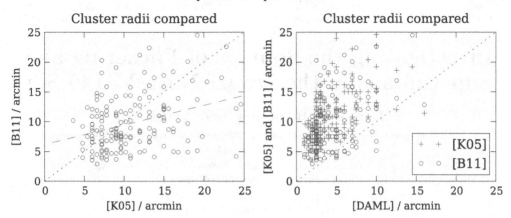

Figure 1. *Left panel:* Cluster radii in [B11] against those in [K05]. Dashed line indicates least squares fit for clarity. [K05] radii are on average larger with large scatter. *Right panel:* Comparison of [K05] and [B11] radii with apparent radii found in [DAML].

Figure 1 (left) compares the radii of 165 clusters common to [K05] and [B11], setting cluster radius < 25′ to avoid border issues arising from the 34′-limited FOV in [B11], while Figure 1 (right) compares radii in the two catalogues with apparent radii from [DAML]. Two important features should be noted: a) there is a large scatter in radii, exceeding a factor of two for a number of objects; b) [B11] radii, though based on 2MASS infrared photometry, are on average smaller than the optically-determined [K05] radii, as can be seen from the least-squares fit indicated by the dashed line. That is very interesting, as it shows the opposite trend to what was noted in [B11]. We suggest the following explanation: Both the fainter magnitude limit of 2MASS photometry and its enhanced sensitivity to red low-mass stars compared to optical photometry used by [K05] increase the field-star density considerably, thereby decreasing the contrast between cluster and field stars. The result is a larger uncertainty in the background density, $\sigma_{\rm bg}$, leading to a smaller radius as per Eq. (3) in [B11].

4. Outlook

Our study is intended to provide a test-bed for stellar-evolution studies, and will be used to investigate the zero-point calibration of the Galactic PLR for Cepheids. Important limitations to this work, not mentioned here, will be discussed in a forthcoming article.

References

Baumgardt, H., Dettbarn, C., & Wielen, R. 2000, *A&AS*, 146, 251
Bukowiecki, L., et al. 2011, *Acta Astronomica*, 61, 231 [B11]
Dias, W. S., et al. 2002, *A&A*, 389, 871 [DAML]
Fernie, J.D., et al. 1995, *IBVS*, 4148, 1
Kharchenko, N.V., et al. 2005, *A&A*, 438, 1163 [K05]
Robichon, N., et al. 1999, *A&A*, 345, 471
Samus N. N., et al. 2011, General Catalog of Variable Stars
Sandage, A. & Tammann, G. A. 2006, *ARA&A*, 44 (1), 93
Turner, D. G. 2010, *AP&SS*, 326 (2), 219
Watson, C. L. 2006, Society for Astronomical Sciences Annual Symposium, 25, 47

New Horizons in Time-Domain Astronomy
Proceedings IAU Symposium No. 285, 2011
R.E.M. Griffin, R.J. Hanisch & R. Seaman, eds.

© International Astronomical Union 2012
doi:10.1017/S1743921312000774

Investigating the Sources of Flickering and Superhumps in the Dwarf Nova V4140 Sgr

R. Baptista[1], B. Borges[2] and A. Oliveira[3]

[1]Universidade Federal de Santa Catarina, 88040-900, Florianópolis - SC, Brazil
email: raybap@gmail.com

[2]Universidade Federal da Grande Dourados, 79804-070, Dourados - MS, Brazil
email: BernardoBorges@ufgd.edu.br

[3]Universidade do Vale do Paraiba, São José dos Campos - SP, Brazil
email: alexandre@univap.br

Abstract. We report the results of maximum entropy eclipse-mapping analysis of an ensemble of light curves of the dwarf nova V4140 Sagitarii (V4140 Sgr) with the objective of studying the spatial distribution of its steady-light and flickering sources in quiescence, and the changing disk structure during an outburst.

Keywords. accretion, accretion disks, stars: dwarf novae, stars: imaging, novae, cataclysmic variables, binaries: eclipsing, stars: individual (V4140 Sagitarii)

1. Context

In dwarf novæ, mass is transferred from a late-type star to a companion white dwarf via an accretion disk. Recurrent outbursts occur on time-scales of days or months, when the disk brightens by a factor of 20–100 for about 1–10 days. Outbursts are explained in terms of either a thermal-viscous disk-instability (DIM; Lasota 2001) or a mass-transfer instability (MTIM; Bath 1975). DIM predicts that matter accumulates in a low-viscosity disk ($\alpha_{quies} \sim 10^{-2}$) during quiescence, whereas in MTIM the disk viscosity is always large ($\alpha \sim 10^{-1}$). Measuring α in a quiescent disk is therefore critical for inferring which model is the more realistic for a given dwarf nova.

"Flickering" refers to the intrinsic brightness fluctuation of 0.01–1 mag on time-scales of seconds to dozens of minutes which are seen in dwarf-novæ light curves (Bruch 2000). Flcikering might arise at the stream-disk impact region (because of unsteady mass inflow, or post-shock turbulence; Warner & Nather 1971) and/or in turbulent inner-disk regions. If the disk-related flickering is caused by magneto-hydrodynamic (MHD) turbulence, it is possible to infer α from the relative flickering amplitude (Geertsema & Achterberg 1992).

V4140 Sgr is an 88-min period eclipsing SU UMa-type dwarf nova showing 1–2 mag, 5 to 10-day outbursts recurring every 80 to 90 days, and longer, brighter super-outbursts when superhumps appear in its light curve (Borges & Baptista 2005).

2. Data Analysis and Results

An ensemble of 22 B-band light curves of V4140 Sgr was obtained with the SOI optical camera on the Southern Astrophysical Research 4.1-m telescope during 2006. The object was observed during two nights during the decline of what seemed to have been a super-outburst in 2006 September 12–24. The outburst light curves show pronounced superhumps, with maximum light occurring at different binary phases at each night.

We applied 3D eclipse-mapping techniques to locate the source of the superhump and to follow the evolution of the disk's surface brightness distibution during the outburst decline. An entropy landscape method was used to derive the disk radius R_d and opening angle β. The disk shrank from $0.34\,R_\odot$ at outburst maximum to $0.21\,R_\odot$ in quiescence, and was geometrically thin both in outburst ($\beta = 1°.0$) and in quiescence ($\beta = 0°.5$). The surface brightness distribution was asymmetric towards the L1 point at outburst maximum, suggesting that the disk was elliptical at that stage. Wide (in azimuth) regions of enhanced emission at the disk rim are responsible for the orbital modulation observed during outburst maximum and decline. The quiescent disk map shows enhanced emission ahead of the stream-disk impact point, and maximum emission along the disk rim coinciding with the predicted azimuth of impact point (bright spot).

The remaining 15 light curves in quiescence were combined to derive the orbital dependency of the steady-light components and of the low- and high-frequency flickering ones (Baptista & Bortoletto 2004). Eclipse mapping of those curves indicate that the steady-light phase is dominated by emission from an extended asymmetric source with negligible contribution from the white dwarf, indicating that in quiescence efficient accretion is taking place through a high-viscosity disk. Flickering maps show an asymmetric source at the disk rim (stream-disk impact flickering) and an extended central source (disk-related flickering) several times larger in radius than the white dwarf at disk centre.

If the disk-related flickering is caused by fluctuations in the energy dissipation rate induced by MHD turbulence in the disk's atmosphere, its relative amplitude will yield a direct measurement of the disk viscosity parameter α and its radial dependency (Geertsema & Achterberg 1992). With that assumption, we find that the inner disk regions of V4140 Sgr have a high viscosity of $\alpha \sim 0.15$–0.3 but which decreases with increasing radius—a similar behaviour to that previously found for the similar dwarf nova HT Cas (Baptista *et al.* 2011). The inferred high disk viscosity is in agreement with the observed surface brightness distribution of the disk, and is inconsistent with the expectations from the DIM.

These results suggest that the outbursts of V4140 Sgr are powered by bursts of enhanced mass transfer rates from the donor star.

References

Baptista, R. & Bortoletto A., 2004. *AJ*, 128, 411

Baptista, R., *et al.*, 2011, in: *Physics of Accreting Compact Binaries*, (Universal Academic Press Inc), in press (arXiv 1105.1382)

Bath, G. T., 1975. *MNRAS*, 171, 311

Borges, B. & Baptista R., 2005. *A&A*, 437, 235

Bruch, A., 2000. *A&A*, 359, 998

Geertsema, G. T. & Achterberg A., 1992. *A&A*, 255, 427

Lasota, J. P., 2001. *New Astronomy Review*, 45, 449

Warner, B. & Nather R. E., 1971. *MNRAS*, 152, 219

New Horizons in Time-Domain Astronomy
Proceedings IAU Symposium No. 285, 2011
R.E.M. Griffin, R.J. Hanisch & R. Seaman, eds.

© International Astronomical Union 2012
doi:10.1017/S1743921312000786

AquEYE and IquEYE, Very-High-Time-Resolution Photon-Counting Photometers

Cesare Barbieri[1], Giampiero Naletto[1], Luca Zampieri[2], Enrico Verroi[1], Serena Gradari[1], Susan Collins[3] and Andy Shearer[3]

[1] University of Padova, Italy   email: `cesare.barbieri@unipd.it`
[2] INAF Astronomical Observatory of Padova, Italy,
[3] NUI Galway, Ireland

Abstract. We describe very high-time-resolution photometers capable of tagging the arrival time of each photon with a resolution and accuracy of few hundred picoseconds, for hours of continuous acquisition, and with a dynamic range of more than 6 orders of magnitude. The final goal is the conceptual definition of a "quantum" photometer for the E-ELT, capable of detecting and measuring second-order correlation effects in photon streams from celestial sources. Two prototype units have been built and operated, one for the Asiago 1.8-m telescope (AquEYE) and one for the 3.5-m NTT (IquEYE).Here we will present results obtained by IquEYE on the Crab Nebula pulsar in simultaneous radio observations with Jodrell Bank in December 2009.

Keywords. instrumentation: photometers pulsars: individual (Crab)

1. Introduction

In 2005 September we completed a study (QuantEYE, the ESO Quantum Eye; Dravins *et al.* 2005) whose main goal was to demonstrate the possibility of reaching picosecond time resolution (the Heisenberg limit) and sustaining a count rate as high as 1 GHz (needed to bring quantum optics concepts into the astronomical domain), with two main scientific aims in mind:

- to measure the entropy of the light beam through the statistics of photon arrival time, and
- to demonstrate the feasibility of astronomical photon correlation spectroscopy and of a modern version of the Hanbury Brown Twiss Intensity Interferometry (HBTII).

Although the project was conceived for the 100-m Overwhelmingly Large (OWL) telescope, the results reached in that study will be of interest for the ESO E-ELT and other 30-m telescopes too.

2. Detector

The most critical point, and the driver for the design of QuantEYE and subsequent prototypes, was the selection of very fast, efficient and accurate photon counting detectors. In order to proceed, we choose Single Photon Avalanche Diodes (SPADs) from the Italian company MPD. These detectors are operated in Geiger, continuous mode, have a 35 ps time resolution through a NIM connector, and sustain count rates as high as 15 MHz. Their main drawbacks are the small sensitive area (50–100 μm diameter), the lack of a CCD-like array, a >77 ns dead time and a 1.5% afterpulsing probability.

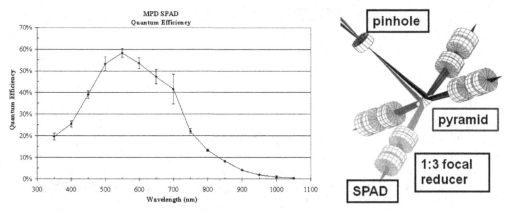

Figure 1. Left: the QE of the MPD SPAD peaks at 55% around 550nm. Right: The opto-mechanical concept of AquEYE. The pyramid splits the light beam into four arms, each feeding a SPAD via a 1:3 focal reducer, inside which filters and polarisers can be inserted.

3. AquEYE

AquEYE, the Asiago Quantum Eye, was designed as the final stage of AFOSC, the imaging spectrophotometer of the Asiago-Cima Ekar (Italy) 182-cm telescope. AFOSC plays the role of a 1:3 focal reducer. The light beam from AFOSC is divided into four parts by means of a pyramidal mirror. Each beam is then focussed on its own SPAD by another 1:3 focal reducer made by a pair of doublets. This configuration provides a possibility for simultaneous multicolour photometry, and allows the cross-correlation of the 4 sub-apertures, in a conceptual proof of HBTII capabilities.

The pulses from each SPAD are fed into a Time-To-Digital Converter board (TDC), together with the time stamps from a Time and Frequency Unit composed of a GPS receiver and a rubidium clock. The time ticks are treated as light pulses, so the delays inside the hardware and software channels are therefore immaterial. The UTC arrival time of each photon is stored separately for each channel, guaranteeing data integrity for subsequent scientific investigations, where one can either sum up the 4 channels or treat them separately, with arbitrary time binning in units of the basic time unit of the system, namely 24 picoseconds.

4. IquEYE

Following the success of AquEYE, IquEYE (Naletto *et al.* 2009) was built for the ESO 3.5-m NTT at La Silla (Chile). The same basic solutions of 4-way pupil splitting and hardware/software configuration were maintained. The main improvement was a new production batch of MPD SPADs with 100 μm diameter, lower dark counts and better engineering. Other improvements were implemented in order to be able to control the instrument remotely from the new LaSilla Control Room.

5. Crab Observations

The Crab pulsar was observed with IquEYE in 2009 January and again in 2009 December. The quality of the data is demonstrated in Fig. 2, which shows a short set of individual pulses, with data binned at 0.001 second intervals.

The optical data were folded using the JB radio ephemeris fitted over the observing dates. A specially fitted ephemeris was required, owing to the rapidly varying dispersion

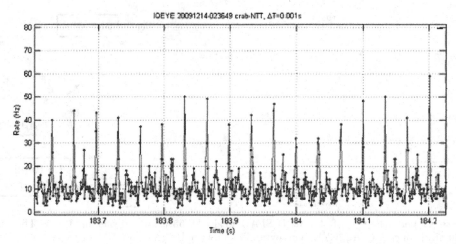

Figure 2. Crab Pulsar Observations: A short section of individual primary and secondary peaks, binned at 0.001 second intervals.

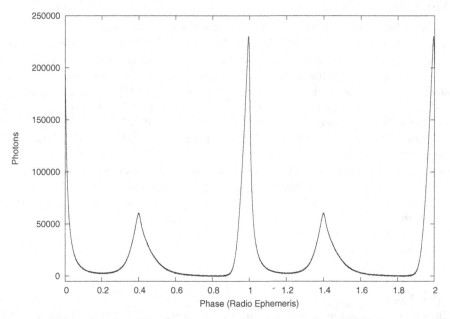

Figure 3. The Crab pulsar light curve, obtained by binning the IquEYE data using 5000 phase bins.

measure at the time. Fig. 3 shows the quality of the light curve obtained. In a few hours we reproduced the JB ephemerides to the picosecond level. Of all optical instruments, IquEYE on the NTT provided the best timing of photon arrival times.

References

Dravins, D., *et al.* 2005, *QuantEYE. Quantum Optics Instrumentation for Astronomy*, ESO document: OWL-CSR-ESO-00000-0162

Naletto, G., *et al.* 2009, *A&A*, 508, 531-539

New Horizons in Time-Domain Astronomy
Proceedings IAU Symposium No. 285, 2011
R.E.M. Griffin, R.J. Hanisch & R. Seaman, eds.

© International Astronomical Union 2012
doi:10.1017/S1743921312000798

The KEPLER Guest Observer Programme

Thomas Barclay[1,2]

[1]NASA Ames Research Center, MS 244-30, Mountain View, CA 94035, USA

[2]Bay Area Environmental Research Institute, Sonoma, CA 95476, USA

email: thomas.barclay@nasa.gov

Abstract. The KEPLER Guest Observer Office is dedicated to the service of the broad science community, with a charter to promote the exploitation of KEPLER data and broaden the scientific impact of the KEPLER mission. There are four routes by which to gain access to KEPLER data: the Guest Observer programme annual call for proposals, the quarterly call for Director's Discretionary Targets, the KEPLER Asteroseismic Science Consortium, and by accessing the KEPLER public data archive. The Guest Observer Office supports all users of KEPLER data no matter which route they took to access data, and to that end have developed software tools which can be used by the community in their data analyses.

Keywords. stars: variables: general, techniques: photometric, methods: data analysis

1. The KEPLER Mission

The KEPLER spacecraft was launched in 2009 March on a mission to determine the frequency of Earth-like planets in or near the habitable zone around Sun-like stars (Koch *et al.* 2010; Borucki *et al.*] 2009; Borucki *et al.* 2010; Caldwell *et al.* 2010). The 0.95-m aperture telescope is in a heliocentric, Earth-trailing orbit and almost continually observes the same 116 deg^2 field of view. KEPLER thus continually observes over 1.6×10^5 targets at a cadence of 30 min with milli- to micro- magnitude precision and has announced the discovery of over 1200 planet candidates, with sizes ranging from smaller than the Earth to twice the size of Jupiter (Borucki *et al.* 2010; Fressin *et al.* submitted) with orbital periods ranging from less than a day to nearly a year (Welsh *et al.*, submitted).

2. Astrophysics with KEPLER

The scientific endeavours accomplished with data from the KEPLER mission encompass a multitude of fields of astrophysics research, from studies of the zodiacal dust in the Solar System (Witteborn *et al.* 2011) to distant AGNs (Mushotzky *et al.* 2011). There are four specific fields which KEPLER has the potential to revolutionise: those of asteroseismology (e.g., Bedding *et al.* 2011; Chaplin *et al.* 2011; Antoci *et al.* 2011; Beck *et al.* 2011), gyrochronology (Meibom *et al.* 2011), stellar activity (e.g., Basri *et al.* 2011; Walkowicz *et al.* 2011) and binary stars (e.g., Carter *et al.* 2011; Derekas *et al.* 2011; Slawson *et al.* 2011).

3. Opportunities to Access KEPLER Data

There are four paths for obtaining KEPLER data on specific targets of interest: the full yearly call for Guest Observer proposals, the quarterly Director's Discretionary Target programme, the KEPLER Asteroseismic Science Consortium and the KEPLER data archive.

3.1. *The annual Guest Observer competition*∗

Once a year there is a call for proposals to observe targets within the KEPLER field of view. The call is open to everyone; proposers need not be based in the United States. There are ∼5000 long-cadence slots (one data point every 29.4 min) and 40 short-cadence ones (one data point every 58.8 s) available at any one time in this programme. The Kepler Guest Observer (GO) Office particularly solicits proposals from the astronomical community for:

(*a*) New sources of astrophysical interest, and

(*b*) Observations of existing targets on the exoplanet list for scientific investigations that are distinct from exoplanetary and related science.

3.2. *Director's Discretionary Targets*

The standard Guest Observer competition occurs on an annual cycle. Because of the proposal review time-line, associated data processing and archive activities, there is currently a 6-month delay between proposal submission and first observations, and a 1-year wait between proposal submission and the delivery of the first Guest Observer data. To provide a faster mechanism for headline KEPLER astrophysics, the Guest Observer Office provides an alternative means for acquiring KEPLER data through the Director's Discretionary Target (DDT) programme.

The DDT program is a quarterly competition. Observations can be proposed at any time, and approved targets will be added during the next spacecraft roll. Up to 100 DDTs are available each quarter. Proposals from all institutions and countries are encouraged. The proposal process is informal, requiring an email request from the proposer to the Guest Observer Office†. The purpose of the DDT programme is to:

- Provide a fast track to KEPLER data so as to yield rapid, high-impact science and thus build a significant sample of KEPLER GO papers rapidly
- Respond to targets of GO interest, newly dropped from the planetary list. Targets can be reinstated onto the observing list via a DDT request
- Respond to "targets of opportunity"
- Permit pilot studies of small samples prior to the next GO solicitation
- Enhance existing GO programmes with additional data.

3.3. *The Kepler Asteroseismic Science Consortium*‡

The main body of asteroseismic investigation of KEPLER data has been performed by members of the KEPLER Asteroseismic Science Consortium (KASC), a collection of over 500 scientists from all over the world who are studying stellar pulsations in order to understand the internal structure of stars. KASC members are guaranteed a minimum of 1700 long- and 140 short-cadence targets, and welcome new members who are able to contribute to the analysis of KEPLER asteroseismic data.

3.4. *Archival data*

The KEPLER archive is hosted by the Multimission Archive at the Space Telescope Sciences Institute (MAST)§. All KEPLER data are archived, and many of them are publicly available for scientific exploitation. MAST hosts not only the archived light-curve data but also the calibrated and uncalibrated pixel-level data and full-frame images which

---

∗For specific details of the call in a given year see http://keplergo.arc.nasa.gov/.
†Email: kepler-go@lists.nasa.gov
‡More information about the KASC is available from http://astro.phys.au.dk/KASC/.
§KEPLER data can be obtained from http://archive.stsci.edu/kepler.

contain all pixels within the KEPLER field of view and not just those within the predefined pixel masks used for long- and short-cadence observations. In 2011 October, over 1.1×10^6 individual light curves for 1.7×10^5 sources are publicly available.

4. Community Software*

The Guest Observer Office has developed a suite of software called PyKE, which can be used in the processing and analysis of KEPLER data. Of particular importance are the tools KEPEXTRACT and KEPCOTREND. The light-curve products at MAST are extracted from the pixel-level data using an aperture which optimises signal to noise—this product contains significant instrumental systematics. The KEPEXTRACT tool allows the user to chose different apertures. The reasons for doing that include:

- Using all pixels in the aperture
- Creating a light curve for certain sources, since the KEPEXTRACT pipeline does not produce a light curve for sources observed with custom or dedicated pixel masks
- Constructing pixel light curves in which the time series for a single pixel can be examined
- Obtaining light curves for extended sources which may be poorly sampled by the optimal aperture.

Astrophysical variability among the KEPLER targets is expected to be uncorrelated, so any apparent correlation is presumed to be due to instrumental systematics. Such correlations can be represented as a linear combination of orthogonal vectors known as cotrending basis vectors. The KEPLER project has released a product containing the cotrending basis vectors and has made them available at MAST†.

The Guest Observer Office has developed a tool called KEPCOTREND for fitting cotrending basis vectors to a light curve. The fitted instrumental signal can then be subtract from the time series, to leave a clean light curve.

References

Antoci, V., *et al.* 2011, *Nature*, 477, 570
Basri, G., *et al.* 2011, *AJ*, 141, 20
Beck, P. G., *et al.* 2011, *Science*, 332, 205
Bedding, T. R., *et al.* 2011, *Nature*, 471, 608
Borucki, W., *et al.* 2009, in: F. Pont, D. Sasselov & M. Holman (eds.), *Transiting Planets*, (IAU S 253) (Cambridge: Cambridge University Press), p. 289
Borucki, W. J., *et al.* 2010, *Science*, 327, 977
Caldwell, D. A., *et al.* 2010, *ApJ* (Letters), 713, L92
Carter, J. A., *et al.* 2011, *Science*, 331, 562
Chaplin, W. J., *et al.* 2011, *Science*, 332, 213
Derekas, A., *et al.* 2011, *Science*, 332, 216
Koch, D. G., *et al.* 2010, *ApJ* (Letters), 713, L79
Meibom, S., *et al.* 2011, *ApJ* (Letters), 733, L9
Mushotzky, R. F., Edelson, R., Baumgartner, W. H., & Gandhi, P. 2011, arXiv:1111.0672
Slawson, R. W., *et al.* 2011, *AJ*, 142, 160
Walkowicz, L. M., *et al.* 2011, *AJ*, 141, 50
Witteborn, F. C., Van Cleve, J. E., Borucki, W., Argabright, V., & Hascall, P. 2011, Proc. SPIE, 8151, 815117

*The community software is hosted at http://keplergo.arc.nasa.gov/Contributed SoftwarePyKEP.shtml.

†Cotrending basis vectors can be downloaded from http://archive.stsci.edu/kepler/ cbv.html.

New Horizons in Time-Domain Astronomy
Proceedings IAU Symposium No. 285, 2011
R.E.M. Griffin, R.J. Hanisch & R. Seaman, eds.

© International Astronomical Union 2012
doi:10.1017/S1743921312000804

Modulated Light Curves of Multiperiodic Stars

József M. Benkő, Róbert Szabó and Margit Paparó

Konkoly Observatory of the Hungarian Academy of Sciences, H-1121 Budapest, Hungary
email: benko@konkoly.hu

Abstract. We modelled the light curves of modulated multiperiodically pulsating stars by a simple double-mode pulsation with combined amplitude and frequency modulations. The synthetic light curves and their spectra show similar features to those we found and discussed for monoperiodic stars. Comparing the synthetic light curves and their spectra with the observed ones helps us to classify the modulations and to distinguish between long-period modulation (Blazhko effect) and the other types of amplitude and/or period changes.

Keywords. methods: analytical, stars: oscillations, stars: variables: other

---

1. Introduction

Benkő *et al.* (2011) described the light variations of Blazhko RR Lyræ stars by a formalism adopted from the theory of electronic signals. In our treatment we assumed monoperiodic light curves to be carrier waves and that they became modulated in either their amplitude (AM), or their frequency (FM), or both. That approach shows numerous advantages compared to the traditional Fourier sums. Since it is general enough, it can be extended to many types of modulated light curves. Here we present some results based on a synthetic light curve generated by a double-periodic carrier signal.

Our present study was motivated by recent publications in which the Blazhko effect was discussed in connection with different multiperiodically pulsating stars. Papers by Henry *et al.* (2005), Moskalik & Kołaczkowski (2009) and Breger (2010) raised the possibility of modulations (Blazhko effect) in γ Dor, beat Cepheids and δ Scuti stars, respectively. A further development is the recent discovery that RR Lyræ stars can pulsate in more than one mode (occasionally or permanently) and can show simultaneous Blazhko effects (Poretti *et al.* 2010, Benkő *et al.* 2010).

2. The Model

To demonstrate how our method works for multiperiodic light curves we chose the simplest multiperiodic (visual double-mode) pulsation as a test case. Let the carrier wave be modelled by a light curve, $c(t)$:

$$c(t) = c_0(t)c_1(t), \quad \text{where} \quad c_k(t) = a_{0k} + \sum_{i_k}^{n_k} a_{ik} \sin\left(2\pi i_k f_k t + \varphi_{ik}\right). \qquad (2.1)$$

$n_k = 1, 2, 3, \ldots$; $k = 0, 1$ integers. In practice, we used a synthetic RRd (= double-mode RR Lyræ) light curve. A signal that is simultaneously modulated in amplitude and frequency can be written formally as

$$m_{\text{comb}}(t) = \left[1 + m_{\text{mod}}^{\text{A}}(t)\right] \tilde{c}(t). \qquad (2.2)$$

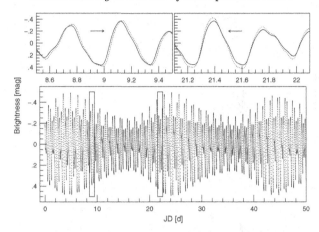

Figure 1. A double-periodic light curve modulated in both amplitude and frequency (bottom). Different non-sinusoidal periodic functions were used for modulations. Manifestation of the phase variation is shown in the top panels, where the modulated light curve (dotted) is compared to the unmodulated carrier one (continuous). The phase shifts are indicated by arrows.

Here $\tilde{c}(t)$ is the same expression as Eq. (2.1), but we substitute functions $\tilde{f}_k(t) = f_k + m_{\mathrm{mod}}^{\mathrm{F}}(t)$ instead of the fixed frequencies f_k for all $k = 0, 1$. The modulation functions $m_{\mathrm{mod}}^{\mathrm{A}}(t), m_{\mathrm{mod}}^{\mathrm{F}}(t)$ are assumed to be periodic and can be represented by finite Fourier sums. Fig. 1 compares the artificial light curve to the carrier light curve of Eq. (2.1).

3. Characteristics of the Light Curve and the Fourier Spectrum

Some features are evident directly from the light curve in Fig. 1. Since the AM is assumed to be non-sinusoidal the envelope curve shows a horizontally asymmetric shape (bottom panel), while the FM indicates phase variations (top panels).

Additional investigations can be carried out in the Fourier spectrum (Fig. 2). The spectrum is dominated by the two main frequencies, f_0, and f_1 and their different number of harmonics ($i_k f_k$, $i_k = 1, 2, \ldots, n_k$; $k = 0, 1$), and their linear combinations constitute the carrier wave spectrum. The appearance of the linear combination frequencies is a well-known observed feature of double-mode RR Lyræ stars. Assumption (2.1) provides those frequencies automatically. When we produced the synthetic light curve we represented the non-sinusoidal AM function $m_{\mathrm{mod}}^{\mathrm{A}}(t)$ as a Fourier sum with two components, so the frequency f_{m} and its harmonic $2f_{\mathrm{m}}$ consequently appear in the resultant spectrum (insert in Fig. 2). FM modulation—even in the sinusoidal case—produces an infinite number of side peaks (see Szeidl & Jurcsik 2009, Benkő et al. 2011) around the main frequencies and their harmonics. Those multiplets ($i_k f_k \pm l f_{\mathrm{m}}$) can be identified in the top panels in Fig. 2. It is important to note that more than two side peaks are seen ($l > 2$) in each multiplet. That shows direct evidence of FM; otherwise, the case can not be separated from a pure non-sinusoidal AM. An additional feature also suggests FM: the higher-order harmonics have more detectable side frequencies than the lower-order ones (compare the top left and right panels of Fig. 2). That distinguishes FM from the very similar phase modulation (PM). Combined AM+FM modulation results in highly non-symmetrical amplitudes of the side peaks. Some side peaks are almost missing, as is shown by the vertical lines at their positions (top right panel, Fig. 2). Those and some additional features have been discussed in detail by Benkő et al. 2011), where single period carrier waves were used. The significant difference between that and the present work is the appearance

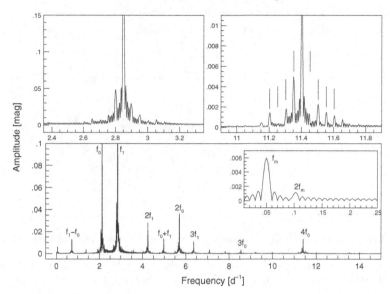

Figure 2. Fourier amplitude spectrum of the synthetic modulated light curve in Fig. 1 (bottom main panel) and its zoomed segments around f_0 (top left panel), $4f_0$ (top right panel—vertical lines show the positions of the side peaks) and f_m (insert in bottom panel).

of linear combination frequencies. By investigating the surroundings of the combination frequencies we find multiplet patterns around all linear combination frequencies that are similar to those around the main frequencies and their harmonics.

All of the phenomena which we have found by this treatment can be explained by the mathematics of the modulation. Conversely, if an observed light curve does *not* show those key features (see also Benkő et al. 2011) we have to find a different origin for the amplitude and/or phase variations.

Acknowledgements

The authors acknowledge the support of ESA PECS projects Nos. 98114 & 4000103541.

References

Benkő, J. M., Kolenberg, K., Szabó, R., Kurtz, D. W., & Bryson, S. *et al.* 2010, *MNRAS*, 409, 1244

Benkő, J. M., Szabó, R., & Paparó, M. 2011, *MNRAS*, 417, 974

Breger, M. 2010, in: Sterken C., Samus N., & Szabados L. (eds.), *Variable Stars, the Galactic Halo and Galaxy Formation* (Moscow: Sternberg Astron. Inst.), p. 95

Chadid, M., Benkő, J. M., Szabó, R., Paparó, M., & Chapellier, E. *et al.* 2010, *A&A*, 510, A39

Guggenberger, E., Kolenberg, K., Chapellier, E., Poretti, E., & Szabó, R. *et al.* 2011, *MNRAS*, 415, 1577

Henry, G. W., Fekel, F. C., & Henry, S. M. 2005, *AJ*, 129, 2815

Moskalik, P. & Kołaczkowski, Z. 2009, *MNRAS*, 394, 1649

Poretti, E., Paparó, M., Deleuil, M., Chadid, M., & Kolenberg, K. *et al.* 2010, *A&A*, 520, A108

Szeidl, B. & Jurcsik, J. 2009, *CoAst*, 160, 17

New Horizons in Time-Domain Astronomy
Proceedings IAU Symposium No. 285, 2011
R.E.M. Griffin, R.J. Hanisch & R. Seaman, eds.

© International Astronomical Union 2012
doi:10.1017/S1743921312000816

Time-Resolved Spectroscopy with SDSS

Steven Bickerton[1], Carles Badenes[2], Thomas Hettinger[3], Timothy Beers[3], and Sonya Huang[1]

[1] Dept. of Astrophysical Sciences, Princeton University, Princeton, NJ 08544, USA
email: bick@astro.princeton.edu

[2] Dept of Physics and Astronomy, University of Pittsburgh, Pittsburgh, PA 15260, USA
email: badenes@pitt.edu

[3] Dept of Physics and Astronomy and JINA: Joint Institute for Nuclear Astrophysics,
Michigan State University, East Lansing, MI 48824, USA
email: hettin12@msu.edu, beers@pa.msu.edu

Abstract. We present a brief technical outline of the newly-formed project, "Detection of Spectroscopic Differences over Time" (DS/DT). Our collaboration is using individual exposures from the SDSS spectroscopic archive to produce a uniformly-processed set of time-resolved spectra. Here we provide an overview of the properties and processing of the available data, and highlight the wide range of time base-lines present in the archive.

Keywords. techniques: spectroscopic, methods:data analysis, surveys, stars: variable: other

1. Introduction

The Sloan Digital Sky Survey (SDSS; York *et al.* 2000) has been in operation for over a decade, and through three separate phases (SDSS I, II, and III) has accumulated an extensive archive of photometric and spectroscopic astrophysical observations. Here we describe a newly-formed data mining collaboration, the "Detection of Spectroscopic Differences over Time" or DS/DT project, to search for variability in SDSS spectra.

Though cadences and exposure times varied, the SDSS spectra were generally observed on three or more occasions, with exposure times of at least 900s. These individual exposures were combined to produce the final spectra released to the community.

Only a handful of groups have examined the sub-spectra for evidence of variability: Hilton *et al.* (2010) examined flaring in M-dwarfs, while Mullally *et al.* (2009) and Badenes *et al.* (2009) used radial-velocity shifts to identify WD–WD binaries. Radial velocities have also been used by Rebassa-Mansergas *et al.* (2010) to identify binaries. With such an enormous diversity of objects, from QSOs and AGNs to stars of all spectral types, there is a vast amount of time-variable data which have never been evaluated.

The primary objective of our work is to produce a uniformly-processed archive of the individual sub-spectra for all SDSS spectroscopic observations. We are developing a data-mining pipeline to target some specific forms of variability (radial velocities, flaring, etc) as well as serendipitous anomalous variability.

2. Data Processing

Our pipeline is an extension of the SDSS spectroscopic pipeline, `spectro`, used to produce the seventh data release of the Sloan Digital Sky Survey (Abazajian *et al.* 2009). The original `spectro` pipeline consists of three principle stages for (1) extraction and calibration of the raw spectroscopic traces, (2) stacking the sub-spectra and stitching the data from the red and blue cameras, and (3) object classification and redshift determination. We have written a modified 2nd stage to perform only the red/blue stitching for the individual exposures. Our new combined step generates individual (per-object) output files containing the full co-added spectra and their associated sub-spectra.

3. Time-Baselines

During the SDSS observing programme, a wide range of time base-lines was sampled. A given plate (640 fibers) was typically observed in three back-to-back 900s exposures, and in most cases the base-line is therefore $\lesssim 30$ minutes. However, plates which were incomplete at the end of an observing session were continued the following night, thus giving many plates a base-line of >12 hours. In some cases, previously-observed plates were later replugged for further observation, generating base-lines of several weeks or sometimes much longer. Finally, some targets were observed on multiple plates, again providing base-lines from weeks to months. Figure 1 shows histograms of the time base-lines sampled (measured between the mid-points of the first and last available exposures) and the number of exposures. The sample includes the SDSS I and II spectra, and the SEGUE-II subset of the SDSS III spectra.

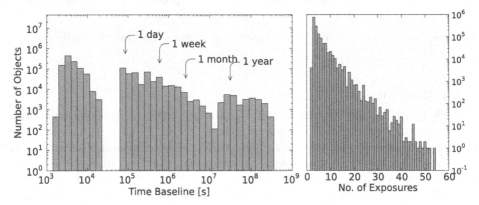

Figure 1. The per-object time base-lines sampled by the SDSS sub-spectra (left), and the number of targets having a given number of exposures available (right).

4. Summary

The DS/DT project is a newly-formed data-mining project to explore time variability in the sub-spectra of the SDSS spectroscopic archive. We have developed a branch pipeline to process and collate the sub-spectra for each target into individual FITS files. It is our intention to make our full data set publicly available. Those interested in testing and providing feedback on our preliminary data are invited to contact us.

References

Abazajian, K. N., *et al.* 2009, *ApJS*, 182, 543
Badenes, C., Mullally, F., Thompson, S. E., & Lupton, R. H. 2009, *ApJ*, 707, 971
Hilton, E. J., West, A. A., Hawley, S. L., & Kowalski, A. F. 2010, *AJ*, 140, 1402
Mullally, F., Badenes, C., Thompson, S. E., & Lupton, R. 2009, *ApJL*, 707, L51
Rebassa-Mansergas, A., *et al.* 2010, *MNRAS*, 402, 620
York, D. G., *et al.* 2000, *AJ*, 120, 1579

New Horizons in Time-Domain Astronomy
Proceedings IAU Symposium No. 285, 2011
R.E.M. Griffin, R.J. Hanisch & R. Seaman, eds.

© International Astronomical Union 2012
doi:10.1017/S1743921312000828

Improved Time-Series Photometry and Calibration Method for Non-Crowded Fields: MMT Megacam and HAT-South Experiences

Seo-Won Chang[1], Yong-Ik Byun[1], and Dae-Won Kim[1,2]

[1]Department of Astronomy, Yonsei University, Seoul, Korea
[2]Harvard-Smithsonian Center for Astrophysics, Cambridge, MA 02138. USA
email: seowony@galaxy.yonsei.ac.kr

Abstract. We present a new photometric reduction method for precise time-series photometry of non-crowded fields that does not need to involve relatively complicated and CPU intensive techniques such as point-spread-function (PSF) fitting or difference image analysis. This method, which combines multi-aperture index photometry and a spatio-temporal de-trending algorithm, gives much superior performance in data recovery and light-curve precision. In practice, the brutal filtering that is often applied to remove outlying data points can result in the loss of vital data, with seriously negative impacts on short-term variations such as flares. Our method utilizes nearly 100% of available data and reduces the rms scatter to several times smaller than that for archived light curves for brighter stars. We outline the details of our new method, and apply it to cases of sample data from the MMT survey of the M37 field, and the HAT-South survey.

Keywords. methods: data analysis—techniques: photometric

1. Multi-Aperture Index Photometry

Our photometric technique is similar to standard aperture photometry, except that we compute the flux in a sequence of several apertures and then determine the optimal aperture (with maximum S/N) individually for each object at each epoch. We also add a new index that isolates peculiar situations where photometry returns misleading information. The method is as follows:

(*a*) Build the master astrometric catalogue. Since the relative centroid positions of all objects are the same for all frames in the time-series, we can easily place an aperture and measure the flux even for the fainter stars.

(*b*) Refine the centroid position by the Gaussian window. The precision in centroiding is very close to that of PSF-fitting on properly sampled point sources.

(*c*) Perform photometry with multiple apertures, then find the optimal aperture size by measuring the S/N values for each object. We achieve the optimal balance between flux loss and noise based on the revised CCD equation (Howell & Everett 2001). For sky background, the annulus value is calculated by a combination of κ-σ clipping and mode estimation. We also use a background map to reduce contamination from nearby sources.

(*d*) Decide the final aperture with our indexing method (Fig. 1). Differential magnitudes between apertures are compared with mean trends for stars of similar brightness, and the difference is indexed. When a signature of contamination occurs, different apertures and an aperture correction term are automatically applied. In that way we can measure correct fluxes even for contaminated stars.

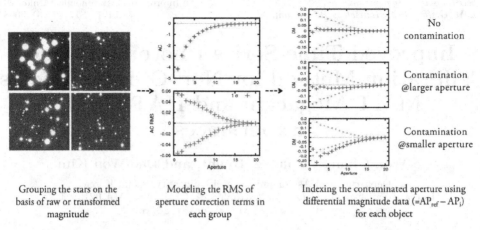

| Grouping the stars on the basis of raw or transformed magnitude | Modeling the RMS of aperture correction terms in each group | Indexing the contaminated aperture using differential magnitude data ($=AP_{ref} - AP_i$) for each object |

Figure 1. Sequence of multi-aperture index photometry. We can see whether, and at what aperture, the differential magnitude begins to diverge from the model curve for each epoch.

2. Spatio-Temporal Photometric De-trending

In order to reduce systematic effects in time-series photometry, several successful methods have been introduced (e.g., Sys-Rem: Tamuz et al. 2005; TFA: Kovacs *et al.* 2005; PDT: Kim *et al.* 2009). Among those, we find that PDT is the most useful for detecting and removing spatially localized patterns in time-series data, as trends in our sample data do show a tendency of localization. Why the trends are different and localized within a single CCD frame is a subject for further study, but is probably related to subtle changes in the PSF and sky conditions within the detector field of view. We apply the PDT algorithm in following steps:

(*a*) Select the template light curves for bright stars. Within the observation span, these light curves have the same temporal ordering of data points. We take a sequence of data points, $L_i(t_{ref})$, as the reference time line.

(*b*) Extract all subset of light curves that show spatially and temporally correlated features such as clusters. The strength of linear dependence between two or more light curves is expressed as the Pearson correlation coefficient. Each cluster is determined by an hierarchical tree clustering algorithm.

(*c*) Construct master trends for each cluster, $T_c(t_{ref})$. Master trends are obtained as the weighted averages of normalized differential light curves, $f_i(t_{ref})$, as described below:

$$T_c(t_{ref}) = \frac{\sum_{i=1}^{N_c} \omega_i f_i(t_{ref})}{\sum_{i=1}^{N_c} \omega_i}, f_i(t_{ref}) = \frac{L_i(t_{ref}) - \bar{L}_i}{\bar{L}_i}, \omega_i = \frac{1}{\sigma_{f_i}^2} \tag{2.1}$$

(*d*) De-trend the light curves of all stars with matching master trend and time lines: we adjust trend sets for each light curve, $f_i(t_i)$, and then determine free parameters (β_{ic}) by minimizing the noise term (ϵ_i).

$$f_i(t_i) = \sum_c \beta_{ic} T_c(t_i) + \epsilon_i(t_i) \tag{2.2}$$

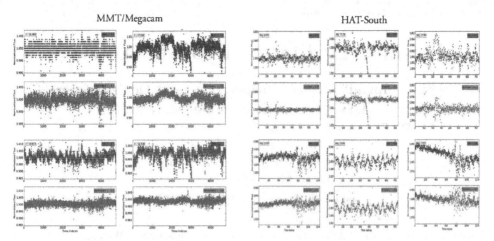

Figure 2. In each subset of MMT/megacam and HAT-South data a multi-aperture indexed light curve (above) is paired with the de-trended light curve (below).

3. Experiments with Time-Series Data

Time-series samples were obtained from two different types of transit survey projects: (a) one month of MMT M37 survey data, using the Megacam instrument (36 2k × 4k CCDs, 24′ × 24′ field of view) (Hartman et al. 2008), (b) HAT-South survey data, from a network of wide-field (one camera ∼ 4°.2 × 4°.2 field of view) telescopes in Australia, Chile and Namibia (Bakos et al. 2009). Fig. 2 shows examples of light curves that demonstrate the effect of our de-trending method. Some systematic effects are clearly present in the light curves of bright stars, yielding artificial flux variations. Our method removes only the systematic variations that are shared by light curves of stars in the adjacent sky regions; all kinds of true variability are preserved.

Acknowledgements

This work is supported by the Korea Institute of Science and Technology Information under the contract of the commissioned research project, Massive Astronomical Data Applications of Cloud Computation (KISTI-P11020). We thank Joel D. Hartman and Matthew J. Holman for their suggestion to use the whole MMT data set. We also thank Daniel Bayliss and the HAT-South team for helping us access HAT-South data.

References

Bakos, G., *et al.* 2009, in: F. Pont, D. Saeelov, & M. Holman (eds.), *Transiting Planets*, IAU Symposium No. 253 (Cambridge: Cambridge Univ. Press), p. 354
Hartman, J. D., *et al.* 2008, *ApJ*, 675, 1233
Howell, S. B, & Everett, M. E. 2001, in: W.J. Borucki & L.E. Lasher (eds.), *3rd Workshop on Photometry*, p. 1
Kim, D.-W., Protopapas, P., Alcock, C., Byun, Y.-I., & Bianco, F. B. 2009, *MNRAS*, 397, 558
Kovács, G., Bakos, G., & Noyes, R. W. 2005, *MNRAS*, 356, 557
Tamuz, O., Mazeh, T., & Zucker, S. 2005, *MNRAS*, 356, 1466

New Horizons in Time-Domain Astronomy
Proceedings IAU Symposium No. 285, 2011
R.E.M. Griffin, R.J. Hanisch & R. Seaman, eds.

© International Astronomical Union 2012
doi:10.1017/S174392131200083X

Fermi LAT Flare Advocate Activity

Stefano Ciprini[1,2], Dario Gasparrini[1,2] and Denis Bastieri[3,4]

[1]ASI Science Data Center, Frascati, Roma, Italy

[2]INAF Observatory of Rome, Monte Porzio Catone, Roma, Italy

[3]University of Padova, Padova, Italy

[4]INFN Padova Section, Padova, Italy

email: stefano.ciprini@asdc.asi.it

(on behalf of the Fermi LAT collaboration)

Abstract. The Fermi Flare Advocate (also known as Gamma-ray Sky Watcher, FA-GSW) service provides a daily quick-look analysis and review of the high-energy gamma-ray sky seen by the Fermi Gamma-ray Space Telescope. The duty offers alerts for potentially new gamma-ray sources, interesting transients and flares. A weekly digest containing the highlights about the GeV gamma-ray sky is published in the web-based Fermi Sky Blog. During the first 3 years of all-sky survey, more than 150 Astronomical Telegrams, several alerts to the TeV Cherenkov telescopes, and targets of opportunity to Swift and other observatories, were realized. That increased the rate of simultaneous multi-frequency observing campaigns and the level of international cooperation. Many gamma-ray flares from blazars (such as extraordinary outbursts of 3C 454.3, intense flares of PKS 1510-089, 4C 21.35, PKS 1830-211, AO 0235+164, PKS 1502+106, 3C 279, 3C 273, PKS 1622-253), short/long flux duty cycles, unidentified transients near the Galactic plane (like J0910-5041, J0109+6134, the Galactic center region), flares associated with Galactic sources (like the Crab nebula, the nova V407 Cyg, the microquasar Cyg X-3), emission of the quiet and active sun, were observed by Fermi and communicated by FA-GSWs.

Keywords. surveys, gamma rays: observations, quasars: general, BL Lacertae objects: general, pulsars: general, stars: novae, Sun: gamma rays

---

The Large Area Telescope (LAT), on board the Fermi Gamma-ray Space Telescope (Atwood *et al.* 2009), is a pair-conversion gamma-ray telescope that is sensitive to photon energies from about 20 MeV up to >300 GeV. The LAT consists of a tracker (two sections, front and back), a calorimeter and an anti-coincidence system to reject the charged-particle background. Fermi LAT, working in all-sky survey mode, is an optimal hunter for high-energy flares, transients and new gamma-ray sources, and is an unprecedented monitor of the variable gamma-ray sky, thanks to its large peak effective area, wide field of view ($\approx$ 2.4 sr), improved angular resolution and sensitivity. This all-sky monitor is complemented by the Flare Advocate service (a.k.a. Gamma-ray Sky Watcher, FA-GSW), belonging to the LAT Instrument Science Operations.

FA-GSWs highlight news of potential interest for LAT science groups and the external community, and offer the first seeds for variability and multiwavelength (MW) investigations; https://confluence.slac.stanford.edu/x/YQw, the LAT MW Coordinating Group page, shows examples. Information and news about the gamma-ray sky are communicated through internal daily reports, Astronomical Telegrams (Fig. 1), notes via the LAT-MW mailing-list (https://lists.nasa.gov/mailman/listinfo/gammamw/), special GCNs for flares, and weekly digests in the Fermi Sky Blog (fermisky.blogspot.com).

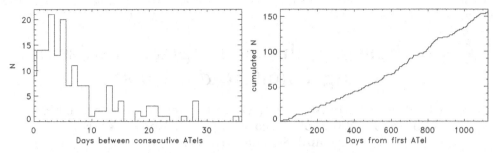

Figure 1. Distribution of the 159 Astronomical Telegrams (ATels), published on behalf of the Fermi LAT Collaboration in about 3 years of Fermi all-sky survey mission, from 2008 July 24 (ATel#1628) to 2011 August 24 (ATel#3580); www.astronomerstelegram.org. ATels and other information about the day-by-day gamma-ray sources detected above 100 MeV are summarized every week in the public Fermi Sky Blog (fermisky.blogspot.com).

The role and activity of the FA-GSW is therefore twofold:

• Gamma-ray Flare Advocate task. Flaring sources approaching a daily flux of 10^{-6} photons cm^{-2} s^{-1} deserve attention (detection, localization, flux, photon index checked, count maps and exposure maps revised). Internal/public notes, ATels, Target of Opportunity (ToO) are submitted; MW observing campaigns are organized when needed.

• Gamma-ray Sky Watcher task. Results from the LAT Automatic Science Processing (ASP) pipeline are checked in 1-day and 6-hour time intervals for transients, brightness trends and new gamma-ray source candidates.

The list of discoveries triggered by FA-GSWs is substantial: many flares from gamma-ray blazars (the extraordinary outburst of 3C 454.3, very bright and large flares, for example from PKS 1510-089, 4C 21.35, PKS 1830-211, AO 0235+164, PKS 1502+106, 3C 279, 3C 273, PKS 1622-253), short/long duty cycles of bright gamma-ray blazars, unidentified transients near the Galactic plane (like J0910-5041, J0109+6134, Galactic centre region) or associated with Galactic sources (like the Crab nebula, the nova V407 Cyg, the micro-quasar Cyg X-3, the binary star system 1FGL J1018.6-5856), intense emission from the quiet and active sun (for example Abdo *et al.* 2011a, Abdo *et al.* 2011b, Abdo *et al.* 2010a, Abdo *et al.* 2010b, Abdo *et al.* 2010c, Ackermann *et al.* 2010).

About a dozen ToOs per year were also submitted to SWIFT following gamma-ray flares detected by the LAT. The all-sky variability monitor run by FA-GSW advocates represents the liaison between the Fermi LAT Collaboration and the MW astrophysical/astroparticle community, and always invites observations of Fermi LAT sources and proposals for MW collaborations.

References

Atwood, W. B., *et al.* 2009, *ApJ*, 697, 1071
Abdo, A. A., *et al.* 2010a, *Science*, 329, 817
Abdo, A. A., *et al.* 2010b, *ApJ*, 710, 810
Abdo, A. A., *et al.* 2010c, *Nature*, 463, 919
Abdo, A. A., *et al.* 2011a, *ApJ*, 734, 116
Abdo, A. A., *et al.* 2011b, *Science*, 331, 739
Ackermann, M., *et al.* 2010, *ApJ*, 721, 1383

New Horizons in Time-Domain Astronomy
Proceedings IAU Symposium No. 285, 2011
R.E.M. Griffin, R.J. Hanisch & R. Seaman, eds.

© International Astronomical Union 2012
doi:10.1017/S1743921312000841

Crab Pulsar: Enhanced Optical Emission During Giant Radio Pulses

Susan Collins[1], Andy Shearer[1], Ben Stappers[2], Cesare Barbieri[3], Giampiero Naletto[3], Luca Zampieri[4], Enrico Verroi[3], and Serena Gradari[3]

[1] NUI Galway, Ireland,
email: susan.collins@nuigalway.ie
[2] Jodrell Bank Center for Astrophysics, UK,
[3] University of Padova, Italy,
[4] INAF Astronomical Observatory of Padova, Italy

Abstract. Although optical pulsar studies have been limited to a few favoured objects, the observation of pulsars at optical wavelengths provides an opportunity to derive a number of important pulsar characteristics, including the energy spectrum of the emitting electrons and the geometry of the emission zone. These parameters will be vital for a comprehensive model of pulsar emission mechanisms. Observations of the Crab pulsar with the high-time-resolution photon-tagging photometer IquEYE show an optical–radio delay of ~178 μs. Incorporating simultaneous Jodrell Bank radio observations suggested a correlation between giant radio pulses and enhanced optical pulses for this pulsar, thus offering possible evidence for the reprocessing of radio photons.

Keywords. pulsars: individual (Crab)

Multiwavelength High-Time-Resolution Astronomy

Observations of the Crab pulsar with the high-time-resolution photometer IquEYE in December 2009 had concurrent radio observations taken on two nights at Jodrell Bank. The radio data were analysed to find so-called 'giant radio pulses' (GRPs): occasional giant pulses with an intensity of up to 1000 times that of a typical pulse (Lorimer and Kramer, 2005.)

To check our reduction and timing correction pipelines, the optical and GRP data were folded using a radio ephemeris specially fitted over the observing dates. The main-pulse GRPs identified for the night of 2009 December 14 are approximately normally distributed about the nominal peak radio phase (Fig. 1). 663 GRPs were identified above a 6.0 σ threshold, which also had concurrent optical observations.

Summary

These results, in combination with those of Shearer *et al.* (2003), have established a clear link between the optical emission and giant radio pulses of the Crab pulsar (Fig. 2.) The delay of 178 μs recorded here between the radio and optical peaks (Fig. 3) points to a phase and spatial separation of the order of 2° and/or 50 kilometres. One possible explanation of the optical emission is the reprocessing of radio photons; see Petrova *et al.* (2009).

We note however that optical polarisation studies (Słowikowska 2009) show an association between the optical polarisation and the arrival phase of the radio precursor. More observations are needed of the linear and circular polarisation during giant radio pulses, as well as a determination of the optical enhancement as a function of GRP phase.

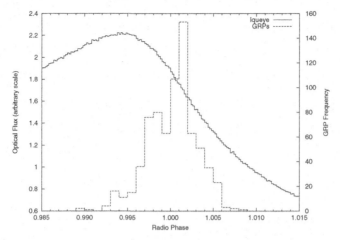

Figure 1. DASHED: Frequency distribution of Crab pulsar Main-Pulse GRPs, with SNR >6.0, observed on 2009 December 14. SOLID: IquEYE optical light curve for the same observing date.

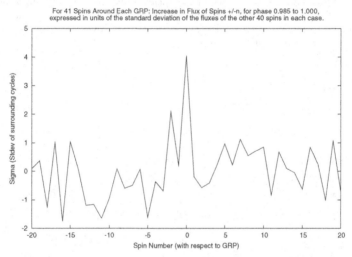

Figure 2. There is a noticeable increase in optical flux at spin0 (which represents the optical photons concurrent with a GRP) up to a 4-σ level. This correlation between increased optical emission and giant radio-pulse emission is in qualitative agreement with previous findings (Shearer *et al.*, 2003).

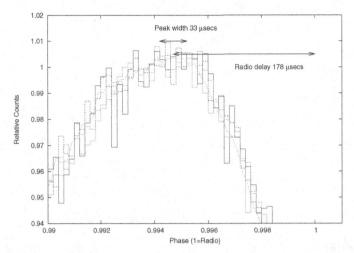

Figure 3. The peak of the optical light curves from 4 nights of data. (For full light curve, see p. 280). There seems to be no night-to-night variation in the average arrival time of the pulses. Furthermore we measured the arrival time of the optical pulse as ∼178 μs before the radio arrival time. That is significantly shorter than recent estimates; for example, Słowikowska *et al.* (2009) estimated the delay to be 235 ± 68 μs. In the combined light over four nights we see a distinct flattening of the peak, with a width of 33 μs.

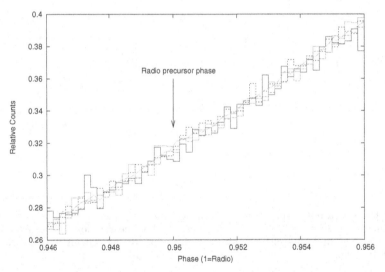

Figure 4. Słowikowska *et al.* (2009) showed that the linear polarisation of the main optical peak occurs in phase with the radio precursor, at phase .95. When we look at the optical intensity (Stokes *I*) we see no change in the slope of the rising edge. We suggest this is indirect evidence for a shift between linear and circular polarisation as observed by McDonald *et al.* (2011) and is consistent with a simple synchrotron model for the optical emission.

References

Lorimer, D. R. & Kramer, M. 2005, *Handbook of Pulsar Astronomy*, Cambridge University Press.

McDonald, J., *et al.* 2011, *MNRAS*, 417 (1) 730

Petrova, S. A. 2009, *MNRAS* 395, 1723

Shearer, A., *et al.* 2003, *Science* 301, 493

Słowikowska, A., Kanbach, G., & Kramer, M. and Stefanescu, A. 2009, *MNRAS* 397, (1), 103

New Horizons in Time-Domain Astronomy
Proceedings IAU Symposium No. 285, 2011
R.E.M. Griffin, R.J. Hanisch & R. Seaman, eds.

© International Astronomical Union 2012
doi:10.1017/S1743921312000853

False-Alarm Probabilities in Period Searches: Can Extreme-Value Distributions be of Use?

Jan Cuypers

Royal Observatory of Belgium, Ringlaan 3, 1180 Brussels, Belgium
email: jan.cuypers@oma.be

Abstract. Results of simulating false-alarm probabilities in irregularly sampled time series are presented. Relations to well-known expressions and earlier-used criteria are shown and tested for applicability. The use of an extreme-values distribution in this context is investigated.

Keywords. methods: statistical

1. Introduction

False-alarm probability (FAP) in period (frequency) search is defined as the probability that a value resulting from a time-series analysis based on some form of periodogram is caused by noise. Since knowing the significance of a peak in periodogram analysis is essential, this probability has to be determined accurately.

False-alarm probabilities for many periodograms can be described by β distributions (Schwarzenberg-Czerny, 1998; Frescura *et al.*, 2008; Baluev, 2008) having the form:

$$1 - (1 - (1 - \frac{z}{[N/2]\sigma_X^2})^{N'})^M \tag{1.1}$$

where z is the power in the periodogram, N the number and σ_X^2 the variance of the data points. In the literature the inner power term N' varies between $(N-3)/2$ and $N/2$.

On the other hand, simulations (Frescura *et al.*, 2008) suggest that formulæ like Eq.(1.1), or some variants, do not fully describe the cumulative distribution functions (CDFs), especially when only a relatively small number of observations is available. For larger numbers the exponential limit expression can be used as well as for the fully equidistant cases. If more accuracy is needed in the tail of the distribution, it would require a huge number of simulations, and ways of avoiding that are always welcome. There is therefore justification in searching for other ways to describe the CDFs and determine the FAP.

2. Extreme-Value Distributions

One option is extreme-value distributions (Coles, 2004; de Haan, 2006)), as was done by Baluev (2008), extending the results of Davies (2002). We can explore a similar but somehow simpler approach: since the peaks of the periodograms are extreme values of a well-known (or not so well-known) distribution, the distribution of the peaks is an extreme-value one. The generalized extreme-value distribution has the form

$$F(x; \mu, \sigma, \xi) = e^{-(1 + \xi \frac{x - \mu}{\sigma})^{-\frac{1}{\xi}}} \tag{2.1}$$

and the Gumbel version, for example ($\xi = 0$) is

$$G(x; \mu, \beta) = e^{-e^{-\frac{x - \mu}{\beta}}} \tag{2.2}$$

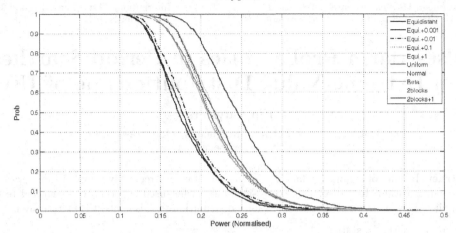

Figure 1. Empirical cumulative distributions for simulated data (50 points, Gaussian noise, time-points distributed as indicated over the same interval; only 1000 simulations are shown).

To construct the CDFs we followed the method outlined by Frescura et al. (2008): we simulated a large number of pseudo-Gaussian random time-series with the sampling times of the actual data. We performed a period analysis for each time series after choosing the appropriate frequency range, constructed the empirical CDF of the highest values of each periodogram and fitted the probability function. How the CDF changes as a function of the distribution of the points in time is illustrated in Fig.1. Remark how large the spread can be for an equal number of data points distributed in the same time interval.

3. Tentative Conclusions

Extreme-value distributions seem in general no better at describing the empirical CDFs than fitted β distributions (Eq. 1.1, with M between $N_f/5$ and N_f, where N_f is the number of scanned frequencies corrected for the oversampling in the periodogram). Only for a few distributions of the time-points are there hints that the tail is closer to an extreme-value distribution.

This is only a first exploration of the field. More research is necessary, and many caveats and questions remain.

Acknowledgements

Thanks to many members of the Gaia CU7 team (Variability Processing) for continued encouragement.

References

Baluev, R. V. 2008, *MNRAS*, 385, 1279
Coles, S. 2001, *An introduction to Statistical Modelling of Extreme values* (London: Springer-Verlag), p. 46
Davies, R. B. 2002, *Biometrika*, 89, 484
Frescura, F. A. M., *et al.* 2008, *MNRAS*, 388, 1693
de Haan, L. & Ferreira, A. 2006, *Extreme Value Theory: An introduction* (New York: Springer-Verlag), p. 9
Lomb, N. R. 1976, *Ap&SS*, 39, 447
Scargle, J. D. 1982, *ApJ*, 263, 835
Schwarzenberg-Czerny, A. 1998, *MNRAS*, 301, 831

New Horizons in Time-Domain Astronomy
Proceedings IAU Symposium No. 285, 2011
R.E.M. Griffin, R.J. Hanisch & R. Seaman, eds.

© International Astronomical Union 2012
doi:10.1017/S1743921312000865

Characterising the Dwarf Nova Population of the Catalina Real-time Transient Survey

Deanne de Budè[1], Patrick Woudt[1] and Brian Warner[1,2]

[1] Department of Astronomy, University of Cape Town, Cape Town, South Africa

[2] School of Physics and Astronomy, Southampton University, Southampton, UK
email: deanne@ast.uct.ac.za

Abstract. Results of a high-speed photometric study of dwarf novæ in the Catalina Real-time Transient Survey are given. A population of faint dwarf novæ near the orbital period minimum is detected. At the shortest periods there is a correlation between orbital period and outburst interval.

Keywords. dwarf novæ, Catalina Real-time Transient Survey, orbital period minimum

Over the past decade the Astronomy Department at the University of Cape Town has carried out high-speed optical photometry of faint southern-hemisphere cataclysmic variables (Woudt & Warner 2010), using candidates taken largely from the Sloan Digital Sky Survey (SDSS). From late 2009 the survey has used CVs from the Catalina Real-time Transient Survey (CRTS: Drake *et al.* 2009), which extends to fainter magnitudes than the SDSS and identifies objects that vary by more than 2 mag in their (approximately) 7-year light curves.

We have observed 56 of these stars, most of which are newly recognised CVs, using the 40-in and 74-in reflectors at the Sutherland site of the South African Astronomical Observatory, equipped with the UCT CCD photometer (O'Donoghue 1995). This has led to the measurement of orbital periods (P_{orb}) for 19 systems with magnitudes $\lesssim 20.5$, almost all of which are near the minimum orbital period, i.e., 80–90 mins. This is in line with the result obtained by (Gänsicke *et al.* 2009), based on SDSS systems, that the high space density of short period CVs predicted from population synthesis models can be found as faint CVs. Supplementing our P_{orb}s with others for CRTS CVs available in the literature, we find the distributions shown in Fig. 1

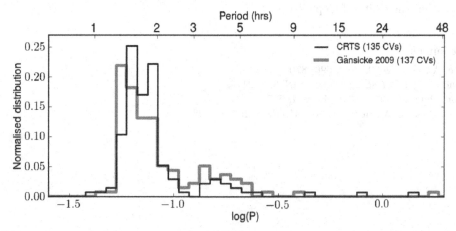

Figure 1. P_{orb} distributions from the SDSS (Gänsicke *et al.* 2009) and CRTS surveys.

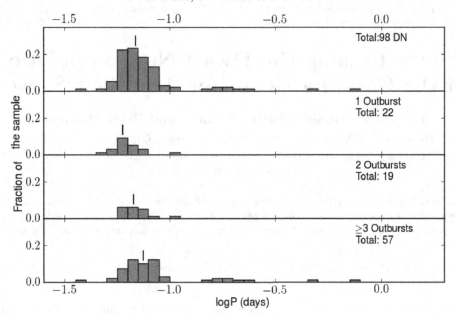

Figure 2. P_{orb} distributions of the CRTS stars according to outburst class. The vertical bars indicate the median values for each class and are given respectively by 1.66 h, 1.43 h, 1.62 h and 1.80 h.

Evolutionary models also predict that most CVs should have evolved past the minimum P_{orb} back towards longer values (post-bounce CVs), with lower mass transfer rates and rarer dwarf nova outbursts. We have classified the CRTS long-term light curves according to whether one, two, or more outbursts were detected; the P_{orb} histograms for the three types are shown in Fig. 2.

There is a clear correlation with P_{orb} in the sense that the least frequent outbursts occur among the dwarf novæ at the shortest orbital periods. Some of these may be post-bounce CVs.

Acknowledgements

DdB is funded by the South African Square Kilometre Array Project and the University of Cape Town, PW and BW's research is supported by the National Research Foundation and by the University of Cape Town.

References

Drake, A., *et al.* 2009, *ApJ*, 696, 870.
Gänsicke, B. T., *et al.* 2009, *MNRAS*, 397, 2170.
O'Donoghue, D. 1995, *Balt. Astr.*, 4, 517.
Woudt, P. A. & Warner, B. 2010, *MNRAS*, 403, p. 398

New Horizons in Time-Domain Astronomy
Proceedings IAU Symposium No. 285, 2011
R.E.M. Griffin, R.J. Hanisch & R. Seaman, eds.

© International Astronomical Union 2012
doi:10.1017/S1743921312000877

Inverse Mapping of Pulsar Magnetospheres: Optical Emission Comes From 300 km Above the Surface

Diarmaid de Búrca, Padraig O'Connor, John McDonald and Andy Shearer

Centre for Astronomy, National University of Ireland, Galway
email: **diarmaiddeburca@gmail.com**

Abstract. In order to determine the emission height of the optical photons from pulsars we present an inverse mapping approach, which is directly constrained by empirical data. The model discussed is for the case of the Crab pulsar. Our method, which uses the optical Stokes parameters, determines the most likely geometry for emission including the magnetic-field inclination angle (α), the observer's line-of-sight angle (χ) and emission height. We discuss the computational implementation of the approach, and the physical assumptions made. We find that the most likely emission altitude is at 20% of the light-cylinder radius above the stellar surface in the open field region.

Keywords. pulsars: general, polarisation

1. Introduction

Even after more than 40 years of observation and theory about pulsar emission, there is still no consensus as to how and where it occurs. There is a general agreement that the energy comes from the slowing down of the pulsar caused by some kind of dipole braking, but there is no accord as to where in the magnetosphere the emission occurs. While recent gamma-ray observations of radio-quiet pulsars have effectively ruled out a polar-cap model (Abdo *et al.* 2009), there are still the outer gap, the slot-gap and the 2-pole caustic models to be considered.

One of the main problems with modelling pulsars is that the emission is heavily dependent on the geometry, and that geometry is usually unknown. In this work we consider the geometry of the pulsar as an additional parameter to be fitted. Light curves are plotted for a range of different viewing angles and inclination angles and are then compared to observation. Once a best fit to observation has been obtained, an inverse mapping approach is used to locate the emission sites.

2. Assumptions

In order to build our model, some simple assumptions were made:
- The optical emission is a synchrotron process
- The magnetic field is in the form of a retarded dipole (Michel & Li 1999)
- The open field region is fully populated with radiating electrons

Starting from those assumptions, we calculated the light curves emitted on the basis of a number of parameters—the orientation of the magnetic-field axis, the electron pitch angle distribution and cut-off, and the observer's viewing angle.

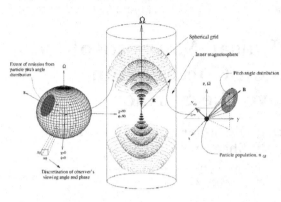

Figure 1. Centre: The concentric spherical Cartesian grid imposed on the pulsar magnetosphere, showing only points in the open magnetosphere. Right: At each individual location (**R**) we calculate the radiation emitted from the local particle distribution, taking the local physical conditions into account (e.g., **B**, v_{co}, the power law distribution, etc). Left: Representation of how emission from a particle pitch-angle distribution (PAD) extends over a range of viewing angles χ and phase Φ. Emission is recorded computationally in discrete bins of length $\Delta\chi$ and $\Delta\Phi$.

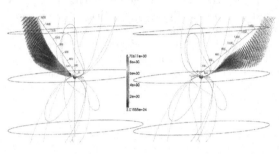

Figure 2. 3-D representation of the regions contributing to the emission in the main peak (left) and to the secondary peak (right), for a model (Crab) pulsar with parameters $(\alpha, \chi, \mathrm{PAD}, \mathrm{PAD}_{co}) = (80, 25, \mathrm{isotropic}, 20)$. The intensity scale shows the emission contributions in the regions mapped. The circles define the light cylinder boundary and the lines are the open volume/closed volume boundary. The emission is seen to be concentrated in the lower magnetosphere from a single pole.

A full description of the model we used can be found in McDonald *et al.* (2011). We constrain our model by comparing the output with observation (all Stokes parameters).

3. Results

Currently we compute all Stokes parameters and compare them against the observations. To date there have been no observations of circular polarisation from the Crab pulsar, so our constraints are limited to intensity and linear polarisation. Inverse mapping results are given for a selected (α, χ) combination, which compared most favourably with the double-peaked structure of the Crab light curve. Values of $(\alpha, \chi) = (70°, 45°)$ with a $(\mathrm{PAD}, \mathrm{PAD}_{co}) = (\mathrm{isotropic}, 45°)$ were chosen. Our preferred values of α and χ were broadly consistent with Ng & Romani (2000), who provided a robust estimate of the Crab inclination angle.

The inverse mapping approach enables us to decompose light curves, with the option of selecting different phase-resolved regions which can then be traced back into the magnetosphere, thereby isolating the emission locations of individual photons. That then localises the magnetospheric emission sites which contribute to the emission at a particular phase. We attempt here to present the results of that inverse mapping in 3 dimensions, though it is not an easy task. In Fig. 3 we present an intensity map of the emission locations. On the left are emission sites for the main pulse while on the right are the emission sites from the secondary pulse. One thing of immediate note is that the emission appears to be coming from one pole (pole one) for the peak, but from the opposite pole for the inter-pulse.

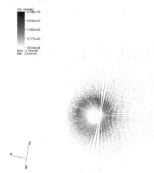

Figure 3. 3-D representation of emission sites from the Crab pulsar for the main peak, at viewing angle of 52° and inclination angle of 70°. As can be seen, there is a ring of high-energy emission. This ring appears to be following the last open field lines, in concordance with the slot-gap model.

4. Conclusions

Optical polarisation studies can be used to determine the local geometry of the emission region. In particular, from this work we see that a simple synchrotron model gives good agreement with observation. The linear polarisation peaks on the rising edge of the main pulse, in agreement with Smith *et al.* (1988) and Słowikowska *et al.* 2009. We find that, for the most part, the radiation is circularly polarised, with an interplay between circular and linear polarisation occurring on the rising edge of the main peak. On that basis, observations of the circularly polarised optical emission from pulsars should provide significant geometrical constraints to the pulsar parameters, though we accept that more detailed work is required in order to define more realistic emission locations.

Our model predicts that the pulsar emission is low in the magnetosphere (roughly 300 km from the star). That is some distance from the regions predicted by most models, with the slot-gap model being the nearest. Our model also predicts a ring of high-energy emission from the region around the last open field lines, again supporting the slot-gap model, as seen in Fig. 3.

Our future work will entail restricting the emission to locations around the last open field lines, thereby removing the inherent symmetry of the model. In that way we hope to make firmer predictions about the optical emission specifically, and to comment more generally on the correlations between optical and radio emission (Słowikowska *et al.* 2009). Furthermore, we will tune our inverse mapping constraints to intensity alone, e.g., for comparison with gamma-ray light curves.

As well as giving the emission location and pulsar orientation, our model can also constrain the PAD and the PAD_{co}. We intend to make a more detailed comparison between our model and circularly-polarised emission. To date there have not been any measurements of optical circular polarisation from any pulsar. A new instrument, the Galway Astronomical Stokes Polarimeter, has that capability, and observations of the Crab pulsar are planned for late 2011.

References

Abdo, A. A., *et al.* 2009, *Sci*, 325, 848

McDonald, J. O'Connor, P., de Búrca, D., Golden, A., & Shearer, A., 2011, *MNRAS*, 417, 730

Michel, F. C. & Li, H. 1999, *Phys. Reports*, 318, 227

Smith, F. G., Jones, D. H. P., Dick, J. S. B. & Pike, C. D. 1988 *MNRAS*, 233, 305

Słowikowska, A., Kanbach, G., Kramer, M., & Stefanescu, A. 2009 *MNRAS*, 397, 103

Ng, C.-T., Romani, R. 2000 *ApJ*, 601, 479

New Horizons in Time-Domain Astronomy
Proceedings IAU Symposium No. 285, 2011
R.E.M. Griffin, R.J. Hanisch & R. Seaman, eds.

© International Astronomical Union 2012
doi:10.1017/S1743921312000889

The Catalina Real-time Transient Survey

A. J. Drake[1], S. G. Djorgovski[1,6], A. Mahabal[1], J. L. Prieto[2], E. Beshore[4], M. J. Graham[1], M. Catalan[3], S. Larson[4], E. Christensen[5], C. Donalek[1] and R. Williams[1]

[1] California Institute of Technology, Pasadena, CA 91125, USA.
email: ajd@cacr.caltech.edu

[2] Dept. of Astrophysical Scinces, Princeton University, NJ 08544, USA.

[3] Depto. de Astronomia y Astrofisica, Pont. Uni. Catolica de Chile, Santiago, Chile.

[4] Lunar and Planetary Lab, University of Arizona, Tucson, AZ 85721, USA.

[5] Gemini South Observatory, c/o AURA, Casilla 603, La Serena, Chile.

[6] Distinguished Visiting Professor, King Abdulaziz Univ., Jeddah, Saudi Arabia.

Abstract. The Catalina Real-time Transient Survey (CRTS) currently covers 33,000 deg^2 of the sky in search of transient astrophysical events, with time base-lines ranging from 10 minutes to ~ 7 years. Data provided by the Catalina Sky Survey provide an unequalled base-line against which $> 4,000$ unique optical transient events have been discovered and openly published in real-time. Here we highlight some of the discoveries of CRTS.

Keywords. (stars:) supernovæ: general, (galaxies:) BL Lacertae objects: general, stars: dwarf novæ stars, stars: flare, galaxies: dwarf

1. Introduction

For the past four years the Catalina Real-time Transient Survey (CRTS; Drake *et al.* 2009a, Djorgovski *et al.* 2011, Mahabal *et al.* 2011) has systematically surveyed tens of thousands of square degrees of the sky for transient astrophysical events. CRTS discovers highly variable and transient objects in real-time, making all discoveries public immediately, thus benefiting a broad astronomical community. Data are leveraged from three telescopes, operated by LPL in a search for NEOs; they cover up to $\sim$2,500 deg^2 per night with four exposures separated by ~ 10 mins. The total survey area is $\sim 33,000$ deg^2 and reaches depth $V \sim 19$ to 21.5 mag (depending on the telescope) during 23 nights per lunation. All data are automatically processed as they are taken, and optical transients (OTs) are immediately distributed using a variety of electronic mechanisms (see http://www.skyalert.org/ and http://crts.caltech.edu/). CRTS has so far discovered $> 4,000$ unique OTs including $> 1,000$ supernovæ and 500 dwarf novæ.

2. Discoveries

Supernovæ and their hosts. Supernovæ are both cosmological tools and probes of the final states of stellar evolution. While many astronomical surveys focus on Type Ia SNe (being standard candles), CRTS uses its wide-area coverage to look for rare types of events that may be missed by many traditional SN surveys. With $> 1,000$ SNe (CRTS published more SN discoveries in both 2009 and 2010 than any other survey), this data set has allowed us to carry out a systematic exploration of supernovæ properties, leading to the discovery of extremely luminous supernovæ and supernovæ in extremely faint host galaxies, with $M_V \sim -12$ to -13, i.e., $\sim$0.1% of L_*.

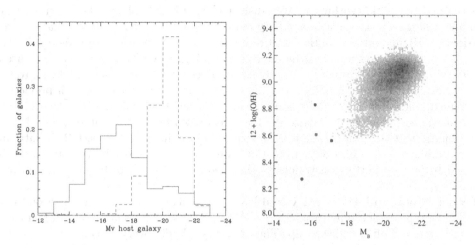

Figure 1. SN Hosts. Left: A comparison of supernovæ host-galaxy absolute magnitudes. Solid line: CRTS. Short-dashed line the Lick Observatory Supernovæ Search (LOSS). Right: Host luminosity and metallicity for four energetic Type IIn SN host galaxies, compared with 53,000 star-forming galaxies from SDSS (Tremonti *et al.* 2004).

Two especially interesting classes of luminous SNe discovered by CRTS, ROTSE and PTF include luminous Type Ic SNe: SN 2005ap (Quimby *et al.* 2007), SN 2009de (Drake *et al.* 2009b, 2010), SN 2009jh (Drake *et al.* 2009, Quimby *et al.* 2011), SN 2010gx (Mahabal *et al.* 2010; Pastorello *et al.* 2010a, 2010b), and CSS110406:135058+261642 (Drake *et al.* 2011b) and ultra-luminous and energetic Type IIn SNe: SN 2008fz, SN 2009jg, etc. (Drake *et al.* 2009c, 2010, 2011a). These supernovæ have been found to favour extremely faint host galaxies (Drake *et al.* 2009a, 2010), suggesting the importance of host-galaxy environment and explaining why more such events have not been discovered previously. In Fig. 1 we contrast the SN host-galaxy absolute magnitudes (from CRTS) with those from the long-running Lick Observatory SN Search (LOSS; Filippenko *et al.* 2001) which concentrates on large nearby galaxies.

The rate of our SN discovery in intrinsically faint galaxies implies phenomenally high specific SN rates (Drake *et al.* 2009a). Although such galaxies are common, a very small fraction of all baryonic matter is expected in them (Kauffmann *et al.* 2003). Evidence suggests that those galaxies include blue compact dwarfs and irregular dwarfs, where excessive star formation rates accelerate SNe rates for the most rapidly evolving massive stars (progenitors of luminous SN). The presence of additional evidence for enhancements in SN Ia rates, of up to 1500%, has been speculated by Della Valle & Panagia (2003).

It is likely that these dwarf galaxies have low metallicities owing to a delayed onset of star formation and expulsion of enriched SN ejecta from their shallow potential wells. According to the galaxy mass-metallicity relationship (Tremonti *et al.* 2004), low-luminosity hosts are expected to be low-metallicity hosts. This prediction was recently confirmed in the work of Neill *et al.* (2011), Stoll *et al.* (2011) and Kozlowski *et al.* (2010) as well as in our recent work shown in Fig. 1. Low metallicities are speculated to lead to a top-heavy IMF, which would account both for an enhanced specific SN rate and for the propensity for highly luminous events (from high-mass progenitors). Low-metallicity host galaxies are also linked to the broad-line type-Ic hypernovæ associated with long-timescale GRBs (Stanek *et al.* 2006).

Another interesting discovery involves a new class of SNe that may be associated with AGN accretion disks. Possibly the most luminous and optically energetic SN ever discovered, CSS100217 (within the AGN disk of a bright NLS1 galaxy), demonstrates that extreme supernovæ can occur in a variety of extreme environments (Drake *et al.* 2011a).

Blazars. Highly variable optical and radio sources, blazars are often targeted for optical follow-up at other wavelengths after their outbursts. CRTS provides an unbiased, statistical optical monitoring of known blazar sources over 75% of the sky. Owing to the erratic nature of blazar variability and the association of those sources with previously catalogued, and often faint, radio sources, we have found several tens of likely blazars based in transient outburst events. We have also discovered variability-selected blazar counterparts to the previously unidentified FERMI gamma-ray sources. CRTS data are being combined with radio data from the Owens Valley Radio Observatory, and will be used to provide better constraints to the theoretical models of blazar emission and variability.

Dwarf Novæ and UV Ceti variables. CRTS has discovered more than 500 new dwarf-nova-type cataclysmic variables (CVs). Since those objects are found in real time, the outbursts are often followed up. Thus far, 132 CV discovery alerts have been sent to users of the VSNET system (www.kusastro.kyoto-u.ac.jp/vsnet/), resulting in successful period determination in dozens of systems. Similarly, CRTS has discovered over 100 UV Ceti variables (flare stars), varying by several magnitudes within minutes. The rate of such flares is still poorly constrained, and must be understood so that future surveys can find rare types of rapid transients. The short cadence of CRTS is well tuned to the discovery of such events. Another class of rapid transients are eclipses of white-dwarf binary systems, which probe the end state of stellar binary evolution. Although first discovered in real time, archival searches have revealed dozens more such eclipsing systems, including some with low mass companions (Drake *et al.* 2011c).

Acknowledgements

CRTS is supported by the NSF grant AST-0909182. We thank the personnel of many observatories involved in the survey and the follow-up observations.

References

Della Valle M. & Panagia, N. 2003, *ApJ*, 587, L71
Drake A. J., *et al.* 2009a, *ApJ*, 696, 87
Drake A. J., *et al.* 2009b, *CBET*, 1958
Drake A. J., *et al.* 2009c, *CBET*, 1766
Drake A. J., *et al.* 2010, *ApJ*, 718, 127
Drake A. J., *et al.* 2011a, *APJ*, 735, 106
Drake A. J., *et al.* 2011b, *ATEL*, 3343
Drake A. J., *et al.* 2011c, *ApJ*, arXiv:1009.308
Djorgovski, S. G., *et al.* 2011, *JAXA*, in press, arXiv:1102.5004
Filippenko, A., Li, W. D., Treffers, R. R., & Modjaz, M. 2001, *PASP Conf Ser.*, 246, 121
Kauffmann G., *et al.* 2003, *MNRAS*, 341, 33
Kozlowski, S., *et al.* 2010, *ApJ*, 722, 1624
Mahabal, A. *et al.* 2009, *ATEL*, 1713, 1
Neill J., *et al.* 2011, *ApJ*, 727, 15
Pastorello, A. *et al.* 2010a, *CBET*, 2413
Pastorello, A. *et al.* 2010b, *ApJ*, 724, 16L
Quimby, R. *et al.* 2007, *ApJ*, 668, 99
Quimby, R. *et al.* 2011, *NATURE*, 474, 487
Stanek, K. Z., *et al.* 2006, *AcA*, 56, 333
Stoll, R., *et al.* 2011, *ApJ*, 730, 34
Tremonti, C. *et al.* 2014, *ApJ*, 613, 898

New Horizons in Time-Domain Astronomy
Proceedings IAU Symposium No. 285, 2011
R.E.M. Griffin, R.J. Hanisch & R. Seaman, eds.

© International Astronomical Union 2012
doi:10.1017/S1743921312000890

Searching for Periodic Variables in the EROS-2 Database

P. Dubath[1,2], I. Lecoeur[1,2], L. Rimoldini[1,2], M. Süveges[1,2], J. Blomme[3], M. López[4], L. M. Sarro[5], J. De Ridder[3], J. Cuypers[6], L. Guy[1,2], K. Nienartowicz[1,2], A. Jan[1,2], M. Beck[1,2], N. Mowlavi[1,2], D. Ordóñez-Blanco[1,2], J. B. Marquette[7], J. P. Beaulieu[7], P. Tisserand[8,9], É. Lesquoy[7,9] and L. Eyer[1]

[1] Observatoire astronomique de l'Université de Genève, Switzerland
email: `Pierre.Dubath@unige.ch`

[2] ISDC Data Centre for Astrophysics, Université de Genève, Switzerland

[3] Instituut voor Sterrenkunde, K.U. Leuven, Belgium

[4] Centro de Astrobiologia (INTA-CSIC), Villanueva de la Canada, Spain

[5] Dpt. de Inteligencia Artificial, UNED, Madrid, Spain

[6] Royal Observatory of Belgium, Brussels, Belgium

[7] Institut d'Astrophysique de Paris, 75014 Paris, France

[8] Research School of Astronomy & Astrophysics, Mount Stromlo Observatory, Weston ACT 2611, Australia

[9] CEA, DSM, DAPNIA, Centre dÉtudes de Saclay, 91191 Gif-sur-Yvette Cedex, France

Abstract. We started a systematic search for periodic variable-star candidates in the EROS-2 database in the context of preparatory work for the GAIA satellite mission. The goal is to evaluate different classification tools and strategies, and to identify a large sample of variable candidates. In this paper we present the results of an assessment study of a three-step identification and classification process. In the study we took a sample of about 80,000 stars from one of the LMC EROS fields.

Keywords. methods: statistical – techniques: photometric – stars: variables: general.

The identification and classification of the periodic variable candidates are achieved through three steps. First, a number of variability criteria are computed to identify variable candidates. Secondly, a supervised classifier is used to isolate periodic variables. Thirdly, another supervised method is employed to classify the periodic stars.

1. Variability Detection

A number of variability criteria are computed from the light curves, and p-values are derived for each criterion. A star is considered to be a variable candidate if the p-value is smaller than a specified threshold.

Fig. 1 shows the number of selected stars as a function of the p-value threshold for 2 of our 6 different criteria. The triangles give the fraction of selected stars which are also identified as variables in the EROS-2 catalogue (Tisserand *et al.* 2007); their total number is indicated by the horizontal line. The vertical dashed lines give the p-value thresholds used in this study. A sample of 11,888 variable candidates was extracted from the total sample of about 80,000 stars.

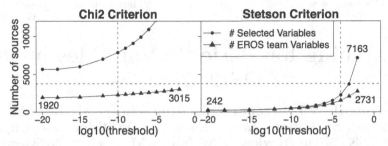

Figure 1. Number of selected stars as a function of p-value thresholds.

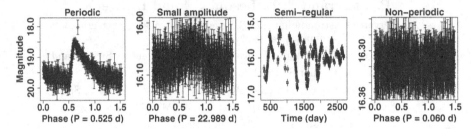

Figure 2. Examples from the four ad-hoc categories used in the visual classification.

	NP	SA	PV	SR	
	1867	13	1		NP (Non–Periodic stars)
	117	54	9		SA (Small Amplitude variables)
		12	87	1	PV (Periodic Variables)
		3	2	15	SR (Semi–regular variables)

Figure 3. Confusion matrix from random forest classification.

The harmonic least-squares analysis method (Zechmeister & Kürster 2009) was used to search for frequencies in the sample of 11,888 variable candidates. Stars with frequencies close to known spurious frequencies (near zero and 1, 2, 3, 4, and 5 cycles per day) were removed, and a sub-sample of 2181 light curves folded using the main period was inspected and visually classified into the four ad-hoc categories illustrated in Fig. 2.

2. Periodicity Detection Training

We followed the classification procedures used in a previous paper (Dubath *et al.* 2011) to identify periodic stars. A supervised random forest classifier (Breiman 2001) was trained on the sub-sample of 2181 visually classified stars by using a large number of attributes.

Fig. 3 shows that the relatively small sample of periodic variables in our data can be identified with good efficiency. The main attributes used in the classification process, ranked according to the importance provided by the random forest methodology, include (1) measures of the model amplitude-to-noise ratio in the B and R bands, (2) the period search false-alarm-probability, (3) the B and R magnitude ranges, (4) the weighted skewness of the magnitude distribution, (5) the point-to-point scatter, (6) the Stetson (1996) variability criterion and (7) the relative difference between the frequencies derived from the B and R bands.

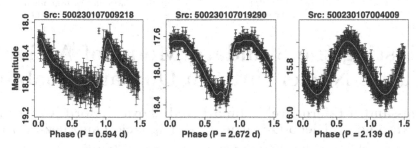

Figure 4. Light curves of three periodic variables.

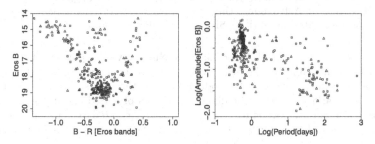

Figure 5. Colour-Magnitude and Period-Luminosity diagrams. Triangles and squares denote the training and test sets, respectively.

The classification model trained with 2181 stars was used to predict types for the sample of 3750 variable candidates with frequencies outside the spurious ranges. That yielded a total sample of 292 periodic variables (100 and 192, from the training and the test sets, respectively). Fig. 4 shows some examples of folded light curves obtained for these sources.

When a similar efficiency for the other EROS fields is assumed, the total number of periodic variables that could be extracted is of the order of 200,000.

3. Classification Plans

The next step in our study is to identify good representatives of the main types of periodic variables included in our sample; they will then be used to train a periodic variable classifier. The structures seen in the colour-magnitude and period-amplitude diagrams displayed in Fig. 5 give us good confidence that it will be possible to classify our set of periodic stars into the main variability types.

References

Breiman, L. 2001, *Machine Learning*, 45(1), 5
Dubath, P., *et al.* 2011, *MNRAS*, 414, 2602
Stetson, P. B. 1996, *PASP*, 108, 851
Tisserand P., *et al.* 2007, *A&A*, 469, 387
Zechmeister, M., Kürster M. 2009, *A&A*, 496, 577

New Horizons in Time-Domain Astronomy
Proceedings IAU Symposium No. 285, 2011
R.E.M. Griffin, R.J. Hanisch & R. Seaman, eds.

© International Astronomical Union 2012
doi:10.1017/S1743921312000907

Testing the Standard Model of Active Galactic Nuclei through Quasar Variability

Alessandro Ederoclite[1,2], Jana Polednikova[1,2], Jordi Cepa[1,2], José Antonio de Diego Onsurbe[1,3], and Ignacio González-Serrano[4]

[1]Instituto de Astrofísica de Canarias, La Laguna, Tenerife, Spain
[2]Departamento de Astronomía, Universidad de La Laguna, Spain
[3]Instituto de Astronomía, Universidad Nacional Autonoma de México
[4]Instituto de Física de Cantabria (CSIC- Universidad de Cantabria), Santander, Spain

Abstract. The standard model of Active Galactic Nuclei (AGNs) predicts that Type 2 AGNs and Type 1 AGNs only differ in the orientation of a dust torus, which does or does not allow one to observe the central region. If the model is correct, the time-scales and the amplitudes of observed temporal variations should be different between obscured and unobscured objects. In order to test this hypothesis, we started a multi-wavelength ($BRIJK_s$) monitoring campaign of a sample of quasars of both types. Here we present the data and preliminary results of that project.

Keywords. quasars: general

1. Introduction

The main idea behind the AGN Standard Model is that the AGN type depends on the inclination of a torus of dust with respect to the line of sight, thus allowing (or not allowing) one to observe directly the accretion on a supermassive black hole. Non-obscured AGNs have typically been identified as showing wide permitted lines as well as narrow permitted and forbidden lines. These objects are usually referred to as Type 1 AGNs. On the other hand, obscured AGNs only show narrow permitted and forbidden lines; they are referred to as Type 2 AGNs. Nevertheless, there are some features of the Type 2 quasars which do not completely fit the unified model, such as the large fraction of moderate radio emitters with a flat spectrum respect to Type 1 quasars (Vir & Ho 2010).

Assuming that the standard model is correct, we would then expect that Type 1 quasars have higher variability, and on shorter time-scales, than Type 2 quasars. Variability in Type 2 quasars could be justified on the assumption of a clumpy torus. If such variability were observed in Type 2 quasars, it would put strong constraints on the structure of the torus. However, since such a systematic study is missing, we took on the challenge to prove the AGN Unified Model by looking for variability in quasars.

2. The Quasar Sample

We selected 5 Type 2 quasars and 5 Type 1 ones of similar brightness, in order to be able to observe them all with a small telescope. Our targets are listed in Table 1. The Type 2 quasars were extracted from the Sloan Digital Sky Survey (Zakamska *et al.* 2003, 2004). The Type 1 quasars were selected from the the 2dF survey (Croom *et al.* 2001, 2004, 2006). Since the main selection criterion is apparent brightness, the Type 1 quasars have a comparatively higher redshift than the Type 2 ones.

Name	RA.	Dec.	mag. (filter)	z
J1045+0046	10:45:34.30	+00:46:17.0	19.558 (g)	1.26
J1211+0049	12:11:18.50	+00:49:25.0	19.88 (V)	1.47
J1334-0120	13:34:11.10	−01:20:53.0	19.68 (V)	1.620
J1449-0120	14:49:48.10	−01:20:42.0	19.76 (B)	0.97
J1402+0026	14:02:50.60	+00:26:07.0	19.95 (B)	0.857
SDSSJ1133+61	11:33:44.02	+61:34:55.7	19.03 (r)	0.426
SDSSJ1157+6003	11:57:18.35	+60:03:45.6	19.48 (r)	0.49
SDSSJ1337-0128	13:37:35.02	−01:28:15.7	18.60 (r)	0.329
SDSSJ1430-0056	14:30:27.66	−00:56:14.9	18.99 (r)	0.318
SDSSJ1501+5455	15:01:17.96	+54:55:18.3	18.57 (r)	0.339

Table 1. Quasars studied for weekly variability. The first five are Type 1 quasars; the second five are Type 2.

Name	RA.	Dec.	g-mag	z
SDSSJ0759+5050	07:59:40.96	+50:50:24.9	16.54	0.054
SDSSJ1430+1339	14:30:29.89	+13:39:12.1	16.75	0.085
SDSSJ1440+5030	14:40:38.10	+53:30:15.9	15.60	0.037

Table 2. Type 2 quasars analyzed for micro-variability.

In order to analyze micro-variability, we included the brightest Type 2 quasars from Reyes *et al.* (2008) and which we list in Table 2. In this case too, the main driver of the target selection was the brightness ($g < 17$), since we wanted to be able to access the objects with the 1.5-m telescope at San Pedro Mártir Observatory. The target sample will be complemented with the sample of Type 1 quasars studied by de Diego *et al.* (1998) and by Ramírez *et al.* (2009).

3. The Observations

Observations of variability on a time-scale of weeks were carried out in BRI with the 80-cm IAC-80 telescope, and in JK_s with the 1.5-m Telescopio Carlos Sanchez (TCS) at Teide Observatory, Tenerife. Taking advantage of service-mode observing, we obtained observations spread over about three months per target. The fields are comparatively sparse, but the relatively large field of view ($\sim 10' \times 10'$) of the camera on the IAC-80 enabled us to identify a number of good comparison stars for differential photometry. (Since our goal was to look for variability only, we did not observe standard stars). The field of view of the near infra-red camera on the TCS is significantly smaller than that of the IAC-80, so we had fewer comparison stars. Those data will be calibrated with stars from 2MASS.

Micro-variability observations were carried out in BVR with the 1.5-m telescope from San Pedro Mártir Observatory during four nights in visitor mode. Individual exposures were 60 seconds, and each quasar was followed for two to four hours.

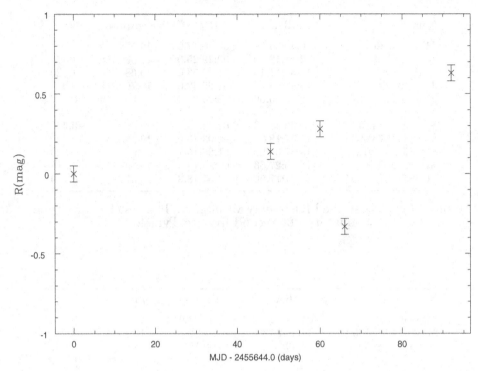

Figure 1. Preliminary R-band light curve of J1334-0120, in days from the beginning of the campaign. The magnitudes are relative to the magnitude of the object during our first observation.

4. Summary and Perspectives

During a pilot programme of a project to test the AGN unification model through quasar variability, we have obtained about three months of observations for ten quasars (5 of Type 1 and 5 of Type 2) in $BRIJK_s$ bands, and four nights of micro-variability (in BVR) for 3 Type 2 quasars (a control sample of Type 1 quasars is already available).

The analysis is under way (see Fig. 1). An extension of the programme to fainter and to more objects is foreseen in order to increase the statistical significance of the results, and to probe a larger volume of Universe. The results of our pilot programme will allow us to choose the best strategy and target selection methods for campaigns on larger (2-m to 10-m) telescopes.

References

Croom, S. M., *et al.* 2001 *MNRAS*, 322, L29
Croom, S. M., *et al.* 2004 *MNRAS*, 349, 1397
Croom, S. M., *et al.* 2009 *MNRAS*, 392, 19
de Diego, J. A., *et al.* 1998, *ApJ*, 501, 69
Rámirez, A., *et al.* 2009 *AJ*, 138, 991
Reyes R., *et al.* 2008 *AJ*, 136, 2373
Schneider, D. P., *et al.* 2010 *AJ*, 139, 2360
Lal, D. V. & Ho, L. C. 2010 *AJ*, 139, 1089
Zakamska, N., *et al.* 2003 *AJ*, 126, 2125
Zakamska, N., *et al.* 2004 *AJ*,128, 1002

New Horizons in Time-Domain Astronomy
Proceedings IAU Symposium No. 285, 2011
R.E.M. Griffin, R.J. Hanisch & R. Seaman, eds.

© International Astronomical Union 2012
doi:10.1017/S1743921312000919

Time-Domain Studies of Gravitationally Lensed Quasars

Luis J. Goicoechea[1] and Vyacheslav N. Shalyapin[1,2]†

[1] Dept. de Física Moderna, Universidad de Cantabria, ES-39005, Santander, Spain
email: goicol@unican.es

[2] Inst. Radiophys. & Elect., Nat. Acad. Sci. Ukraine, UA-61085, Kharkov, Ukraine
email: vshal@ukr.net

Abstract. We present an overview and current results of an ongoing optical/NIR monitoring of seven gravitationally-lensed quasars (GLQs) with the 2-m Liverpool Robotic Telescope. The photometric data from the first seven years (2005–2011) of this programme are leading to high-quality light curves, which in turn are being used as key tools for different standard and novel studies. While brightness records of non-lensed distant quasars may contain unrecognized extrinsic variations, one can disentangle intrinsic from extrinsic signals in certain GLQs. Thus, some GLQs in our sample allow us to assess their extrinsic and intrinsic variations; we then discuss the origin of both kinds of fluctuations. We also demonstrate the usefulness of GLQ time-domain data for obtaining successful reverberation maps of the inner regions of accretion disks around distant supermassive black holes, and for estimating the redshifts of distant lensing galaxies.

Keywords. gravitational lensing, black hole physics, accretion, galaxies: general, quasars: general.

An overview of our ongoing Liverpool Quasar Lens Monitoring (LQLM) project is presented in Table 1. The data collection are being carried out in different phases: LQLM I (from 2005 January to 2007 July), LQLM II (from 2008 February to 2010 July) and LQLM III (from 2010 October to the present), and is using available optical/NIR instrumentation. The relevant instruments on the Liverpool Robotic Telescope are the RATCam CCD camera and its associated Sloan *griz* filter set, the RINGO2 optical polarimeter, and the FRODOspec spectrograph (3900–9400 Å). Some astrophysical results and expectations for each target GLQ are given here; a more complete and updated information can be found on the GLENDAMA website http://grupos.unican.es/glendama.

SBS 0909+532. The LQLM I light curves in the r band led to a robust time-delay between its two images of $\Delta t_{AB} = -49 \pm 6$ days and $\Delta t_{ij} = t_j - t_i$, B leading (Goicoechea *et al.* 2008a). In addition, the optical flux ratio A/B changed little in the first 10 years of observations, i.e., between the identification as a quasar pair in 1996 and our LQLM I campaign (see Dai & Kochanek 2009 and references therein). For example, the r-band light curve of the A image and the properly shifted r-band light curve of B were consistent with each other throughout the LQLM I period, so the variability over this time segment was basically intrinsic to the distant quasar (Goicoechea *et al.* 2008a). However, the LQLM III light curves indicate that the r-band flux ratio had evolved in 2010–2011. Gravitational microlensing by stars within the main lensing galaxy could account for the detected extrinsic variation.

FBQ 0951+2635. Gravitational microlensing seems to be an important variability mechanism for this GLQ (Paraficz *et al.* 2006; Shalyapin *et al.* 2009). We are taking a few frames per year in the r band to trace the long-term behaviour of A/B, and thus to

† On behalf of the GLENDAMA Project Team.

Table 1. Current status of the LQLM project.

GLQ (redshift)	Comments[1]	Main lens (redshift)	Observation phases[2]	Instruments[2]	Outputs[3] (status)
SBS 0909+532 ($z = 1.38$)	2 images: A-B size $\sim 1.11''$	early-type galaxy ($z = 0.83$)	I+III	RATCam gr filters	LC (final reduction)
FBQ 0951+2635 ($z = 1.25$)	2 images: A-B size $\sim 1.10''$	early-type galaxy ($z = 0.26$)	I+II+III	RATCam ri filters	LC+DI (final reduction)
QSO 0957+561 ($z = 1.41$)	2 images: A-B size $\sim 6.17''$	cD galaxy ($z = 0.36$)	I+II+III	RATCam $griz$ filters FRODOspec RINGO2	LC+S+PM (LC=final reduction, S=first reduction, PM=pending)
SDSS 1001+5027 ($z = 1.84$)	2 images: A-B size $\sim 2.86''$	early-type galaxy ($z \sim 0.2$–0.5)	II	RATCam g filter	LC (final reduction)
SDSS 1339+1310 ($z = 2.24$)	2 images: A-B size $\sim 1.69''$	early-type galaxy ($z \sim 0.4$)	II+III	RATCam ri filter	LC+DI (first reduction)
HE 1413+117 ($z = 2.56$)	4 images: A-D size $\sim 1.35''$	? ($z = 1.9$)	II	RATCam r filter	LC+DI (final reduction)
QSO 2237+0305 ($z = 1.69$)	4 images: A-D size $\sim 1.78''$	face-on Sb galaxy ($z = 0.04$)	II	RATCam gr filters	LC (first reduction)

Notes:
[1] See the CASTLES (http://www.cfa.harvard.edu/castles/) and
SQLS (http://www-utap.phys.s.u-tokyo.ac.jp/ sdss/sqls/) websites.
[2] See main text.
[3] LC = light curves, DI = deep images, S = spectra, PM = polarization measurements.

obtain information about the structure of the source and the lensing galaxy (Wambsganss 1990; Kochanek 2004).

QSO 0957+561. We did not find evidence of extrinsic variability in the LQLM I light curves in the g and r bands. These initial brightness records were used to measure time delays between images and optical bands (Shalyapin *et al.* 2008), and to analyse the structure function of the rest-frame UV variability (Goicoechea *et al.* 2008b; Goicoechea *et al.* 2010). Later, LQLM II fluxes in the *griz* bands, together with concurrent space-based observations from Swift/UVOT and Chandra, unveiled details of the accretion flow and its jet connection in a distant radio-loud quasar for the first time (see Gil-Merino *et al.* 2011). Our global database in the gr bands is also providing surprising results on the chromaticity in Δt_{AB} and the long-term evolution of B/A, which are probably related to the presence of a dense cloud within the cD lensing galaxy along the line of sight to the A image. We are also exploring the spectro-polarimetric evolution of this fascinating first GLQ.

Two new GLQs. SDSS 1001+5027 was discovered in 2005 (Oguri *et al.* 2005). The first monitoring campaign in the R band did not produce any time delay between its two images (Paraficz *et al.* 2009). We have recently observed this double GLQ in the g band (February–May 2010), since we expect to see more variability at shorter wavelengths. If A leads B, and there are no significant extrinsic variations, the LQLM II g-band fluxes suggest a time delay ranging from 12 to 22 days. The other new GLQ (SDSS 1339+1310; Inada *et al.* 2009) was monitored in the r band just after its discovery (February–July 2009; LQLM II). Although Fig. 1 shows prominent flux variations, our LQLM II r-band light curves do not reveal any conclusive delay. Additional data during LQLM III are required in order to decide on the time delay and other properties of this system.

Two famous quads. We followed up the r-band variability of the four images A-D of the Cloverleaf quasar (HE 1413+117) in order to measure its time delays for the first time. The LQLM II fluxes of this GLQ (February–July 2008) enabled us to obtain

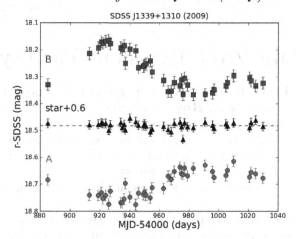

Figure 1. LQLM II r-band light curves of SDSS 1339+1310.

$\Delta t_{AB} = -17 \pm 3$ days, $\Delta t_{AC} = -20 \pm 4$ days and $\Delta t_{AD} = 23 \pm 4$ days (B-C leading, D trailing), which we then used to estimate the redshift of the main lensing galaxies: $z = 1.88^{+0.09}_{-0.11}$ (Goicoechea & Shalyapin 2010). Although useful spectroscopic data are not yet available, we derived an accurate value of z via gravitational lensing. We also monitored the Einstein Cross (QSO 2237+0305) in the g and r bands; the light curves and microlensing analyses will be presented soon. QSO 2237+0305 is the most emblematic target for microlensing studies (Shalyapin *et al.* 2002; Kochanek 2004; Gil-Merino *et al.* 2006).

References

Dai, X. & Kochanek, C. S. 2009, *ApJ*, 692, 677

Gil-Merino, R., González-Cadelo, J., Goicoechea, L. J., Shalyapin, V. N., & Lewis, G. F. 2006, *MNRAS*, 371, 1478

Gil-Merino, R., Goicoechea, L. J., Shalyapin, V. N., & Braga, V. F. 2011, *ApJ*, in press (arXiv:1109.3330)

Goicoechea, L. J. *et al.* 2008a, *New Astron.*, 13, 182

Goicoechea, L. J., Shalyapin, V. N., Gil-Merino, R., & Ullán, A. 2008b, *A&A*, 492, 411

Goicoechea, L. J., Shalyapin, V. N., & Gil-Merino, R. 2010, *The Open Astron. J.*, 3, 193 (see http://www.benthamscience.com/open/toaaj/)

Goicoechea, L. J. & Shalyapin, V. N. 2010, *ApJ*, 708, 995

Inada, N., *et al.* 2009, *AJ*, 137, 4118

Kochanek, C. S., 2004, *ApJ*, 605, 58

Oguri, M., *et al.* 2005, *ApJ*, 622, 106

Paraficz, D., Hjorth, J., Burud, I., Jakobsson, P., & Elíasdóttir, Á. 2006, *A&A*, 455, L1

Paraficz, D., Hjorth, J., & Elíasdóttir, Á. 2009, *A&A*, 499, 395

Shalyapin, V. N. *et al.* 2002, *ApJ*, 579, 127

Shalyapin, V. N., Goicoechea, L. J., Koptelova, E., Ullán, A., & Gil-Merino, R. 2008, *A&A*, 492, 401

Shalyapin, V. N., *et al.* 2009, *MNRAS*, 397, 1982

Wambsganss, J., 1990, *PhD thesis*, Munich University (also available as report MPA 550)

New Horizons in Time-Domain Astronomy
Proceedings IAU Symposium No. 285, 2011
R.E.M. Griffin, R.J. Hanisch & R. Seaman, eds.

© International Astronomical Union 2012
doi:10.1017/S1743921312000920

The VAO Transient Facility

Matthew J. Graham[1], S. G. Djorgovski[1,3], Andrew Drake[1], Ashish Mahabal[1], Roy Williams[1] and Rob Seaman[2]

[1] California Institute of Technology, Pasadena CA USA
email: `mjg, ajd, ashish, roy@caltech.edu, george@astro.caltech.edu`

[2] National Optical Astronomical Observatory, Tucson AZ USA
email: `seaman@noao.edu`

[3] Distinguished Visiting Professor, King Abdulaziz Univ., Jeddah, Saudi Arabia

Abstract. The time-domain community wants robust and reliable tools to enable the production of, and subscription to, community-endorsed event notification packets (VOEvent). The Virtual Astronomical Observatory (VAO) Transient Facility (VTF) is being designed to be the premier brokering service for the community, both collecting and disseminating observations about time-critical astronomical transients but also supporting annotations and the application of intelligent machine-learning to those observations. Two types of activity associated with the facility can therefore be distinguished: core infrastructure, and user services. We review the prior art in both areas, and describe the planned capabilities of the VTF. In particular, we focus on scalability and quality-of-service issues required by the next generation of sky surveys such as LSST and SKA.

Keywords. Standards, methods: miscellaneous, astronomical data bases: miscellaneous

1. Introduction

The endorsement of the Large Synoptic Survey Telescope (LSST) in the National Research Council's Decadal Survey of Astronomy and Astrophysics (Decadal Survey 2010) identified transient astronomy as a vital area of astronomical research for the next decade and beyond. With its stated goals of enabling and facilitating science, the VAO is well placed to deliver the tools and services required by the community to exploit that new domain. Yet it is not a trivial undertaking. The transient astronomy community is a broad church—a unique, burgeoning, multi-wavelength group consisting of both professional and non-professional astronomers. Fortunately, it is already VO cognizant, with ~47000 VOEvents having been sent at the time of writing, including ~26000 from the Catalina Real-Time Transient Survey (Drake *et al.* 2009). However, it faces daunting data challenges, with LSST projecting event production rates of between ~10^5 and ~10^7 notifications per night and predictions that SKA will be capable of detecting 1 core-collapse supernova per second over the whole sky originating somewhere in the redshift range $0 < z < 5$.

Over the past 6 years two NSF-funded projects—VOEventNet and SkyAlert—have already contributed many aspects of the current event infrastructure. The time is now ripe to define the next-generation service—the VAO Transient Facility—which will continue that work and put it onto a sound operational footing, capable of addressing the needs of the community. Below we describe a review study which we have carried out concerning the development of such a facility within the VAO, built upon the existing SkyAlert product. We have identified the short- to mid-term requirements of the community and a programme to transition the prototype SkyAlert to a production-level service—with all that that entails.

2. Community Engagement

A VAO Transient Facility aims to provide the transient astronomy community with robust and reliable tools to produce event notifications, to subscribe to event notifications, and to leverage VAO capabilities for data discovery/analysis. Responses were solicited from a cross-section of the community—LSST, LOFAR, GAIA, SKA Pathfinders (MeerKAT, ASKAP), AAVSO, Zooniverse and LIGO—to identify specific issues within those areas that require attention and to assist in planning and determining priorities. Specific questions were asked about the mode and rates of event delivery, brokering requirements including levels of service and functionality, current standards and infrastructure, repository holdings and performance, and other areas of potential interest such as EPO opportunities. The responses from the community can be broadly grouped into two categories: simplicity and programmatic access.

It is all too easy to try to make everything as feature-rich as possible in order to meet all possible user cases. There is also a strong desire to keep things manageable. Event notifications should just report the basics of a transient event. References can always be used to link to extra data such as images, light curves, etc., but the current version (2.0) of the VOEvent standard (Seaman *et al.* 2011) is perfectly adequate. The ability to retrieve stored events based solely on their sky positions (RA, Dec) and timings meets most anticipated queries. At this stage, querying just any aspect of the full VOEvent data model is not required. There are also concerns as to whether the existing infrastructure will be able to scale to the event rates of some projects, but the existing protocols (Jabber/XMPP and TCPV) seem satisfactory.

A browser interface is great for trying out things and getting a feel for what something does, but for operational usage having an API to code against is much more preferable. In particular, APIs were requested so that it should be possible to define subscription filters using an expression syntax to limit the events received to those of interest. Something akin to the current Python-based syntax used by SkyAlert would be broadly acceptable. Users would also like to be able to provide specific pieces of code ("plugins") to execute when matching events are received by the broker. They might add extra information to an event report before it is passed on to the subscriber.

3. SkyAlert

SkyAlert (`http://www.skyalert.org`) is an excellent prototype of the kind of requested broker service. Instances have been deployed by various projects around the world, and it has a strong user base (~300 subscribers). The VAO has its own set of requirements to which products and services must comply, covering development aspects such as version control and testing as well as operational and user support issues. A review of SkyAlert has been carried out to ensure that it can be integrated into the VAO with minimal effort, and also to identify potential problems that we can address: points of failure and bottlenecks. Two main activities have been identified as being required: the development of tests, and documentation. Additional tasks are then required for a production-level transient facility, such as addressing scalability issues—migrating the code base away from a single server instance—and potential integration with VAO security mechanisms.

4. Event Infrastructure

The current event infrastructure is an *ad hoc* mixture of unwritten agreements, ideology, and hacks that work. There are no inherent conceptual issues with this framework

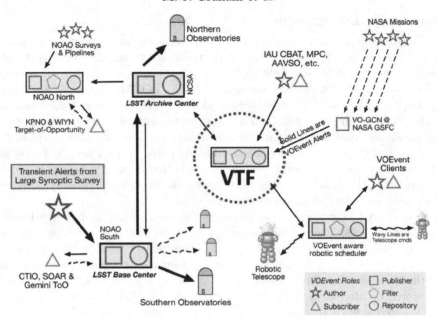

Figure 1. This shows a possible architecture for the dissemination of LSST events.

(see Fig. 1 as to how LSST events could be disseminated by it), but with its increasing uptake and usage by new projects, its various components need to be standardized to ensure interoperability. Specifically, the format for describing infrastructure metadata, a protocol for accessing and querying stored event notifications, and the transport mechanisms used to distribute event notifications need to be specified.

5. Recommendations and Schedule

The VAO Transient Facility can meet the specific needs of the transient community, in terms of event notification and brokering capabilities, by offering services based on the SkyAlert system and by ensuring that the event infrastructure is completely formalized. The development schedule for such a venture is driven by consideration of a number of potential external science collaborations, specifically the LSST 2012 Data Challenge (June 2012), in which event notifications will be circulated at a 25% expected data rate, and the GAIA Event Challenge (March 2013). Obviously such activities would serve as a good test and demonstrator of the capabilities of the facility. To meet them, we propose to have a cluster-based fully-tested installation by mid-late 2012, compliant with a fully-specified event infrastructure.

Acknowledgements

This work has been funded through NSF grants AST-0834235, AST-0909182 and OCI-0915473.

References

Drake, A. J., *et al.* 2009, *ApJ*, 696, 870
NAS Committee, *New Worlds, New Horizons*, Decadal Survey of Astronomy and Astrophysics (Washington, DC: NAS), 2010
Seaman, R., *et al.* 2011, *IVOA Recommendation*, arXiv:1110.0523

New Horizons in Time-Domain Astronomy
Proceedings IAU Symposium No. 285, 2011
R.E.M. Griffin, R.J. Hanisch & R. Seaman, eds.
© International Astronomical Union 2012
doi:10.1017/S1743921312000932

Searching for Fast Optical Transients using a VERITAS Cherenkov Telescope

Sean C. Griffin

Physics Department, McGill University,
3600 University Street, Montreal, QC, H3A2T8, Canada
email: griffins@physics.mcgill.ca

Abstract. Astronomical transients are intrinsically interesting objects to study. However, fast optical transients (μs time-scales) are a largely unexplored field of optical astronomy. Most optical observations use instruments that have integration times of the order of seconds and are thus unable to resolve fast transients. Current-generation atmospheric Cherenkov gamma-ray telescopes such as VERITAS, which consists of four 12-m telescopes, have huge collecting areas, much larger than those of any existing optical telescopes. This paper outlines the benefits of using a Cherenkov telescope to detect optical transients, and the implementation of the VERITAS Transient Detector (TRenDy), a dedicated multi-channel photometer based on field-programmable gate arrays which can be used on VERITAS for such studies without interfering with gamma-ray observations.

Keywords. accretion, instrumentation: photometers, X-rays: binaries

1. Introduction

In optical astronomy, fine time-resolution observations on a μs time-scale are rare, owing to the fact that most optical detectors (such as CCDs) have long integration times, on the scale of seconds or longer. Moreover, because large optical telescopes are oversubscribed, few attempts have been made to search for fast optical transients.

Atmospheric Cherenkov detectors such as VERITAS (Holder *et al.* 2006) have huge collecting areas (at the cost of angular resolution), and the nature of gamma-ray observations make observing under moonlight difficult or impossible. Thus, time is usually available for other studies. H.E.S.S. has carried out a search for fast optical transients using a purpose-built camera built specifically for this search (Deil *et al.* 2009). This paper describes the VERITAS Transient Detector (TRenDy): a high time-resolution, multi-channel, photon counting rate-meter. TRenDy was designed to be integrated into the standard VERITAS telescope readout system, thus making the transition from gamma-ray to optical observing rapid. The limits of the sensitivity of TRenDy due to background light can be traded against gains made by extended observations.

The primary source candidates are X-ray binary systems: the emission mechanism for fast optical transients is probably the accretion of matter onto compact objects. Compact objects such as black holes and neutron stars have sizes of 10s of kilometres, which suggests typical time-scales of 10–100 μs. Millisecond time-scale optical variability has already been observed in X-ray binaries (Kanbach *et al.* 2001; Baratolini *et al.* 1994) and pulsars (Shearer *et al.* 2003). The X-ray binary Cygnus X-1 has been observed by Dolan (2001) to emit pulses in the UV with durations of the order of $\leqslant 10$ ms, a time-scale which requires the production mechanism to be somewhere in the system's accretion disk. That agrees with measurements of X-ray sources GX339-4 (Fabian *et al.* 1982) and V4641 Sagittarii (Uemura *et al.* 2002). Fluctuations in the optical afterglow of gamma-ray

Figure 1. *Left:* The TRenDy rate meter. The design is based on the Xilinx ML402 FPGA board and contains a Virtex-4 XC4VSX35-FF668-1 FPGA (centred). The custom mezzanine board (right of the FPGA) acts as an interface between the VERITAS CFDs and the FPGA. The LEMO cables to the right connect to the standard VERITAS readout chain. *Right:* The VERITAS camera is made up of 499 photomultiplier tubes with attached light cones (not shown here) to minimize dead space and block light not coming from the VERITAS reflector. The hexagonal packing leads to a natural subset of the central seven PMTs being used by TRenDy.

bursts, on time-scales of less than an hour, have also been detected (Bersier *et al.* 2003). It may be possible to see fine structure in the light curves of such transients.

2. The VERITAS Transient Detector

The TRenDy rate-meter is built from the commercially available Xilinx ML402 Evaluation Platform FPGA board. It also contains a custom-built mezzanine board for loading NIM signals from the constant-fraction discriminators (CFDs) of the standard VERITAS readout into the FPGA. A photo of TRenDy is shown in Fig. 1 (left).

The device has seven channels corresponding to the centre seven pixels of the VERITAS camera (Fig. 1, right) and is designed to use the centre pixel to collect light from the target source and six channels around it to form a "veto ring". The point spread function of a VERITAS telescope is less than the field of view of a single camera pixel ($0°.15$): all light from a point source will be well contained within the centre pixel. The six channels of the ring provide a measure of the background rates, but also the ability to veto non-astronomical events: if a signal is seen in both the veto ring and the central pixel simultaneously it must be from a terrestrial source, such as lightning. Similarly, if a signal is seen to move from pixel to pixel, the event can also be vetoed; possible sources of this type are aeroplanes, meteors or spacecraft.

3. Tests and Results

One test performed to verify the functionality of TRenDy was the observation of the transit of the image of a bright star across the PMTs of the Whipple 10-m telescope (Kildea *et al.* 2007). The telescope was slewed to a position ahead of the target star in right ascension. The telescope's tracking system was then disengaged, causing the star's image to drift across the face of the camera. The effect is clearly seen in Fig. 2 (left). A second test was the observation of the Crab Optical Pulsar. The light curve shown in Fig. 2 (right) was reconstructed from TRenDy data taken on a VERITAS telescope using standard barycentering codes and a radio ephemeris from Jodrell Bank Observatory (Lyne *et al.* 1993).

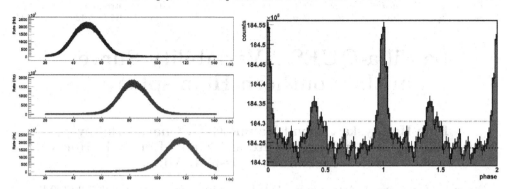

Figure 2. *Left:* A star is allowed to pass through the field of view of the camera. Three of the seven TRenDy channels are shown here. The count rates from three neighbouring pixels rise and fall as the star passes through their field of view. The scatter is statistical. *Right:* Phase profile of the Crab Optical Pulsar; two periods are shown to aid the eye. The bottom line is the mean number of counts; the top line represents 5σ. The sensitivity of TRenDy is $31\sigma/\sqrt{\text{hour}}$. This data set represents about 45 minutes of data.

4. Conclusions

The design for the TRenDy, a multi-channel high-time-resolution photometer, has been outlined, as has its use on VERITAS. Following the success of tests done over the past few months, searches for microsecond optical flares will now take place. Currently under investigation is the possibility of measuring the PMT photocurrents rather than counting pulses as a method of working around the constraints imposed by observing in moonlight.

5. Acknowledgements

Special thanks go to my supervisor, David Hanna, for proposing this project, and to Adam Gilbert for his invaluable help in much of the hardware design. TRenDy is supported by grants from NSERC and FQRNT. We acknowledge the excellent work of the technical support staff at the FLWO and at the collaborating institutions in the construction and operation of the instrument.

References

Bartolini, C., *et al.* 1994, *ApJS* 92, 455
Bersier, D., *et al.* 2003, *ApJ* 584, L43
Deil, C., *et al.* 2009, *Astropart. Phys.* 31, 156
Dolan, J. F. 2001, *PASP* 113, 974
Fabian, A. C., *et al.* 1982, *A&A* 111, L9
Holder, J., *et al.* 2006, *Astropart. Phys.* 25, 391
Kanbach, G., *et al.* 2001, *Nature*, 414, 180
Kildea, J., *et al.* 2007, *Astropart. Phys.* 28, 182
Lyne, A. G., *et al.* 1993, *MNRAS* 265, 1003
Shearer, A., *et al.* 2003, *Science* 301(5632), 493
Uemura, M., *et al.* 2002, *PASJ* 54, L79

New Horizons in Time-Domain Astronomy
Proceedings IAU Symposium No. 285, 2011
R.E.M. Griffin, R.J. Hanisch & R. Seaman, eds.

© International Astronomical Union 2012
doi:10.1017/S1743921312000944

La Silla-QUEST Variability Survey in the Southern Hemisphere

Ellie Hadjiyska[1], David Rabinowitz[1], Charles Baltay[1],
Nancy Ellman[1], Peter Nugent[2], Robert Zinn[3], Benjamin Horowitz[1],
Ryan McKinnon[1], Lissa R. Miller[3]

[1] Center for Astronomy & Astrophysics, Yale University, New Haven, CT 06520-8120 USA
email: ellie.hadjiyska@yale.edu

[2] Lawrence Berkeley National Laboratory, Berkeley, CA, 94720-8139 USA

[3] Dept. of Astronomy, Yale University,
P.O. Box 208101, New Haven, CT 06520-8101 USA

Abstract. We describe the La Silla-QUEST (LSQ) Variability Survey. LSQ is a dedicated wide-field synoptic survey in the Southern Hemisphere, focussing on the discovery and study of transients ranging from low redshift ($z < 0.1$) SN Ia, Tidal Disruption events, RR Lyræ variables, CVs, Quasars, TNOs and others. The survey utilizes the 1.0-m Schmidt Telescope of the European Southern Observatory at La Silla, Chile, with the large-area QUEST camera, a mosaic of 112 CCDs with field of view of 9.6 square degrees. The LSQ Survey was commissioned in 2009, and is now regularly covering ~1000 square deg per night with a repeat cadence of hours to days. The data are currently processed on a daily basis. We present here a first look at the photometric capabilities of LSQ and we discuss some of the most interesting recent transient detections.

Keywords. surveys, techniques: photometric, supernovae: Type Ia, stars: variables

1. Introduction

After the completion of the Palomar-QUEST northern sky survey in September 2008 the QUEST Large Field Camera (Baltay et al. 2007) was moved and installed on the 1.0-m ESO Schmidt in La Silla and had first light on April 24, 2009. Since September 2009 the southern survey has been routinely observed (Andrews et al. 2008), and the telescope and camera are controlled from Yale and are fully robotic. We have 90% of the time on the telescope with 10% allocated to Chile. The QUEST camera consists of 112 CCDs of 600×2400 Sarnoff thinned pixels, back illuminated devices with 13 μm $\times$ 13 μm pixel pitch. The camera covers an area of $4°.6 \times 3°.6$ on the sky and a plate scale of 0.86 arcsec/pixel. The survey covers ~1000 square degrees per night, primarily between $\pm 25°$ to allow for follow-up from both hemispheres. The LSQ variability (SN and transient) survey uses 60-sec exposures (and the TNO survey 180 sec) taken twice a night with a cadence of 2 nights in one broad-band filter of 4000 to 7000 Å (Qst*-band). The seeing at La Silla for the 60-sec exposures is 1.7 arcsec FWHM, reaching a depth of 20.5 mag. The LSQ survey subtraction pipeline has started producing between 400 and 900 transient candidates each night (Fig. 1).

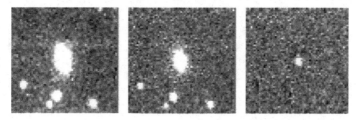

Figure 1. A possible supernova. From left – reference image, night1, subtracted image.

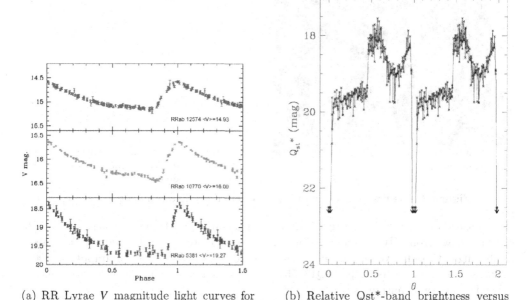

(a) RR Lyrae V magnitude light curves for RRab 12574, 10770 and 5381.

(b) Relative Qst*-band brightness versus orbital phase θ, from LSQ observations.

Figure 2. LSQ Example Transient Detections.

2. Transient Detections

2.1. *RR Lyræ Variables*

The LSQ RR Lyræ star (RRLS) survey is searching the galactic halo for RRLS that have V magnitudes between roughly 14 and 20. Because RRLS are excellent standard candles, they provide a powerful probe of the density distribution of the halo, which is being examined for halo substructure. Plotted are the V magnitude light-curves for three type ab RRLS (Fig. 2a), which illustrate the typical photometric precisions at these magnitudes. From their mean V magnitudes, we estimate that RRab 12574, 10770, and 5381 lie 7, 13, and 52 kpc from the Sun, respectively.

2.2. *A Deep Eclipsing CV*

A deep eclipsing cataclysmic variable (Rabinowitz *et al.* 2011a) was discovered with eclipse depths >5.7 magnitudes, orbital period 94.657 min, and peak brightness $V \sim 18$ at J2000 position 17h 25m 54.8s, -64 deg 38 min 39 sec. Light curves in B, V, R, I, z and J were obtained with SMARTS 1.3-m and 1.0-m telescopes at Cerro Tololo and spectra from 3500 to 9000 Å with the SOAR 4.3-m telescope at Cerro Pachon. The optical light curves (Fig. 2b) show a deep, 5-min eclipse immediately followed by a shallow 38-min eclipse and then sinusoidal variation. No eclipses appear in J. During the deep eclipse

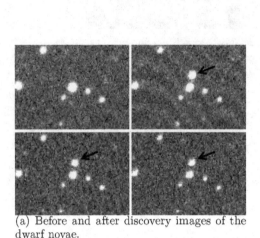
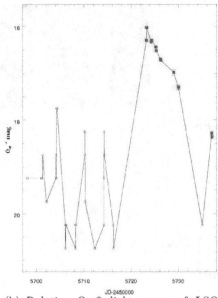

(a) Before and after discovery images of the dwarf novae.

(b) Relative Qst* light curve of LSQ J16531857-1617542, upper limits and error barred detections.

Figure 3. The first followed up LSQ subtraction pipeline transient candidate.

the measure of $V-J > 7.1$ corresponds to a spectral type M8 or later secondary. The spectra show strong Hydrogen emission lines, Doppler broadened by 600 – 1300 km s-1, oscillating with radial velocity that peaks at mid deep eclipse with semi-amplitude 500 ± 22 km s^{-1}. It is suggested that LSQ172554.8-643839 is a polar with a low-mass secondary viewed at high inclination.

2.3. *A Dwarf Nova*

An apparent dwarf nova was discovered (Rabinowitz *et al.* 2011b) on June 11.046 (when the variable was at magnitude $R = 16.3$) and June 11.063 UT (at R = 16.0) (Figs. 3a and 3b). A faint source is reported at this position in the Guide Star Catalog V2.3.2 (with $Bj = 20.76$). Simultaneous visible and J-band observations were taken with ANDICAM on the 1.3-m SMARTS telescope at Cerro Tololo. A spectrum (range 0.350–0.966 nm) taken on June 26 with GMOS on the Gemini South telescope reveals strong H-α and H-β emission lines, with H-alpha clearly double-peaked, indicating the presence of an accretion disk with a rotational velocity of 1000 km s^{-1}.

This work has been made possible with the help of NERSC computer resources and funding from DOE and NASA.

References

Andrews, P., *et al.* 2008, *PASP*, 120, 703A
Baltay, C., *et al.* 2007, *PASP*, 119, 1278B
Rabinowitz, D., *et al.* 2011, *ApJ*, 732, 51
Rabinowitz, D., Baltay, C., Ellman, N., Hadjiyska, E., & Tourtellote, S. 2011, *CBET*, 2757, 1R

New Horizons in Time-Domain Astronomy
Proceedings IAU Symposium No. 285, 2011
R.E.M. Griffin, R.J. Hanisch & R. Seaman, eds.

© International Astronomical Union 2012
doi:10.1017/S1743921312000956

Inferring Rotation Periods of Young Stars from Synoptic Observations

Patrick Hartigan[1], Christopher M. Johns-Krull[1], and Paul Scowen[2]

[1] Dept. of Physics and Astronomy, Rice University, Houston, TX, USA
emails: hartigan@sparky.rice.edu; cmj@rice.edu

[2] School of Earth and Space Exploration, Arizona State University, Tempe, AZ, USA
email: paul.scowen@asu.edu

Abstract. Using known distributions for the periods, amplitudes and light-curve shapes of young stars, we examine how well one could measure periods of these objects in the upcoming era of large synoptic surveys. Surveys like the LSST should be able to recover accurate rotation periods for over 90% of targets of interest in regions near to massive-star formation. That information will usher in a new era in our understanding of how the angular momentum of a young star/disk system evolves with time.

Keywords. stars: pre-main-sequence, stars: rotation

1. Introduction

One of the main goals of star-formation research is to learn how individual stars, binaries and clusters of various masses form, evolve and disperse within a cloud. To accomplish those goals, one must first find all the young stars within a given region, including any that may have drifted away from their birthplaces, and determine their spectral types and luminosities in order to place them in an H–R diagram and to infer masses and ages of each source from a set of pre-main-sequence models. With that information one can begin to recover the history of star formation within a given region, to assess whether or not winds from the most massive stars trigger collapse within molecular cores, to observe clustering behaviour, and to understand the degree to which massive stars affect initial mass functions in star-forming regions. The addition of rotational periods for each of the objects opens up a wealth of research related to the changes in the angular momentum of the objects as they contract towards the main sequence—an area of great importance for studies of young solar systems and early stellar winds. The photometric databases required to measure rotation periods also support other areas of research, such as testing pre-main-sequence evolutionary tracks through newly-discovered eclipsing binaries, and correlating observed accretion events with the ejection of knots in stellar jets.

Unfortunately it can be quite difficult to identify all the young stars within a given area. Low-mass ($\lesssim 1.2~M_\odot$) young stars are classified either as classical T Tauri stars (cTTs), or as weak-lined T Tauri stars (wTTs), according to the strength of their Hα emission lines. Although cTTs are surrounded by dense accretion disks that produce strong infrared excesses, and readily stand out in near-IR colour–colour diagrams and in Hα emission surveys (Guieu *et al.* 2009), the wTTs, which dominate for clusters older than about 5 Myr (Fedele *et al.* 2010), resemble field stars in those surveys. One can identify both wTTs and cTTs via their active chromospheres and large starspots, which give rise respectively to X-ray emission and a low-amplitude quasi-periodic variability in the optical. X-ray surveys uncover many wTTs, but those efforts are limited in both areal coverage and depth.

Optical synoptic surveys such as LSST will usher in a new era in which variability should become the new standard for discovering young stars. That has the important benefit of determining periods as well, provided that the sensitivities and cadence are sufficiently high to detect the variables and to recover their periods accurately. Optical surveys are particularly attractive because the amplitudes of the optical light curves are higher than they are in the near-IR. The light curves of both wTTs and cTTs have non-periodic components, but the phases of the periodic parts of the light curves generally remain stable over the course of a year.

In this contribution we create an ensemble of typical T-Tauri light curves, and assess how well one can extract periods of those systems from a survey that has a cadence of once every three days, with an uncertainty of 0.02 mag. The parameters approximate those of LSST for $V < 19$–20 mag. Since a typical T Tauri star in a nearby star-forming region such as Orion has $V \sim 17$ mag, the parameter range is critical for studies of nearby star-forming regions as well as for more massive clusters such as Carina that are four times more distant (3 magnitudes fainter).

2. Model Parameters

To determine how well a synoptic survey will recover periods for a given class of variable stars, we must first define the period and amplitude distribution of the variables and the shapes of the light curves. For the survey we adopt a cadence of once every three days, with a standard deviation of 0.2 days, in order to mimic typical observing variations (and to help eliminate artificial aliases). We adopt a Gaussian noise of 0.02 mag in each data point, and use a Scargle method to recover periods with some false-alarm probability (FAP) (Horne & Baliunas 1986).

2.1. Period distribution

Classical T Tauri stars rotate somewhat more slowly than weak-lined T Tauri stars do, probably owing to the influence of the surrounding circumstellar disk. A recent survey of rotational properties of both types of T Tauri stars by Affer *et al.* (2011) shows that the combined sample of the objects exhibits a period distribution analogous to a Poisson one, with a mean of roughly 4 days. We therefore adopt a Poisson distribution with a mean of 3.5 days, and then add 0.5 days to the distribution to ensure that all stars have periods longer than 0.5 days, as defined by the observations.

2.2. Amplitude distribution

Amplitudes of T Tauri light curves calculated by the ROTOR programme (Grankin *et al.* 2008) peak at around 0.1 magnitude, with an extended tail to higher variability. Because the limit of the ROTOR survey was ~ 0.1 mag, we anticipate that lower-amplitude variables also exist. For the purposes of this study we adopt for the amplitudes a Poisson distribution that has a mean of 0.05 mag, and then add 0.05 mag to ensure that the mean variability is 0.1 mag, with a minimum variability of 0.05 mag.

2.3. Light-curve shapes

Photometric variations in weak-lined T Tauri stars are caused by starspots, with the additional complication of accretion variability present in classical T Tauri stars. By observing sources continuously from space-based platforms over the course of several weeks, the MOST (Siwak *et al.* 2011) and CoRoT satellites (Alencar *et al.* 2010) have collected dozens of high-precision light curves of T Tauri stars. The shapes tend to be either sinusoidal, or flat with a positive or negative bowl-like feature that extends over

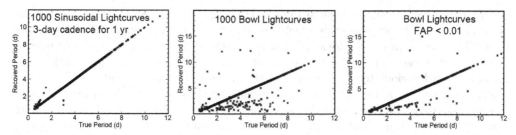

Figure 1. Recovered period vs. true period for a sample of sinusoidal (left), bowl-shaped (middle) and bowl-shaped with False Alarm Probability < 0.01 (right), assuming a 3-day cadence and one year of observing. The bowl-shaped curves are more difficult to recover than the sinusoids, but the method is highly successful in both cases.

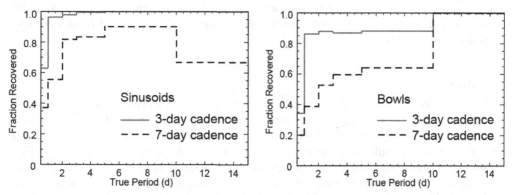

Figure 2. Fraction of periods recovered correctly for sinusoidal (left) and bowl light curves (right) for 3-day (solid line) and 7-day (dashed line) cadences over an observing period of one year. A 3-day cadence is significantly better than a 7-day one. Over 98% of sinusoidal, and 86% of bowl, light curve periods are recovered successfully with the 3-day cadence. The percentages drop to about 82% and 59% respectively, for the 7-day cadence.

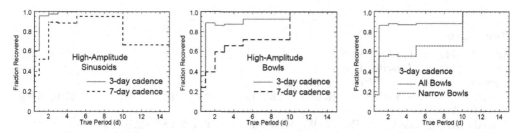

Figure 3. Left and centre: Same as Fig. 2 but restricting the sample to amplitudes greater than 0.1 mag. The method is only marginally more successful with the larger amplitude objects than it is with the entire sample. Right: The narrowest 278 bowls have a significantly higher error rate than the entire sample does.

some range of phase. We consider both sinusoids and bowl-shapes in our models. For the latter shape we use a Gaussian distribution with a FWHM chosen uniformly within the interval of 0–0.75 units of period for the bowl.

3. Results

The results of our simulations are shown in Figs. 1–3. We define a period to be recovered correctly if it is less than 1% in error. Using the parameters described above, the method recovers over 98% of the sinusoidal periods (Fig. 1, left), and 86% of the bowl curves (Fig. 1, centre). The few sinusoids which are missed tend to have periods less than about a day. The bowl light curves are somewhat more difficult to extract, although restricting the sample to those with FAP < 0.01 eliminates over half the outliers (Fig. 1, right). The erroneous bowl periods tend to be an alias that is twice the true period.

Fig. 2 shows that reducing the cadence to once every seven days significantly reduces the success rate to about 82% for the sinusoids and to about 59% for the bowls. For both types of light curve the shorter periods are more difficult to extract correctly. Comparing the large-amplitude (> 0.1 mag) curves in Fig. 3 with the entire sample in Fig. 2 shows little difference, implying that reduced amplitude is not the main source of error. However, when comparing the 278 narrowest bowls with the entire bowl sample (Fig. 3, right) we see a marked reduction in the success rate of the method at all periods. A narrow bowl resembles a shorter period, and those bowls are the most difficult to interpret correctly because the number of points with useful period information is reduced compared with sinusoids or wider bowls.

4. Summary

A survey like LSST should be extremely successful in recovering accurate periods of young stars. Using a sampling rate of about 3 days and uncertainties of 0.02 mag (easily within reach for $V \lesssim 19$ with LSST), we were able to recover the correct periods for over 90% of T-Tauri-like variable stars. The most difficult targets will be the few that have periods less than about a day, or light curves that are largely flat over a large phase interval, so the time-scale of variability is correspondingly shorter than it would be for more smoothly varying systems. A limit of $V \sim 20$ mag would reach most young stars in nearby star-forming regions like the Orion Nebula, and the all-sky nature of the survey will ensure that the distributed populations of young objects are included. The resulting dataset should revolutionize our understanding of how stellar angular momentum evolves in young clusters and associations.

References

Affer, L., Micela, G., Favata, F., & Flaccomio, E. 2011, Poster contribution to 2nd CoRoT Symposium, Marseille, France

Alencar, S. H. P., et al. 2010, A&A, 519, A88

Fedele, D., van den Ancker, M., Henning, Th., Jayawardhana, R., & Oliveira, J. 2010, A&A, 510, A72

Grankin, K. N., Bouvier, J., Herbst, W., & Melnikov, S. Y.u. 2008, A&A, 479, 827

Guieu, S., et al. 2009, ApJ, 697, 787

Horne, J. H. & Baliunas, S. 1986, ApJ, 302, 757

Siwak, M., et al. 2011, MNRAS, 415, 1119

New Horizons in Time-Domain Astronomy
Proceedings IAU Symposium No. 285, 2011
R.E.M. Griffin, R.J. Hanisch & R. Seaman eds.

© International Astronomical Union 2012
doi:10.1017/S1743921312000968

Proposal for Multi-Messenger Observations of Radio Transients by Nasu and Ligo-Virgo

Kazuhiro Hayama[1,3] **and Kotaro Niinuma**[2] **and Tomoaki Oyama**[3]

[1] Max-Planck-Institut für Gravitationsphysik, D-30167 Hannover, Germany
email: kazuhiro.hayama@ligo.org
[2] Graduate School of Science and Engineering, Yamaguchi University,
Yamaguchi 753-8511, Japan
[3] National Astronomical Observatory of Japan, Mitaka, Tokyo 181-8588, Japan

Abstract. The gravitational waves radiated from explosions of compact objects are expected to be accompanied by other astronomical radiation such as radio flares, neutrinos and gamma-rays. In order to improve the detection of gravitational waves, mutual follow-up observations have been proposed between gravitational wave telescopes and other astronomical telescopes, known as "multi-messenger" observations. We propose multi-messenger observations using the LIGO, Virgo and Nasu Radio Telescopes.

Keywords. multi-messenger, gravitational waves, radio transients

1. Introduction

The Nasu radio telescope, located in Nasu, 160 km north of Tokyo, consists of eight spherical dish antennæ each of 20 m diameter, and a spherical dish antenna of 30 m diameter. Its observation frequency is 1.42 GHz. It is designed to detect transient radio signals and variable radio objects, and has detected more than 10 events since 2004. Although the origin of these transient events is not known, one potential source is the radio remnants of compact binary mergers; they are prime sources of gravitational waves. A merger is expected to produce significant energetic outflows, whose optimal frequency is $\sim$ 1.4 GHz (Naka & Piran 2011)—the observing frequency of the Nasu telescope.

Searches for gravitational waves have been performed by km-scale interferometetric gravitational-wave telescopes such as LIGO (Abbott *et al.* 2009a) and Virgo (Accadia *et al.* 2011). One of the main searches carried out is a blind search for transient gravitational waves, including binary neutron-star coalescence (Abbott *et al.* 2009b, 2009c). That type of search does not use astrophysical information from electro-magnetic observation. On the other hand, so-called "multi-messenger" observations, which use both gravitational-wave observations and other astronomical observations, have been performed to search for gravitational waves associated mainly with gamma-ray bursts (Abbott *et al.* 2010a, 2010b).

In this paper we propose multi-messenger observing of radio transients using the Nasu Radio Telescope and LIGO-Virgo.

2. Overview of Waseda Nasu Pulsar Observatory

2.1. *Specification of the observatory*

The Waseda Nasu Pulsar Observatory is located in Nasu, Tochigi Prefecture, Japan, 160 km north of Tokyo, at latitude $36° 55' 41''.3$ north and longitude $139° 58' 54''.3$ east.

The transients monitoring array is chiefly composed of eight 20-m spherical dish antennæ (Daishido *et al.* 2000; Takeuchi *et al.* 2005). In order to search for radio transients or to monitor the radio variability of AGNs, those antennæ utilize four pairs of interferometers with an 84-m base-line. Four different declinations are monitored simultaneously with four fringe beams. Observations are performed as an earth-rotation drift-scan (Kuniyoshi *et al.* 2006). For any direction within $\pm 5°$ of the zenith, a declination of $+32° - 42°$ is observable by moving a symmetrical Gregorian sub-reflector. The half-power beam-width is approximately $0°.8$, and the fringe spacing approximately $10' - 11'.5$, depending on the declination and the projected base-line (Kida 2008). The field of view is $0°.8$ in latitude, $3°.2$ in declination (Niinuma 2007b). The frequency of the observations is 1.42 ± 0.01 GHz.

2.2. *Method for detecting radio transients*

If a point source passes through the fringe beam of the Nasu interferometers, the beam pattern (defined as the product of the primary beam and the fringe pattern, spaced at ~ 40–46 seconds), appears in the data (Kuniyoshi *et al.* 2006; Takefuji 2007). The method of searching for radio transients is the same as for normal radio sources: the fringe signal is identified as coming from a transient source if the beam pattern appears clearly. The duration of all the radio transients detected at Nasu is at least 4 minutes. The fringe cycle provides further evidence that the signal originated from a celestial object.

2.3. *Potential of Nasu Observatory*

The spatial FFT observing system for direct imaging was developed in Daishido laboratory (Daishido *et al.* 2000; Nakajima 1993; Otobe 1994), and we have been carrying out test observations with this system at Nasu Observatory. When one can carry out a stable observation, it will be possible to

• obtain the dispersion measure within a bandwidth of 20 MHz; we can estimate the distance to the origin by assuming the column density of electrons, and

• detect radio sources with a time-resolution of 50 micro-seconds by observing the same object eight times in a day, about 25 seconds apart.

3. Nasu and LIGO-Virgo: linked operations

Multi-messenger observations with the Nasu radio telescope and LIGO-Virgo relax the detection threshold for gravitational-wave searches, and should improve the efficiency of detections. Fortunately, since several radio transients were detected by the Nasu instruments while LIGO and Virgo were also making observations in 2005–2007, we could carry out multi-messenger observations.

The multi-messenger observation has two modes. One is to follow up observations triggered by other astronomical detections. When a transient event is detected and recognized, information about the position and arrival time of the event is recorded. The follow-up observation of a gravitational wave associated with that event then incorporates that information in its search parameters. By noting any coincidences between the two observations, a low-threshold search can be performed that involves archived data, thus enabling the detection of a fainter gravitational wave signal than would result from the blind search. The other is a follow-up observation of an astronomical wave triggered by a gravitational wave. In typical explosions of compact objects such as merging binary neutron stars, electromagnetic waves are radiated for a duration of a few milliseconds to a few weeks, after any gravitational radiation (Hansen *et al.* 2001). When a gravitational-wave candidate is found the alert—including information about sky position and arrival time—is sent to other astronomical telescopes and follow-up observations are requested. The difference in the arrival times will vary according to frequency. In the case of a radio flare, the time difference is a few weeks, but as the durations of radio transients are a few days to a few weeks that is long enough to arrange follow-up observations. Since Nasu is observing continuously in a direction that is within ±5 degrees of the zenith, the discovery of radio transients is more likely from the region within the sensitivity of the Nasu telescope.

References

Abadie, J., *et al.* (2011), *Physical Review D*, 82, 102001
Abbott, B. P., *et al.* (2009a), *Reports on Progress in Physics*, 72, 7
Abbott, B. P., *et al.* (2009b), *Physical Review D*, 80, 062001
Abbott, B. P., *et al.* (2009c), *Physical Review D*, 80, 062002
Abbott, B. P., *et al.* (2010a), *ApJ*, 715, 1438
Abbott, B. P., *et al.* (2010b), *ApJ*, 715, 1453
Accadia, T., *et al.* (2011), *Classical and Quantum Gravity*, 28, 11
Daishido, T., *et al.* (2000), *Proc. SPIE*, 4015, 73
Hansen, B. M. S., *et al.* (2001), *Mon. Not. R. Astron. Soc.*, 322, 695
Kida, S., *et al.* (2008), *New Astronomy*, 13, 519
Kuniyoshi, M., *et al.* (2006), *PASP*, 118, 901
Kuniyoshi, M., *et al.* (2007), *PASP*, 119, 122
Nakajima, J., *et al.* (1993), *PASJ*, 45, 477
Nakar, E. & Piran, T. (2011), *Nature*, 478, 82
Niinuma, K., *et al.* (2007), *PASP*, 119, 112
Niinuma, K., *et al.* (2007), *ApJ*, 657, L37
Niinuma, K., *et al.* (2009), *ApJ*, 704, 652
Otobe, E., *et al.* (1994), *PASJ*, 46, 503
Takefuji, K., *et al.* (2007), *PASP*, 119, 1145
Takeuchi, H., *et al.* (2005), *PASJ*, 57, 815

New Horizons in Time-Domain Astronomy
Proceedings IAU Symposium No. 285, 2011
R.E.M. Griffin, R.J. Hanisch & R. Seaman, eds.
© International Astronomical Union 2012
doi:10.1017/S174392131200097X

Variability with WISE

Douglas Hoffman, Roc Cutri, John Fowler, and Frank Masci

IPAC, California Institute of Technology, Pasadena, CA 91125, USA
email: dhoffman@ipac.caltech.edu

Abstract. WISE mapped the entire sky in four bands during its approximately 7-month cryo-genic mission. The number of exposures for each point on the sky increased with ecliptic latitude, and ranged from ∼12 on the ecliptic to over 1000 at the ecliptic poles. The observing cadence is well suited to studying variable objects with periods between ∼2 hours to ∼2 days on the ecliptic, with the maximum period increasing up to several weeks near the ecliptic poles. We present the method used to identify several types of variables in the WISE Preliminary Release Database, and the mid-IR light curves of several objects. Many of these objects are new, and include RR Lyr, Algol, W UMa, Mira, BL Lac and YSO-type variables, as well as some unknown objects.

Keywords. surveys, stars: variables

1. Introduction

The Wide-field Infrared Survey Explorer (WISE) mapped the entire sky in four bands centred at wavelengths of 3.4, 4.6, 12 and 22μ (referred to as W1, W2, W3, and W4, respectively). WISE conducted its survey using a 40-cm cryogenically cooled telescope equipped with four 1024 × 1024 infrared array detectors that simultaneously imaged the same 47′ × 47′ field-of-view on the sky. WISE flew in a 531-km sun-synchronous polar orbit and employed a freeze-frame scanning technique whereby the telescope scans the sky continually at a rate of approximately 3.8 arcmin/sec and a scan mirror freezes the sky on the focal plane detectors while 7.7 sec (W1 and W2) and 8.8 sec (W3 and W4) exposures are acquired. The FOV of each successive exposure set overlaps the previous one by 10%, and the scan paths on adjacent orbits overlap by approximately 90% on the ecliptic (see Fig. 1). The WISE survey strategy alternated stepping the scan path forward and backward on subsequent orbits, in an asymmetric pattern that approximately matched the orbital precession rate. In this way, each point near the ecliptic was observed on every other orbit, approximately each 191 minutes, and typically 12 independent exposures were accumulated for each point near the ecliptic. The number of exposures increases with ecliptic latitude, reaching over 1000 at the ecliptic poles.

2. Variable Source Identification

Flux variables were identified through the statistics made while generating WISE multi-frame coadditions. The primary identification method uses the cross-correlation coeffi-cients between the flux measurements in adjacent bands. Objects with band correlation significance > 70% and with no artifact flags are considered strong candidates. Variable candidates are also found by using the standard deviation of the flux measurements and the maximum difference between uncontaminated flux measurements. The χ^2 statistic was computed for those two quantities, yielding the probability that the source was in-consistent with the reference distribution. The majority of the sources identified by that method are periodic variables. A Lomb-Scargle periodogram was computed for them,

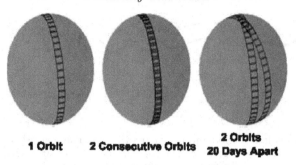

Figure 1. The WISE survey strategy.

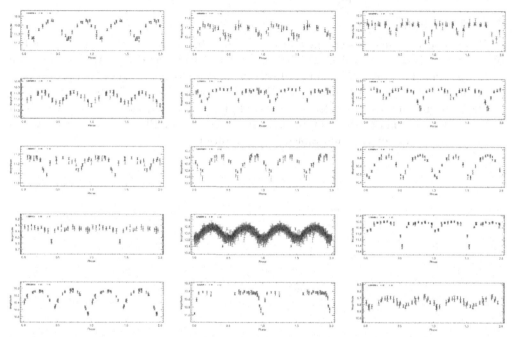

Figure 2. Various types of periodic variables in the WISE data set, phased to the peak power of the Lomb-Scargle periodogram. Example types include W Uma, RR Lyr, Algol, and β Lyr.

and the light curve was phased to twice that period (assuming that most of the periodic variables are eclipsing/rotational variables). Non-periodic variables were also found, but were less numerous.

3. Statistics of Variables

At a preliminary estimate, about 0.5% of all WISE sources are classified as significantly variable in flux. Similar surveys in the optical produce variability rates of 2.0–4.0%. The lower rate is probably due to the fact that the mid-infrared emission from common variables such as W UMa stars is low, as well as having relatively few observations near the ecliptic. The amplitudes of flux variations also tend to decrease at longer wavelengths for many pulsating variables. Nonetheless, a wide variety of variables has been identified with WISE, with W UMa, RR Lyr and Algol systems being the most common. Fig. 2 shows a sample of phased light curves, while Fig. 3 shows the light curves of five long-period variables. The next step is to classify those objects into variability class according to colour, period and light-curve morphology.

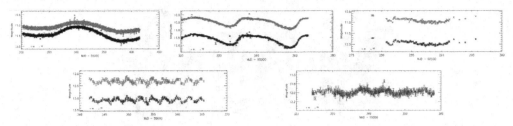

Figure 3. Examples of long-period and irregular variables. These mostly come from the ecliptic polar regions where coverage is greater. The top row features a Mira variable, a Cepheid and a BL Lac object. The bottom row features a young stellar objects with semi-regular variations, and a variable of unknown type.

4. Acknowledgements

This publication made use of data products from the Wide-Field Infrared Survey Explorer, which is a joint project of the University of California, Los Angeles, and the Jet Propulsion Laboratory/California Institute of Technology, funded by the National Aeronautics and Space Administration. Long-term archiving and access to the WISE single-exposure database is funded by NEOWISE, which is a project of the Jet Propulsion Laboratory/California Institute of Technology, funded by the Planetary Science Division of the National Aeronautics and Space Administration.

New Horizons in Time-Domain Astronomy
Proceedings IAU Symposium No. 285, 2011
R.E.M. Griffin, R.J. Hanisch & R. Seaman, eds.

© International Astronomical Union 2012
doi:10.1017/S1743921312000981

Hottest Superfluid and Superconductor in the Universe: Lessons from the Cooling of the Cassiopeia A Neutron Star

Wynn C. G. Ho[1,6], Craig O. Heinke[2], Daniel J. Patnaude[3], Peter S. Shternin[4,5,7], and Dmitry G. Yakovlev[4,5]

[1] School of Mathematics, University of Southampton, Southampton SO17 1BJ, UK

[2] Department of Physics, University of Alberta, Edmonton, AB T6G 2G7, Canada

[3] Smithsonian Astrophysical Observatory, Cambridge, MA 02138, USA

[4] Ioffe Physical Technical Institute, Politekhnicheskaya 26, 194021 St. Petersburg, Russia

[5] St. Petersburg State Polytechnical Univ., Politekhnicheskaya 29, 195251 St. Petersburg, Russia

[6] email: wynnho@slac.stanford.edu [7] email: pshternin@gmail.com

Abstract. The cooling rate of young neutron stars gives direct insight into their internal makeup. Using CHANDRA observations of the 330-year-old Cassiopeia A supernova remnant, we find that the temperature of the youngest-known neutron star in the Galaxy has declined by 4% over the last 10 years. The decline is explained naturally by superconductivity and superfluidity of the protons and neutrons in the stellar core. The protons became superconducting early in the life of the star and suppressed the early cooling rate; the neutron star thus remained hot before the (recent) onset of neutron superfluidity. Once the neutrons became superfluid, the Cooper pair-formation process produced a splash of neutrino emission which accelerated the cooling and resulted in the observed rapid temperature decline. This is the first time a young neutron star has been seen to cool in real time, and is the first direct evidence, from cooling observations, of superfluidity and superconductivity in the core of neutron stars.

Keywords. dense matter, equation of state, neutrinos, stars: evolution, stars: interiors, stars: neutron, supernovae: individual (Cassiopeia A), X-rays: stars

Neutron stars are created in the collapse and supernova explosion of massive stars. They begin their lives very hot ($T > 10^{11}$ K) but cool rapidly through the emission of neutrinos. Neutrino emission depends on uncertain physics at the supra-nuclear densities ($\rho \gtrsim 2.8 \times 10^{14}$ g cm^{-3}) of the neutron star core (Tsuruta 1998; Yakovlev & Pethick 2004; Page *et al.* 2006; Yakovlev *et al.* 2011). Current theories indicate that the star may contain exotica such as hyperons and deconfined quarks, and matter may be in a superfluid/superconducting state (Migdal 1959; Lattimer & Prakash 2004; Haensel *et al.* 2007). Observing cooling neutron stars and comparing their temperatures to theoretical models allow one to constrain the (uncertain) physics that governs the stellar interior.

The compact object at the centre of the Cassiopeia A supernova remnant was discovered in CHANDRA first-light observations (Tananbaum 1999) and subsequently identified as a neutron star (Ho & Heinke 2009). The supernova explosion is estimated to have occurred in 1681 ± 19 (Fesen *et al.* 2006); that makes the Cassiopeia A neutron star the youngest-known neutron star: its age is ~ 330 yr. Heinke & Ho (2010) analyzed CHANDRA ACIS observations taken during the last 10 years and found a steady temperature decline of 4%. If the rapid decline is due to passive cooling, then that is evidence for superfluidity and superconductivity in the core of a neutron star.

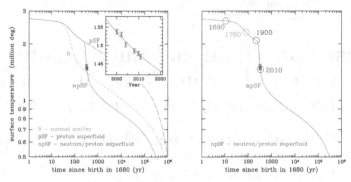

Figure 1. Theoretical models of neutron-star cooling with superfluid neutrons and protons (npSF – solid), normal neutrons and superfluid protons (pSF – long-dashed), and normal neutrons and protons (N – short-dashed). Left: Data (inset) are from CHANDRA observations of the Cassiopeia A neutron star (crosses). Right: Calculated temperatures at particular ages (circles; denoted by the year) corresponding to the four sets of panels in Fig. 2

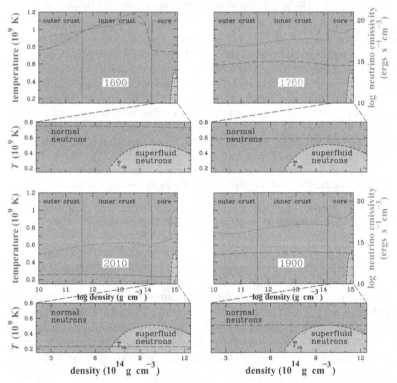

Figure 2. Profiles of neutron-star interior temperature (long-dashed) and neutrino emissivity (solid) as a function of density. The short-dashed lines indicate the critical temperature T_{cn} for superfluid formation: neutrons become superfluid at densities where $T < T_{cn}$. Each set of (upper and lower) panels is at a particular age/year (see also right panel of Fig. 1), starting with the top left (in year 1690) and continuing clockwise until the bottom left (in year 2010). Upper set: outer parts of the star are at lower densities and labelled outer/inner crust, and inner parts at supra-nuclear densities are labelled core; the core has a radius of $\sim$10 km, while the crust thickness is $\sim$1 km. Lower set: close-up of the core and neutron superfluid transition region.

The left panel of Fig. 1 shows surface temperatures for three theoretical models of neutron star cooling: "N – normal matter" corresponds to neutron star matter that does not contain any sort of superfluid, "pSF–proton superfluid" is for a proton superfluid in the core, and "npSF–neutron/proton superfluid" is for superfluid protons and neutrons in the core. The inset shows CHANDRA surface temperature measurements of the Cassiopeia A neutron star from 1999 to 2010 (Heinke & Ho 2010; Shternin *et al.* 2011). Note the clear difference between the cooling behaviour of models with normal matter (N) and matter containing superfluids (pSF or npSF) after time ~40 yr. A proton superconductor forms soon after neutron star formation, and that suppresses neutrino emission so the cooling rate is weaker than for normal matter. The neutron star is thus able to stay relatively warm, but precipitating a rapid temperature drop once neutrons become superfluid (Gusakov *et al.* 2004; Page *et al.* 2004). The model with superfluid neutrons and protons (npSF) fits the data at an age of a few hundred years. The right panel of Fig. 1 shows the npSF model of neutron star cooling; the four circles trace the cooling curve predicted by that model from about 10 years after the supernova explosion (SN in ~1680) to near the present date.

The four sets of panels of Fig. 2 show the interior temperature and neutrino emissivity at the four ages/dates corresponding to the circles in Fig. 1. At early ages (top left), the neutron-star core cools so rapidly by neutrino emission that the crust does not have time to react. The crust was thus hotter than the core in 1690 (age ~10 yr; protons were superconducting by that time), and the surface temperature declined very slowly. The surface temperature eventually reacted to the "cooling wave" that swept through the crust and started to drop more quickly. After 1760 (top right) the temperature became almost constant throughout the star. Then in ~1900 (bottom right) the interior temperature dropped below the critical value for a neutron superfluid to form, and a spike in neutrino emission occurred in the core as neutron Cooper pairs formed. Energy was lost as the neutrinos were emitted, causing the core to cool off and another cooling wave to travel outwards. As neutrons in large regions of the core became superfluid the surface temperature dropped quickly, beginning in ~1930 and continuing to the present date (bottom left). Further details can be found in Shternin *et al.* (2011).

Very similar results and conclusions were obtained independently by Page *et al.* (2011). Continued monitoring will allow tests of our model, and will also reveal important fundamental physics that cannot be accessed in laboratories on Earth.

References

Fesen, R. A., *et al.* 2006, *ApJ*, 645, 283
Gusakov, M. E., Kaminker, A. D., Yakovlev, D. G., & Gnedin, O. Y. 2004, *A&A*, 423, 1063
Haensel, P., Potekhin, A. Y., & Yakovlev, D. G. 2007, *Neutron Stars 1. Equation of State and Structure* (New York: Springer)
Heinke, C. O. & Ho, W. C. G.. 2010, *ApJ*, 719, L167
Ho, W. C. G. & Heinke, C. O. 2009, *Nature*, 462, 71
Lattimer, J. L. & Prakash, M. 2004, *Science*, 304, 536
Migdal, A. B. 1959, *Nucl. Phys.*, 13, 655
Page, D., Geppert, U., & Weber, F. 2006, *Nucl. Phys. A*, 777, 497
Page, D., Lattimer, J. M., Prakash, M., & Steiner, A. W. 2004, *ApJS*, 155, 623
Page, D., Prakash, M., Lattimer, J. M., & Steiner, A. W. 2011, *Phys. Rev. Lett.*, 106, 081101
Shternin, P. S., *et al.* 2011, *MNRAS* (Lett.), 412, L108
Tananbaum, H. 1999, *IAU Circ.* 7246
Tsuruta, S. 1998, *Phys. Rep.*, 292, 1
Yakovlev, D. G. & Pethick, C. J. 2004, *ARAA*, 42, 169
Yakovlev, D. G., *et al.* 2011, *MNRAS*, 411, 1977

New Horizons in Time-Domain Astronomy
Proceedings IAU Symposium No. 285, 2011
R.E.M. Griffin, R.J. Hanish & R. Seaman, eds.

© International Astronomical Union 2012
doi:10.1017/S1743921312000993

Fast Transient Detection as a Prototypical "Big Data" Problem

Dayton L. Jones, Kiri Wagstaff, David Thompson, Larry D'Addario, Robert Navarro, Chris Mattmann, Walid Majid, Umaa Rebbapragada, Joseph Lazio, and Robert Preston

Jet Propulsion Laboratory, California Institute of Technology, Pasadena, CA 91109, USA
email: dayton.jones@jpl.nasa.gov

Abstract. The detection of fast (< 1 second) transient signals requires a challenging balance between the need to examine vast quantities of high time-resolution data and the impracticality of storing all the data for later analysis. This is the epitome of a "big data" issue—far more data will be produced by next generation-astronomy facilities than can be analyzed, distributed, or archived using traditional methods. JPL is developing technologies to deal with "big data" problems from initial data generation through real-time data triage algorithms to large-scale data archiving and mining. Although most current work is focused on the needs of large radio arrays, the technologies involved are widely applicable in other areas.

Keywords. methods: data analysis, astronomical data bases: miscellaneous

Fast transient signals are predicted from a number of astronomical objects and processes, and are expected to be associated with extreme physical conditions. Several technologies are critical for large-scale dedicated or commensal searches for fast transients.

Power consumption by digital electronics is likely to be a major operating cost for future radio facilities in searches for fast transient signals, particularly for wide-band digital beam-forming and correlation. We must be able to afford the cost of generating high-rate data before the analysis of high-rate data becomes relevant. JPL has investigated ASIC architectures that can reduce power consumption in cross-correlation dramatically for arrays with large numbers of antennæ (D'Addario 2011). This has obvious application to the SKA and other large instruments.

A second area of development at JPL is adaptive algorithms to perform real-time processing in high-volume data flows, when storage of data for later analysis is not an option (data triage). Machine-learning algorithms that provide transient triggers and automated processing of buffered high-time-resolution data are now being tested by the VLBA V-FASTR project (Brisken *et al.* 2011; Thompson *et al.* 2011; Wayth *et al.* 2011). Similar techniques can be used to detect time-varying interference (self-induced or external), anomalies in array performance-monitoring data, and some aspects of time-varying calibration. Data-adaptive algorithms could also control front-end data collection based on the statistical properties of event detections, allowing optimized sampling.

We are also working on highly scalable data archive frameworks for astronomy. Archive design will determine how much real-time data triage will be needed, and how much analysis can be done off-line. JPL has developed a Process Control System based on Object Oriented Data Technology (OODT), components of which are being evaluated for use at several observatories. OODT is open source software, and is the first NASA software to become a top-level project at the Apache Software Foundation.

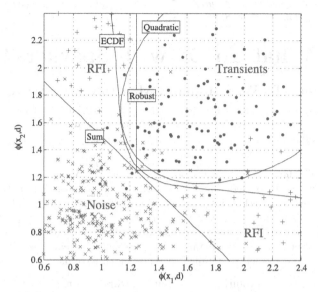

Figure 1. Ability of different multi-station transient detection methods to separate noise, RFI, and true transients. The axes show signal strength from each of two separate VLBA antennæ.

The JPL-funded technology development described here is part of an evolving end-to-end approach to "big data" problems. Fast transient searches provide an excellent test case for that work, in addition to the possibility of exciting near-term scientific results.

Low-Power Digital Signal Processing. JPL has completed a systematic study of the effect of architecture choices on the power consumption of large cross-correlators used for all aperture synthesis radio arrays. Architectures that minimize memory use and I/O data rates can significantly reduce power consumption, sometimes making the difference between a particular facility or instrument being feasible to operate or not.

Real-Time Machine-Learning Algorithms. As an example of the value of machine-learning algorithms for real-time data triage, Fig. 1 shows an application in fast (< 1 s) transient signal detection. The quadratic discriminant algorithm provides greater robustness against false alarms, and greater sensitivity.

Scalable Data Archives and Data Mining. The JPL Process Control System is a set of reusable components from the Object Oriented Data Technology (OODT) framework developed by Dan Crichton. It is scalable, hardware- and database-independent, and is interoperable, with a flexible plug-in capabilities for user data processing tools and algorithms.

Acknowledgements

This work was carried out at the Jet Propulsion Laboratory, California Institute of Technology, under contract with the U.S. National Aeronautics and Space Administration.

References

D'Addario, L. 2011 *Square Kilometre Array Memo*, 133, http://www.skatelescope.org/
Brisken, W., *et al.* 2011, *Proc. National Radio Science Meeting*, Boulder, CO, 2011 January 6
Thompson, D. R., *et al.* 2011, *ApJ*, 735, 98
Wayth, R. B., *et al.* 2011, *ApJ*, 735, 97

New Horizons in Time-Domain Astronomy
Proceedings IAU Symposium No. 285, 2011
R.E.M. Griffin, R.J. Hanisch & R. Seaman, eds.

© International Astronomical Union 2012
doi:10.1017/S1743921312001007

What To Do with Sparkers?

E. F. Keane[1], B. W. Stappers[2], M. Kramer[1,2] and A. G. Lyne[2]

[1]Max Planck Insitut für Radioastronomie, 53121 Bonn, Germany.
email: ekeane@mpifr-bonn.mpg.de

[2]School of Physics & Astronomy, University of Manchester, Manchester M13 9PL, UK.

Abstract. In 2007, the discovery of the so-called "Lorimer Burst" was announced—a single radio pulse that was so dispersed that it could only have originated outside our Galaxy. The apparently unique event, together with the large inferred distance (a redshift $z \sim 0.2$ is required to explain its high dispersion) implies a very high luminosity. Suggested progenitors include a supernova, a binary neutron-star merger, and a black-hole annihilation event. Crude estimates of the rates of such events predict that many such bursts should already be detectable in archived pulsar-survey data, and has led to detailed searches which have had some success.

Keywords. Surveys, stars: evolution, stars:neutron, stars:supernovae:general, ISM:general

1. Pulsar Searches and the Lorimer Burst

Fourier-domain searches are the ones most commonly used for identifying new pulsars. However, searches for the brightest individual pulses are the best way to discover certain highly-modulated pulsars (the so-called "RRATs"; Keane & McLaughlin 2011), and perhaps the *only* way of finding extra-galactic pulsars. In 2007 Lorimer *et al.* (2007) reported the discovery of a single 5-ms 30-Jy pulse with a very high DM (dispersion measure) of $375 \, \mathrm{cm^{-3} \, pc}$, of which only $25 \, \mathrm{cm^{-3} \, pc}$ is attributable to material in the Galaxy for that line of sight (Cordes & Lazio 2002). The pulse obeyed the theoretical dispersion law and showed frequency-dependent broadening consistent with interstellar scattering. It was detected in 3 of the 13 beams of the telescope receiver, as expected for a boresight signal. Despite many tens of hours of follow-up a second pulse was never observed. The interpretation of the burst was of a very bright, single event originating at a cosmological distance, with suggested progenitors including a supernova, a binary neutron-star merger, and black hole annihilation.

2. "Perytons"

Recently Burke-Spolaor *et al.* (2011) have identified a number of radio bursts in data from the on-going High Time Resolution Universe Survey; they are relatively weak, and are seen in all 13 beams of the receiver that is used. These are characteristic of radio-frequency interference (RFI), but uncharacteristically for RFI they are dispersed (although with certain "kinks" in their dispersive sweeps). It was initially reported that all of these "peryton" events were at the same DM as the Lorimer burst, raising (by association) an air of suspicion over that particular value of DM. However, the search only considered pulses in the DM range $200–500 \, \mathrm{cm^{-3} \, pc}$, and as more such signals have since been identified they are now seen to span that range. Furthermore, recent work shows that these signals, when detected, arise $\sim 80\%$ through each integer second of time (UTC), with characteristic 22-second gaps (Kocz *et al.* 2011). These signals are therefore clearly terrestrial.

3. J1852−08: A Second "Sparker"?

A recent successful search of the Parkes Multi-beam Pulsar Survey for bright bursts with DM from 0–2000 cm^{-3} pc has, amongst numerous discoveries, resulted in the identification of J1852−08 (Keane *et al.* 2011). This 7-ms pulse has a lower flux density than the Lorimer burst but is strongly detected and, like the Lorimer burst, shows the theoretically-expected dispersion sweep with a DM of 745 cm^{-3} pc. According to the NE2001 model for the Galactic electron density distribution (Cordes & Lazio 2002), 222 cm^{-3} pc of that must be explained by the intergalactic medium and any putative host galaxy. The inferred redshift is $z \sim 0.1$ (or a distance of $\sim 0.5h^{-1}$ Gpc in the standard cosmological model). There seem to be two possible explanations for the nature of J1852−08. The first is that the NE2001 model is incorrect for that line of sight and so the event is actually a "giant pulse" from a Galactic pulsar. The second is that the NE2001 model is reliable, so the distance, and hence the luminosity, is too large to be due to the brightest giant pulse known from any pulsar, and that the observation was a "single event". The first possibility can be investigated readily through observation. If J1852−08 were a giant pulse-emitting pulsar (like the Crab), then since the pulse was well above our detection threshold we would expect to have observed many weaker pulses in the original 35-minute survey pointing. Those were not seen, but to exhaust the possibility a total of 16 hours of follow-up observations were performed at the Parkes Observatory, resulting in no detection. Thus, if NE2001 is incorrect then J1852−08 would also necessarily have the most extreme pulse amplitude distribution of any known pulsar. The second possibility therefore seems more likely, as even though we cannot prove that the event was unique the probability increases the longer we do not see any repetition of it. For further proof we might look for a gravitational-wave signal. However, as the event actually occurred on the 2001 June 21 it was a pre-LIGO and pre-GEO600 event so no gravitational-wave counterpart can be sought. The same holds for the Lorimer burst on 2001 August 24.

4. Outlook—What to Do with Sparkers?

The lag (of ~ 10 years!) between occurrence and discovery of these two bursts means that there is no more that can be done to investigate their actual progenitors. However, as we enter the era of "real-time all-sky" monitoring, detection rates should increase dramatically. Every next-generation radio telescopes has Transients as a key science goal. With future instruments like LOFAR and (in the next decade) the SKA, it is vital to plan for "multi-messenger" confirmations; they may take the form of cross-community automated alerts and/or ATels. Such moves are essential for solving the mystery of sparkers.

References

Burke-Spolaor, S., *et al.* 2011, *MNRAS*, 416, 2465.

Cordes, J. M. & Lazio, T. J. W. 2002, astro-ph/0207156.

Keane, E. F., Kramer, M., Lyne, A. G., Stappers, B. W., & McLaughlin, M. A. 2011, *MNRAS*, 415, 3065.

Keane, E. F. & McLaughlin, M. A. 2011, *BASI*, 39, 1.

Kocz, J., Bailes, M., Barnes, D. Burke-Spolaor, S., & Levin, L. 2011, *MNRAS*, in press.

Lorimer, D. R., Bailes, M., McLaughlin, M. A., Narkevic, D. J., & Crawford, F. 2007, *Science*, 318, 777.

New Horizons in Time-Domain Astronomy
Proceedings IAU Symposium No. 285, 2011
R.E.M. Griffin, R.J. Hanisch & R. Seaman, eds.

© International Astronomical Union 2012
doi:10.1017/S1743921312001019

A Refined QSO Selection Method Using Diagnostics

Dae-Won Kim[1,2,3], Pavlos Protopapas[1,3], Markos Trichas[1], Michael Rowan-Robinson[4], Roni Khardon[5], Charles Alcock[1] and Yong-Ik Byun[2,6]

[1] Harvard-Smithsonian Center for Astrophysics, Cambridge, MA, USA

[2] Department of Astronomy, Yonsei University, Seoul, South Korea

[3] Institute for Applied Computational Science, Harvard University, Cambridge, MA, USA

[4] Astrophysics Group, Imperial College, London, UK

[5] Department of Computer Science, Tufts University, Medford, MA 02155, USA

[6] Yonsei University Observatory, Yonsei University, Seoul, South Korea

Abstract. We present 663 QSO candidates in the Large Magellanic Cloud (LMC) that were selected using multiple diagnostics. We started with a set of 2,566 QSO candidates selected using the methodology presented in our previous work based on time variability of the MACHO LMC light curves. We then obtained additional information for the candidates by cross-matching them with the Spitzer SAGE, the 2MASS, the Chandra, the XMM, and an LMC *UBVI* catalogues. Using that information, we specified diagnostic features based on mid-IR colours, photometric redshifts using SED template fitting, and X-ray luminosities, in order to discriminate more high-confidence QSO candidates in the absence of spectral information. We then trained a one-class Support Vector Machine model using those diagnostics features. We applied the trained model to the original candidates, and finally selected 663 high-confidence QSO candidates. We cross-matched those 663 QSO candidates with 152 newly-confirmed QSOs and 275 non-QSOs in the LMC fields, and found that the false positive rate was less than 1%.

Keywords. Magellanic Clouds, methods: data analysis, quasars: general

---

1. Introduction

In previous work (Kim *et al.* 2011a) we developed a QSO selection method using a supervised classification model trained on a set of variability features extracted from the MACHO light curves, and including a variety of variable stars, non-variable stars and QSOs. The trained model showed a high efficiency of 80% and a low false positive rate of 25%. Using that method, we selected 2,566 QSO candidates from the light-curve database. We then developed and employed a decision procedure on the basis of diagnostics using (1) mid-IR colours, (2) photometric redshifts and (3) X-ray luminosities for those candidates in order to separate *high confidence* QSO candidates (hereinafter hc-QSOs). We thus chose in total 663 hc-QSOs out of 2,566. Those 663 candidates are very probably QSOs; if confirmed, they will increase the number of known QSOs in the MACHO LMC database by a factor of ∼12.

2. Selection Methods

We selected 2,566 QSO candidates from the MACHO light-curve database using the QSO selection method developed by Kim *et al.* (2011a), and applied multiple diagnostics tests on them.

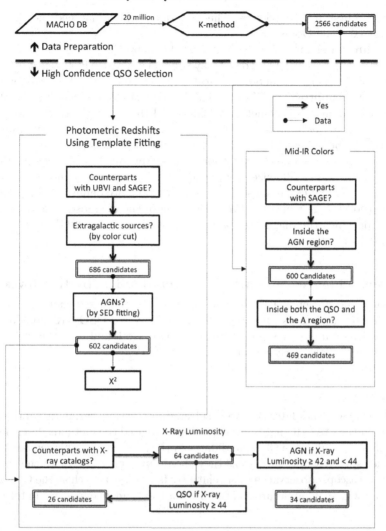

Figure 1. The rectangles with bold borderlines are the diagnostics. For most of the diagnostics, we determined whether the candidates are likely to be QSOs (solid line arrows). The thin arrows show the data flow. The double-lined rectangles show the number of candidates.

- **Spitzer mid-IR properties.** It is known that the mid-IR colour is an efficient discriminator between AGNs and stars/galaxies because their spectral energy distributions are substantially different from one another (Laurent *et al.* 2000; Lacy *et al.* 2004). Lacy *et al.* (2004) introduced a mid-IR colour cut to separate AGNs in the Spitzer SAGE catalogue (Surveying the Agents of a Galaxy's Evolution; Meixner *et al.* 2006). We used those mid-IR colour selections as the first diagnostic.
- **Photometric redshifts using SED fitting.** We cross-matched the 2,566 QSO candidates with the *UBVI* catalogue for the LMC (Zaritsky *et al.* 2004) and the 2MASS catalogue (Skrutskie *et al.* 2006), to extract *UBVI* and *JHK* magnitudes. We next separated stars from Galaxies plus AGNs (i.e., extragalatic sources) using a criterion proposed by Eisenhardt *et al.* (2004) and Rowan-Robinson *et al.* (2005). The 686 extragalatic sources thus identified were then fitted with galaxy templates in order to derive photometric redshifts (Rowan-Robinson *et al.* 2008). The template objects included three

QSO, one starburst and 10 galaxies. The candidates which matched the QSO templates were considered to be QSOs.

- **X-ray luminosities.** We cross-matched the 2,566 QSO candidates with two X-ray point source catalogues, the Chandra X-ray source catalogue (Evans *et al.* 2010) and the XMM-Newton 2nd Incremental Source catalogue (Watson *et al.* 2009). We found 88 counterparts. 64 of them fitted the SED templates and therefore had estimated photometric redshifts; we used those photometric redshifts and the X-ray fluxes from the catalogues to calculate their X-ray luminosities. The candidates showing high X-ray luminosity were deemed likely to be QSOs (Elvis *et al.* 1994; Persic *et al.* 2004).

We then employed the one-class Support Vector Machine (SVM) classification method to select just the high-confidence QSO candidates because we do not have negative examples (i.e., a non-QSO training set). To train a model, we define the diagnostics results as feature vectors. Fig. 1 outlines the calculation of the diagnostics, and the number of candidates for which the diagnostics are available. Kim *et al.* (2011b) give details about the selection method.

3. Crossmatching with Newly Discovered QSOs by Kozłowski

To estimate the efficiency and the false positive rate of our selection method, we cross-matched the 663 candidates with the 152 newly-discovered QSOs (Kozłowski *et al.* 2011) and 275 confirmed non-QSOs (i.e., false positives). We found that the yield was higher than 43%, and the false positive rate was less than 1%.

4. Summary

From 2,566 QSO candidates that were selected by the time variability of their MACHO light curves in the MACHO light-curve database, we used the methods described above to identify 663 high-confidence QSO candidates in the LMC fields. This set can be used as a target set for spectroscopic surveys as they should maximize the yield; that is important because spectroscopic observations for relatively faint objects such as the QSO candidates in dense- and wide-field areas around the LMC are extremely expensive in telescope time.

References

Eisenhardt, P. R., *et al.* 2004, *ApJS*, 154, 48
Elvis, M., *et al.* 1994, *ApJS*, 95, 1
Evans, I. N., *et al.* 2010, *ApJS*, 189, 37
Kim, D.-W., *et al.* 2011a, *ApJ*, 735, 68
Kim, D.-W., *et al.* 2011b, *ArXiv e-prints*
Kozłowski, S., *et al.* 2011, *ArXiv e-prints*
Lacy, M., *et al.* 2004, *ApJS*, 154, 166
Laurent, O., *et al.* 2000, *A&A*, 359, 887
Meixner, M., *et al.* 2006, *AJ*, 132, 2268
Persic M., *et al.* 2004, *A&A*, 419, 849
Rowan-Robinson, M., *et al.* 2008, *MNRAS*, 386, 697
Rowan-Robinson, M., *et al.* 2005, *AJ*, 129, 1183
Skrutskie, M. F., *et al.* 2006, *AJ*, 131, 1163
Watson, M. G., *et al.* 2009, *A&A*, 493, 339
Zaritsky, D., Harris J., Thompson, I. B., & Grebel, E. K. 2004, *AJ*, 128, 1606

New Horizons in Time-Domain Astronomy
Proceedings IAU Symposium No. 285, 2011
R.E.M. Griffin, R.J. Hanisch & R. Seaman, eds.

© International Astronomical Union 2012
doi:10.1017/S1743921312001020

Interstellar Scintillation as a Cosmological Probe: Prospects and Challenges†

J. Y. Koay[1]*, **J.-P. Macquart**[1], **B. J. Rickett**[2], **H. E. Bignall**[1], **D. L. Jauncey**[3], **J. E. J. Lovell**[4], **C. Reynolds**[1], **T. Pursimo**[5] **L. Kedziora-Chudczer**[6] **and R. Ojha**[7]

[1]International Centre for Radio Astronomy Research, Curtin University, Australia
*email: kevin.koay@icrar.org

[2]Department of Electrical & Computer Engineering, University of California, San Diego, USA

[3]CSIRO Astronomy & Space Science, Australia Telescope National Facility, Australia

[4]School of Mathematics & Physics, University of Tasmania, Australia

[5]Nordic Optical Telescope, E-38700 Santa Cruz de La Palma, Spain

[6]School of Physics & Astrophysics, University of New South Wales, Australia

[7]Institute for Astrophysics & Computational Sciences, Catholic University of America, USA

Abstract. The discovery that interstellar scintillation (ISS) is suppressed for compact radio sources at $z \gtrsim 2$ has enabled ISS surveys to be used as cosmological probes. We discuss briefly the potential and challenges involved in such an undertaking, based on a dual-frequency survey of ISS carried out to determine the origin of this redshift dependence.

Keywords. intergalactic medium, galaxies: active, radio continuum: ISM, scattering

1. Prospects

The radio variability observed in many compact active galactic nuclei (AGN) at intra-hour and intra-day time-scales has been linked to scintillation caused by scattering in the interstellar medium (ISM) of our own Galaxy (Kedziora-Chudczer *et al.* 1997; Jauncey & Macquart 2001; Rickett *et al.* 2001; Dennett-Thorpe & de Bruyn 2002). Two factors make interstellar scintillation (ISS) an ideal probe of the physics of the ISM and of the background sources themselves. (1) The amplitude of scintillation decreases with increasing source angular size (thus, in the case of atmospheric scintillation at optical wavelengths, stars twinkle but planets do not), so ISS can be utilized to probe source sizes down to μas scales (Lazio *et al.* 2004), which is orders of magnitude better than any ground-based radio telescope can achieve. (2) The phenomenon is pervasive; the 5 GHz Microarcsecond Scintillation Induced Variability (MASIV) Survey found that $\sim 60\%$ of flat-spectrum AGNs exhibit 2–10% rms flux density variations due to scintillation (Lovell *et al.* 2008).

Both the fraction of scintillating sources and their scintillation amplitudes have been found to decrease for sources with redshifts $z \gtrsim 2$ (Lovell *et al.* 2008), thereby imbuing ISS surveys with a cosmological significance. What causes that redshift dependence of ISS? Possible explanations include (1) a $(1+z)^{0.5}$ scaling of angular size due to cosmological expansion (Rickett *et al.* 2007) in a flux-limited and brightness temperature-limited sample of sources, (2) selection effects due to the redshift dependence of source luminosities, Doppler boosting factors or spectral indices, (3) evolution of AGN morphology, (4)

† Presented on behalf of J.Y. Koay by J.-P. Macquart.

scatter broadening in the ionized intergalactic medium (IGM) due to multi-path propagation (Rickett *et al.* 2007), and (5) weak gravitational lensing by foreground objects.

There are great prospects for ISS as a cosmological probe at μas resolution. Investigations into its redshift dependence can potentially probe the angular size–redshift relation of flat-spectrum AGNs, AGN evolution, turbulence in the IGM, and/or the distribution of matter in the Universe; at the very least it will place strong constraints on them.

2. Observations and Challenges

We observed 140 compact, flat-spectrum AGNs at $0 \lesssim z \lesssim 4$ to investigate the origin of the redshift dependence of ISS. The sources were observed simultaneously at 4.9 GHz and 8.4 GHz on two VLA sub-arrays; each was observed for one minute at 2-hour intervals for a total duration of 11 days, scheduled in sidereal time. The main motivation for the observations was to determine whether the redshift dependence of ISS scales with frequency, since scatter broadening, intrinsic source size effects and space-time curvature each scale differently with frequency. The observations also function as a precursor for future larger-scale surveys with next-generation radio telescopes, providing a platform for exploring the challenges (discussed below) involved in such an experiment.

• *Instrumental and Systematic Errors.* Possible errors in the time domain are flux-dependent errors (variable calibrators, residual gain and pointing errors) and flux-independent errors (system noise, resolved sources, and confusion). Many of them appear as flux-density variations that repeat daily, since our observations were scheduled in sidereal time. We developed various methods to estimate them—see Koay *et al.* (2011).

• *Intrinsic Source Variability.* Owing to their compact nature, some of the sources may exhibit intrinsic variations that are difficult to deconvolve from ISS. The variability amplitudes of our sources were found to be significantly correlated with line-of-sight Galactic Hα intensities (Koay *et al.* 2011), confirming that ISS dominates their variability on a statistical level at least.

• *Source Selection Biases.* Since our survey was flux-limited and involved relativistically beamed sources, it is likely that the high-redshift sources are more luminous and have larger Doppler boosting factors. The fact that the sources are observed at increasing rest-frame emission frequencies with increasing redshift adds to the bias. We devised a means of mitigating those effects by examining the ratio of the variability amplitudes at the two frequencies for each source individually, and then comparing the distribution of the ratios at high and low redshift.

• *Complexity of ISM and Source Properties.* The anisotropy and inhomogeneity of the ISM, coupled with the complexity of source structures, complicate the interpretation of the redshift dependence. We used Monte-Carlo simulations to model the statistics of ISS amplitudes for comparisons with observed distributions.

References

Dennett-Thorpe, J. & de Bruyn, A. G. 2002, *Nature*, 415, 57
Jauncey, D. L. & Macquart, J.-P. 2001, *A&A*, 370, L9
Kedziora-Chudczer, L. *et al.* 1997, *ApJ*, 490, L9
Kedziora-Chudczer, L. *et al.* 2001, *MNRAS*, 325, 1411
Koay, J. Y., *et al.* 2011, *AJ*, 142, 108
Lazio, T. J. W., *et al.* 2004, *New Astron. Revs*, 48, 1439
Lovell, J. E. J., *et al.* 2008, *ApJ*, 689, 108
Rickett, B. J., *et al.* 2001, *ApJ*, 550, L11
Rickett, B., *et al.* 2007, *Proc. Science*, 46, PoS(MRU)046

New Horizons in Time-Domain Astronomy
Proceedings IAU Symposium No. 285, 2011
R.E.M. Griffin, R.J. Hanisch & R. Seaman, eds.

© International Astronomical Union 2012
doi:10.1017/S1743921312001032

An Extremely Luminous Outburst from a Relativistic Tidal Disruption Event

A. J. Levan, on behalf of a larger collaboration

Department of Physics, University of Warwick, Coventry, CV4 7AL, UK
email: `A.J.Levan@warwick.ac.uk`

Abstract. We present the discovery and monitoring observations of *Swift* 1644+57, a luminous outburst from the nucleus of a galaxy at $z = 0.35$. Precise astrometry ties the source to within a few hundred parsecs of the nucleus of its host, and suggests a link to the massive black hole that probably resides there. The high luminosity and rapid variability are strongly indicative of a beamed source. We suggest that this event is best explained by the tidal disruption of a passing star by the supermassive black hole, which simultaneously created a powerful panchromatic explosion. However, it has also been proposed that such events may be related to the core collapse of massive stars. Future observations of a sample of similar events, focussing on their locations within the hosts, should distinguish in a straightforward manner between the two proposals.

Keywords. gamma-rays: bursts, galaxies: active, accretion

1. Introduction

Swift 1644+57 was discovered on 2011 March 28, and triggered the gamma-ray burst (GRB) detectors onboard the SWIFT satellite on a total of four occasions over a 48-hour period (Levan *et al.* 2011; Burrows *et al.* 2011). Such behaviour is unheard of for GRBs, which normally trigger the detector only once and then decline rapidly over the following hours to days, fading into invisibility (at least to the SWIFT X-ray Telescope, XRT) within a few weeks. In contrast, the light curve of Swift 1644+57 (shown in Fig. 1) remained bright for weeks and months after the initial trigger, and many months after the initial outburst it was still a bright X-ray source ($F_X \sim 3 \times 10^{-12}$ ergs s^{-1} cm^{-2}). Such behaviour immediately marked Swift 1644+57 as a new class of astrophysical source, especially once its cosmological nature was established.

In addition to bright X-ray emission Swift 1644+57 also exhibited bright emission in the IR (Levan *et al.* 2011, see also Figs. 1 & 2) and in radio (Zauderer *et al.* 2011). No optical variability was seen, suggesting that the source is highly obscured. However, the presence of the IR and radio counterparts enabled precise astrometry from VLBI observations and HST imaging. Those place the location of Swift 1644+57 to within $0''.03$ of the nucleus of its host galaxy (and fully consistent with a nuclear origin). While the galaxy is compact, and doubtless contains many stars within that compact radius (approximately 5% of the galaxy light lies within the region formally allowed at 1σ by our astrometry), such positional coincidence is strongly suggestive that Swift 1644+57 is associated with the black hole in the nucleus of its host galaxy.

The rapid rise from quiescence, the fast variability in the X-ray light curve and the absence of any clear AGN activity in the optical spectra clearly demonstrate that Swift 1644+57 is not a variable active galactic nucleus (Levan *et al.* 2011; Bloom *et al.* 2011a). The observed luminosity of Swift 1644+57 is $L_X = 3 \times 10^{48}$ ergs s^{-1} at peak, and corresponds to the Eddington limit of a 10^{10}-M$_\odot$ black hole. In contrast, our late-time IR observations, combined with the bulge mass–black hole mass relation, imply that the

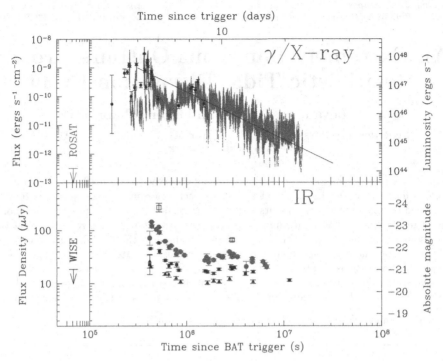

Figure 1. The high-energy (gamma-ray: black, X-ray: dense grey) and IR (*L*: open squares, *K*: grey triangles, *H*: solid squares and *J*: solid triangles) light curves of Swift 1644+57, taken from Levan *et al.* (2011) and Burrows et al. (2011). The light cuves show strong variability in X-rays, with increasing outburst intensity at redder IR wavelengths (probably because of heavy extinction within the host galaxy).

mass is unlikely to be greater than 10^7 $M_\odot$; variability arguments also lead to a low mass for the black hole (Bloom *et al.* 2011a, Miller *et al.* 2011). All that strongly suggests that the emission is relativistically beamed (Bloom *et al.* 2011a), and indeed the radio expansion implies a moderately relativistic source (Zauderer *et al.* 2011).

A natural possibility, first suggested by Bloom *et al.* (2011b) prior to the precise astrometric tie to the host galaxy, is that Swift 1644+57 represents a tidal disruption event, which simultaneously produces a relativistic outflow. In contrast to most TDE candidates, where we do not view down the jet, in the case of Swift 1644+57 we were viewing along the jet axis, causing us to see a very different observational phenomenon. While the properties of Swift 1644+57 were clearly unique, the presence of jets with TDEs has been suggested by van Velzen *et al.* (2011) and Giannios & Metzger (2011), who considered such jets and the likely impact they would have on the observed radio luminosity of "standard" TDEs at late times.

While the properties of Swift 1644+57 are broadly consistent with those of a relativistic TDE, it has also been suggested that such events could arise from core-collapse events, in which material far out in the collapsing star has sufficient angular momentum that it forms a centrifugally-supported disk (Quataert & Kasen 2011; Woosley & Heger 2011). That model differs from the one invoked for GRBs, where the disk is formed from material close to the core at the time of collapse and naturally leads to a longer-lived event, which may resemble Swift 1644+57.

The location within the nucleus of its host galaxy, the prediction of such jets from TDEs (van Velzen *et al.* 2011; Giannios & Metzger 2011) and the apparent late-time fading

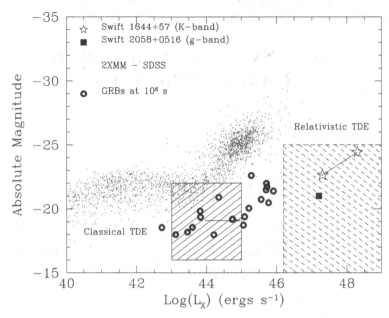

Figure 2. Comparative X-ray (x-axis) and optical (y-axis) properties of various X-ray emitting sources, compared to the properties of Swift 1644+57 (Levan *et al.* 2011) and Swift 2058+0516 (Cenko *et al.* 2011). The majority of sources represent manifestations of active nuclei in galaxies, while we also show the late-time properties of GRB afterglows and the expected locations of classical tidal disruption events (see Levan *et al.* 2011, and references therein for more details). Swift 1644+57 is offset well from any of those sources (the two stars for Swift 1644+57 represent its evolution from peak to 10^6s), at a location in parameter space where no sources have been seen before. Future examples of such events may therefore be identified by their location within that parameter space.

at $t^{-5/3}$ (expected for a TDE) all lend credence to a model in which Swift 1644+57 is created in a TDE event. However, from the available data we cannot rule out the collapsar hypothesis. Future observations of a sample of events should easily make that distinction. We would expect that any stellar event would occur across its host galaxy and not exclusively in the nucleus. In contrast, TDE events should be exclusively nuclear, and hence even a small sample (~ 5) of such events should provide a strong diagnostic of their true nature. Indeed, a second candidate event has already been uncovered (Cenko *et al.* 2011), and late-time observations of its hosts and the subsequent astrometry will provide a strong test of those models.

References

Bloom, J. S., *et al.* 2011a, *Science*, 333, 203
Bloom, J. S. *et al.* 2011b, *GCN* 11847
Burrows, D. N., *et al.* 2011, *Nature*, 476, 421
Cenko, S. B., *et al.* 2011, arXiv:1107.5307
Giannios, D. & Metzger, B. D. 2011, *MNRAS*, 416, 2102
Levan, A. J., *et al.* 2011, *Science*, 333, 199
Miller, J. M. & Gületekin, K. 2011, *ApJ*, 738, L13
Quataert, E. & Kasen, D. 2011, *MNRAS*, L360
van Velzen, S., Körding, E., & Falcke, H. 2011, *MNRAS*, 417, L51
Woosley, S. E. & Heger, A. 2011, arXiv:1110.3842
Zauderer, B. A., *et al.* 2011, *Nature*, 476, 425

New Horizons in Time-Domain Astronomy
Proceedings IAU Symposium No. 285, 2011
R.E.M. Griffin, R.J. Hanisch & R. Seaman, eds.

© International Astronomical Union 2012
doi:10.1017/S1743921312001044

Solar System Science with Robotic Telescopes

T. A. Lister

Las Cumbres Observatory Global Telescope Network (LCOGT), Goleta, CA 93117, USA
email: tlister@lcogt.net

Abstract. An increasing number of sky surveys is already on-line or soon will be, leading to a large boost in the detection of Solar System objects of all types. For Near-Earth Objects (NEOs) that could potentially hit the Earth, timely follow-up is essential. I describe the development of an automated system which responds to new detections of NEOs from Pan-STARRS and automatically observes them with the LCOGT telescopes. I present results from the first few months of operation, and plans for the future with the 6-site, 40-telescope global LCOGT Network.

Keywords. minor planets, asteroids, astrometry, celestial mechanics, telescopes

1. Introduction

Scientific interest in asteroids is motivated by their status as remnants from the solar-system formation process. Their study can provide insights into the assembly of terrestrial planets and to planetary formation mechanisms in general. Near-Earth Objects (NEOs) are the closest neighbours of the Earth-Moon system, and research into them is important not only for Solar System science but also for understanding and protecting human society from potential impact hazards. NEOs originated in collisions between bodies in the main asteroid belt, and have found their way into near-Earth space via complex dynamical interactions. This transport of material from the main belt into the inner Solar System has shaped the histories of the terrestrial planets.

Moving objects such as asteroids, which have typical rates of motion between about $1''$–$10''$/min for NEOs and $2''$–$3''$/hr for Kuiper Belt Objects, add an extra complexity to the population of transients being discovered in time-domain surveys. New sky surveys such as Pan-STARRS (PS1; Kaiser 2004) and the Palomar Transient Factory (PTF; Law *et al.* 2009) which have recently started operating have the capability of discovering large numbers of new solar-system objects and thus to increase greatly our understanding of the solar-system formation and evolution processes. They will be joined by upcoming surveys such as LSST (Ivezić *et al.* 2008) and SkyMapper (Keller *et al.* 2007). There is a coupled growth in the number of robotic telescopes which can respond automatically and follow up new detections from those sky surveys.

PS1 uses a 1.8-m telescope coupled with a 1.4-Gigapixel CCD camera, and images a 7-deg.^{2} field with a variety of survey cadences which in turn yield NEO detections. The PS1 cadence is not enough by itself to confirm newly-detected NEOs; that is where the follow-up network comes in. In order to cope with the large number of NEO candidates that PS1 produces, I have developed an automated system to retrieve new PS1 NEOs, compute their orbits, plan observations and automatically schedule them for follow-up on the robotic telescopes of the LCOGT Network (Shporer *et al.* 2010; also p. 408). The current follow-up telescope are the two 2-m Faulkes Telescopes (FTN: Maui, Hawaii; FTS: Siding Spring, Australia) and the 0.8-m Sedgwick telescope (Sedgwick Preserve, CA).

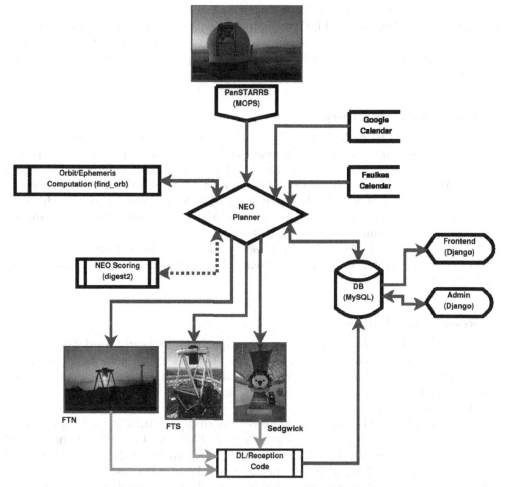

Figure 1. Overall layout of the automated follow-up system.

2. Automated Follow-Up Planning

The overall layout of the automated follow-up systems is shown in Fig. 1 above. The planning process proceeds by:

- retrieving new PS1 NEO candidates from the Moving Object Processing System,
- calculating new orbits and ephemerides for the selected candidates,
- computing the subset of an object's ephemeris while it is dark, the object is $> 30°$ altitude and telescope is unbooked (valid slots are scored based on the target altitude and distance from the Moon),
- estimating exposure times from the object's rate of motion and the pixel scale of the particular detector, balanced by the need for at least 3 exposures within the chosen block length,
- submitting observations to the telescope and the recording the details in a `Django` database (DB),
- fetching frames of successfully observed targets and recording them in the DB.

The code makes extensive use of Open Source software: the main code is `Python`, with Starlink `SLALIB` (Wallace 1994) for astronomical calculations. The Google Data API (for Sedgwick) and `MySQL` (for the two Faulkes telescopes) are used for fetching telescope bookings. The writing of the front-end uses the `Django` web framework, with a `MySQL` back-end.

3. The LCOGT Network

Over the next few years the LCOGT Network will undergo a major expansion from the existing three telescopes to 12 1-m telescopes and 24 0.4-m ones. They will be deployed over six sites; those in the Southern Ring (CTIO, Chile; SAAO, South Africa; Siding Spring, Australia) will be completed first, followed by the northern sites (McDonald Obs., TX; Tenerife, Canary Islands; Asian site TBD). All the 1-m telescopes, which are most suitable for NEO follow-up, will be equipped with CCD cameras and a standardized set of 21 filters (including Sloan and Bessell sets).

A worldwide network of robotic telescopes capable of continuous coverage and greater availability, when combined with dynamic scheduling, will create a very powerful resource for NEO follow-up.

4. Results and Future Work

The automated follow-up system has been in operation from early April 2011, and has successfully queued and observed ~ 200 objects. Of those, a high percentage ($\sim 75\%$) of groups that were actually observed were successfully reported to the Minor Planet Center. During 2011 those observations resulted in over 60 Minor Planet Electronic Circulars announcing new objects of interest.

Follow-up with LCOGT telescopes has given rise to the confirmation of 65 new objects from PS1 and other surveys. Those objects include 27 NEOs, 4 comets, 2 Centaurs and 32 other objects (primarily Main-Belt asteroids). In the course of the survey follow-up I have discovered 7 new Main-Belt objects and 2 new Jupiter Trojans.

Observations with the 0.8-m Sedgwick telescope have recovered objects as faint as $V \sim 21$. The 1-m network therefore has substantial potential for NEO follow-up as the instruments will have greater collecting area, better CCD cameras (with higher QE), and will be situated at better-quality sites.

References

Ivezić, Z. *et al.* 2008, arXiv:0805.2366
Kaiser, N. 2004, *SPIE Proceedings*, 5489, 11
Keller, S., *et al.* 2007, *PASA*, 24, 1
Law, N., *et al.* 2009, *PASP*, 121, 1395
Shporer, A., *et al.* 2010, *IAUS* 276, arXiv:1011.6394
Wallace, P. T. 1994, in: D. R. Crabtree, R. J. Hanisch and J. Barnes (eds.), *ADASS III*, ASPC, 61 (San Francisco: ASP), p. 481

New Horizons in Time-Domain Astronomy
Proceedings IAU Symposium No. 285, 2011
R.E.M. Griffin, R.J. Hanisch & R. Seaman, eds.

© International Astronomical Union 2012
doi:10.1017/S1743921312001056

Real-Time Classification of Transient Events in Synoptic Sky Surveys

Ashish A. Mahabal[1], C. Donalek[1], S. G. Djorgovski[1,2], A. J. Drake[1], M. J. Graham[1], R. Williams[1], Y. Chen[1], B. Moghaddam[3], and M. Turmon[3]

[1] California Institute of Technology, Pasadena, CA 91125, USA
email: aam@astro.caltech.edu

[2] Distinguished visiting professor, King Abdulaziz Univ., Jeddah, Saudi Arabia.

[3] Jet Propulsion Laboratory, Pasadena, CA 91109, USA

Abstract. An automated rapid classification of the transient events detected in modern synoptic sky surveys is essential for their scientific utility and effective follow-up when resources are scarce. This problem will grow by orders of magnitude with the next generation of surveys. We are exploring a variety of novel automated classification techniques, mostly Bayesian, to respond to those challenges, using the ongoing CRTS sky survey as a testbed. We describe briefly some of the methods used.

The increasing number of synoptic surveys is now generating tens to hundreds of transient events per night, and the rates will keep growing, possibly reaching millions of transients per night within a decade or so. Generally, follow-up observations are needed in order to exploit scientifically these data streams to the full. In optical surveys, for instance, all transients look the same when discovered—a starlike object that has changed its brightness significantly—and yet between them they could represent vastly different physical phenomena. Which ones are worthy of a follow-up? This is a critical issue for the massive event streams such as LSST, SKA, etc., and the sheer volume demands an automated approach (Donalek *et al.* 2008; Mahabal *et al.* 2010; Djorgovski *et al.* 2011a).

The process of scientific measurement and discovery operates typically on time-scales from days to decades after the original measurements, feeding back to a new theoretical understanding. However, that clearly will not work when changes occur on time-scales that are shorter than those needed to set up a new round of measurements. It demands real-time systems incorporating a computational analysis and decision engine, and optimized follow-up instruments that can be rapidly deployed with immediate analysis and feedback, and implies automated classification and decision-making systems.

The classification process for a given transient involves: (1) obtaining available contextual archival information, and combining it with the measured parameters from the discovery pipeline, (2) determining (relative?) probabilities or likelihoods of it belonging to some class of transient, (3) obtaining follow-up observations to disambiguate competing classes, (4) using those as a feedback and repeating for an improved classification.

We describe below a few techniques that help in this process. Our principal data-set is the transient event stream from the Catalina Real-time Transient Survey (CRTS; http://crts.caltech.edu; Drake *et al.* 1999; Djorgovski *et al.* 2011b; Mahabal *et al.* 2011), but the methodology we are developing is more universally applicable.

Bayesian Networks. The available data for any given event would generally be heterogeneous and incomplete. That is difficult to accommodate in the standard machine-

355

learning feature vector approach, but can be naturally accommodated in a Bayesian approach, such as Bayesian Networks (BN) (Mahabal *et al.* 2008).

We have used three colours obtained from the Palomar 60-inch telescope from follow-up observations of CRTS transients, and two contextual parameters: Galactic latitude and proximity to a galaxy. Priors for 6 classes have been used: CVs, SNe, Blazars, other AGN, UV Ceti, and the "Rest" (everything else). We are currently adding more parameters and classes. About 300 objects each have been used for SN and CV, and ∼100 for blazars. The number statistics for other AGN and UV Ceti are still too small. 82% of the objects classified as SN are indeed SN (79% for CVs, 69% for Blazars); the contamination is ∼10–20%. Since a single set of observations achieved that result, the potential for extending the BN and combining its output with other techniques is very promising.

Light-Curve Based Classification. Structure in sparse and/or irregular light curves (LC) can be exploited by automated classification algorithms. Our procedure is to collect LCs for different objects belonging to a given class and representing and encoding the characteristic structure probabilistically in the form of an empirical probability distribution function. That can then be used for subsequent classification of a LC that covers only a few epochs. Moreover, the comparison can be made incrementally over time as new observations become available, the final classification scores improving with each additional set of observations. This forms the basis for a real-time classification methodology. Since the observations come in the form of flux at a given epoch, for each point after the very first one we can form a $(\delta m, \delta t)$ pair. We focus on modelling the joint distribution of all such pairs of data points for a given LC. By virtue of being increments, the empirical probability density functions of these pairs are invariant to absolute magnitude and time shifts—a desirable feature. Upper limits can also be encoded in this methodology, for example, forced photometry magnitudes at a SN location in images taken before the star exploded. We currently use smoothed 2-D histograms to model the distribution of elementary (dm, dt) sets. In our preliminary experimental evaluations with a small number of object classes (single outburst like SNe, periodic variable stars like RR Lyræ and Miras, as well as stochastic variables like blazars and CVs) we have been able to show that the density models for these classes are potentially a powerful method for object classification from sparse/irregular LC data.

Currently we are using the (dm, dt) distributions for classification in a binary mode: successive two-class classifiers in a tree-structure SNe are first separated from non-SNe (the easiest bit, currently performing at ∼99% completness); next, non-SNe are separated into stochastic versus non-stochastic variables, and each group is then further separated into more branches. The most difficult so far has been the CV-blazar node, which is based on just the (dm, dt) density, i.e. without bringing in the proximity to a radio source since we are also interested in discovering blazars that were not active when the archival radio surveys were done. This classifier is currently performing at ∼71% completeness. We are also exploring Genetic Algorithms to determine the optimal (dm, dt) bins for different classes. Those will in turn help us optimise follow-up observing intervals for specific classes (Mahabal *et al.* 2011 or Djorgovski *et al.* 2011a).

Incorporating Contextual Information. Contextual information can be highly relevant to resolving competing interpretations; for example, the light curve and observed properties of a transient might be consistent with it being a cataclysmic variable star, a blazar or a SN. If it is subsequently known that there is a galaxy in close proximity, the SN interpretation becomes much more plausible. Such information, however, can be characterized by high uncertainty and absence, and by a rich structure: if there were two galaxies nearby instead of one then details of galaxy type and structure and native stellar populations become important (for instance, is this type of SN more consistent with it being in the extended halo of a large spiral galaxy or in close proximity to a faint dwarf galaxy?) The ability to incorporate such contextual information in a quantifiable fashion is highly desirable. We have been compiling priors for such information as well; they then get incorporated into the Bayesian network mentioned earlier.

We are also investigating the use of crowd-sourcing (citizen science) as a means of harvesting human pattern-recognition skills, especially in the context of capturing relevant contextual information, and turning them into machine-processable algorithms. A methodology employing contextual knowledge forms a natural extension to the logistic regression and classification methods mentioned above. It will be necessary for larger future surveys where the data flow exceeds available human resources, and moreover it would make such classification objective and repeatable. It also represents an example of a human-machine collaborative discovery process.

Transients can also be found with the technique of image subtraction that employs a matched older observation or a deeper co-added image (Drake *et al.* 1999). If the images are properly matched, a transient stands out as a positive residual, though when used with white light (as is the case with CRTS) the difference images tend to have bipolar residuals, thus leading to false detections. We have been experimenting with this technique to look for supernovæ in galaxies via citizen science, whereby a few amateur astronomers regularly look at the galaxy images along with the residuals presented to them. A large number of SNe have been found in that fashion (see Prieto *et al.* 2011, for an example, and http://nesssi.cacr.caltech.edu/catalina/current.html for a list).

A given classifier may not be optimal for all classes, nor to all types of input, and is the primary reason why multiple types of classifiers have to be employed in the complex task of classifying transients in real time. The presence of different bits of information can trigger different classifiers. In some cases more than one classifier can be used for the same kinds of input. An essential task, then, for handling input from a diverse set of classifiers such as those described above is to derive an optimal event classification. However, combining different classifiers with a different number of output classes and in the presence of error bars is a non-trivial task, and is still under development.

Acknowledgements

This work was supported in part by NASA grant 08-AISR08-0085 and NSF grants AST-0909182 and IIS-1118041.

References

Djorgovski, S. G., *et al.* 2011a, in: A. Srivasatva & N. Chawla (eds.), *Stati. Anal. Data Mining* (CIDU 2011 conf.), in press.

Djorgovski, S. G., *et al.*, 2011b, in: T. Mihara & N. Kawai (eds.), *The First Year of MAXI: Monitoring Variable X-ray Sources* (Tokyo: JAXA Special Publ.), in press

Donalek, C., *et al.* 2008 , in: Bailer-Jones (ed.), *Classification and Discovery in Large Astronomical Surveys*, AIPC, 1082, 252,

Drake, A. J., *et al.* 1999, *ApJ*, 521, 602

Drake, A. J., *et al.* 2009, *ApJ*, 696, 870

Mahabal, A. A., *et al.* 2008, *AN*, 329, 3, 288

Mahabal, A. A., *et al.* 2010, *ASPCS*, 434, 115, in: Y. Mizumoto, K. I. Morita & M. Ohishi (eds.), *ADASS XIX*

Mahabal, A. A., *et al.* 2011, *BASI*, 39,387

Prieto, J., *et al.* 2011, *ApJ*, submitted (arXiv:1107.5043)

New Horizons in Time-Domain Astronomy
Proceedings IAU Symposium No. 285, 2011
R.E.M. Griffin, R.J. Hanisch & R. Seaman, eds.

© International Astronomical Union 2012
doi:10.1017/S1743921312001068

Towards Improving the Prospects for Coordinated Gravitational-Wave and Electromagnetic Observations

Ilya Mandel[1], Luke Z. Kelley[2] and Enrico Ramirez-Ruiz[3]

[1] School of Physics & Astronomy, University of Birmingham, Edgbaston, B15 2TT, UK
email: imandel@star.sr.bham.ac.uk

[2] Harvard-Smithsonian Center for Astrophysics, Cambridge, MA 02138, USA

[3] Dept. of Astronomy & Astrophysics, University of California, Santa Cruz, CA 95064, USA

Abstract. We discuss two approaches to searches for gravitational-wave (GW) and electromagnetic (EM) counterparts of binary neutron-star mergers. The first approach relies on triggering archival searches of GW detector data based on detections of EM transients. Quantitative estimates of the improvement to GW detector reach due to the increased confidence in the presence and parameters of a signal from a binary merger gained from the EM transient suggest utilizing other transients in addition to short gamma-ray bursts. The second approach involves following up GW candidates with targeted EM observations. We argue for the use of slower but optimal parameter-estimation techniques and for a more sophisticated use of astrophysical prior information, including galaxy catalogues to find preferred follow-up locations.

Advanced ground-based gravitational-wave detectors LIGO (Harry & the LIGO science collaboration 2010) and Virgo (Virgo Collaboration 2009) are expected to begin taking data around 2015, operating at a sensitivity approximately a factor of 10 better than their initial counterparts. Gravitational waves (GWs) emitted during the late inspirals and mergers of compact-object binaries will be one of the main sources for these detectors. Although predictions for merger rates of binary neutron stars (BNSs) and neutron-star–black-hole (NSBH) binaries are highly uncertain (Mandel & O'Shaughnessy 2010), we may expect detections at a rate between one per few years and a few hundred per year, with perhaps a few tens of detections per year being most likely (Abadie *et al.* 2010a).

These detections would usher in an era of genuine gravitational-wave astronomy, with GWs being used as another tool to observe the sky. There has been a lot of discussion in the literature of the promise of multi-messenger observations (e.g., Bloom 2009). In fact, the recent focus on multi-messenger astronomy has been so exclusive that it is worthwhile to recall that a significant amount of astrophysics can be extracted from GW observations alone, since the GW signal encodes the masses and spins of the binary components, and can be used to probe astrophysics, strong-field gravity, and cosmology even in the absence of electromagnetic (EM) observations of counterparts to GW events. Nonetheless, there is no doubt that observing both EM and GW counterparts of the same event would be of great astrophysical significance, and would allow us to settle crucial questions such as the origins of short GRBs. Here we describe some thoughts about (1) future possibilities for triggered searches of archival GW data based on EM transients observed during surveys, and (2) possible improvements to recently started efforts to follow up GW triggers with targeted EM observations.

GW searches triggered on EM transients. The LIGO Scientific Collaboration and the Virgo Collaboration have previously used short, hard Gamma-Ray Bursts (SGRBs)

as triggers to search for compact binary coalescences in GW detector data (Abadie *et al.* 2010b). SGRBs are believed to be associated with relativistic jets from BNS or NS–BH mergers (Lee & Ramirez-Ruiz 2007). However, the impact of the SGRB observation on gaining confidence in the presence of a GW signature from a binary merger has not been quantitatively estimated. In fact, most of the observed SGRBs are too distant for us to detect the associated GW signal; the average luminosity distance for the 16 SGRBs with confident host identifications and redshift measurements compiled by Berger (2010) is approximately 5 Gpc, which exceeds the anticipated event horizon of Advanced LIGO by more than a factor of 10. Hence, even when an SGRB is detected, it is unlikely that a gravitational-wave counterpart is observable, so the detectability threshold is not lowered by as much as could be anticipated.

We have developed a Bayesian framework to incorporate accurately the information from an EM transient in order to quantify the benefits of triggered searches for improving the detector reach. The detection of an EM transient which may originate in a compact-object binary merger will increase the *a priori* probability that a given stretch of data from the LIGO-Virgo ground-based gravitational-wave detector network contains a signal from a binary coalescence. That increase depends on the confidence that the transient is associated with a binary coalescence, the probability that the coalescence occurs within the detection volume, and the accuracy of the merger-time reconstruction from the transient. Additional information contained in the EM signal, such as the sky location, inclination or distance to the source, can further rule out false alarms and thus lower the necessary threshold for a detection. For a fixed false-alarm probability, the EM transient can reduce the signal-to-noise ratio threshold for detection by up to ∼60–80% for optimal searches, depending on the assumptions made. If the untriggered search is incoherent while the triggered search is optimized because of the known sky location (Harry & Fairhurst 2011), the improvement can be even more significant.

Preliminary results suggest that optical signatures from r-processes in the tidal tails (also known as "kilonovæ") could be one of the most promising triggers. They are likely to be sufficiently numerous in all-sky surveys like the LSST, differentiable from other transients based on the light-curve profiles if the observational cadence is sufficiently high, and could yield moderate improvements in the detectability thresholds.

EM followups of GW candidate events. The benefits of following up GW candidates with targeted EM observations have long been recognized (Finn *et al.* 1999). Recently the first pilot programme for following up GW candidate events has been activated (Abadie *et al.* 2011). There are two significant complications in this effort: (a) the prompt analysis of GW candidates, including detection confidence and sky location; and (b) the large uncertainty in the positional reconstruction of GW detections, which could encompass tens or even a few hundred square degrees on the sky depending on the candidate and the detector network configuration (Van Der Sluys *et al.* 2008b).

A significant amount of effort has been expended to allow for very rapid processing of GW data with the aim of achieving latencies of only a few seconds or tens of seconds to identify GW candidates (Cannon *et al.* 2011). Timing triangulation between different GW interferometers comprising the network has so far been used to localize rapidly the source on the sky (Fairhurst 2009). Such incoherent use of detector data permits rapid analysis, but is likely to yield sub-optimal results. The "need for speed" can be over-stated: it is simply not necessary for many of the possible EM counterparts. For example, off-axis optical afterglows outside the jet opening angle will only peak on time scales greater than ∼1 day (van Eerten & MacFayden 2011). We may hope that if we are inside the jet opening angle the GRB will be picked up by a mission like SWIFT; but even if not, the optical signal has decay time-scales of at least several hours. Time-scales for lower-frequency follow-ups such as radio are even longer. We thus have the luxury of applying slower parameter-estimation techniques that perform a coherent analysis of all of the detector data (Van Der Sluys *et al.* 2008b), and should be able to improve sky localization and quantify better the confidence regions.

Even with slower, optimal data-analysis techniques, the confidence regions are much larger than the fields of view of typical follow-up instruments. Efforts to reduce the GW sky error box have relied on a galaxy catalogue (Nuttall & Sutton 2010). In particular, pixels within the error box have been re-weighted by the blue-light luminosity of the galaxies contained in them (Abadie *et al.* 2011). While that approach is a reasonable first cut, it could be problematic for three reasons. First, blue-light luminosity is a proxy for the star-formation rate, but mergers could be significantly delayed relative to star-formation episodes, and red elliptical galaxies may substantially contribute to present-day merger rates (O'Shaughnessey *et al.* 2010). Secondly, galaxy catalogues are unlikely to be complete to ~400 Mpc (the horizon distance of Advanced LIGO for optimally located and oriented coalescing neutron-star binaries). Finally, some fraction of mergers may happen outside host galaxies altogether, if the progenitor binary experienced large supernovæ kicks (Kelley *et al.* 2010). An ongoing study by Vousden *et al.* (2012) aims to account for these shortcomings by employing more sophisticated astrophysical priors.

Despite the improvements mentioned above, finding an electromagnetic counterpart of a GW candidate will remain extremely challenging, as discussed by Metzger & Berger (2011). Coordinated observing among several facilities may be required to cover the large uncertainty region with smaller field-of-view instruments. Another alternative that may be worth investigating is the deployment of a network of inexpensive robotic telescopes specifically with the goal of following up GW candidates, though it will be difficult to detect any but the closest afterglows. While it is impossible to guarantee that an EM counterpart to given candidate would be found, we can still strive to maximize the probability of a successful follow-up with a view to ensuring that a sufficient fraction of triggers are followed up to make at least some multi-messenger observations.

References

Abadie, J., *et al.* 2010a, *Classical and Quantum Gravity*, 27, 173001
Abadie, J., *et al.* 2010b, *ApJ*, 715, 1453
Abadie, J., *et al.* 2011, ArXiv e-prints, 1109.3498
Berger, E. 2010, *ApJ*, 722, 1946
Bloom, J. S., *et al.* 2009, ArXiv e-prints, 0902.1527
Cannon, K., *et al.* 2011, ArXiv e-prints, 1107.2665
Fairhurst, S. 2009, *New Journal of Physics*, 11, 123006
Finn, L. S., Mohanty, S. D., & Romano, J. D. 1999, *Phys. Rev. D*, 60, 121101
Harry, G. M., the LIGO Scientific Collaboration. 2010, *Class. Quant. Grav.*, 27, 084006
Harry, I. W. & Fairhurst, S. 2011, *Phys. Rev. D*, 83, 084002
Kelley, L. Z., *et al.* 2010, *ApJ (Letters)*, 725, L91
Lee, W. H. & Ramirez-Ruiz, E. 2007, *New Journal of Physics*, 9, 17
Mandel, I. & O'Shaughnessy, R. 2010, *Class. Quant. Grav.*, 27, 114007
Metzger, B. D. & Berger, E. 2011, ArXiv e-prints, 1108.6056
Nuttall, L. K. & Sutton, P. J. 2010, *Phys. Rev. D*, 82, 102002
O'Shaughnessy, R., Kalogera, V., & Belczynski, K. 2010, *ApJ*, 716, 615
van der Sluys, M. V., *et al.* 2008, *ApJ*, 688, L61
van Eerten, H. J. & MacFadyen, A. I. 2011, *ApJ* 733, L37
Virgo Collaboration. 2009, Technical Report VIR-0027A-09

New Horizons in Time-Domain Astronomy
Proceedings IAU Symposium No. 285, 2011
R.E.M. Griffin, R.J. Hanisch & R. Seaman, eds.

© International Astronomical Union 2012
doi:10.1017/S174392131200107X

The NOAO Variable-Sky Project

T. Matheson, R. Blum, B. Jannuzi, T. Lauer, D. Norman, K. Olsen, S. Ridgway, A. Saha, R. Shaw, and A. Walker

National Optical Astronomy Observatory, Tucson, AZ 85719, USA
email: `matheson@noao.edu`

Abstract. Modern time-domain surveys have demonstrated that finding variable objects is relatively straightforward. The problem now is one of selecting and following up discoveries. With even larger-scale surveys on the horizon, the magnitude of the problem will inevitably increase. One way to prepare for the coming deluge is to have realistic estimates of the numbers of potential detections so that resources can be developed to meet that need. To that end, astronomers at the National Optical Astronomy Observatory (NOAO) have begun a project to characterize the variable sky in terms of type of objects, distribution on the sky and range of variation.

Keywords. Surveys

1. Introduction

The transient alert rate from LSST will be tremendous (LSST Science Book 2009). Estimates of the number of detections per night have ranged over many orders of magnitude, and knowing the scale of the problem is critical to anticipating future observing requirements. If the science to be done with variables requires follow-up observations on rapid (or even not so rapid) time-scales, then we will need to have those facilities in place to handle the flow of alerts from LSST. That need will include photometric and spectroscopic instruments for virtually all accessible wavelengths. Having reliable estimates of the numbers of objects will make for more efficient allocations of resources to develop those facilities and observing strategies.

In our approach we include both theoretical models and empirical studies to predict the number of variables that can be detected in any given pointing of a time-domain survey. We test them against prior and on-going surveys, and also with our own on-sky experiments. The ultimate goal is an easily accessible tool that any astronomer can use to predict numbers and distributions of variables for any general time-domain project.

2. Solar System

A large source of transient objects in any time-domain survey is the Solar System itself. There are many classes of objects, not all confined to the ecliptic plane. The Grav *et al.* (2011) model of Solar System bodies contains over 14 million objects from 10 broad categories (main-belt asteroids, near-Earth objects, Trojans, Centaurs, trans-Neptunian objects, scattered disk objects, potential Earth impactors, short-period comets, long-period comets and hyperbolic comets). From the orbital elements provided by that model we can predict the number of objects for a given field of view and their magnitude distribution for any particular pointing (see Fig. 1).

There are some subtleties involved in using this model. The epoch for which coordinates are derived is an important consideration. For Solar System objects, the relative

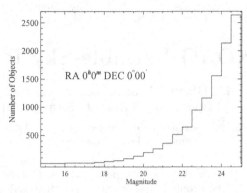

Figure 1. Magnitude distribution of the Solar System objects from the Grav model in the LSST field of view (9.6 deg^2) pointed at the LSST reference field (l=96°, b=−60°, which is also RA 0^h, Dec 0° and in the ecliptic plane). There are 10752 objects brighter than V=25 (the approximate depth of single-epoch LSST images). If we go off the ecliptic by 23°, there are 926 objects.

positions of the Sun, Earth and the individual body can dramatically affect the visibility of the body. This includes not only the relative distance of the main belt from the Earth (depending on whether you are looking at opposition or conjunction), but also the illumination of the surface visible from Earth. In addition, the orbital solutions do not include perturbations, but for the purposes of this exercise the position of any individual object is not as important as the overall number of expected objects in the given field. Finally, any model has limitations in completeness, but missing elements will in general be rare.

3. Galactic Variables

To estimate the number of Galactic variables, we first constructed a simulation of the stellar content in a given field of view using the Besançon Galaxy model (Robin *et al.* 2003). This model generates stars based on the Galactic components in the field of view. We plan to explore other Galactic models as well. We then selected only those stars that would be detected in a single-epoch visit by LSST. As we know the basic parameters of each star, we can determine which stars are in the instability strip (e.g., Gautschy & Saio 1995) and may thus appear as variable. In the LSST reference field (see Fig. 2) there are ∼100 instability-strip variables. Other types of stellar variables will require further analysis.

A different approach is to look at variability by stellar type as derived from previous studies. We used the results from Kepler (Ciardi *et al.* 2011) and HATNet (Hartman *et al.* 2011) to predict the total number of stellar variables based on their empirical calibration. Here we define variability as changing enough to meet the 5σ level of LSST. Combining the probability of variation with the number of stars of that class in magnitude bins enables us to derive the numbers of expected variables. For the LSST reference field, that total is ∼800 stellar variables. This technique has some limitations (especially long-period variability because of the duration of the Kepler mission), but it does provide a realistic estimate for those variables that were accessible in the prior studies.

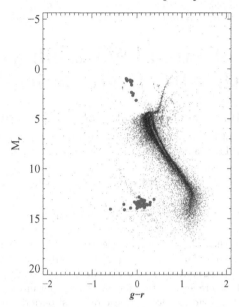

Figure 2. Colour–magnitude diagram of the stars generated from the Besançon Galaxy model for a 10-deg^2 field of view at the LSST reference field (l=96° b=−60,° RA 0^h, Dec 0°). Note that the nominal LSST FOV is 9.6 deg^2. There are 105982 stars brighter than r=25, and 93 instability-strip variables (large dots). Most of the instability-strip variables are pulsating white dwarfs.

4. Future Work

Although this paper focusses on results applicable to LSST, the techniques can be used for any time-domain survey. Given the field of view and depth of each single-epoch visit, the same results can be generated to match other surveys.

This is work in progress. The next steps will be to check further the results described here, both with models and on-sky calibrations. We will explore models to predict stellar variables outside the instability strip, in order to tie together theoretical predictions with the empirical variability results. We will incorporate models of local galaxy distributions to accommodate stellar variables detectable in nearby galaxies. We will also add the next major component—extragalactic variables. Finally, for this programme to be of general use we will develop an interface that allows users to enter their own survey parameters and thence to obtain realistic predictions of expected numbers of variables.

References

Ciardi, D. R., von Braun, K., Bryden, G., van Eyken, J., Howell, S. B., Kane, S. R., Plavchan, P., Ramirez, S. V., & Stauffer, J. R. 2011, *AJ*, 141, 108

Gautschy, A. & Saio, H. 1995, *ARAA*, 33, 75

Grav, T., Jedicke, R., Denneau, L., Chesley, S., Holman, M. J., & Spahr, T. B. 2011, *PASP*, 123, 423

Hartman, J. D., Bakos, G. A., Noyes, R. W., Sipocz, B., Kovacs, G., Mazeh, T., Shporer, A., & Pal, A. 2011, *AJ*, 141, 166

LSST Science Collaborations & LSST Project 2009, *LSST Science Book, Version 2.0*, arXiv:0912.0201, http://www.lsst.org/lsst/scibook

Robin, A. C., Reyle, C., Derriere, S., & Picaud, S. 2003, *A&A*, 409, 523

New Horizons in Time-Domain Astronomy
Proceedings IAU Symposium No. 285, 2011
R.E.M. Griffin, R.J. Hanisch & R. Seaman, eds.

© International Astronomical Union 2012
doi:10.1017/S1743921312001081

Statistics of Stellar Variability in Kepler Data with ARC Systematics Removal

Amy McQuillan[1], Suzanne Aigrain[1] and Stephen Roberts[2]

[1]Department of Physics, University of Oxford, Oxford, OX1 3RH, UK
email: amy.mcquillan@astro.ox.ac.uk, suzanne.aigrain@astro.ox.ac.uk

[2]Department of Engineering Science, University of Oxford, Oxford, OX1 3PJ, UK
email: sjrob@robots.ox.ac.uk

Abstract. We investigate the variability properties of main-sequence stars in the first month of Kepler data, using a new astrophysically robust systematics correction. We find that 36% appear more variable than the Sun, and confirm the trend of increasing variability with decreasing effective temperature. We define low- and high-variability samples, with a cut at twice the level of the active Sun, and compare properties of the stars belonging to each sample. We find tentative evidence that the more active stars have lower proper motions. The frequency content of the variability shows clear evidence for periodic or quasi-periodic behaviour in 16% of stars, and highlights significant differences in the nature of variability between spectral types. Most A and F stars have short periods (< 2 days) and highly sinusoidal variability, suggestive of pulsations, whilst G, K and M stars tend to have longer periods (> 5 days, with a trend towards longer periods at later spectral types) and show a mixture of periodic and stochastic variability, indicative of activity. Finally, we use autoregressive models to characterise the stochastic component of the variability, and show that its typical amplitude and time-scale increase towards later spectral types, which we interpret as an increase in the characteristic size and life-time of active regions. Full details will be published shortly.

Keywords. stars: activity, stars: rotation, stars: statistics, stars: spots, Galaxy: stellar content

1. Systematics Correction

The Kepler pipeline (PDC) is unsuitable for the study of stellar variability so a new Astrophysically Robust Correction for systematics (ARC) was developed. The key feature of our reduction is the removal of a set of basis functions that are found to be present in small amounts across many light curves, therefore effectively removing systematics while leaving the true variability signal unchanged. Full details of this method can be found in Roberts *et al.* (in prep.). The improvement can be seen in Fig. 1, which shows an apparent bimodality in variability where the PDC has removed intrinsic stellar signals at medium variability levels, while the ARC preserves them.

2. Variability Statistics

Using the ARC data we revisit and confirm many of the relationships presented by Basri *et al.* (2010), Basri *et al.* (2011) and Ciardi *et al.* (2011) and extend that work to a more thorough study of the periodic and stochastic nature of the variability. We divide the targets into high- and low-variability groups by comparison to the active Sun (top lines in Fig. 1), allowing us to examine the stellar properties of each sample. Using this method, we find that 36% of dwarf stars observed by Kepler appear more variable than the active Sun on a 33-day timescale. We determine the variability statistics and

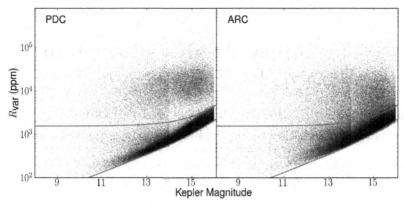

Figure 1. Variability (defined as $5^{th} - 95^{th}$ percentile of normalised flux) for the PDC and ARC data, with the photometric uncertainty (lower line) and twice the solar value (upper line).

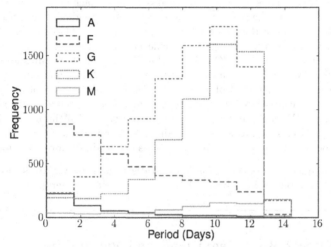

Figure 2. Period distribution for each spectral type, for stars displaying a significant period, showing a clear trend towards longer periods for later type stars.

characteristics of each spectral type, and show tenuous evidence that that the high-variability sample typically has lower proper motion. There is evidence for periodic or quasi-periodic behaviour in 16% of dwarf stars, with varying fractions and typical periods for each spectral type (Fig. 2). Specific caveats apply to this distribution, based on dataset length and period selection method.

The stochastic component of variability associated with each spectral type was parameterised by fitting autoregressive models to the periodograms (Aigrain *et al.* 2004). This reveals a trend of increasing amplitude and time-scale towards later types. The spectral density distribution of A and F stars implies small scale, short-lived active regions, whereas for G, K and M stars the active regions appear larger and more stable.

References

Aigrain, S., Favata, F., & Gilmore, G. 2004, *A&A*, 414, 1139
Basri, G., *et al.* 2010, *ApJ*, (Letters) 713, L155
Basri, G., *et al.* 2011, *ApJ*, 141, 20
Ciardi, D. R., *et al.* 2011, *ApJ*, 141, 108
McQuillan, A., Aigrain, S., & Roberts, S. 2011, submitted to *A&A*

New Horizons in Time-Domain Astronomy
Proceedings IAU Symposium No. 285, 2011
R.E.M. Griffin, R.J. Hanisch & R. Seaman, eds.

© International Astronomical Union 2012
doi:10.1017/S1743921312001093

Variability Analysis based on POSS1/POSS2 Photometry

Areg M. Mickaelian[1], Alain Sarkissian[2] and Parandzem K. Sinamyan[1]

[1]Byurakan Astrophysical Observatory (BAO), Byurakan 0213, Aragatzotn Province, Armenia
email: `aregmick@aras.am, spk7711@gmail.com`

[2]Laboratoire Atmosphères, Milieux et Observations Spatiales, 78280 Guyancourt, France
email: `alain.sarkissian@latmos.ipsl.fr`

Abstract. We introduce accurate magnitudes as combined calculations from catalogues based on accurate measurements of POSS1- and POSS2-epoch plates. The photometric accuracy of various catalogues was established, and statistical weights for each of them have been calculated. To achieve the best possible magnitudes, we used weighted averaging of data from APM, MAPS, USNO-A2.0, USNO-B1.0 (for POSS1-epoch), and USNO-B1.0 and GSC 2.3.2 (for POSS2-epoch) catalogues. The r.m.s. accuracy of magnitudes achieved for POSS1 is 0.184 in B and 0.173 mag in R, or 0.138 in B and 0.128 in R for POSS2. By adopting those new magnitudes we examined the First Byurakan Survey (FBS) of blue stellar objects for variability, and uncovered 336 probable and possible variables among 1103 objects with POSS2–POSS1 $\geqslant 3\sigma$ of the errors, including 161 highly probable variables. We have developed methods to control and exclude accidental errors for any survey. We compared and combined our results with those given in Northern Sky Variability Survey (NSVS) database, and obtained firm candidates for variability. By such an approach it will be possible to conduct investigations of variability for large numbers of objects.

Keywords. techniques: photometric, methods: statistical, stars: variables: other, stars: early-type, cataclysmic variables, white dwarfs, BL Lacertae objects: general

1. POSS1/POSS2 Best Magnitudes and Variability

Variability is one of the key parameters for investigating stellar physics and evolution; the discovery and study of variable stars and understanding their changes is therefore extremely important. However, there are only 80,671 known variables listed in the General Catalogue of Variable Stars (GCVS; Samus *et al.* 2011), and compared to the total of ~ 1 billion stars with available photometric data is is of course a rather small number. Accurate photometry is needed to compare brightness and reveal candidate variables. Photometric catalogues of the required accuracy contain relatively small numbers of objects, and cannot be used to check the possible variability of any object over the whole sky. More than 1 billion objects have been measured photometrically from the Palomar Observatory Sky Surveys (POSS) epochs 1 and 2: (1948–1958 and 1986–2000, respectively) and their Southern Sky extensions at ESO and AAO, together covering the whole sky and at present giving $\sim 0.2^m$–0.3^m r.m.s. photometry for objects down to at least $\sim 21^m$. In this paper we use "POSS1" and "POSS2" to refer to the whole sky. The data are given in United States Naval Observatory catalogue (USNO-A2.0; Monet *et al.* 1998), Automated Plate Measurement catalogue (APM; McMahon *et al.* 2000), Minnesota Automated Plate Scanner catalogue (MAPS; Cabanela *et al.* 2003), USNO-B1.0 (Monet *et al.* 2003), and the HST Guide Star Catalogue (GSC 2.3.2; Lasker *et al.* 2008). As those catalogues are all based on the same observations (POSS1 and POSS2 photographic plates), we may expect to derive similar results for POSS1 and POSS2 photometry, respectively.

Table 1. Photometric accuracy and statistical weights of individual catalogs.

Catalogue	Epoch	Photometric bands	Number of objects	Catalogues r.m.s.	r.m.s.B/weight	r.m.s.R/weight
MAPS	POSS1	O, E	89,234,404	0.2-0.3	0.308 / 0.409	0.271 / 0.389
APM	POSS1	b, r	166,466,987	0.5	0.413 / 0.228	0.318 / 0.283
USNO-A2.0/B1.0	POSS1	B1, R1	526,280,881	0.3	0.327 / 0.363	0.295 / 0.328
USNO-B1.0	POSS2	B1, R1, B2, R2, I	1,045,913,669	0.3	0.206 / 0.414	0.153 / 0.408
GSC 2.3.2	POSS2	j, V, F, N	945,592,683	0.13-0.22	0.173 / 0.586	0.127 / 0.592
Tycho-2	1989-1993	BT, VT	2,539,913	0.01-0.10		
SDSS	2000-2008	u, g, r, i, z	357,000,000	0.03		
2MASS	1997-2001	J, H, Ks	470,992,970	0.02		
GCVS	1900-2011	V or other	80,671	0.1		
NSVS	1999-2000	R (VROTSE)	14,000,000	0.04		
Combined POSS1	1948-1958	O/b/B1, E/r/R1			0.184	0.173
Combined POSS2	1986-2000	B2/j, R2/F			0.138	0.128

Table 2. Variability classes and the numbers based on POSS2/POSS1 comparison.

Variability classes	POSS2/POSS1 σ	B	R	B+R
Extremely variable	$\geqslant 10.00$	15	9	1
Strongly variable	5.00-9.99	46	52	28
Probable variable	3.00-4.99	115	101	132
Possible variable	2.00-2.99	200	177	175
All objects	$\geqslant 2.00$	376	339	336

Unfortunately, during automatic object identification and measurements on different plates, accidental errors can happen, and automatic cross-matching then leads to erroneous results. To derive best magnitudes from all the POSS1 and POSS2 measurements, we

- corrected for systematic differences between MAPS–APM–USNO-A2.0/B1.0 and USNO-B1.0–GSC 2.3.2 magnitudes,
- combined POSS1-based magnitudes (MAPS, APM, USNO) onto POSS1-best,
- combined POSS2-based magnitudes (USNO and GSC 2.3.2) onto POSS2-best,
- eliminated accidental errors (measurements with large individual errors),
- established statistical weights for each catalogue:
  MAPS : APM : USNO = 0.409 : 0.228 : 0.363 and USNO : GSC = 0.414 : 0.586,
- calculated O/b/B1, E/r/R1 and B2/j, R2/F by using the derived statistical weights.

Table 1 gives the previous and newly-obtained photometric accuracy and statistical weights of individual catalogues, together with the accuracy for the combined POSS1- and POSS2-based magnitudes, which we call POSS1-best and POSS2-best.

We have also derived transformation formulæ from POSS2 to POSS1, POSS to Johnson B and V, POSS to SDSS g and r, and POSS-R to NSVS-R. The differences between POSS2-best and POSS1-best then reveal objects which are variables. Table 2 summarizes our counts and classifications of those objects.

There were 294 NSVS variables among those objects. The 336 objects which we newly identified as variable included 37 that were already in the NSVS, so our analysis of our analysis of POSS1/POSS2 photometry has identified 299 new possible variables.

2. Multiwavelength Data and the Physical Nature of Objects

Possible variable objects among FBS Blue Stellar Objects (BSOs) might include blazars (large variability, radio, X-ray, polarization), QSOs/Seyferts (small variability), CVs (DN, NL, etc.; optical variability, X-ray), WDs (pulsating WDs, ZZ Ceti variables), flare stars (dMe: dwarf M stars at flare stage), and some others. Cross-correlation with radio and X-ray catalogues and proper-motion data revealed:

• 19 radio/IR variable sources (18 are radio and 6 are IR sources). Of those, 12 are extragalactic objects (11 AGN and 1 galaxy) and 2 are planetary nebulae (PN). One source appears to be an unknown object and 4 are associated with stars (WDs and subdwarfs; chance associations with large positional differences).

• 36 X-ray sources having variability (9 of them are the same radio AGN). Out of those, 21 are AGN, including 9 radio sources (among them 3 blazars). But X-ray is also useful for identifying CVs and WDs; 5 CVs and 8 variable WDs have thus been identified, while 2 objects have no classification but are also probable AGN/CV/WD.

• Altogether, 46 out of 336 variable FBS BSOs are either radio, IR or X-ray sources. Among the other 290 objects 4 are AGN, 7 are other galaxies, 10 are CVs, 53 are WDs, 66 are subdwarfs, 7 are NHB/HBB, 26 are stars without a classification, and 117 are unknown objects. We have thus established the variability of 336 objects, including 117 of unknown nature.

• If we take only the 161 objects with extreme/strong/probable variability, then the total numbers are 13 AGN, 4 other galaxies, 1 PN, 9 CVs, 27 WDs, 32 subdwarfs, 6 NHB/HBB, 11 stars without a classification, and 58 unknown objects. Special attention must be paid to variable WDs as ZZ Ceti-type objects may appear among them.

3. Conclusions

POSS1/POSS2 photometry for over 1 billion objects provides a means to examine the objects for variability when deriving the most accurate magnitudes possible from POSS1 and POSS2 measurements. By applying such a method we have revealed 336 variables in the catalogue of FBS blue stellar objects; they include blazars, cataclysmic variables, white dwarfs and other recognized types. A similar approach is being applied to the FBS late-type stars (among which Mira type and semi-regular variables dominate), the Second Byurakan Survey (SBS) objects (blazars, QSOs/Seyferts, cataclysmic variable, white dwarfs, etc.), and to the AGN catalogue for blazars and variable QSOs.

References

Abazajian, K. N., *et al.* 2009, *ApJS*, 182, 543
Cabanela, J. E., *et al.* 2003, *PASP*, 115, 837
Høg, E., *et al.* 2000, *A&A*, 355, L27
Lasker, B., *et al.* 2008, *The GSC, V. 2.3.2, Astron. J.*, 136, 735
McMahon R. G., Irwin, M. J., & Maddox, S. J. 2000, *APM-North Catalogue* (Cambridge, UK: IoA)
Mickaelian, A. M. 2008, *AJ*, 136, 946
Mickaelian, A. M. & Sinamyan, P. K. 2010, *MNRAS*, 407, 681
Mickaelian, A. M., Mikayelyan, G. A., & Sinamyan, P. K. 2011, *MNRAS*, 415, 1061
Monet D., *et al.* 1998, *USNO-A V2.0, USNO Flagstaff St. & Univ. Space Res. Ass.*
Monet, D. G., *et al.* 2003, *AJ*, 125, 984
Samus N. N., *et al.* 2010, *Combined GCVS (Vizier catalog II/250, version 2010)*
Wozniak, P. R., Vestrand, W. T., & Akerlof, C. W. 2004, *AJ*, 127, 2436

New Horizons in Time-Domain Astronomy
Proceedings IAU Symposium No. 285, 2011
R.E.M. Griffin, R.J. Hanisch & R. Seaman, eds.

© International Astronomical Union 2012
doi:10.1017/S174392131200110X

Optical Pulsations from Isolated Neutron Stars

Roberto P. Mignani[1,2]

[1] Mullard Space Science Laboratory, Dorking, Surrey, RH5 6NT, UK
email: `rm2@mssl.ucl.ac.uk`

[2] Kepler Institute of Astronomy, University of Zielona Góra, Zielona Góra, Poland

Abstract. Because they are fast rotating objects, isolated neutron stars (INS) are obvious targets for high-time-resolution observations. With the number of optical/UV/IR INSs detections now at 24, timing observations become increasingly important in INS astrophysics.

Keywords. stars: neutron; pulsars: general; radiation mechanisms: general

---

24 INSs have been detected in the UV, optical, and IR (Mignani 2011). As well as pulsars, they include the magnetars (Mereghetti 2008) and the X-ray Dim INSs (XDINSs; Turolla 2009). Optical timing (see Mignani 2010a) yields the direct INS identification, input for models of neutron-star magnetospheres through comparisons of multi-wavelength light curves, evidence of debris disks, the spin-down parameters of radio-silent INSs, and measurements of giant pulses—only detected in the radio and optical (Mignani 2010b, c). This paper focusses on classes for which counterparts have been detected; it describes the optical pulsation emission mechanisms and the observational challenges in timing studies of INSs, and outlines the characteristics of pulsations from pulsars and magnetars.

1. Optical Pulsations: Mechanisms and Observations

The production of optical pulsations from INSs depends on the underlying emission process. In some cases, the optical emission is the result of energy irradiation from relativistic particles in the neutron-star magnetosphere through synchrotron losses or other non-thermal processes. Optical pulsations are, then, expected from INSs which have a strong magnetospheric activity, i.e., pulsars and magnetars. In those cases, the optical emission is produced near the magnetic poles, yielding a small beaming factor. Indeed, optical pulsations are mostly characterised by sharp, double-peaked profiles. Phase shifts with respect to the X-ray and gamma-ray light curves are usually observed if the emission comes from different regions in the magnetosphere with (for instance) the gamma-ray emission being produced in the outer magnetosphere.

In some other cases, the optical emission is of thermal origin and is produced by the cooling of the neutron-star surface. Optical pulsations are therefore expected from INSs with dominant thermal emission components, e.g., the XDINSs, if the optical emission is associated with a non-isotropic temperature distribution on the neutron-star surface. In that case, optical pulsations are expected to have shallow profiles, while phase shifts between the optical and X-ray light curves are a natural consequence of the emission being produced from regions of different temperature on the neutron-star surface.

Finally, it is possible that optical pulsations do not originate directly from the neutron star's magnetosphere or surface but from the reprocessing of the pulsed X-ray radiation in a circumstellar debris disk, formed out of fall-back material after the supernova explosion. In that case, optical pulsations can be produced from any type of INS with a disk. The reprocessing affects the optical light curve, with wider profiles expected with respect to the X-ray one owing to the smearing of the X-ray pulse by the disk material. Moreover, phase shifts are expected between the optical and X-ray light curves, and are due to the radiation travel-time between the neutron star and the disk and typically depend on the size of the disk inner radius and geometry. Depending on the disk viscosity, a continuum emission component, produced by the X-ray reprocessing, can be present, and might be stronger than the pulsed one.

2. Optical Pulsars: Challenging Targets

Only 8 of the 24 INSs detected at optical wavelengths (Mignani 2011) are also detected as optical pulsars. There are several reasons for this paucity. One is related to characteristics of their optical emission and to their intrinsic faintness, which limits the search for a periodicity: only three INSs are brighter than $V \sim 25$. Moreover, the value of the Pulsed Fraction (PF) depends on the underlying emission process and it is difficult to determine it *a priori* without knowing the nature of the optical emission, i.e., without adequate spectral information. The latter is usually obtained through multi-band photometry measurements which are sometimes collected over several years. At the same time, the value of the PF measured at other wavelengths, e.g., in the X-rays, cannot be taken as an absolute reference since light-curve profiles vary significantly as a function of wavelength, very much like the INS spectrum. The slope of the optical spectrum and/or the extinction along the line of sight also biasses the choice of the observing wavelengths which, in turns, has to cope with the availability of a detector/instrument working in that wavelength range. Moreover, as observed (e.g., in the magnetars), the PF depends very much on the source brightness, varying significantly from active to quiescent states.

Difficulties of running periodicity-search algorithms also contribute. The low number of photons which can be collected over integrations that are a few hours long makes it impossible to analyse the time series through a Fast Fourier Transform (FFT). Thus, one needs to fold the time series around a reference period available from radio, X- or gamma-ray observations. In this case, the INS must be a stable rotator: it must now feature sudden variations of the spin-down rate (glitches), otherwise the search for pulsations would require contemporary ephemeris. A further difficulty is in the *a priori* estimate of an expectation value for the PF, and thence of the required integration time.

A further reason is related to the availability and characteristics of the instruments used for optical timing. For instance, for a given object one can expect a higher flux (e.g., in the IR than in the UV) depending on the spectrum and extinction. Pulsations might thus be undetectable outside an optimised wavelength range, which implies an instrument/detector selection effect. Moreover, the timing of fainter INSs became feasible only in the late 1990s with the advent of 8-m-class telescopes like the VLT or Gemini. Even in those cases, however, the search for optical pulsars clashed with the paucity of on-site instruments for high-time-resolution observations, either photon counters, time-resolved imagers or fast read-out windowed CCD devices. HST was equipped with instruments for high-time-resolution observations in the optical/UV but for various reasons they have not been available to users. Most timing facilities on ground-based telescopes are guest instruments, built, maintained and operated by private consortia and not directly available to the community for open-time proposals. Moreover, they are not easily portable, have to be properly interfaced to different telescope structures and hardware, and the instrument shipping adds non-negligible costs to travel expenses both for equipment and manpower.

3. Pulsars and Magnetars

So far, five of the 12 identified pulsars also pulsate in the optical. In general, pulsars are the best target INSs for optical timing since they usually count on accurate radio ephemeris, spin-down parameters, distances and positions, with many potential targets routinely discovered in radio and gamma-ray surveys. Moreover, many pulsars are observed in X-rays, which gives the interstellar extinction via the N_H and thence an estimate of the brightness and of the most-suited observing wavelength. In general, optical light curves of pulsars are all double-peaked, with a phase separation $\Delta\phi = 0.4$–0.6, the only exception being B0540–69 for which $\Delta\phi \sim 0.2$. The peaks in the optical light curve are not always in phase with the gamma/X/radio ones, as expected if the pulse originates in different regions of the magnetosphere. All pulsars except B0540–69 are also detected as optical/UV pulsars, but only the Crab is detected as an IR pulsar as well. Interestingly, one of the very few measurements of a pulsar braking index has been obtained from the optical timing of B0540–69 (Gradari *et al.* 2011), while the Crab is the only pulsar where giant optical and radio pulses have been observed simultaneously (see p. 296).

Three out of 6 magnetars identified in the optical/IR pulsate. Their emission is either of magnetospheric origin, perhaps powered by the magnetic field, or is produced by X-ray reprocessing in a debris disk (Mignani 2011 and references therein). In all cases, the profile of the optical pulsation reproduces the X-ray one. For 1E 1048.1–5937, the optical PF is $\sim 70\%$ of the X-ray one, providing evidence for disk reprocessing. However, there is also a marginal evidence (2σ) for X-ray lags, which is not expected in the reprocessing model. For 4U 0142+61 the optical PF is larger than the X-ray one and there is evidence (2σ) of optical lags. For SGR 0501+4516 the optical PF is a larger than the X-ray one, and the optical light curve is in phase with the X-ray one.

4. Future Perspectives

The wealth of pulsating INSs detected in X-rays (~ 60) and gamma-rays (~ 80) highlights a quantitative gap between the UV/optical/IR (8) and high-energy domains. The huge collecting areas of the Extremely Large Telescope (ELT) together with new generation instruments is needed to start a new era in optical timing (Mignani 2010b, c) and to match the potentials of the LOFT X-ray mission (p. 372). QuantEYE, the first pilot study for the OWL 100-m telescope, was based on quantum detector technology to reach pico-s time resolution (p. 280). QuantEYE was father to prototypes for the Asiago 182-cm telescope (AqEYE) and for the 3.6-m NTT (IquEYE), which produced the best measurements of pulsar light curves. A new prototype (EquEYE) is now being studied for the VLT as a possible precursor for a new quantum photometer for the E-ELT. That will open a new era in optical-timing studies of isolated neutron stars.

References

Gradari, S., *et al.* 2011, *MNRAS*, 412, 2689
Mereghetti, S., 2008, *A&AR*, 15, 225
Mignani, R. P., 2010a, in: *High Time Resolution Astrophysics IV*, *PoS*, arXiv:1008.0605
Mignani, R. P. 2010b, in: *Astronomy with Megastructures*, arXiv:1008.5037
Mignani, R. P. 2010c, in: *Astrophysics of Neutron Stars 2010*, arXiv:1009.3378
Mignani, R. P. 2011, *ASR*, 47, 1281
Turolla, R. 2009, *ASSL*, 357

New Horizons in Time-Domain Astronomy
Proceedings IAU Symposium No. 285, 2011
R.E.M. Griffin, R.J. Hanisch & R. Seaman, eds.

© International Astronomical Union 2012
doi:10.1017/S1743921312001111

LOFT: Large Observatory For X-Ray Timing

R. P. Mignani[1,2], S. Zane[1], D. Walton[1], T. Kennedy[1], B. Winter[1], P. Smith[1], R. Cole[1], D. Kataria[1], and A. Smith[1] (for the *LOFT* team)

[1]Mullard Space Science Laboratory, Dorking, Surrey, RH5 6NT, UK
email: rm2@mssl.ucl.ac.uk

[2]Kepler Institute of Astronomy, University of Zielona Góra, Zielona Góra, Poland

Abstract. High-time-resolution X-ray observations of compact objects provide direct access to strong-field gravity, black-hole masses and spins, and the equation of state of ultra-dense matter. LOFT†, the Large Observatory for Xray Timing, is specifically designed to study the very rapid X-ray flux and spectral variability that directly probe the motion of matter down to distances very close to black holes and neutron stars. A 10-m^2-class instrument in combination with good spectral resolution (<260 eV @ 6 keV) is required to exploit the relevant diagnostics, and has the potential of revolutionising the study of collapsed objects in our Galaxy and of the brightest supermassive black holes in active galactic nuclei. LOFT will carry two main instruments: a Large Area Detector (LAD), to be built at MSSL/UCL in collaboration with the Leicester Space Research Centre, and a Wide Field Monitor (WFM). The ground-breaking characteristic of the LAD (it will work in the energy range 2–30 keV) is a mass per unit surface in the range ∼10 kg/m^2, giving an effective area of ∼10 m^2 (@10 keV) at a reasonable weight—an improvement by ∼20 over all predecessors. This will allow timing measurements of unprecedented sensitivity, providing the capability to measure the mass and radius of neutron stars with ∼5% accuracy, or to reveal blobs orbiting close to the marginally stable orbit in active galactic nuclei. We summarise the characteristics of the LOFT instruments and give an overview of its expected capabilities.

Keywords. space vehicles: instruments; stars: neutron; pulsars: general; radiation mechanisms: general; equation of state; gravitation

1. A New X-Ray Mission

LOFT is one of four M3 missions that have been selected by ESA for an Assessment Phase and to be considered for a possible launch in 2020–2022. The LOFT Consortium includes institutes across the UK, Europe, Israel, Turkey, Canada, the US and Brazil. In addition to MSSL, the UK participation includes the Space Research Centre (SRC) in Leicester and the Universities of Southampton, Durham, Manchester and Cambridge. The UK participation is sponsored by the UK Space Agency. MSSL/UCL will lead the LOFT Large Area Detector (LAD) instrument within the consortium (S. Zane and D. Walton), and will also have a major role in hardware/software development and system engineering (thermal, mechanical, electronics and software). Those efforts will be supported by Leicester SRC (G. Fraser) in leading the development of the collimators.

LOFT (Feroci *et al.* 2011a,b) is a 10-m^2-class telescope specifically designed to study the very rapid X-ray flux and spectral variability that directly probe the motion of matter down to distances very close to black holes and neutron stars. High-time-resolution X-ray observations of compact objects provide direct access to strong-field gravity, black-hole masses and spins, and the equation of state of ultra-dense matter. They provide

† http://www.isdc.unige.ch/loft/index.php/the-loft-mission

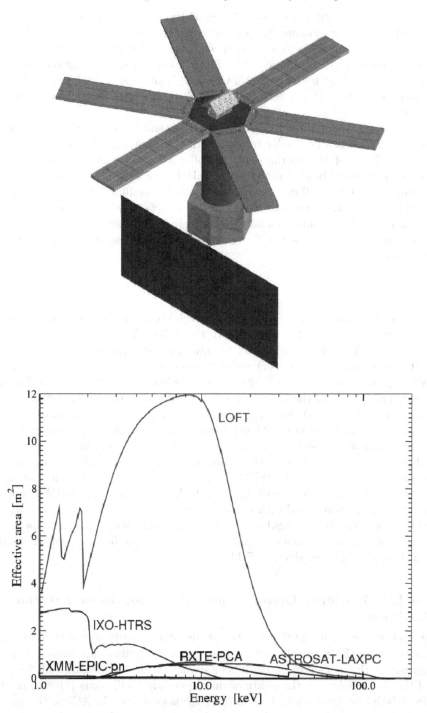

Figure 1. Upper: Conceptual scheme of the *LOFT* satellite. From top to bottom along the satellite axis: Wide Field Monitor, optical bench, the six Large Area Detector petals, structural tower, bus, and solar array. Lower: LAD effective area plotted against energy in linear scale, as compared to that of other satellites for X-ray astronomy (from Feroci *et al.* 2011a)

unique opportunities to reveal for the first time a variety of general relativistic effects, and to measure fundamental parameters of collapsed objects. They offer unprecedented information on strongly curved space-times and matter at supra-nuclear densities and in supercritical magnetic fields. That in turn bears directly on several fundamental questions raised both by ESA's Cosmic Vision Theme, "Matter under extreme conditions", and the STFC road map, "What are the laws of physics under extreme conditions?"

A 10-m^2-class telescope like LOFT requires the combination of good spectral resolution (<260 eV @ 6 keV) in order to exploit the relevant diagnostics. It will then have the potential to revolutionise the study of collapsed objects in our Galaxy and of the brightest supermassive black holes in active galactic nuclei (AGNs). The time-scales and phenomena that LOFT will investigate range from sub-millisecond, quasi-periodic oscillations to year-long transient outbursts, and the objects to be studied include many that flare up and change state unpredictably. Relatively long observations, flexible scheduling and continuous monitoring of the X-ray sky are therefore essential elements for success in this project.

2. Payload

LOFT will be launched in a ∼600 km equatorial orbit. It will carry two instruments: a Large Area Detector (LAD), operating in the 2–50 keV range with energy resolution <260 eV (@ 6 keV), and a Wide Field Monitor (WFM). The LAD consists of 6 panels deployable in space (Fig. 1, upper) which provide a total effective area of ∼10 m^2 (@ 10 keV), improving by a factor of ∼20 over its predecessors (Fig. 1, lower). The groundbreaking characteristic of the LAD is a mass per unit area of ∼10 kg/m^2, a factor of 10 lower than the *RXTE*/PCA, enabling a ∼10-m^2-area payload at reasonable weight. The ingredients for a sensitive but light experiment are the large-area Silicon Drift Detectors, and a collimator based on lead-glass micro-channel plates. An unprecedently large throughput (∼3 × 10^5 cts/s from the Crab) will be achieved, while making pile-up and dead-time secondary issues. The WFM is a coded-mask telescope mounted at the top of the structural tower at the centre of the LAD deployable array. The WFM will operate in the energy range 2–50 keV and with a field of view of 3 steradians, corresponding to ∼1/4 of the whole sky. The WFM angular resolution (5 arcmin) will enable it to locate sources with a 1-arcmin accuracy, with a 5σ sensitivity of 2 mCrab (50 ks). Some characteristics of the instrumentation are given in Table 1.

3. The LOFT Science Driver: Study of Matter under Extreme Conditions

The science drivers for LOFT are the study of the neutron-star structure and the equation of state (EOS) of ultra-dense matter (mass, radius and crustal properties of neutron stars), the motion of matter under strong gravity conditions and the mass and spin of the black holes via the study of quasi-periodic oscillations (QPOs) in the time domain, relativistic precession, Fe-line reverberation studies in AGNs, the measure of small-amplitude periodicities in X-ray transients, millisecond pulsars, etc., discovery of new X-ray transients, early trigger of jets over many astronomical scales, X-ray flashes, and many others.

The LAD ∼10-m^2 effective area in the 2–50 keV energy range will allow timing measurements of unprecedented sensitivity, leading for instance to measuring the mass and radius of neutron stars with ∼5% accuracy, or to reveal blobs orbiting close to the

Table 1. Overview of the LOFT instrument performances.

Item	Requirement	Goal
Large Area Detector (LAD)		
Energy Range	2–50 keV	1–50 keV
Effective Area (2–10 keV)	10 m^2 @ 8 keV	12 m^2 @ 8 keV
Energy Resolution (@ 6 keV)	260 ev @ 6 keV	200 ev @ 6 keV
Field of View (FWHM)	$< 1°$; transparency $<1\%$ @ 20 keV	30 arcmin
Time Resolution	10μs	7μs
Dead Time	$<1\%$ (@1 Crab[1])	$<0.5\%$ (@1 Crab)
Background	< 10 mCrab	< 5 mCrab
Maximum source flux (steady,peak)	>500mCrab; >15 Crab	>500mCrab; >30 Crab
Wide Field Monitor (WFM)		
Energy Range 2–50 keV	1–50 keV	
Energy Resolution (FWHM)	500 eV	300 eV
Field of View	50% of the accessible LAD sky coverage	Same, with improved sensitivity
Angular Resolution	5 arcmin	3 arcmin
Point Source Localisation	1 arcmin	0.5 arcmin
Sensitivity (5σ, 50 ks)	5 mCrab	2 mCrab

Notes: [1] Flux values are in units of the Crab pulsar flux in the energy range of interest.

marginally stable orbit in active galactic nuclei. The LAD energy resolution will also allow the simultaneous exploitation of spectral diagnostics, in particular from the relativistically broadened 6–7 keV Fe-K lines. The WFM will monitor a large fraction of the sky and constitute an important resource in its own right. The WFM will discover and localise X-ray transients and impulsive events and monitor spectral state changes with unprecedented sensitivity. It will then trigger follow-up pointed observations with the LAD as well as with other multi-wavelength facilities.

References

Feroci, M., *et al.* 2011a, *Experimental Astronomy*, in press
Feroci, M., *et al.* 2011b, *Proceedings of the SPIE*, 7732, 57

New Horizons in Time-Domain Astronomy
Proceedings IAU Symposium No. 285, 2011
R.E.M. Griffin, R.J. Hanisch & R. Seaman, eds.

© International Astronomical Union 2012
doi:10.1017/S1743921312001123

Search for Turbulent Gas through Interstellar Scintillation

M. Moniez[1], R. Ansari[1], F. Habibi[1], and S. Rahvar[2,3]

[1]LAL, IN2P3-CNRS, Université de Paris-Sud, 91898 Orsay Cedex, France
email: moniez@lal.in2p3.fr

[2]Department of Physics, Sharif University of Technology, Tehran, Iran

[3]Perimeter Institute for Theoretical Physics, Waterloo, Ontario N2L 2Y5, Canada

Abstract. Stars twinkle because their light propagates through the atmosphere. The same phenomenon is expected when the light of remote stars crosses a Galactic—disk or halo—refractive medium such as a molecular cloud. We present the promising results of a test performed with the ESO–NTT, and consider its potential.

Keywords. Galaxy:structure, dark matter, ISM

1. What is Interstellar Scintillation?

Refraction through an inhomogeneous transparent cloud ("screen") distorts the wavefront of incident electromagnetic waves (Fig. 1). For a *point-like* source the intensity in the observer's plane is affected by interference which, in the case of stochastic inhomogeneities, causes it to take on a speckle appearance that is characterized by at least two distance scales:

• The diffusion radius $R_{diff}(\lambda)$ of the screen, defined as the transverse separation for which the root mean square of the phase difference at wavelength λ is 1 radian, and

• The refraction radius

$$R_{ref}(\lambda) = \frac{\lambda z_0}{R_{diff}} \sim 30860\,km \left[\frac{\lambda}{1\,\mu m}\right] \left[\frac{z_0}{1\,kpc}\right] \left[\frac{R_{diff}(\lambda)}{1000\,km}\right]^{-1}, \qquad (1.1)$$

where z_0 is the distance to the screen. This is the size, in the observer's plane, of the diffraction spot from a patch of $R_{diff}(\lambda)$ in the screen's plane.

After crossing a fractal cloud described by the Kolmogorov turbulence law (Fig. 1, left), the light from a *monochromatic point* source produces an illumination pattern on Earth consisting of speckles of size $R_{diff}(\lambda)$ within larger structures of size $R_{ref}(\lambda)$ (Fig. 1, right). The illumination pattern from a stellar source of radius r_s is the convolution of the point-like intensity pattern with the projected intensity profile of the source (Fig. 2, upper right).

A cloud moving with a transverse velocity V_T relative to the line of sight will induce stochastic intensity fluctuations with amplitude predicted by Fig. 3 at the characteristic time-scale

$$t_{ref}(\lambda) = \frac{R_{ref}(\lambda)}{V_T} \sim 5.2\,minutes \left[\frac{\lambda}{1\mu m}\right] \left[\frac{z_0}{1\,kpc}\right] \left[\frac{R_{diff}(\lambda)}{1000\,km}\right]^{-1} \left[\frac{V_T}{100\,km/s}\right]^{-1}. \quad (1.2)$$

Signature of the scintillation signal. The first two signatures point to a propagation effect, which is incompatible with any type of intrinsic source variability.

• Chromaticity: Since R_{ref} depends on λ, one expects a variation in the characteristic time-scale $t_{ref}(\lambda)$ between the red side of the optical spectrum and the blue side.

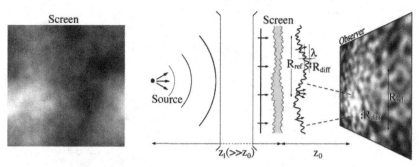

Figure 1. Left: A 2D stochastic phase screen (grey scale), from a simulation of gas affected by Kolmogorov-type turbulence. Right: The illumination pattern from a point source (left) after crossing such a phase screen. The distorted wavefront produces structures at scales $\sim R_{diff}(\lambda)$ and $R_{ref}(\lambda)$ in the observer's plane.

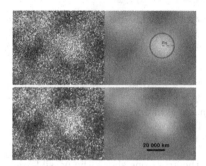

Figure 2. Simulated illumination map at $\lambda = 2.16\mu m$ on Earth from a point source (upper left) and from a K0 V star ($r_s = 0.85 R_\odot$, $M_V = 5.9$) at $z_1 = 8\,kpc$ (right). The refracting cloud is assumed to be at $z_0 = 160$ pc with a turbulence parameter $R_{diff}(2.16\mu m) = 150\,km$. The circle shows the projection of the stellar disk (with radius $R_S = r_s \times z_0/z_1$). The bottom maps are illuminations in the K_s wide band ($\lambda_{central} = 2.162\mu m$, $\Delta\lambda = 0.275\mu m$).

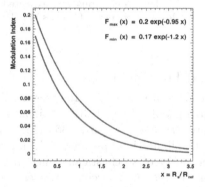

Figure 3. The intensity modulation index $m_{scint.} = \sigma_I/\bar{I}$ decreases when the ratio of the projected stellar disk R_S to the refraction scale $R_{ref}(\lambda)$ increases. The modulation index is contained between the curves represented by functions $F_{min}(x)$ and $F_{max}(x)$.

In figure 3: $F_{max}(x) = 0.2\,exp(-0.95\,x)$; $F_{min}(x) = 0.17\,exp(-1.2\,x)$; axes: Modulation Index vs $x = R_s/R_{ref}$.

• **Spatial de-correlation:** We expect a de-correlation between the light-curves observed at different telescope sites, increasing with their distance.

• **Correlation between the stellar radius and the modulation index:** Big stars scintillate less than small stars through the same gaseous structure.

• **Location:** The probability for scintillation is correlated with the foreground gas column density. Extended structures may therefore induce clusters of neighbouring scintillating stars.

Foreground effects, background to the signal. Atmospheric *intensity* scintillation is negligible through a large telescope (Dravins *et al.* 1997, 1998). Any other atmospheric effect should be easy to recognize as it affects all stars. Asteroseismology, granularity of the stellar surface, spots or eruptions produce variations of very different amplitudes and time-scales. A rare type of recurrent variable star exhibits emission variations on the minute scale, but such objects could be identified from their spectra.

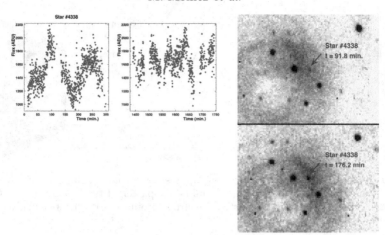

Figure 4. Above: Light-curves for the two nights of observation. Right: images of the candidate found toward B68 during a low-luminosity phase (upper) and a high-luminosity phase (lower); North is up, East is left.

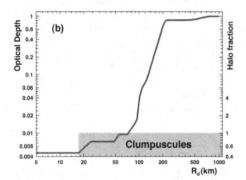

Figure 5. The $95\% \, CL$ maximum optical depth of structures with $R_{diff} < R_d$ towards the SMC. The right scale gives the maximum contribution of structures with $R_{diff}(1.25\mu m) < R_d$ to the Galactic halo (in fraction); the grey zone shows the possible region for the hidden gas clumpuscules expected from the model of Pfenniger & Combes (1994).

2. Preliminary Studies with the NTT, and Future Prospects

During two nights in 2006 June, 4749 consecutive exposures of $T_{exp} = 10 \, s$ were made with the infra-red SOFI detector in K_s and J through nebulae B68, cb131, Circinus and towards the SMC (Habibi *et al.* 2011). A candidate was found towards B68 (Fig. 4), but the poor photometric precision in K_s and other limitations prevented definitive conclusions. Nevertheless, we can surmise from the rarity of stochastically fluctuating objects that there is no significant population of stars that can mimic scintillation effects, and future searches should not be overwhelmed by a background of fakes.

From the observed SMC light-curves we also established upper limits (not yet competitive) on invisible gaseous structures as a function of their diffusion radius (Fig. 5). Those constraints are limited by the statistics and by the photometric precision.

Prospects. LSST will be an ideal set-up to search for this signature of gas, thanks to the fast readout and to the wide and deep field. Scintillation signal would provide a new tool to measure the inhomogeneities and dynamics of nebulæ, and to probe the molecular hydrogen contribution to the baryonic hidden matter of the Milky Way.

References

Dravins, D., Lindegren, L., Mezey, E., & Young, A. T. 1997, *PASP*, 109, 173; 109, 725
Dravins, D., Lindegren, L., Mezey, E., & Young, A. T. 1998, *PASP*, 110, 610
Habibi F., Moniez M., Ansari R., & Rahvar S. 2011, *A&A*, 525, 108
Pfenniger, D. & Combes, F. (1994) *A&A*, 285, 94

New Horizons in Time-Domain Astronomy
Proceedings IAU Symposium No. 285, 2011
R.E.M. Griffin, R.J. Hanisch & R. Seaman, eds.

© International Astronomical Union 2012
doi:10.1017/S1743921312001135

Optical Polarimetry of the Crab Nebula

Paul Moran[1], Andy Shearer[1], and Roberto Mignani[2]

[1]Centre for Astronomy, NUI Galway, Newcastle, Galway, Ireland
email: `p.moran4@nuigalway.ie`
[2]Mullard Space Science Laboratory, Dorking, Surrey, RH5 6NT, UK

Abstract. Time-resolved polarisation measurements of pulsars provide an unique insight into the geometry of the emission regions. Hubble Space Telescope (HST) polarisation data of the Crab Nebula were obtained from the Multimission Archive at STScI (MAST). The data are composed of a series of observations of the Crab Nebula with the HST and ACS camera system taken in three different polarisation filters (0°, 60° and 120°) between 2003 August and 2005 December. Polarisation vector maps of the Nebula were produced with the polarimetry software IMPOL. The degree of polarisation (P.D.) and the position angle (P.A.) of the pulsar's integrated pulse beam were measured, and also that of the nearby Synchrotron Knot, yielding P.D. = 4.90 ± 0.33 %, P.A. = 106°.46 ± 1°.9 for the pulsar, and P.D. = 61.70 ± 0.72 %, P.A. = 126°.86 ± 0°.23 for the Synchrotron Knot. These results are consistent with those of obtained by others using INTEGRAL.

Keywords. Pulsar, Polarimetry, Crab Nebula

1. Introduction

The raw HST ACS polarisation science frames of the Crab Nebula (M1) were obtained from the Multimission Archive at STScI (MAST) (See Fig. 1). The data are composed of a series of observations of the Crab Nebula with the HST and ACS camera system taken in three different polarisation filters (0°, 60° and 120°) between 2003 August and 2005 December. The images had already been flat-fielded; they were then geometrically aligned, combined, and averaged with cosmic-ray removal using IRAF. For each set of observations the images taken in the 0°, 60° and 120° polarisers were combined to give a single Stokes intensity image. The intensity images were analysed by the IMPOL software which produces polarisation vector maps.

Determining the polarisation in a crowded field such as the Crab Nebula is complex. In order to determine the Crab pulsar's polarisation profile we need to know the level of background polarisation. The problem is compounded by possible variations in the Crab gamma-ray flux, such as flaring phenomena, in the the surrounding pulsar wind nebula. Our work is intended to map accurately the polarisation of the Cab Nebula, and to act as a guideline for future time-resolved polarisation measurements of the Crab pulsar using the Galway Astronomical Stokes Polarimeter (GASP). GASP is an ultra-high-speed, full Stokes, astronomical imaging polarimeter based on the Division of Amplitude Polarimeter (DOAP). It has been designed to resolve extremely rapid variations in objects such as optical pulsars and magnetic cataclysmic variables.

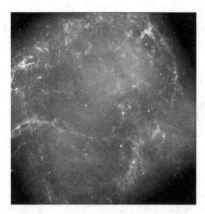

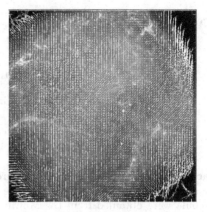

Figure 1. HST image of the Crab Nebula

Figure 2. Polarisation Map of the Crab Nebula

Table 1. Overall Results for the Polarisation Degree and Position Angle.

	Polarisation Degree (%)	Position Angle(°)
Pulsar	4.90±0.33	106.46±1.90
Synchrotron Knot	61.70±0.72	126.86±0.23

2. Polarimetry

In order to determine the polarimetry, aperture photometry was first performed on the pulsar and its synchrotron knot in each image. The Stokes vectors were then calculated using the following formulæ:

$$I = \tfrac{2}{3}\left[r(0) + r(60) + r(120)\right]$$
$$Q = \tfrac{2}{3}\left[2r(0) - r(60) - r(120)\right]$$
$$U = \tfrac{2}{\sqrt{3}}\left[r(60) - r(120)\right],$$

where r(0), r(60) and r(120) are the calibrated count rates in the 0°, 60° and 120° polarised images, respectively.

2.1. Computing the fractional polarisation (P.D.) of the target

Included is a factor which corrects for cross-polarisation leakage in the polarising filters. This correction is useful for the POLUV filters; values of T_{par} and T_{perp} can be found in Figure 5.4 of the ACS Instrument Handbook. The instrumental polarisation of the WFC ($\sim 2\%$) must be subtracted from it.

2.2. Computing the position angle (P.A.) on the sky of the polarisation E-vector

The parameter PAV3 is the roll angle of the HST spacecraft, and is called PA_V3 in the data headers. The parameter χ contains information about the camera geometry which is derived from the design specifications; for HRC, $\chi = -69°.4$, and for the WFC $\chi = -38°.2$.

$$\text{P.D.} = \frac{\sqrt{Q^2 + U^2}}{I} \frac{T_{par} + T_{perp}}{T_{par} - T_{perp}}$$

$$\text{P.A.} = \tfrac{1}{2}\tan^{-1}\left(\tfrac{U}{Q}\right) + \text{PAV3} + \chi$$

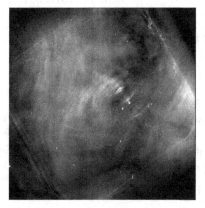

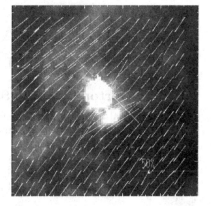

Figure 3. Polarised Flux Map of the Crab Nebula

Figure 4. Polarisation Vector Map of region around pulsar and synchrotron knot

3. Discussion

The polarisation maps show the variation of the polarisation throughout the nebula and particularly in the vicinity of the pulsar itself. One can distinctly see the overall structure of the nebula, the degree of polarisation of the knots and the synchrotron emission (see Figs. 2, 3 & 4). The Crab Nebula was observed by AGILE in September 2010 to flare in the gamma-ray spectrum. This was the first ever discovery of such an event. The flare emission is synchrotron from a small region close to the pulsar and possibly the nearby knot. The April 2011 flare was roughly 30 times brighter than the average pulsar and nebula flux. These observations challenge emission models of the pulsar wind interaction, particle acceleration processes, as well as our understanding of the Crab system and pulsars themselves.

4. Conclusion

The results (listed in Table 1) are in good agreement with those of Słowikowska *et al.* (2009). These measurements will be used as a background measure of the pulsar and nebula contribution for future time-resolved polarisation measurements of the Crab pulsar. We plan to use GASP to measure both the linear and circular polarisation from the Crab on time-scales of <100 microseconds. In order to do this we have to subtract the polarisation of the background. Polarisation measurements give a unique insight into the geometry of the emission region see (McDonald *et al.* 2011), and enable us to determine the pulsar geometry, emission altitude, pulsar inclination, field inclination and pitch angle.

References

McDonald, J., *et al.* 2011, *MNRAS, arXiv:1106.5207M*
Słowikowska, A., *et al.* 2009, *MNRAS, arXiv:0901.4559v1* [astro-ph.SR]

New Horizons in Time-Domain Astronomy
Proceedings IAU Symposium No. 285, 2011
R.E.M. Griffin, R.J. Hanisch & R. Seaman, eds.

© International Astronomical Union 2012
doi:10.1017/S1743921312001147

Time-Domain Astrophysics with SuperWASP

A. J. Norton[1] and the SuperWASP Consortium[2]

[1] Department of Physical Sciences, The Open University, Walton Hall, Milton Keynes, U.K.
email: a.j.norton@open.ac.uk

[2] The SuperWASP project is funded and operated by Queen's University Belfast, the Universities of Keele, St. Andrews and Leicester, the Open University, the Isaac Newton Group, the Instituto de Astrofisica de Canarias, the South African Astronomical Observatory and by STFC

Abstract. SuperWASP is the world's leading ground-based survey for transiting exoplanets. Its database now contains over 300 billion data points covering 30 million unique objects from 10 million images obtained over 1700 nights since 2004. In addition to having discovered 70 transiting exoplanets, SuperWASP enables long-baseline, high-cadence studies of variable stars to be performed. We summarise some of the studies already carried out, and look ahead to the prospects for studying periodic variables with varying periods. The science which is thus supported will include studies of the Blazhko effect in RR Lyræ stars, migrating starspots in rotational variables, third bodies in eclipsing binaries, and coalescing binary stars.

Keywords. Surveys, Catalogues, Stars: Binaries: Eclipsing, Stars: Rotation, Stars: Variables

1. Introduction

The Wide Angle Search for Planets (WASP) is the world's leading ground-based survey for transiting exoplanets (Pollacco *et al.* 2006), having discovered 70 to date (Enoch *et al.* 2011). It is comprised of two installations, on La Palma (Canary Islands) and at Sutherland (South Africa), and consists of 8 cameras per instrument on a single robotic mount. Each camera has a Canon lens of 111 mm aperture and 200 mm focal length, that is backed by a 2048 × 2048-pixel Andor CCD. The combination provides a 7°.8 × 7°.8 field of view and an image scale of 13″.7 per pixel. Light curves are generated for all stars within the field within a magnitude range of $8 < m_v < 15$.

The SuperWASP database, whose public archive may be found at the web address http://wasp.le.ac.uk/public, holds over 300 billion data points covering 30 million unique objects, from 10 million images obtained over 1700 nights since 2004. The coverage of the survey is now virtually the entire sky, with the exception of the Galactic plane where the stellar density is too high to permit useful aperture-photometry of objects on account of the instrument's large pixel size.

In order to investigate stellar variability, positions of SuperWASP objects are first matched with the USNO-B1 catalogue. That enables light curves to be constructed for individual objects across the entire observing period. The stellar light curves so produced are sensitive to variability on time-scales from minutes to years, with a cadence as short as 30 seconds. Period searching on SuperWASP light curves uses a combination of a CLEANed power spectrum and a phase dispersion minimisation technique to identify periods in common between the two. Many systematic noise periods at $1/n$ fractions of a sidereal day are apparent, and are caused by events such as temperature-dependent focus shifts which vary across the image plane. In addition, several sets of multiple objects display identical periods because of the large pixel size and issues with blending. Such effects complicate things, but nonetheless over 1 million light curves which are found display genuine periodic variability. Automated Neural Network classification of folded light curves based on their shape is on-going.

2. Periodic Variable Stars Coincident with ROSAT Sources

In a first attempt to investigate the variability of a manageable but interesting subset of the SuperWASP objects, the positions of the objects were cross-correlated against ROSAT X-ray all sky survey catalogues (1RXS & 2RXP catalogues). That initial study (Norton *et al.* 2007) found 428 unique objects displaying periodic variability in their SuperWASP light curves and which were also coincident with ROSAT X-ray sources. We determined that there is a roughly 5% chance alignment between ROSAT sources and SuperWASP objects, and that at least 80% of those matches are likely to be real.

68 of the 428 objects just mentioned were previously-identified variable stars, 66 having known periods. They included 47 objects in the General Catalog of Variable Stars (GCVS), 17 objects discovered by the Robotic Optical Transient Search Experiment (ROTSE), and 2 discovered by the Semi-Automatic Variability Search (SAVS). They were classified as follows: 13 pre-main-sequence stars, 10 EA (Algol type) binaries, 5 EB (Beta Lyræ type) binaries, 10 EW (W Ursa Majoris type) binaries, 6 BY Dra systems, 5 RS CVn systems, 2 RR Lyræ stars, 15 Cepheid variables, 3 cataclysmic variables, 1 SuperSoft X-ray Source and 1 low-mass X-ray binary. (The Cepheid classifications given by ROTSE were probably mistaken).

Amongst the 360 newly-identified periodic variables, we demonstrated that most were likely to be rapid rotators of some sort (e.g., pre-main sequence stars, RS CVn stars, BY Dra stars), but the sample included many eclipsing binaries too. The project thus demonstrated the effectiveness of the SuperWASP survey for detecting photometric modulation over a range of time-scales amongst these moderately bright stars.

3. Short-Period Eclipsing-Binary Candidates

A second exploratory project was an investigation of some of the shortest-period variable objects revealed by the SuperWASP survey. Low-mass dwarf stars are very common, but their evolution in close binaries is poorly understood. In fact, a short period cut-off for eclipsing binaries with main-sequence components exists around a period of ~0.22 days, but few are known close to that limit. The cause of that cut-off is unclear; it is possibly linked to the finite age of the population.

To investigate this problem further, Norton *et al.* (2011) identified 5,600 periodic signals in the SuperWASP database in the period range 125–167 minutes, plus a further 17,300 signals close to 1/9 d (~160 minutes), 14,500 signals close to 1/10 d (144 minutes), and 9,300 signals close to 1/11 d (~131 minutes)—most of which were false for the reasons outlined earlier. Contact binaries will generally be detected by our search technique at half their true orbital period, so the range searched corresponded to orbital periods between 250 minutes and 333 minutes. Visual examination of the light curves revealed 53 candidate eclipsing binaries (EW type) with broad maxima and narrow minima, only 5 of which were previously known. The rest of the objects display sinusoidal modulation (because they are rotational variables) or narrow maxima and broad minima (characteristic of pulsating variables). The process also allowed us to identify the shortest-period binary known that has dM components: GSC2314–0530 = 1SWASP J022050.85+332047.6, with a period of 277.4 minutes (or 0.1926 days).

The orbital-period distribution produced by that sample increased the number of known short-period binaries (i.e., with orbital periods less than 0.23 days) by a factor of 6. Follow-up spectroscopic observations of these relatively bright candidates will allow the theory of the evolution of low-mass binary stars and their proposed mass–radius–period relationships to be tested further.

4. Periodic Variables with Varying Periods

Future research on the variable-star population revealed by SuperWASP will focus on those variable stars for which evidence of a varying period may be detected. That work is forming the basis of an STFC-funded PhD project. Four types of variability will be considered and sought:

4.1. *Varying period and/or amplitude of pulsating (RR Lyræ) stars*

The cause of this phenomenon in RR Lyræ stars, known as the Blazhko effect, is presently unknown. The long time-base and continuous coverage of the $> 10,000$ RR Lyræ stars which SuperWASP will observe will enable the phenomenon to be investigated in unprecedented detail, and should offer new insights into its origin.

4.2. *Varying periods in stars displaying rotational modulation*

The surfaces of cool stars are often dominated by star spots, whose presence imparts a photometric variability to the stellar light curve as the star rotates. Star spots are expected to migrate in latitude with time. Many stars rotate differentially, giving rise to a variation in the observed rotation period and hence allowing investigation of the outer structure of the star. SuperWASP detects hundreds of thousands of stars displaying rotational modulation, and offers the prospect of statistical analysis of this behaviour across a variety of stellar spectral types.

4.3. *Third bodies in eclipsing binaries*

The $> 10,000$ eclipsing binaries detected by SuperWASP offer an accurate "clock" by which to measure the orbital periods of binary motion. Some of these binary stars will be part of hierarchical triple systems and some may host circumbinary exoplanets. By searching for eclipsing binary periods which themselves vary cyclically with longer periods, the presence of third objects may be identified and characterized.

4.4. *Coalescing binaries*

Close eclipsing binary stars will lose angular momentum via a combination of magnetic braking and gravitational radiation. The very closest binaries will eventually merge, resulting in a single stellar core. Such a merger has been observed only once (Tylenda *et al.* 2011), in the star V1309 Sco. The long time-base offered by SuperWASP will allow us to search for other eclipsing binaries whose periods are systematically decreasing, and which may therefore be in or near the final stages of merger.

References

Enoch, B., *et al.* 2011, *AJ*, 142, 86
Norton, A. J., *et al.* 2007, *A&A*, 467, 785
Norton, A. J., *et al.* 2011, *A&A*, 528, A90
Pollacco, D. L., *et al.* 2006, *PASP*, 118, 1407
Tylenda, R., *et al.* 2011, *A&A*, 528, A114

New Horizons in Time-Domain Astronomy
Proceedings IAU Symposium No. 285, 2011
R.E.M. Griffin, R.J. Hanisch & R. Seaman, eds.

© International Astronomical Union 2012
doi:10.1017/S1743921312001159

ARCONS: a Highly Multiplexed Superconducting UV-to-Near-IR Camera

Kieran O'Brien[1], Ben Mazin[1], Sean McHugh[1], Seth Meeker[1] and Bruce Bumble[2]

[1] Department of Physics, University of California, Santa Barbara, CA 93106, USA
email: kobrien@physics.ucsb.edu

[2] Jet Propulsion Laboratory, Pasadena, California 91107, USA

Abstract. ARCONS, the Array Camera for Optical to Near-infrared Spectrophotometry, was recently commissioned at the coudé focus of the 200-inch Hale Telescope at the Palomar Observatory. At the heart of this unique instrument is a 1024-pixel Microwave Kinetic Inductance Detector (MKID), exploiting the Kinetic Inductance effect to measure the energy of the incoming photon to better than several percent. The ground-breaking instrument is lens-coupled with a pixel scale of 0″.23/pixel, each pixel recording the arrival time ($< 2\,\mu$ sec) and energy of a photon ($\sim 10\%$) in the optical to near-IR (0.4–1.1 microns) range. The scientific objectives of the instrument include the rapid follow-up and classification of transient phenomena.

Keywords. instrumentation: detectors, instrumentation: spectrographs, pulsars: individual (Crab)

---

1. Background

(a) *The Kinetic Inductance Detector.* The working principle of the Kinetic Inductance Detector was described in detail in Day *et al.* (2003), and is summarized here. Photons with energy $h\nu$ are absorbed in a superconducting film, producing a number of excitations, called "quasiparticles" (Fig. 1a). To measure those quasiparticles sensitively, the film is placed in a high-frequency planar resonant circuit (Fig. 1b). Figs. 1c, d show the effect on the amplitude and phase respectively of a microwave excitation signal sent through the resonator. The change in the surface impedance of the film following a photon absorption event pushes the resonance to lower frequency and changes its amplitude. If the detector (resonator) is excited with a constant on-resonance microwave signal, we can measure the degree of phase and/or amplitude shift caused by a single incident optical photon.

(b) *Energy resolution.* Since the energy of the incoming photon is many times that necessary to generate a quasiparticle, several thousand quasiparticles are generated by each photon (in contrast to a semiconductor, where the incident photon has an energy only slightly above the band-gap). The degree of phase-shift is related to the number of quasiparticles, and hence to the energy of the incoming photon. The maximum theoretical resolution, $R\,(=E/\delta E)$ is given by,

$$R = \frac{1}{2.355}\sqrt{\frac{\eta h\nu}{F\Delta}} \;, \tag{1.1}$$

where η is an efficiency factor, F the Fano factor and Δ the superconducting energy gap.

(c) *Multiplexing scheme.* By engineering each resonator to have a slightly different resonant frequency, a large number (a few thousand) of resonators (pixels) can be probed

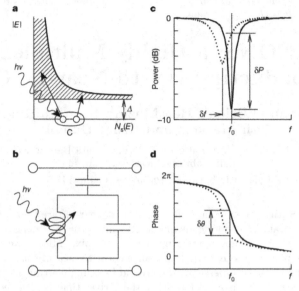

Figure 1. An illustration of the detection principle, from Day *et al.* 2003.

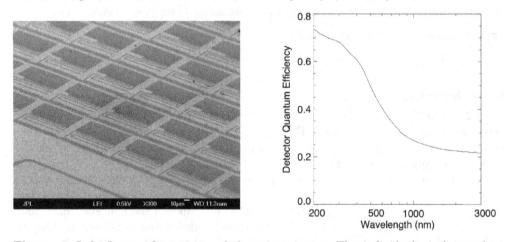

Figure 2. Left: Image of a section of the science array. The individual pixels can been seen to have a slightly different length meandered section in order to tune the resonant frequency, enabling the highly multiplexed read-out. Right: the measured quantum efficiency of the TiN-lumped element detector.

simultaneously by a comb of frequencies sent down a single coaxial line. In the current scheme we use two coaxial cables with 512 resonators on each line in the range 4–5 GHz.

2. The Instrument

ARCONS (Mazin *et al.* 2010) uses a cryogen-free ADR to cool an array of Titanium-Nitride (TiN)-lumped element MKIDs to a temperature of 85 mK (which is well below the superconductor T_c of ~800 mK). The instrument has a hold time of ~12 hrs before it needs to be regenerated; the latter step takes ~2 hrs. The 32 × 32 array (shown in Fig. 2) is on a 100-μm pitch, which is behind a 200-μm focal-length micro-lens array, increasing the fill factor to 64% and increasing the uniformity by concentrating the light on a small region of the inductor. While the MKIDs are sensitive from the UV to mid-IR

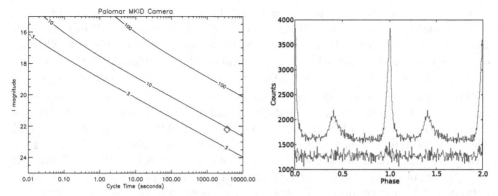

Figure 3. Left: predicted sensitivity of ARCONS at Palomar, with contours for signal-to-noise ratios of 3, 10 and 100. The diamond shows a measurement from the 2011 commissioning run. Right: the phase-folded Crab Pulsar light curve (repeated for 2 cycles of the 33 msec spin period for clarity) from a 20-sec twilight observation. The sky variability is also shown (offset).

(see Fig. 2), the choice of glass cuts them off below 400 nm and a filter (ASAHI "supercold") sets the red limit at 1100 nm, limiting the total count-rate (dominated by the sky). The pixels are 0″.23 on the sky, giving a field of view of ∼7″.5 × 7″.5. Each pixel has a R∼12 at 400 nm, but was strongly affected by "substrate events" where the photon is absorbed in the Si substrate, leading to a breakdown in the relationship between photon energy and phase shift.

The pixels are read out using a custom-built software-defined radio (SDR) system (McHugh *et al.* 2011), in which a frequency comb is created in software, and up-converted to the required frequency range. That signal is sent along the coax and passes through the MKIDs. The transmitted signal is amplified and then down-converted and digitized via onboard A/D converters. It is then "channellized", and the pulse heights (a direct measure of the photon energy) and pulse start-times (photon arrival times) are measured in a powerful FPGA.

3. First Light

ARCONS was successfully commissioned during four nights in 2011 July/August. The throughput was as expected. We observed a broad range of "science demonstration" targets, including interacting binaries (AM Cvns, LMXBs, short-period eclipsing sources), QSOs (for low-resolution redshift measurements), supernovæ (Type Ia and Type II) and the Crab pulsar.

4. Science Goals

(*a*) *Rapid characterization of transients.* ARCONS will allow us to determine the spectrum of a transient source as it evolves. The small field, readout-noise-free IFU will allow point-and-shoot deep spectroscopic observations. That will enable better marshalling of follow-ups on large-aperture telescopes.

(*b*) *Time-resolved Opt/IR observations of pulsars.* Detailed time-resolved spectra, especially if combined with simultaneous radio (GBT) and gamma-ray (Fermi) observations, should reveal details about pulsar emission and the structure of the magnetosphere that will allow us to differentiate between models of optical emission.

(*c*) *Characterization of short-period variables.* There is a growing number of short-period variables being discovered by sky surveys, and that number will increase with future large surveys, most notably LSST. These discoveries require detailed follow-up observations in order to characterize the components of the variables. ARCONS is ideally suited for such observations, offering zero dead-time, low-resolution spectroscopy.

(*d*) *Redshift determination via spectro-imaging.* We will use the intrinsic energy resolution of the MKIDs to determine the redshift of galaxies out to $z \sim 4$ with a high degree of accuracy. A spectral resolution of 20 in the UV with ARCONS translates into 14 wavelength resolution elements in the wavelength range of the camera. The advantage of ARCONS over multi-filter photometry, such as that employed by COMBO-17, is that all wavelengths are observed simultaneously (thus reducing the exposure time considerably) and through the same conditions (seeing, sky transparency), making the analysis less complicated.

5. The Future

With the information and experience we have gained from the recent commissioning run, we have identified a number of upgrades which, when combined, will improve the instrument significantly and promote it from a demonstration instrument to a front-line scientific one. The improvements include increasing the pixel-scale to match better the median seeing at Palomar and to give a wider field of view, increasing the number of pixels and bandwidth of the read-out electronics for reading out a 2048 pixel array, and improving our calibration scheme. In the longer term, we are developing an anti-reflection scheme to improve the QE of the detectors. MKIDs promise to revolutionize many areas of astronomy, not limited to the time domain, as they are capable of producing large area, read-noise-free integral-field spectroscopy without the need for complicated optical systems. In the time domain, they will offer deep, low-resolution integral-field spectroscopy with the advantage of time-tagging the arrival time of each photon.

6. Acknowledgements

We would like to thank the management and staff of Palomar Observatory for their hard work and support during the commissioning of ARCONS. Material in this paper is based upon work supported by the National Aeronautics and Space Administration under Grant NNX09AD54G, issued through the Science Mission Directorate, Jet Propulsion Lab's Research & Technology Development Program, and a grant from the W.M. Keck Institute for Space Studies. Part of the research was carried out at the Jet Propulsion Laboratory, California Institute of Technology, under a contract with the National Aeronautics and Space Administration.

References

Day, P., Leduc, H., Mazin, B., Vayonakis, A., & Zmuidzinas, J. 2003, *Nature*, 425, 817
Mazin, B., *et al.* 2010, in: *Proc. SPIE*, Vol. 7735, 773518
McHugh, S. 2011, *Review of Scientific Instruments*, submitted.

New Horizons in Time-Domain Astronomy
Proceedings IAU Symposium No. 285, 2011
R.E.M. Griffin, R.J. Hanisch & R. Seaman, eds.

Photographic Archives of Ukrainian Observatories: Digitizing a Heritage

Ludmila Pakuliak[1], Lilia Kazantseva[2], Natalia Virun[3] and Vitaly Andruk[1]

[1] Main Astronomical Observatory, NAS of Ukraine, Kyiv, 03680 Ukraine
email: pakuliak@mao.kiev.ua

[2] Astronomical Observatory of Kyiv National University, Kyiv, 04053 Ukraine
email: likaz@observ.univ.kiev.ua

[3] Astronomical Observatory of Lviv National University, Lviv, 79005 Ukraine
email: virun@astro.franko.lviv.ua

Abstract. We describe a key project of the Ukrainian Virtual Observatory (UkrVO), namely, a collaboration to digitize the large collections of photographic plates that had been exposed during more than 100 years at Ukrainian observatories, and to combine the digitized images with CCD archives to form the UkrVO Joint Digital Archive. The application of flatbed scanners for digitizing plates is discussed.

Keywords. astronomical data bases: miscellaneous

1. Plate Collections of Ukrainian Observatories

The archives of photographic observations at three observatories of Ukraine—at Kyiv, Lviv and Odessa Universities—are significant through both their size and their age; the collections are very extensive, and cover the period from the late-19th century to the mid-20th century. A small part of the plates from that period may have been dispersed when observing activities were severely disrupted during WWI and WWII and several eastern European observatories lost major parts of their plate collections. The loss of a collection also means the loss of information about the observing methods used.

The history of progress in observational techniques has a number of dimensions, cultural as well as scientific. Observing involved Hartmann or hexagonal diaphragms, tubular, wedge or Fesenkov photometers, hypersensitizing in special chemical solutions, and many other methods and processes that have now been relegated to the status of little-known techniques of historic interest only. Unfortunately, it is difficult to recover complete information about the images that were observed, as logs and other records were often lost. Nevertheless, each series of historic photographic observations is worthy of careful attention and study. We therefore included the preservation of historical observational archives within the framework of the Ukrainian Virtual Observatory project without segregation into scientific and historical sections.

1.1. The AO LNU collection

The Astronomical Observatory of Lviv National University (AO LNU) made photographic observations from 1939 to 1976; just a few were also made between 1936 and 1939. The instruments that were used included a camera with a Zeiss triplet lens (D/F= 100/500 mm), a Mertz refractor, an astro-camera (D/F=140/700 mm) and a Zeiss refractor (D/F = 130/2400 mm). Up to the 1950s the plates were measured for photovisual photometry with a Schilt photometer. Part of the collection had never been reduced at all, or processed only partially.

1.2. *The AO KNU collection*

The photographic collection of the Astronomical Observatory of Kyiv National University (AO KNU) has over 20,000 plates; it covers the period 1898–1996, and is sub-divided into more than 200 series. 65% of the collection is on glass; the rest is on large-format film. Most of the series include images of photometric standard fields obtained on the same dates as target images, or other exposures allowing different methods of calibration (out-of-focus images for use in conjunction with a photo-tube, a photometric wedge, the Sabattier effect, or the method of equi-densities for photometry of extended objects). In 2010 the first attempts were made to provide open access to the historical (pre-1950) section of the archive. Test digitization of about 100 plates gave very encouraging precisions: $0''.1$ in position and 0.07 in magnitude.

2. The Joint Digital Archive of the UkrVO

The Joint Digital Archive (JDA) of photographic and CCD observations was conceived as a key project of the UkrVO (Vavilova *et al.* 2010). Digitizing Ukraine's photographic plates started in 2008. The JDA is being required to include digitized photographic images from at least six Ukrainian observatories. The plates had been exposed in more than 20 different instruments, so it means that every "publisher" of digitized images has to transform the data to a common standard before placing them in a shared database.

Flatbed scanners have recently become popular as digitizers for photographic plate archives. The quality of the output from digitization by commercial scanners has caused, and still causes, considerable concern. Nevertheless, commercial scanners are the most widely available appliances for relatively rapid digitization work, and applying the proper scan procedures and algorithms to the digitized images yields the best accuracy that can be achieved for a given appliance and given observational material.

Processed files from the digital archives of MAO NASU, AO LNU and AO KNU are mounted on the computer cluster of the Main Astronomical Observatory NAS, which are shared resources. A database of the Golosiiv plate archive has been available there via open access (DBGPA, http://www.gua.db.ukr-vo.org) since 2003, and serves as a test area for JDA development and software upgrades for data access. To date that JDA prototype holds digitized data from four plate collections, including their historical sections.

Digitizing the plates is carried out with two models of flatbed scanner (Table 1). Two types of digitized images are derived. For the highest-quality plates, high-resolution scans are made twice, one with the plate rotated clockwise through 90°. The TIFF output images have a dynamic range of 16 bits in grey scale and a resolution of 1200 dpi. The maximum linear dimension is 13,000 pixels. A different type of scanned image is made for "previews"—rapid preliminary visualization of the content and quality of the material. Most of the plates scanned in that way are wide-field plates, like the ones included in the WFPDB (Tsvetkova *et al.* 2008). In the preview images any ink marks or writing are preserved as they may have historic value (especially when the plates are nearly a century old), and are only subsequently cleaned off. JPEG previews are scanned with a grey-scale of 8 bits, or 24 bits for a coloured image, and a resolution of 300–1200 dpi. The maximum linear dimension is 1,200 pixels along the larger side for plates of any size.

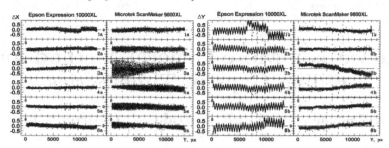

Figure 1. Trends of individual deviations of coordinates ΔX, ΔY from the average values

Table 1. Comparing accuracies of performance for different image resolutions.

	Epson Expression 10000XL					Microtek ScanMaker 9800XL				
dpi	300	600	900	1200	1600	300	600	900	1200	1600
$\sigma_m\,(^m)$	0.18	0.22	0.24	0.22	0.22	0.22	0.34	0.40	0.35	0.36
$\sigma_\alpha\,('')$	1.12	0.69	0.62	0.49	0.41	1.84	1.27	0.90	0.78	0.63
$\sigma_\delta\,('')$	1.17	0.73	0.63	0.54	0.46	2.04	1.52	1.18	1.07	0.98

3. Comparing Scanners to Estimate their Accuracy

In order to estimate the photometric and positional accuracy of the scanners, a set of six sequential scans was made of the same plate at 1200 dpi (21.17 μm/pixel). Averages of the six scans were derived, for each scanner and every object. The total number of recorded objects was $\sim$100,000, though selecting objects with $6 < V < 13$ from the TYCHO-2 Catalogue reduced that number to 6000. Special attention was paid to (a) systematic trends (and their elimination) produced by the scanner, (b) correct separation of images on multi-exposure plates, and (c) the effects of different resolution (for which another set of images with resolutions of 300–1600 dpi was obtained).

The results showed that the optimum resolution for both instruments was 1200 dpi. Trends of individual positional deviations (ΔX, ΔY) from the average values are illustrated in Fig. 1. After removing systematic trends, the rms errors were $\sigma_{X,Y} = \pm\,0.03$–0.06 pixels, meaning that $\sigma_{\alpha,\delta} \leqslant \pm\,0''.1$ for both scanners. The values of (σ_m), the rms photometric errors, were $\leqslant \pm\,0.015$ mag, and $\sigma_B \leqslant \pm\,0.03$ mag. for both digitizers too. Details of the accuracy and precision for different resolutions are given in Table 1. In the light of those results, an Epson Expression model was selected for making high-resolution images, and a Microtek one for preview digitization.

At the time of writing, the number of digitized images in the MAO NASU archive is approaching 4000, and there are nearly 2500 in the AO LNU archive. Digitization of the AO KNU archive has just commenced.

The rationale and methods adopted for digitizing these historical components of an archive are described on our Website at `http://ukr-vo.org/history/index.php?b1&4` and `http://ukr-vo.org/history/index.php?b2&4`. Placing these descriptions in the public domain provides a resource that can be used for educational and museum activities, in particular in relation to the history of the science of astronomy in the Ukraine.

References

Vavilova, I. B., Pakuliak, L. K., & Protsyuk, Yu.I. 2010, *Kosm. Nauka Tekhn.*, 16, 62
Tsvetkova, K. P., Tsvetkov, M. K., Sergeeva, T. P., & Sergeev, A. V. 2009, *Kin. & Phys.of Celest. Bodies*, 25, 402

New Horizons in Time-Domain Astronomy
Proceedings IAU Symposium No. 285, 2011
R.E.M. Griffin, R.J. Hanisch & R. Seaman, eds.

© International Astronomical Union 2012
doi:10.1017/S1743921312001172

Towards a More General Method for Filling Gaps in Time Series

J. Pascual-Granado[1], R. Garrido[1], J. Gutirrez-Soto[1,2] and S. Martín-Ruiz[1]

[1] Instituto de Astrofsica de Andaluca (CSIC), Granada, Spain.
email: javier@iaa.es

[2] Universidad Internacional Valenciana - VIU, 12006 Castelln de la Plana, Spain

Abstract. The need for a proper interpolation method for data coming from space missions like CoRoT is emphasized. A new gap-filling method is introduced which is based on auto-regressive moving average interpolation (ARMA) models. The method is tested on light curves from stars observed by the CoRoT satellite, filling the gaps caused by the South Atlantic Anomaly.

Keywords. methods: data analysis, stars: oscillations

1. Introduction

Recent space missions like CoRoT (Baglin *et al.* 2006) and Kepler (Borucki *et al.* 2010) provide photometric data with very high duty cycle and unprecedented resolution. However, the light curves always have some invalid flux measurements owing to operational procedures such as (among others) mask changes or reorientation of the satellite, or (as in the case of CoRoT) the impact of energetic particles when the satellite passes through the South Atlantic Anomaly. Those gaps produce aliases in the frequency spectra, and a gap-filling procedure is normally applied to avoid such aliases. The most commonly used one is linear interpolation, which is very simple but very unrealistic. It is the interpolation used for CoRoT data.

We have developed a new gap-filling technique (ARMA) based on auto-regressive moving average models, as a means of avoiding the errors introduced by linear interpolation methods. Our technique models the data points around the gaps by auto-regressive processes and interpolates the gaps with a forward-backward predictor, thence yielding a corrected light curve with regular sampling.

In this contribution we apply our gap-filling technique to the CoRoT light curves of two stars, HD 51193 and HD 172189.

2. Results

The first case is the Be star HD 51193 (Gutirrez et al. 2007) observed in the LRa02 target field (see Fig. 1). Between the two vertical dashed lines the improvement of the ARMA interpolation (crosses) is compared to the linear interpolation (dots).

In the right panel of Fig. 1 the grey peaks appearing on the right-hand side are produced by the spectral window caused by the gaps. In the ARMA-interpolated (black) spectra those peaks have disappeared. About 1.5% of the power spread over those peaks is thus recovered.

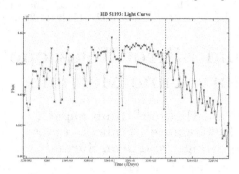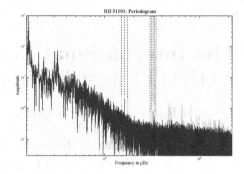

Figure 1. Left: a section of the HD 51193 light curve. Between the two vertical lines the dots represent linear interpolation and crosses the ARMA one. Right: a Scargle periodogram of the light curve. Linear interpolation is in grey and ARMA interpolation in black.

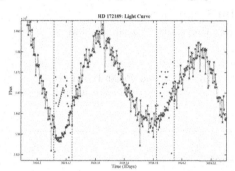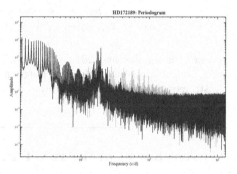

Figure 2. The same as Fig. 1, but for the star HD 172189.

The second case is the binary system HD 172189, (Martín-Ruiz et al. 2005) which shows classical δ Scuti pulsations (see Fig. 2). As the Figure shows, in this case the error introduced by linear interpolation is more pronounced and more critical. Again, the spectral window is removed much more efficiently with ARMA interpolation than with linear interpolation.

These two cases reveal the importance of introducing a reliable gap-filling process. Owing to the size of the gaps in CoRoT data, a greater improvement is expected in the case of δ Scuti stars.

A more detailed description of the method, together with further results, will be published shortly.

References

Baglin, A., Michel, E., & Auvergne, M., The COROT Team 2006, in: K. Fletcher & M. Thompson (eds.), *Beyond the spherical Sun* (Noordwijk: ESA), SP-624
Borucki, W. J. *et al.* 2010, *Science*, 327, 977
Gutirrez-Soto, J., *et al.* 2007, *A&A*, 476, 927
Martín-Ruiz, S., *et al. A&A*, 440, 711

New Horizons In Time-Domain Astronomy
Proceedings IAU Symposium No. 285, 2011
R.E.M. Griffin, R.J. Hanisch & R. Seaman, eds.

© International Astronomical Union 2012
doi:10.1017/S1743921312001184

The International Liquid Mirror Telescope (ILMT) as a Variability Time Machine

Joël Poels[1], Ermanno Borra[2], Paul Hickson[3], Ram Sagar[4],
Przemyslaw Bartczak[5], Ludovic Delchambre[1], François Finet[1],
Serge Habraken[1], Jean-Pierre Swings[1] and Jean Surdej[1]

[1] AEOS, University of Liège, B-4000, Sart Tilman, Belgium
email: poels@astro.ulg.ac.be

[2] Dept. of Physics, Laval University, Canada

[3] Dept of Physics and Astronomy, UBC, Canada

[4] Dept of Astronomy, ARIES, Nainital, India

[5] Dept of Astronomy of A. Mickiewicz University, Poznan, Poland

Abstract. During the year 2012 the International Liquid Mirror Telescope (a collaboration between astronomical institutions in Belgium, Canada, India and Poland) wil see first light. The instrument will provide substantial, in-depth sky coverage and make an unprecedented number of nightly observations.

1. Current Status of the ILMT Project

The science achievable with this unique instrument is exciting in terms of variability studies, and includes possible cosmological inference. Details of the ILMT in particular can be found on our Web pages, whose URLs are given below. The Website offers access to many public documents featuring pioneer papers dealing with Liquid-Mirror technology, as well as images and videos which we are proud to share. One will also find there some more didactic documents for those interested in this upcoming and promising technology.

Large sections of the ILMT equipment have been shipped to India recently (2011 December), though other parts of the assembly are still in a building phase. Many issues and technical problems were solved during the past two years. Of those, the chief one was related to the quality of the mercury surface. As the objective of the project is to achieve a surface quality close to that of a glass mirror with similar dimensions, we carried out extensive tests and found that if we can avoid surface waves (concentric waves due to vibrations, spiral waves due to rotation, wind, etc.) we achieve a surface accuracy of $\lambda/2$ when the mercury layer is 1 mm thick (or even less); see Fig. 1. In order to carry out those measurements and correct such dynamic surface defects we thought that shooting an incident laser beam and capturing its reflection with a dedicated camera could help. We then analysed the signal recorded by the camera using using algorithms based on Fourier transforms to disentangle and characterize the waves. The analysis then guided the tweaking and fine-tuning of a few parameters, and finally we arrived at the expected performance figures. It was also recognised that the mirror container (Fig. 2) is a key component of the system. By meeting all the stated specifications for properties such as rigidity and temperature stability we could be sure that the mirror surface would be as perfect as possible and without distorting wavelets.

The ILMT is erected vertically (Fig. 3), a design which admittedly incorporates both advantages and drawbacks. The latter mainly stem from the fact that objects passing above the ILMT FOV (field of view) do not follow a straight line, and as a consequence

Figure 1. The high quality of the mirror surface.

Figure 2. The mirror container is a key component of the system. It must fulfil strict specifications of rigidity and temperature stability in order to guarantee a final shape that is as perfect as possible and to avoid the formation of wavelets on the surface of the mercury.

the image of a point-like source will be degraded and cannot be fitted with a gaussian "seeing" profile. Furthermore, the integration time across the FOV is not the same, and will depend on the declination of the object. In practical terms that effect will depend on the terrestrial latitude where the ILMT is installed. To correct such effects, we designed and built a dedicated corrector that enabled us to eliminate the (well known) Time Delay Integration (TDI) effect. Since the correction is latitude dependent, a dedicated corrector is needed to compensate for the latitude correction of each installation site.

The cost of building such a telescope is roughly 1/50 that of building a conventional instrument of the same class. Even if the ILMT points only to the zenith, that is nevertheless an ideal observing mode since the airmass stays roughly the same, and in addition, pointing to the zenith guarantees the best air transparency. Regarding the choice of the filters, the main one (i') allows observations for a maximum number of nights because its spectral range is less sensitive to the bright phases of the moon. The camera will be equipped with a set of additional filters (g', r', shutter), which can be selected according to need. Since the travel time of an object over the FOV is constant, the CCD camera could be dazzled by bright objects; however, such eventualities can obviously be predicted in advance and the shutter activated for as long as it is needed according to the brightness of the source.

Figure 3. Vertical fixed structure. Focal length = 8m.

If we were to observe on each night the same 30'-wide strip of sky and apply a systematic image subtraction method to the current and the reference image, there is little doubt that interesting extragalactic objects, as well as other variable ones, will be discovered, and can then be monitored on a daily basis.

We designed an analysis routine so as to preserve data integrity. A strip of night sky was sliced into smaller strips of 4K × 4K pixels, and we applied a software framework which (among other tasks) allows the registration of the location of local/remote rough images into a Relational Database Management System. That database is the heart of a clustering architecture which permits the reduction of multiple images in parallel, using tools such as an Object Oriented (OO) C++ code that relies on CORBA middleware. Then, when a new ILMT image is stored, various pipelines can be triggered. C++ pipelines can run independently and can achieve many different scientific tasks; some can compute accurately the astrometry and photometry for a detected image feature, while others can provide light curves of objects of interest that are signalled through a Web-browser request which is triggered by users. The potential of this system is vast, within obvious hardware limitations.

Further details of this project can be found on the International Liquid Mirror Telescope Homepage: http://www.aeos.ulg.ac.be/LMT

New Horizons in Time-Domain Astronomy
Proceedings IAU Symposium No. 285, 2011
R.E.M. Griffin, R.J. Hanish & R. Seaman, eds.

© International Astronomical Union 2012
doi:10.1017/S1743921312001196

Classification of ASKAP VAST Radio Light Curves[†]

Umaa Rebbapragada[1], Kitty Lo[2], Kiri L. Wagstaff[1], Colorado Reed[3], Tara Murphy[2] & David R. Thompson[1]

[1] Jet Propulsion Laboratory, Pasadena, CA, 91109 USA
email: Umaa.D.Rebbapragada@jpl.nasa.gov
[2] Sydney Institute for Astronomy, University of Sydney, Sydney, NSW 2006, Australia
[3] Department of Physics, University of Iowa, Iowa City, IA 52242, USA

Abstract. The VAST survey is a wide-field survey that observes with unprecedented instrument sensitivity (0.5 mJy or lower) and repeat cadence (a goal of 5 seconds) that will enable novel scientific discoveries related to known and unknown classes of radio transients and variables. Given the unprecedented observing characteristics of VAST, it is important to estimate source classification performance, and determine best practices prior to the launch of ASKAP's BETA in 2012. The goal of this study is to identify light-curve characterization and classification algorithms that are best suited for archival VAST light-curve classification. We perform our experiments on light-curve simulations of eight source types and achieve best-case performance of approximately 90% accuracy. We note that classification performance is most influenced by light-curve characterization rather than classifier algorithm.

Keywords. ASKAP, VAST, radio astronomy, classification, data analysis

1. Introduction

The Australian Square Kilometer Array Pathfinder (ASKAP) will observe the entire visible radio sky, including previously unexplored regions of phase space, in a single day with sub-mJy sensitivity at a 5-second cadence. Because no other telescope in operation has those capabilities, ASKAP has the potential of advancing significantly the study of known transients and variables, while also enabling the discovery of new objects and object classes. The Variables and Slow Transits (VAST) survey science project of ASKAP is focused on the development of new algorithms for the detection of transients with time-scales as short as 5 seconds (Murphy & Chatterjee 2009). Source types of interest include X-ray binaries, supernovæ, Extreme Scattering Events, Intra-Day Variables, novæ, and dMe flare stars, and RSCVn. Source classification is a prerequisite for scientific study of radio transients and variables.

The overall goal of our study is to evaluate state-of-the-art machine-learning methods for archival classification of VAST light curves. Because VAST has no existing counterpart with data to use for empirical evaluation, we simulate light curves for each of the sources discussed above plus background sources, and study performance using different classifiers, observing strategies, signal-to-noise ratios, and light-curve characterizations.

Here we present just a summary of results using a daily observational strategy (VAST Wide; Murphy & Chatterjee 2009) which observes with an r.m.s. of 0.5mJy. Our results show that we achieve approximately 90% classification accuracy using a Support Vector

---

† Presented on behalf of U. Rebbapragada by K. Lo

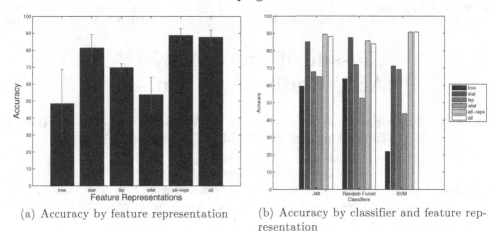

(a) Accuracy by feature representation   (b) Accuracy by classifier and feature representation

Figure 1. Classification accuracy.

Machine and a concatenation of different feature representations. These results and others will be published as a VAST Memo in early 2012.

2. Classifiers and Light-Curve Characterizations

We selected standard classification algorithms that have proved successful in other light-curve classification tasks (Richards *et al.* 2011 and Wachman *et al.* 2009). Specifically, we evaluated Support Vector Machine (SVM; Cortes & Vapnik 1995), Decision Tree (J48; Quinlan 1986), and Random Forest (Breiman 2001) classifiers using implementations provided by the Weka data mining package (Hall *et al.* 2009). We have also worked with other types of classifiers, including probabilistic classifiers such as Naive Bayes and Logistic Regression. However, we found that SVMs, Decision Trees and Random Forests produced superior results on these data.

Machine-learning methods for classification presume the existence of a structured data set, where each example is a vector of "features." Real light-curve observations may not meet this requirement because light curves may contain different numbers of observations taken at differing sampling rates. Thus, in a real setting, one must create representations of the data that meet these requirements.

Our first feature set extracts statistics from the flux measurements of each light curve. These are a subset of the "non-periodic statistical features" used by Richards *et al.* (2011) and include moment statistics (e.g., mean, standard deviation, skew, kurtosis), flux percentile ratios, and shape statistics. We refer to this feature set as *stat*. Our second feature set, *lsp*, extracts coefficients from the Lomb-Scargle Periodogram representation of each light curve (Scargle 1982). We actually extract power information from the top 20 frequencies. Our third feature set, *wlet*, extracts wavelet coefficients using the discrete wavelet transform (DWT).

From the original time-domain observations, the statistical and two frequency space characterizations, we create the following six feature sets for our experiments: *tme* (time-domain flux measurements), *stat, lsp, wlet, all-reps* (concatenation of stat, lsp, and wlet), and *all* (concatenation of *tme, stat, lsp* and *wlet*).

3. Experimental Setup and Results

We simulated 200 400-day light curves per source type at signal-to-noise ratios (SNR) of 3, 5, 7 and 10 (each is a unit of standard deviation). For source types Intra-day Variables and Extreme Scattering Events, SNR is defined in relation to the source's quiescent flux. For all other (transient) source types, SNR is defined with respect to the source's peak flux. For transient source types, the event occurs at time 0.

Our first result, in Fig. 1, shows accuracy by feature representation, averaged across all other parameters (SNR and classifier). We measure accuracy using 10-fold cross validation. The results show that the time domain observations alone yield the weakest performance on average, and combining the feature representations (*all* and *all-reps*) yields the best performance.

Fig. 1 shows accuracy per classifier and feature, averaged across SNR. SVM seems to have the largest variability in performance, recording the lowest performance for the *tme* feature, but the highest performance for the *all-reps* feature set. For the higher-performing *all-reps* and *all* feature representations, the three classifiers perform similarly. We conclude that feature representation more strongly informs performance than classifier selection.

4. Conclusions

These results are part of ongoing work to estimate the classification performance of VAST data prior to the arrival of commissioning data from ASKAP's BETA. We have also studied the impact of different VAST observational strategies, and estimated classification performance per source type. We plan to publish those results along with the results in this paper in a VAST Memo in early 2012. Our future plans are to refine our methods and feature representations in order to optimize classification performance in the archival setting.

5. Acknowledgements

This work was carried out in part at the Jet Propulsion Laboratory, California Institute of Technology, under contract with the National Aeronautics and Space Administration. Government sponsorship is acknowledged.

References

Breiman, L. 2001, *Machine Learning*, 45, p. 5
Cortes, C. & Vapnik, V. 1995, *Machine Learning* 20, p. 273
Hall, M., *et al.* 2009, *SIGKDD Explorations* 11, 1
Murphy, T. & Chatterjee, S. 2009,
    http://www.physics.usyd.edu.au/sifa/vast/index.php/Main/Documents
Quinlan, J. R. 1986, *Machine Learning*, 1, 1
Richards, J. W., *et al.* 2011, *ApJ* 733, 1
Scargle, J. D. 1982, *ApJ* 263, p. 835
Wachman, G., Khardon, R., Protopapas, P., & Alcock, C. R. 2009, in: W.L. Buntine, M. Grobelnik, D. Mladenic & J. Shawe-Taylor (eds.), *Kernels for Periodic Time Series Arising in Astronomy*, Proc. ECML (Bled, Slovenia), p. 489

New Horizons in Time-Domain Astronomy
Proceedings IAU Symposium No. 285, 2011
R.E.M. Griffin, R.J. Hanisch & R. Seaman, eds.

© International Astronomical Union 2012
doi:10.1017/S1743921312001202

The Importance of Timing Metadata

Arnold H. Rots

Smithsonian Astrophysical Observatory, Cambridge, MA 02138, USA.
email: arots@head.cfa.harvard.edu

Abstract. We emphasize the crucial importance for authors and researchers to attach good and adequate metadata to their time-domain data. We provide pointers to existing and emerging standards, and provide guidance for labelling time in publications.

Keywords. Time

1. When Time is of the Essence

When accurate timing really counts it is crucial to have good, complete and self-consistent metadata. Considering that often one cannot predict unambiguously how data will be used in the future, it is good practice and proper stewardship to attach good metadata to all data objects. To do so when the objects are created is really not a lot of trouble, and it makes life much easier later on. It will also make the data much more useful:

> if you want your data to be used 1, 2, 5, 10, 20, 50, 100 years from now, attach all the metadata you can, as accurately as you know them

There are three metadata items that simply have to be there:
- **Time-Scale**: TCB, TCG, TDB, TT, TAI, GPS, UTC ...
- **Observation Location**: The precise location from where the observation was made
- **Time Reference Location**: Location to which the (arrival) time is referenced, such as Geocenter, Barycenter, Topocenter (i.e. the Observation Location)

There are three more items that may be required:
- **Time Zero Point**, in cases where time is recorded in the data as relative (or elapsed) time (such as a standard reference point for a mission, or the start of a time-series)
- **Time Reference Direction**, where the Time Reference Position is not the Observatory Location (Topocenter), in order to reconstruct path-length compensation
- **Solar System Ephemeris**, for cases in which the Time Reference Position, Observatory Location, and/or Time Scales are not tied to the earth (e.g., Barycenter)

Finally, there are two metadata items which may be implied by the Solar System Ephemeris or may need to be supplied separately:

- **State vectors** of Observatory Location and/or Time Reference Position
- **Gravitational potential** at Observatory Location and/or Time Reference Position

Concerning the Time-Scales, TT and TCG are earth-bound, while TDB and TCG are tied to the solar-system barycenter; TT and TDB are dynamical times, while TCG and TCB are co-ordinate times. TT, TAI, and GPS run synchronously at constant offsets from each other (TT–TAI = 32.184 sec; TAI–GPS = 19 sec), while UTC is offset by leap seconds (currently 34) from TAI to keep it within 0.9 sec of UT1. For further details on Time-Scales, see (for instance) Wallace (2011) or Rots (2010).

2. Standards for these Metadata

There are two standards that are, or will soon be, available to the astronomical community to guide data publishers in making proper metadata available:

2.1. *IVOA*

Time metadata standards for the Virtual Observatory are included in the standard for Space-Time Coordinate metadata (STC): http://ivoa.net/Documents/latest/STC.html

2.2. *FITS*

A FITS World Coordinate System paper (FITS WCS V) on Time is in preparation; the authors are Rots, Bunclark, Calabretta, Allen, Manchester & Thompson. A second draft was circulated in October 2011.

3. Proper Labelling of Time in Publications

It is vitally important to exercise care when referring to time in publications, especially in the labelling of axes in figures. There are three officially accepted ways to denote time: an ISO-8601 string (yyyy-mm-ddThh:mm:ss, where decimals may be added to the seconds field), Julian Date (JD), and Modified Julian Date (MJD, where MJD = JD−2400000.5). Note that none of these implies any particular Time-Scale or any particular Time Reference Position; that information needs to be provided separately. We strongly recommend labels such as

JD 2452300.487 (TT; Geocenter)
MJD 53245.964 (TDB; Barycenter)
2011-10-10T11:56:34 (UTC; Topocenter)

The above constitute an extension (by adding the Time Reference Position) of the IAU's recommended notation. What should be avoided at all cost are creative but utterly confusing labels like "BJD" or "BJD-240000". The former case appears to indicate a time, somehow related to the barycenter, but it is not clear whether it represents TDB, or just TT or UTC reduced to the barycenter. The latter case is even more confusing still since it leaves the reader wondering whether the author literally meant "2400000" or forgot to add the half day to it. Please use standard time designations; don't invent your own.

Acknowledgement

This work has been supported by NASA under contract NAS 8-03060 to the Smithsonian Astrophysical Observatory for operation of the Chandra X-ray Center.

References

Rots, A. H. 2010, *ASP-CS*, 434, 107
Wallace, P. T. 2011, *Metrologia*, 48, S200

New Horizons in Time-Domain Astronomy
Proceedings IAU Symposium No. 285, 2011
R.E.M. Griffin, R.J. Hanisch & R. Seaman, eds.

© International Astronomical Union 2012
doi:10.1017/S1743921312001214

Using the Gregory-Loredo Algorithm for the Detection of Variability in the Chandra Source Catalog

Arnold H. Rots

Smithsonian Astrophysical Observatory, Cambridge, MA 02138, USA.
email: arots@head.cfa.harvard.edu

Abstract. In searching for a reliable variability indicator for Chandra X-ray event data, we implemented the Gregory-Loredo (G–L) algorithm to assess variability and to generate light curves for the Chandra Source Catalog. A test was performed on 118 sources detected in a single observation, spanning the intensity range 5–24,000 photons over a total observing time of 102,000 sec. We conclude that the G–L algorithm is extremely robust, yielding both a reliable variability indicator and light curves with optimal resolution, while requiring a very modest amount of CPU time. The algorithm can also work on binned data, and is capable of handling data gaps as well as variations in effective area.

Keywords. Methods: statistical, catalogs, X-rays: general

1. Introduction

Determining whether a source of discrete (X-ray) photons exhibits variability is a nontrivial problem. We have tested the algorithm published by Gregory & Loredo (1992), modified for aperiodic variability, for use in the production of the Chandra Source Catalog (CSC; Evans *et al.* 2010). In brief, N events are binned in histograms of m bins, where m runs from 2 to m_{max}. The algorithm is based on the likelihood of the observed distribution $n_1, n_2, ..., n_m$ occurring. Out of a total number of mN possible distributions, the multiplicity of this particular one is $N!/(n_1!.n_2!.....n_m!)$. The ratio of the latter to the former provides the probability that this distribution came about by chance. Hence, its inverse is a measure of the significance of the distribution. In that way we calculate an odds ratio for m bins versus a flat light curve. The odds are summed over all values of m to determine the odds that the source is time-variable. For more details, see Gregory & Loredo (1992) and Loredo (p. 87).

2. Evaluation and Usage

We tested the algorithm on all 118 sources found by the programme WAVDETECT in Chandra ObsId 635. The total time-span of the observation was 102 ks, and the sources varied between 5 and 24000 counts. The average time to run the programme was 1.5σ per source. 71 sources were found to be variable with an odds ratio > 1.0 (probability > 0.5). Visual inspection of the light curves of all 118 sources found 54 that are variable, though there were a few borderline cases on either side of the divide.

Our tests proved that the method works very well on event data, and is capable of dealing with data gaps. We added a capability of taking into account temporal variations in effective area. As a byproduct, a light curve with optimal resolution is delivered. That light curve is effectively the sum of the binnings weighed by their odds ratios, and as such it represents the most optimal binning for the curve. The standard deviation σ is provided for each point of the light curve.

The odds ratio yields a probability for variability. The tests showed that there is an ambiguous range of probabilities: $0.5 < P < 0.9$, and in particular the range between 0.5 and 0.67 (above 0.9 all is variable, below 0.5 all is non-variable). For that range we have developed a secondary criterion based on the light curve, its average σ and the average count rate. We calculated the fractions f_3 and f_5 of the light curve that are within 3σ and 5σ, respectively, of the average count rate. If $f_3 > 0.997$ **and** $f_5 = 1.0$ for cases in the ambiguous range, the source is deemed to be non-variable.

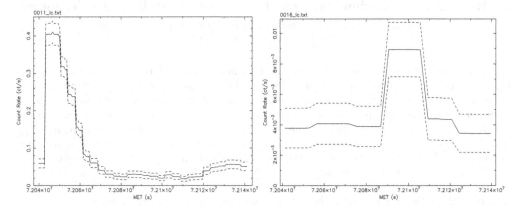

Figure 1. Two examples of light curves produced by our implementation, derived from sources with 8697 counts (left) and 14 counts (right). The dashed lines indicate the 3σ levels.

3. Conclusion

G–L provides a robust algorithm for detecting temporal variability which is insensitive to the type and shape of the variability and which takes properly into account the uncertainties in the count rate, requiring a statistically significant departure from a flat count rate for it to declare variability. The light curves provided by the programme appear to be near optimal for what we intended to present to users.

The addition of the secondary criterion results in a reliable test, though careful users may want to inspect the light curves of all sources with a non-zero variability index.

The programme is integrated into the production pipeline of the CSC. Fig. 1 provides two examples of light curves that are produced. Further information may be found at

http://cxc.cfa.harvard.edu/csc/
http://cxc.cfa.harvard.edu/csc/why/gregory_loredo.html
http://hea-www.harvard.edu/~arots/GL/

Acknowledgement

This work has been supported by NASA under contract NAS 8-03060 to the Smithsonian Astrophysical Observatory for operation of the Chandra X-ray Center.

References

Rots, A. H., *et al.* 2010, *ApJS*, 189, 37
Gregory, P. C. & Loredo, T. J. 1992, *ApJ*, 398, 146

New Horizons in Time-Domain Astronomy
Proceedings IAU Symposium No. 285, 2011
R.E.M. Griffin, R.J. Hanisch & R. Seaman, eds.

© International Astronomical Union 2012
doi:10.1017/S1743921312001226

On Our Multi-Wavelength Campaign of the 2011 Outburst of T Pyx†

L. Schmidtobreick, A. Bayo, Y. Momany, V. Ivanov, D. Barria,
Y. Beletsky, H. M. J. Boffin, G. Brammer, G. Carraro, W.-J. de Wit,
J. Girard , G. Hau, M. Moerchen, D. Nuernberger, M. Pretorius,
T. Rivinius, R. Sanchez-Janssen, F. Selman, S. Stefl, and I. Yegorova

European Southern Observatory, Casilla 19001, Santiago 19, Chile
email: lschmidt@eso.org

Abstract. The well-known recurrent nova T Pyx has brightened by 7 magnitudes, starting on 2011 April 14, its first eruption since 1966. T Pyx is unique amongst recurrent novæ in being surrounded by a nebula formed of material ejected during previous eruptions. The latest eruption therefore offers the rare opportunity to observe a light echo sweeping through the existing shell, and a new one forming. The sudden exposure of the existing shell to high-energy light is expected to result in a change of the dust morphology as well as in the part destruction of molecules. We observe this process in the near- and mid-IR during several epochs using ESO's VLT instruments SINFONI, VISIR and ISAAC. Unfortunately, in the data analysed so far we only have a tentative detection in Brα from the shell, so might in the end have to be content with upper limits for the emission from the various molecular bands and ionised lines.

Keywords. novae, ISM: molecules, dust

---

1. Introduction

Recurrent novæ (RNe) are a class of cataclysmic variables that erupt at intervals of several decades. The eruptions are caused by a thermonuclear runaway (TNR) on the surface of a white dwarf accreting hydrogen-rich matter from a close companion star. The same mechanism is thought to be responsible for classical nova eruptions, whose recurrence times are, however, of the order of thousands of years instead of decades (Patterson 1984; Shara *et al.* 1986).

T Pyx—a recurrent nova—had five previous recorded eruptions (in 1890, 1902, 1920, 1944, 1966), and started the sixth on 2011 April 14. The accretion rate of the white dwarf is at least 100 times higher than expected for its orbital period (Uthas *et al.* 2010) and it is suggested that it has a white-dwarf mass close to the Chandrasekhar limit, thus making it a particularly good candidate as a type Ia SN progenitor (see also Knigge *et al.* 2000). It is surrounded by at least five dust and gas shells that were expelled during previous eruptions. Part of the nova remnant has been imaged both from the ground (where it appears as a roughly spherical shell) and with *HST*, where it is resolved into many individual knots (Shara *et al.* 1997; Schaefer *et al.* 2010).

2. Data

Although the shell is mostly invisible in the optical and near-IR continuum, we expect to find emission in certain frequencies in the IR centred on ionised lines. To our knowledge, none of the known nova shells has ever been observed in the mid-IR, but reference data

† Based on data of the UT3/UT4-team observatory project with program ID 287.D-5021.

from similar objects like SN 1987A and V838 Mon indicate the formation of molecules and dust. Furthermore, the formation of dust has been observed during nova outbursts themselves: see, e.g., Nielbock & Schmidtobreick (2003) for the outburst of V4745 Sgr. As the ionisation front from the current eruption sweeps through the shell we expect it to brighten in the UV and optical regions, amd in the Balmer lines and lines of ionised metals. The evolution of the shell is being monitored with the ESO's VLT instruments VISIR, SINFONI and ISAAC.

2.1. VISIR

Imaging for our first epoch was carried out in early May in PAH1, PAH2, Ne II, Ar III and S IV filters. The central source (T Pyx) was clearly visible in all regions. Several faint features are present in the surroundings but a careful analysis showed them all to be detector artifacts. So far, no detection of the shell itself can be claimed in any of the observations.

2.2. SINFONI

SINFONI integral-field spectroscopy was carried out at two epochs, the first directly after the nova outburst and the second about two months later. Even after a careful reduction of the data and a thorough search through the spectra we did not find any emission that could be attributed unambiguously to a shell around T Pyx. The images are instead dominated by stray light and remnants of the central source reaching out as far as 7″, and thus make a clear distinction of the shell impossible.

2.3. ISAAC

ISAAC LW imaging was carried out with two narrow-band filters centred on $4.07 \, \mu m$ (Brα) and $3.80 \, \mu m$ (H$_2$), and with the broad-band L filter. So far only one epoch (directly after the nova explosion) has been reduced and analysed. The central source is present in all three filters. We have a tentative detection of the brighter parts of the shell in Brα. The shell was not detected in the L-band continuum or in molecular hydrogen.

3. Conclusion

It appears that the old shell is still very weak in the early epochs after the nova outburst. That is to be expected, since the ionising light of the new explosion has not yet reached the old shell. Depending on which distance is adopted for T Pyx, the light will take between 8 and 20 weeks to reach the ring of material at a distance of about 5″. The data taken at later epochs have not yet been analysed, but we are working on them and will discuss the new information together with the non-detections at early epochs in a forthcoming paper.

References

Knigge, C., King, A. R., & Patterson, J. 2000, *A&A*, 364, 75
Nielbock, M. & Schmidtobreick, L. 2003, *A&A*, 400, 5
Patterson, J. 1984, *ApJS*, 54, 533
Schaefer, B., Pagnotta, A., & Shara, M. 2010, *ApJ*, 708, 381
Shara, M., Zurek, D. R., & Williams, R. E., *et al.* 1997, *ApJ*, 114, 258
Shara, M., Livio, M., Moffat, A. F. J., & Orio, M. 1986, *ApJ* 311, 163
Uthas, H., Knigge, C., & Steeghs, D. 2010, *MNRAS* 409, 237

New Horizons in Time-Domain Astronomy
Proceedings IAU Symposium No. 285, 2011
R.E.M. Griffin, R.J. Hanisch & R. Seaman, eds.
© International Astronomical Union 2012
doi:10.1017/S1743921312001238

Multi-wave Monitoring of the Gravitational Lensed Quasar Q0957+561

Vyacheslav N. Shalyapin[1,2], Luis J. Goicoechea[1] and Rodrigo Gil-Merino[1]

[1]Universidad de Cantabria, 39005 Santander, Spain
email: vshal@ire.kharkov.ua

[2]Institute for Radiophysics and Electronics, 61085 Kharkov, Ukraine

Abstract. We present X-ray (CHANDRA), ultraviolet (SWIFT/UVOT; U band) and optical-infrared (Liverpool Telescope; $griz$ bands) continuum light curves of Q0957+561 observed in the first half of 2010. A cross-correlation analysis of the light curves shows that the U-band fluctuation leads the other variations at higher and lower energies. The study constrains the geometry of the continuum emission regions in a distant radio-loud AGN for the first time. We note that our work opens a new window in echo-mapping of high-z AGNs with the use of lensed quasars, since the variability of some of the images of a given multiply-imaged quasar can be predicted in advance, provided there is a modest optical follow-up of the system.

Keywords. Gravitational lensing: strong, black hole physics, quasars: individual (Q0957+561).

1. Motivation

Optical monitoring of the double imaged quasar Q0957+561 ($z = 1.41$) enabled us to detect a very prominent increase in the brightness of the leading 'A' image between late 2008 and the middle of 2009. From the observed delay of ∼14 months between the two images, we predicted successfully that a repetition of the event would occur in the trailing 'B' image in the first half of 2010 (Goicoechea & Shalyapin 2009).

2. Observations

A multi-wavelength monitoring campaign was organized in the first half of 2010 as a ground-based and space-based project, and involved X-ray, NUV, optical and NIR facilities. We obtained 6 light curves covering a large part of the electromagnetic spectrum:

- X-ray, 2–10 KeV CHANDRA/ACIS, 12 exposures, each for 3000 secs;
- U-band, SWIFT/UVOT, 35 exposures for 155–1092 secs;
- $griz$-bands, Liverpool Telescope: 55, 58, 50 and 45 exposures respectively per band, each for 120 secs.

The prominent flash of the 'A' image (2008/2009) is indeed repeated in the 'B' image in 2010. This fact demonstrates the intrinsic origin of the variability. The amplitudes of the variations significantly exceed their photometric errors. Moreover, the amplitude of the variability increases toward shorter wavelengths (higher energies).

In addition, we also detect, by chance, a clear U-shaped variation in image 'A'. It has a signal-to-noise ratio that is lower than the very prominent fluctuations in 'B' (Fig. 1).

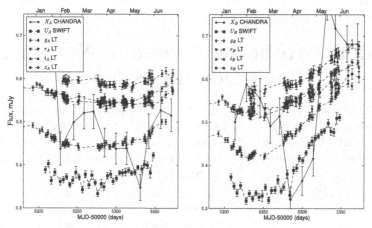

Figure 1. Variability in the 'A' (left) and 'B' (right) images of Q0957+561 in 2010.

3. NUV/Optical/NIR Inter-Band Time Delays

From the large variability of the 'B' image in all bands we can estimate accurate inter-band time delays. The variability of the 'A' image can also be used to check our estimates from the 'B' image data. We use both the D^2 and χ^2 methods (Shalyapin *et al.* 2008 and references therein) to estimate time delays with respect to the g-band record (1944 Å in the source rest-frame). The inter-band delays from 'A' agree with the delays from 'B'. As expected, the delays based on 'A' data have larger errors. The deviations between the two delay curves from 'A' and 'B' deserve more attention. The NUV fluctuations in 'A' and 'B' lead the variations at other wavelengths. Our measurements suggest the presence of a central driving source that produces delays $\tau(\lambda) \sim \lambda^{4/3}$ for standard accretion disk rings around the central supermassive black hole (Collier *et al.* 1999). The driving source would emit at $\lambda < 1438$ Å (source rest-frame); its flares would be reprocessed in the disk to trigger the observed NUV/optical/NIR variations.

4. Origin of X-Ray Emission

There is a ~ 32-day delay between the X-ray radiation and the U-band photons, so we can reject a standard corona-disk radiative coupling in which the disk variability is driven by a corona just above the black hole (Kazanas & Nayakshin 2001). That result is also supported by X-ray reprocessing simulations and by the absence of X-ray reflection features in the spectrum of 0957+561. The X-ray emitting region seems to be at the base of the Q0957+561 jet, at a height of $\sim 200 R_g$, or 0.05 pc above the accretion disk. A central EUV source would drive the variability of Q0957+561 (Gil-Meruno *et al.* 2011).

References

Collier, S., Horne, K., Wanders, I., & Peterson, B. M. 1999, *MNRAS*, 302, L24

Gil-Merino, R., Goicoechea, L. J., Shalyapin, V. N., & Braga, V. F. 2011, *ApJ*, in press (arXiv:1109.3330)

Goicoechea, L. J. & Shalyapin, V. N. 2009, *The Astronomer's Telegram*, no. 2228

Kazanas, D. & Nayakshin, S. 2001, *ApJ*, 550, 655

Shalyapin, V. N., Goicoechea, L. J., Koptelova, E., Ullán, A., & Gil-Merino, R. 2008, *A&A*, 492, 401

New Horizons in Time-Domain Astronomy
Proceedings IAU Symposium No. 285, 2011
R.E.M. Grifin, R.J.. Hanisch & R. Seaman, eds.

© International Astronomical Union 2012
doi:10.1017/S174392131200124X

A Global Robotic Telescope Network for Time-Domain Science

R. A. Street[1] and T. A. Lister[1], Y. Tsapras[1,2], A. Shporer[1], F. B. Bianco[1], B. J. Fulton[1], D. A. Howell[1], B. Dilday[1], M. Graham[1], D. Sand[1], J. Parent[1], T. Brown[1], K. Horne[3], M. Dominik[3]†, P. Browne[3], C. Snodgrass[4], N. Kains[5], D. Bramich[5], N. Law[6] and I. Steele[7]

[1]LCOGT, 6740 Cortona Drive, Suite 102, Goleta, CA 93117, USA,
email: rstreet@lcogt.net

[2]School of Mathematical Sciences, Queen Mary College, London E1 4NS, UK

[3]SUPA, School of Physics & Astronomy, University of St. Andrews, Fife, KY16 9SS, UK,

[4]Max Planck Institute for Solar System Research, 37191 Katlenburg-Lindau, Germany

[5]ESO, 85748 Garching bei München, Germany

[6]Dept. of Astronomy and Astrophysics, University of Toronto, Ontario, M5S 3H4, Canada

[7]Astrophysics Research Institute, Liverpool JMU, Birkenhead, CH41 1LD, UK.

Abstract. Las Cumbres Observatory Global Telescope Network (LCOGT) is currently building a new kind of general-purpose astronomical facility: a fully robotic network of telescopes of 2m, 1m and 0.4m apertures and homogeneous instrumentation. A pan-network approach to scheduling (rather than per individual telescope) offers redundancy in the event of poor weather or technical failure, as well as the ability to observe a target around the clock. Here we describe the network design and instrumentation under development, together with the main science programmes already being lead by LCOGT staff.

Keywords. telescopes, instrumentation: photometers, instrumentation: spectrographs,stars: planetary systems, stars: supernovae

1. Introduction

At any given location, ground-based observations of time-domain phenomena suffer from gaps in their coverage caused by the diurnal cycle. The gaps can be supplemented with data from longitudinally-spaced sites, but at the cost of combining data from different instruments taken under very different conditions. While a number of such telescope networks like GONG and BiSON have been built to date, most have been designed for, and operated by, specific projects, with limited instrumentation. The Las Cumbres Observatory Global Telescope Network (LCOGT) will be a new, general-purpose facility for optical/NIR photometry and spectroscopy, which we describe in Sections 2 & 3. Designed from the outset for time-domain astronomy, the main science drivers for LCOGT staff are exoplanet and supernova science; we highlight these programmes in Sections 4 & 5.

2. Robotic Observing Network

Telescopes in the LCOGT network will be distributed longitudinally, at ∼6 sites in both hemispheres, making it possible to observe around the clock. All telescopes will

† Royal Society University Research Fellow

operate robotically as a cohesive whole, taking advantage of different sites to compensate for weather losses elsewhere. The network will incorporate 3 classes of telescope aperture: 2m, 1m and 0.4m. Each site will consist of a cluster of 2 or 3 1-m telescopes accompanied by about 3 0.4-m telescopes. There will be at least 6 sites. The "Southern Ring" includes Siding Spring (Australia), the South African Astronomical Observatory and the Cerro Tololo Inter-American Observatory (Chile), while the "Northern Ring" will link sites at Maui (Hawai'i), Sedgwick (California), McDonald (Texas), Tenerife (Canary Islands) and a site in Asia yet to be selected. In addition to the 1-m and 0.4-m instruments we operate the existing two 2-m Faulkes Telescopes North (Hawai'i) and South (Australia). All telescopes of a given aperture class in the network will have an homogeneous instrument set and a consistent set of filters and calibration facilities. At present we have imaging cameras available at the Faulkes Telescopes; our plans for new instrumentation are described in Section 3. In partnership with SUPA/St. Andrews, LCOGT will be deploying its Southern Ring facilities first; construction is underway at CTIO and SAAO.

3. New Instrumentation

Our standard imaging instrument for the 1-m telescopes will be a SiNiSTRO camera, which consists of a Fairchild Imaging CCD 486 Bl plus an LCOGT-designed controller. It will have a field of view (FOV) of $\sim26'$ with a pixel scale of $\sim0''.4$/pixel.

To explore variable phenomena at very short time-scales by diffraction-limited imaging we are currently commissioning Lucky Imager, High Speed Photometer (LIHSP) cameras for the 2-m telescopes. They house an Andor iXon+ DU888, 1k × 1k back-illuminated frame-transfer EMCCD, and offer a FOV of up to $2'.2 \times 2'.2$ and observing cadences as fast as several tens of Hz. The instruments will be exploited primarily for microlensing-event follow-up (Dominik 2010) and occultations, and are currently being tested. We will shortly begin testing an Andor Neo 2560 × 2160 sCMOS detector on a 1-m telescope, off-axis port; it has $0''.19$ pixels, a FOV of $360'' \times 416''$ ($\sim6' \times 7'$), and is capable of 100 fps at full frame.

The need for globally-networked spectographic facilities is even more acute than for imaging. Accordingly, LCOGT is developing spectrographs for our 2-m and 1-m telescopes. FLOYDS (Folded Low Order whYte-pupil Double-dispersed Spectrograph) will produce low-resolution ($R\sim300$–500) spectra between $\lambda\,350$–1100 nm once mounted on the 2-m telescopes. The science drivers include supernova classification, GRB follow-ups and AGN reverberation mapping.

Our 1-m telescopes will be equipped with MRES, our medium-resolution ($R\sim25000$) spectrograph which will cover a wavelength range of ~390–860 nm and aims for a radial-velocity precision of 0.5 km s^{-1} for a $V = 15$ G0 star with $v\sin i < 3$ km s^{-1}.

4. Exoplanet Science

LCOGT runs a number of programmes to confirm exoplanet discoveries and characterise the systems. As members of the Palomar Transient Factory (PTF), and with team members collaborating with the WASP, Qatar Exoplanet Survey, HATNet and the CoRoT, Kepler teams, we have provided targeted high-precision photometry of discoveries from those surveys, contributing to the growing catalogue of exoplanets (Street *et al.* 2010; Lister *et al.* 2009). In targets of special interest we monitor transit timings long-term to search for variations (Fulton *et al.* 2011), and characterise long period/rare transit events such as those of HD 80606 b (Hidas *et al.* 2010; Shporer *et al.* 2010a).

We operate the only fully-robotic follow-up programme for microlensing events, RoboNet-II (Tsapras *et al.* 2009), combining our own facilities with those of the Liverpool Telescope, Canary Islands. This is a highly challenging programme: the ~1500 events discovered throughout a 6-month season have to be prioritised dynamically as new data come in, and trigger observations in response to light-curve anomalies which betray the presence of planets. However, the planets are worth the effort, as this is the fastest way to find cool exoplanets in the outer reaches of their systems (Muraki 2011; Sumi *et al.* 2010).

5. Supernova Science

The supernova program at LCOGT obtains light curves of SNe discovered at the PTF and Pan-STARRS–1. Follow-up spectroscopy is obtained elsewhere, including Lick, Keck, Gemini and VLT, and (soon) with the low-resolution FLOYDS spectrographs on the Faulkes Telescopes. The SN programme includes:

- building a new sample of low-redshift SNe Ia to anchor the Hubble diagram measuring Dark Energy (Sullivan *et al.* 2011)
- investigating the progenitors of SNe Ia (Bianco *et al.* 2011)
- developing a better understanding of SNe Ia as cosmological probes (Conley *et al.* 2011; Cooke *et al.* 2011)
- following new classes of transients such as those too faint (Kasliwal *et al.* 2010) or too bright (Quimby *et al.* 2011) to fit into normal SN paradigms.

6. Additional Science

In addition to the main science programmes, LCOGT scientists collaborate widely and support a variety of other scientific programmes ranging from tracking Near-Earth Asteroids discovered by Pan-STARRS (resulting in 66 MPECS in 2011 to date) to the detection of the first detached double white-dwarf eclipsing binary (Steinfadt *et al.* 2010) and a ground-based detection of the beaming effect in that system (Shporer *et al.* 2010b).

References

Bianco, F. B., *et al.* 2011, *ApJ*, 741, 20
Conley, A., *et al.* 2011, *ApJS*, 192, 1
Cooke, J., *et al.* 2011, *ApJ* (Letters), 727, L35
Corsi, A., *et al.* 2011, arXiv:1101.4208
Dominik, M., 2010, *General Relativity & Gravitation*, 42, 2075.
Fulton, B.J., *et al.* 2011, *AJ*, 142, 84.
Hidas, M. G., Tsapras, Y., Mislis, D., *et al.*, 2010, *MNRAS*, 406, 1146.
Kasliwal, M. M., *et al.* 2010, *ApJ* (Letters), 723, L98
Lister, T. A., *et al.* 2009, *ApJ*, 703, 752.
Muraki, Y., Han, C., & Bennett, D. P. 2011, *ApJ*, 741, 22.
Quimby, R. M., *et al.* 2011, *Nature*, 474, 487
Shporer, A., *et al.* 2010, *ApJ*, 722, 880.
Shporer, A., *et al.* 2010,*ApJ* (Letters), 725, L200.
Street, R. A., *et al.* 2010, *ApJ*, 720, 337.
Steinfadt, J.D.R., *et al.* 2010, *ApJ* (Letters), 716, L146.
Sullivan, M., *et al.* 2011, *ApJ*, 737, 102
Sumi, T., *et al.* 2010, *ApJ*, 710, 1641.
Tsapras, Y., *et al.* 2009, *AN*, 330, 4.

New Horizons in Time-Domain Astronomy
Proceedings IAU Symposium No. 285, 2011
R.E.M. Griffin, R.J. Hanisch & R. Seaman, eds.

© International Astronomical Union 2012
doi:10.1017/S1743921312001251

FRATs: Searching for Fast Radio Transients in Real Time with LOFAR

S. ter Veen, P. Schellart, and H. Falcke for the LOFAR Transients and Cosmic Ray Key Science Projects

Department of Astrophysics, IMAPP, Radboud University Nijmegen, 6500 GL Nijmegen, NL
email: s.terveen@astro.ru.nl

Abstract. The aim of the FRATs project is to detect single dispersed pulses from Fast Radio Transients with LOFAR in real time. The pulses can originate from pulsars, RRATS and other classes of known or unknown objects. To detect the pulses a trigger algorithm is run on an incoherent beam from the different LOFAR stations. The beam has a wide field of view and can be formed parallel to other observations. A precise localisation is achieved by storing and processing off-line the data from each dipole, giving all-sky coverage with a spatial resolution of the order of arc-seconds. The source is identified by making high-time-resolution images. The method has been tested by detecting and identifying a giant pulse from the Crab pulsar.

Keywords. techniques: interferometric, (stars:) pulsars: general

1. Introduction

The most well-known radio transients are pulsars. Detecting more pulsars will assist in studies of the neutron-star population in the Galaxy and in globular clusters, and in understanding the emission mechanism of these extreme objects. In the pioneering years pulsars were found by looking for single pulses, but later on periodicity searches were usually used because they offered higher sensitivity. Single-pulse searches have once again become popular, and have led to the discovery of Rotating Radio Transients (RRATS) (McLaughlin *et al.* 2006)—pulses that repeat only occasionally. What objects give those sporadic pulses, and how many more are there? If the pulses originate from pulsars they can be giant pulses from distant normal pulsars, or can originate in almost dead, nearby pulsars that only emit rarely. Other suggestions regarding the origin of sporadic pulses include exoplanets with an emission mechanism similar to the Jupiter-IO system. To hunt for those (and for any other kind of pulses) we developed a real-time detection mode for LOFAR (the LOw Frequency Array; see Stappers *et al.* 2011).

2. Method

The difficulty of detecting strong single pulses is that they occur only occasionally, so a large instantaneous sky coverage and observing time are therefore required. Furthermore, when only a single pulse is found, one wants to be absolutely sure that it is from an astrophysical source and not a terrestrial one. For example, it is still undecided what was the origin of the Lorimer burst (Lorimer *et al.* 2007). A precise position and correction for dispersion are also required. We explain how LOFAR meets those requirements within the Fast Radio Transients (FRATs) project.

2.1. *LOFAR*

LOFAR is a new-generation distributed digital radio telescope. Instead of using large dishes, it uses stations with fields of small dual polarization antennæ. Most of those stations are located in the Netherlands, but there are also ones in Germany, France, the UK and Sweden. There are two types of antennæ: Low Band Antennæ (10–90 MHz) and High Band Antennæ (110–240 MHz). The stations are placed on base-lines from 100 m to 1000 km. There are three observating modes: imaging, beam-forming, and storing the data per dipole. Data are stored in the memory ring-buffers of the Transient Buffer Boards (TBBs) at a sample time of 5 ns. Currently 1.3 seconds of raw data can be stored, but an increase to 5.2 seconds is planned. Buffer-time can be traded for bandwidth or, by storing fewer dipoles, for sensitivity. If an interesting event is found, the buffers can be stopped and read out, partially or entirely.

The beam-forming mode corrects for the delays between the different receiving elements, in the first stage in each station and in the second stage between the stations. The latter can be done either coherently (added in phase, small field of view (FoV), high sensitivity) or incoherently (added in power, large FoV (10–18 deg.^{2} @150 MHz), lower sensitivity). The total bandwidth in this mode is 48 MHz, but it can be split for observing in many different directions simultaneously. The beam-formed data consist of many channels with a high frequency- and time-resolution (700 Hz–12 kHz; 0.08–1.3 ms).

2.2. *Real-time detection*

To detect pulses we use the incoherently summed beam-formed data from all active stations, taken in parallel to an imaging observation. That meets the requirements of a large FoV and a long observation time. The trigger algorithm that processes the data first has to correct for dispersion, i.e. the delay caused by the interstellar medium and making signals of a lower frequency arrive later, according to the relation $\Delta t_{DM} = 4.15\,\mathrm{ms}\,\mathrm{DM}(\nu_{1,\mathrm{GHz}}^{-2} - \nu_{2,\mathrm{GHz}}^{-2})$; that is done in several frequency bands (∼MHz bandwidth). Narrow-band radio-frequency interference (RFI) may trigger only one band but a real signal will trigger multiple bands. It corrects for the time delay between the ∼kHz channels in each band for several dispersion measure trials (DM), and checks for peaks using a simple $P > S_{sys} + N * \Delta S$ algorithm, where P is the power of the signal, S_{sys} is calculated by taking the mean, and ΔS is the standard deviation of the de-dispersed data. The algorithm then checks for a coincidence between peaks found in the different bands corresponding to the expected delay for the given DM (Fig. 1). If such an event is found, a trigger message is sent to obtain the TBB data.

2.3. *Offline localisation*

The TBB data are used to find the direction of the pulse; they contain the raw data from each available dipole, and beams can be formed from them in any direction at the full resolution capabilities of LOFAR. They can, for instance, be used to image the whole sky (LBA), or 2000 deg^2 (HBA). Near-field beam-forming is used to discriminate against terrestrial signals (and also to study cosmic-ray air showers and lightning).

To investigate where the pulse came from, incoherent beams are formed in all directions to check whether it originated in the pointing direction or in a sidelobe. Coherent beams are formed in the actual direction of the pulse to find its precise location; a precision of < 1 arcsecond resolution is expected. Fig. 1 shows the de-dispersed data in four frequency bands of a giant pulse for which a trigger was issued and TBB data were obtained. Fig. 2 shows the de-dispersed data of a pulse imaged from those TBB data.

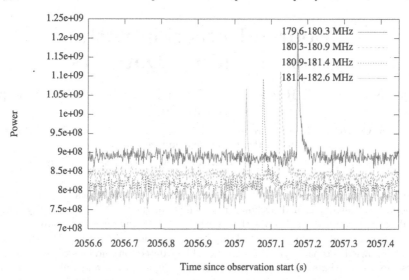

A pulse from the Crab pulsar in 4 frequency bands

Figure 1. A pulse from the Crab pulsar, de-dispersed in each of 4 frequency bands. The time delay between the different pulses is caused by dispersion.

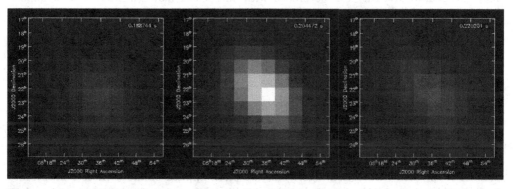

Figure 2. A de-dispersed image in the direction of the Crab pulsar before, during and after the pulse. (Made from the TBB data from only one LOFAR station, hence the inferior resolution).

3. Conclusion

A real-time search for Fast Radio Transients is feasible with LOFAR. By using an incoherent beam, parallel to other observations, large sky and time coverage are obtained. To refine the direction and discriminate against RFI the raw data from each dipole are obtained and analyzed. A prototype of this pipeline has run succesfully, proving the concept and pointing the way to a new horizon in time-domain radio astronomy.

References

McLaughlin, M. A., *et al.* 2006, *Nature*, 439, 817
Stappers, B., *et al.* 2011, *A&A*, 530, A80
Lorimer, D. R., Bailes, M., McLaughlin, M. A., Narkevic, D. J., & Crawford, F. 2007, *Science*, 318, 777

New Horizons in Time-Domain Astronomy
Proceedings IAU Symposium No. 285, 2011
R.E.M. Griffin, R.J. Hanisch, R. Seaman, eds.

© International Astronomical Union 2012
doi:10.1017/S1743921312001263

Source Detection with Interferometric Datasets

Cathryn M. Trott[1,2], Randall B. Wayth[1], Jean-Pierre R. Macquart[1], and Steven J. Tingay[1]

[1] International Centre for Radio Astronomy Research, Curtin University, WA 6845, Australia
email: `cathryn.trott@curtin.edu.au`

[2] ARC Centre of Excellence for All-Sky Astrophysics (CAASTRO)

Abstract. The detection of sources in interferometric radio data typically relies on extracting information from images, formed by Fourier transform of the underlying visibility dataset, and CLEANed of contaminating sidelobes through iterative deconvolution. Variable and transient radio sources span a large range of variability timescales, and their study has the potential to enhance our knowledge of the dynamic universe. Their detection and classification involve large data rates and non-stationary PSFs, commensal observing programs and ambitious science goals, and will demand a paradigm shift in the deployment of next-generation instruments. Optimal source detection and classification in real time requires efficient and automated algorithms. On short time-scales variability can be probed with an optimal matched filter detector applied directly to the visibility dataset. This paper shows the design of such a detector, and some preliminary detection performance results.

Keywords. methods: statistical, techniques: interferometric, radio continuum: general

1. Introduction

Source detection with radio interferometric datasets typically relies on detection of point sources in deconvolved (CLEANed) images, generated by the Fourier Transform of the underlying complex visibility dataset, observed on each base-line and at each spectral channel of the array. Detection algorithms are often applied offline with user input, and are limited by the weighting applied to the visibility data at the time of image generation; an example is SFIND, Hopkins *et al.* (2002).

Systematic study of slowly-varying transient and variable sources (slow transients, with variability time-scales >5 seconds) present an opportunity to explore the universe along a different dimension than has historically been the case (variability time-scale rather than electromagnetic spectrum), with the aim to characterize large samples of object classes, and the potential to discover new classes of objects, based on their variability properties. The next-generation of wide-field, high-sensitivity interferometric arrays have the capability to survey the sky with high cadence, yielding appropriate datasets for detecting and classifying sources varying with time-scales ranging from seconds to months. The real-time, automated detection and classification of slow transient sources from commensal observing requires statistically robust algorithms to extract information optimally, without the benefits of user input and optimal image formation. In addition, one may wish to sample the data at higher rates than the imaging time-scale, to isolate short time-scale variability. Detection of variable sources using the underlying visibility dataset is an option to interrogate the data at higher cadence, and with simpler noise properties, making optimal detection feasible. We explore visibility-based source detection in this note.

2. Visibility-Based Detection: the Matched Filter

For a known signal in multivariate Gaussian noise, the matched filter detector yields optimal detection performance (probability of detection maximized for a given false-positive rate). This detector can be demonstrated to be equivalent to image-based techniques. The matched filter detection performance is given by:

$$\text{SNR} = \frac{<T(x)>}{\sqrt{\text{var}[T(x)]}}, \tag{2.1}$$

where $T(x)$ is the detector test statistic, and is the output of applying the detector. Comparison of the test statistic to a threshold forms the basis of the binary signal-present/signal-absent decision. For a source of strength I_0 located at source position (l_0, m_0) embedded in white Gaussian noise, the visibility space performance has the properties:

$$
\begin{aligned}
<T(x)> &= S^\dagger S \tag{2.2}\\
&= \frac{1}{N_{\text{vis}}} \sum_u \sum_v I_0 \exp{-2\pi i(ul_0 + vm_0)} I_0 \exp{2\pi i(ul_0 + vm_0)} \\
&= I_0^2 N_{\text{vis}} \\
\text{var}[T(x)] &= <(N^T S)^2> \\
&= <S^\dagger N N^T S> \\
&= S^\dagger <N N^t> S \\
&= S^\dagger S \sigma^2/2.
\end{aligned}
$$

Therefore, the signal-to-noise ratio is given by:

$$\text{SNR}_{\text{Vis}} = I_0^2 N_{\text{vis}} / \sqrt{I_0^2 N_{\text{vis}} \sigma^2/2} = I_0 \sqrt{2 N_{\text{vis}}/\sigma^2}. \tag{2.3}$$

Image-space performance relies on pixel-based detection. For a source at (l_0, m_0), a visibility is given by:

$$V(u, v) = I_0 \exp{-2\pi i(ul_0 + vm_0)} + \text{noise}. \tag{2.4}$$

The image is formed by Fourier Transform, $\vec{I} = \mathbf{H}\vec{V}$, where $\mathbf{H}$ is the Fourier operator. Therefore:

$$I(l, m) = \frac{1}{N_{\text{Vis}}} \sum_u \sum_v V(u, v) \exp{2\pi i(ul + vm)} + \mathbf{H C}_N \mathbf{H}^\dagger. \tag{2.5}$$

At image position, (l_0, m_0):

$$I(l_0, m_0) = \frac{1}{N_{\text{Vis}}} \sum_u \sum_v I_0 + \mathcal{N}(0, \sigma^2/2N_{\text{vis}}). \tag{2.6}$$

Then, the signal-to-noise ratio is given by:

$$\text{SNR}_{\text{Image}} = I_0 \sqrt{2 N_{\text{vis}}/\sigma^2} = \text{SNR}_{\text{Vis}}. \tag{2.7}$$

The visibility-space matched filter is therefore equivalent to pixel-based source detection in the image plane (such as that performed by SFIND), assuming that the source is located at the pixel centre, and that the visibility data have been weighted naturally (equally) in the image-formation process. These theoretical results have been presented in Trott *et al.* (2011), which also discusses source localization and optimal weighting of data for realistic noise.

3. Improved Performance

There are cases where applying a matched filter detector in visibility space will yield superior detection performance compared to traditional techniques in the image plane. Two such cases are given below.

- Image formation requires gridding the underlying visibility dataset and sampling the image plane at high spatial resolution in order to characterize adequately the instrument point spread function (synthesized beam). At the true source location and with naturally-weighted data, the amplitude of the image-space pixel contains all the information about the source. For a source not co-located with a pixel, the flux is spread across pixels and pixel-based detection no longer contains all the information. Visibility-based detection does not suffer from such discretization errors;

- Commensal observing modes for detecting and classifying slow transients will be routine for next-generation instruments. In that mode, the image-formation process will be designed to be optimal for the science case being studied. The images may not be weighted for optimal point-source sensitivity. In addition, if the underlying noise in the visibility dataset is not white Gaussian (i.e., it is coloured or correlated), failure to include that in the image-formation process will lead to degraded performance. Operating directly on the visibility data allows optimal weighting of the data in the source detector to be achieved.

4. Summary

The detection of previously unknown sources, whether static, variable or transient, is one of the key goals of future wide-field, high-sensitivity, large interferometric radio instruments. To realise the potential to explore the unknown parameter space of variable radio objects and to discover new classes of such objects, we require sophisticated source detection, variability detection and source-localization techniques. The large data rates and advanced scientific demands of these instruments will require those detections and classifications to be performed in real time, automatically, and often in a commensal mode of observing. Those constraints demand novel thinking about how to design optimal pipelines to extract as much information before data are destroyed. Working in visibility space allows high-time-resolution access to the raw data in a form with maximal information content and simple noise statistics. In this paper we have explored the theoretical possibility of achieving optimal source detection with visibility data, and demonstrate its equivalence to image-based techniques, for the case of simple noise properties. Details of the source-localization algorithm and visibility-space detection performance will appear in a forthcoming publication.

References

Hopkins, A. M., *et al.* 2002, *AJ*, 123, 1086
Trott, C. M., Wayth, R. B., Macquart, J.-P., & Tingay, S. J. 2011, *MNRAS*, 731, 81

New Horizons in Time-Domain Astronomy
Proceedings IAU Symposium No. 285, 2011
R.E.M. Griffin, R.J. Hanisch & R. Seaman, eds.

© International Astronomical Union 2012
doi:10.1017/S1743921312001275

Wide-Field Plate Database:
Latest Results

Milcho Tsvetkov[1] and Katya Tsvetkova[2]

[1] Institute of Astronomy, Bulgarian Academy of Sciences, Sofia 1784, Bulgaria
email: milcho@skyarchive.org

[2] Institute of Mathematics and Informatics, Bulgarian Acad. of Sciences, Sofia 1113, Bulgaria

Abstract. We present the Wide-Field Plate Database, a basic source of data and meta-data for astronomy's more than 2.4 million wide-field photographic images obtained with professional telescopes worldwide. The technology developed in Sofia for plate digitization with commercial high-quality flatbed scanners yields low-resolution digital images for quick visualization and easy online access, and optimal high-resolution ones for photometric and astrometric investigations.

1. Introduction

The Wide-Field Plate Database (WFPDB, http://www.skyarchive.org; Tsvetkov 1992) was started at the Institute of Astronomy of the Bulgarian Academy of Sciences in 1991. It collects data from, and meta-data for, astronomy's wide-field plates that are stored worldwide, and includes not only index catalogues extracted from logbooks but also digitized logbooks, digitized plate images and relevant research papers. Along with each catalogue, under the appropriate WFPDB identifier, can be found specific details about the location of the archive, the observatory, telescope parameters and period of operation, and essential meta-data for each observation such as the coordinates of the plate centre in epoch 2000.0, the date and beginning of the observation in UT, object name and type, method of observation, the exposure duration and number (if multiple), the type of emulsion, the filter and spectral band, the size of the plate and its quality, plus the identity of the observer, the location of the plate, and the status of plate digitization and the name of astronomer in charge of the archive.

The WFPDB offers the option to select plates from the various plate catalogues. The search can be done either by object or field co-ordinates, or by constraints on observational parameters. From the results page the user can display an additional page showing details about the archive to which a selected plate belongs, with a map of the all-sky distribution of the observations from that archive and an additional page with details about the selected observation including (if available) any notes, the name of the observer, and information about the plate's availability and its digitization. The same page may also link to the plate preview ("thumbnail"), if available. The image can then be examined in some detail by zooming in on the preview.

In 1997 a static version of the WFPDB, giving details of about 323,000 plates, was copied to Strasbourg (http://vizier.u-strasbg.fr/cats/ VI.htx), where an online search is provided via VizieR (see Search in the WFPDB Catalogue number VI/90). In parallel, an enlarged and daily-updated version has been running in Sofia since 2001 (http://www.skyarchive.org); that Catalogue was later mirrored at Potsdam in 2007 (http://vodata.aip.de/WFPDBsearch) and at the Institute of Mathematics and Informatics of the Bulgarian Academy of Sciences (http://www.wfpdb.org) in 2010, both also offering options for data searches.

2. Catalogue of Wide-Field Plate Archives

The most recent version of the Catalogue of Wide-Field Plate Archives (CWFPA, January 2011) gives, in table form, descriptive information for each plate archive that is listed. It uses an archive identifier, which is composed of the name of the observatory, the respective instrument aperture (and an instrument aperture suffix in the case of instruments with the same aperture). One can also find details about the observatories where the plates were exposed, the parameters of the telescopes used and their period of operation, as well as the present location of the archive and the name of astronomer in charge to contact. The number of the known archives rose from 68 in 1989 (as in the first lists prepared by B. Hauck and C. Jaschek) to 476 at the present day, the numbers of direct and objective prism plates listed correspondingly rising from 1,500,000 to 2,475,636.

3. Catalogue of Wide-Field Plate Indexes

The current (2011 September) version of the Catalogue of Wide-Field Plate Indexes (CWFPI) contains the parameters of 563,600 plates from 133 archives, with online access and data search possibilities, together with quick previews (low-resolution JPEG files) of some of the plates and an option for complete images at high resolution (FITS files) upon request. The all-sky distribution of the centres of the plates in the CWFPI in equatorial co-ordinates is shown in Fig. 1.

4. Plate Digitization

The aims of the preview scans at low resolution (600 or 1200 dpi) in JPEG format are to provide quick visualization and examination of the images, with easier Web accessibility, and to store any marks which the observer may have made on the plate. The high-resolution images are carried out at optimal resolution (1600 or 2400 dpi, in FITS file format). The latter scans are only made accessible upon request. Systematic digitization of plates takes considerable funds and yields a huge volume of data which then has to be stored. One solution to that problem is to make digital archives of selected plates only from a given observing programme (Tsvetkova & Tsvetkov 2009)—observations

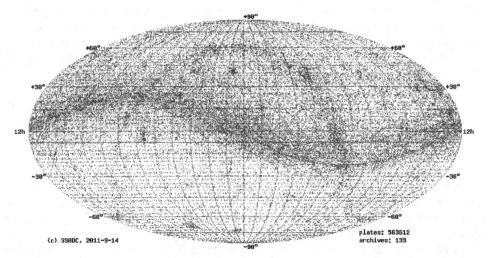

Figure 1. All-sky distribution of the plate centres, in equatorial co-ordinates (J2000).

of the Pleiades, supernova search (Tsvetkova *et al.* 2008), flare stars, Potsdam Carte du Ciel plates (Tsvetkova *et al.* 2009), installed on the German Astrophysical Virtual Observatory Potsdam server (`http://vo.aip.de/plates`).

Plate digitization includes an estimate of the quality of the digitized data, a link to the WFPDB for online access, the implementation of Wavelet transformation methods for compression, and a digitization of the relevant logbooks and notes to link the plate to its description in the appropriate logbook.

At present about 250,000 plates currently stored in European observatories (Sonneberg, Tautenburg, Potsdam, Bamberg, Heidelberg, Hamburg, Byurakan, Brussels, Rozhen, Bucharest, Cluj, Belgrade, Konkoly, Pulkovo, Moscow, Kiev, Asiago and Vatican) have been digitized, mainly with professional flatbed scanners, and are awaiting the implementation of online access. A future goal of the WFPDB is to accomplishing the high-resolution digitization (in standardized FITS file format) together with and digitization of the catalogues, logbooks, and any observers' notes that have come to light.

5. References and Online Services

The project also registers links between specific archived plates and any papers based on them. In the case of flare stars, for instance, publications in the electronic Information Bulletin on Variable Stars are linked to the scanned plates containing the relevant images (e.g. Holl *et al.* 2006), and efficient representation of image scans at different resolutions.

A basic scheme for further development is to establish Photographic Multimedia Plate Libraries at astronomical observatories. Such libraries, which are based on the latest computer-science methods, unify the efforts of astronomers, librarians, and networking and information technology specialists and give good storage, access, effective search possibilities and data optimization, digital curation and similar services.

Acknowledgement

This work is supported by the BG NSF DO-02-273 and BG NSF DO-02-275 grants. We thank the Symposium organizers for an IAU grant.

References

Holl, A., Kalaglarsky, D., Tsvetkov, M., Tsvetkova, K., & Stavrev, K. 2006, in: M. Tsvetkov, V. Golev, F. Murtagh, & R. Molina (eds.), *Virtual Observatory, Plate Content Digitization, Archive Mining, Image Sequence Processing*, Heron Press Science Series, Sofia, p. 374
Tsvetkov, M. 1992, *IAU WGWFI Newsletter*, 2, 51
Tsvetkova, K., Holl, A., & Balazs, L. G. 2008, *BaltA*, 17, 405
Tsvetkova, K. & Tsvetkov, M. 2009, in: D.N. Arabelos, M.E. Contadakis, Ch. Kaltsikis, & S.D. Spatalas (eds.), *Terrestrial and Stellar Environment*, Ziti Press, Thessaloniki, p. 302
Tsvetkova, K., Tsvetkov, M., Boehm, P., Steinmetz, M., & Dick, W. R. 2009, *AN*, 330, 879

New Horizons in Time-Domain Astronomy
Proceedings IAU Symposium No. 285, 2011
R.E.M. Griffin, R.J. Hanisch & R. Seaman, eds.
© International Astronomical Union 2012
doi:10.1017/S1743921312001287

Period Analyses of 100+ Years of RR Lyræ Data

E. N. Walker

The Stargazers Trust, Herstmonceux, E. Sussex, BN27 1PU
email: enw07@btinternet.com

Abstract. The GEOS data base for RR Lyr stars has been analysed to search for secondary (Blazhko) periods in several stars for which over 100 years of measurements are available. The results indicate that, even when there is a secondary period present, its behaviour can be dramatically different from star to star or even over time in the same star. In one star there is a clear correlation between the O–C diagrams for both the pulsation and Blazko periods; that has implications for the origins of the Blazhko periods.

1. Introduction

At a time when consideration is being given to instruments which will produce terabytes of data per day, it can be instructive to look at data where only kilobytes of data have been obtained over more than 100 years. Examples of the latter are the times of maxima in RR Lyr-type variable light curves in the GEOS data-base, and we present here analyses of some of them. For RR Lyr itself, whose pulsation period has shown both 5,000-day and 10,000-day variations, we find that the pulsation period varies and yet the Blazhko phase diagram with different periods has similar shapes. For RW Dra (Blazhko's star) there have been at least six different pulsation periods over the last 100 years; the shape of its Blazhko phase diagram varies and there is a correlation between the O–C diagram for both the pulsation and Blazhko periods. These and similar results impose constraints on models for both the pulsation and Blazhko variations.

1.1. *RR Lyr*

Fig. 1 shows the O–C diagram for over 100 years of light-maxima timing. Approximately the first half of the data have a longer period than does the later half, and the variations in the later data show evidence of both 5,000- and 10,000-day variations.

1.2. *RW Dra, Blazhko's star*

Fig. 2 shows the shape of the Blazhko phase diagram for three different values of the period which have occurred over the last 100 years. The shape has stayed similar although the period has change by about 12%.

Fig. 5 shows a clear correlation between the normalised O–C values for both the pulsation and Blazhko periods. This is consistent with models, which require the Blazhko effect to be produced by a beat between the pulsation period and another similar period.

More details are given in a series of papers by Walker 2010 and references therein.

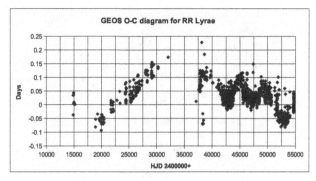

Figure 1. The GEOS O–C diagram for RR Lyræ, obtained using an ephemeris of 2442923.41930 + 0.566837800 x E days

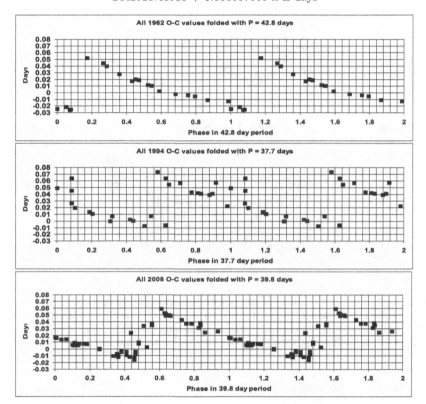

Figure 2. The GEOS O–C diagram for RW Dra, obtained using an ephemeris of 2442923.41930 + 0.566837800 x E days

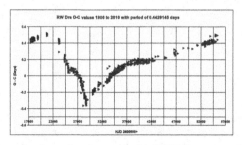

Figure 3. Changes in the pulsation period of RW Dra over the last 100 years.

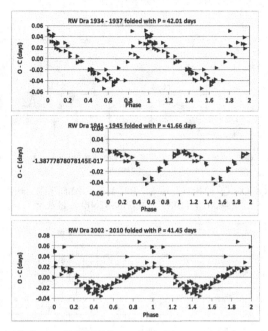

Figure 4. Period changes in RW Dra after the removal of two linear gradients

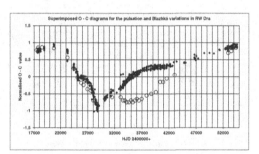

Figure 5. Normalised O–C values for the pulsation period (filled squares). The large triangles are the normalised O–C values for the Blazhko variations, and are plotted inverted with respect to the O–C values for the pulsation. In both cases the amplitudes were normalised with respect to the values of the underlying variation, either pulsation or Blazhko.

Reference

Walker, E. N. 2010, *Observatory*, 130, 225

New Horizons in Time-Domain Astronomy
Proceedings IAU Symposium No. 285, 2011
R.E.M. Griffin, R.J. Hanisch & R. Seaman, eds.

© International Astronomical Union 2012
doi:10.1017/S1743921312001299

The VLBA Fast Radio Transient Experiment: Progress and Early Results

Randall B. Wayth[1], Walter F. Brisken[2], Adam T. Deller[3], Walid A. Majid[4], David R. Thompson[4], Steven J. Tingay[1] and Kiri L. Wagstaff[4]

[1] International Centre for Radio Astronomy Research, Curtin University, WA 6845, Australia
email: r.wayth@curtin.edu.au

[2] NRAO, PO Box O, Socorro, NM 87801, USA

[3] ASTRON, P.O. Box 2, 7990 AA Dwingeloo, The Netherlands

[4] Jet Propulsion Laboratory, California Institute of Technology, Pasadena, CA 91109, USA

Abstract. Motivated by recent discoveries of isolated, dispersed radio pulses of possible extragalactic origin, we are performing a commensal search for short-duration (ms) continuum radio pulses using the Very Long Baseline Array (VLBA). The geographically separated antennæ of the VLBA make the system robust to local RFI and allow events to be verified and localised on the sky with milli-arcsec accuracy. We report sky coverage and detection limits from the experiment to date.

Keywords. radio continuum: general – methods: data analysis

1. Introduction

Recent discoveries of isolated, dispersed pulses of radio emission in pulsar surveys (Lorimer *et al.* 2007; Burke-Spolaor *et al.* 2011; Keane *et al.* 2011) hint at the possibility of a population of extragalactic sources of fast transient radio emission. The possible sources of that emission are somewhat speculative, but include some exotic objects and physics such as annihilating black holes, and merger events generating gravitational waves (see Macquart *et al.* 2010).

The isolated radio bursts discovered to date have been found with the Parkes radio telescope, which is a large single dish. Big dishes have great sensitivity but poor resolution, making the localisation of an event on the sky, and hence follow-up, difficult. In addition, single dishes cannot easily distinguish between genuine astronomical events and the terrestrial radio-frequency interference (RFI) that mimics the frequency-swept "chirp" expected, owing to the dispersion of radio waves in the ionised interstellar medium.

To overcome these issues when searching for fast radio transients, we are conducting the V-FASTR experiment (Wayth *et al.* 2011). V-FASTR is a commensal search for fast transients that uses the Very Long Baseline Array (VLBA)—a long-baseline interferometer with baseline lengths ranging from hundreds to thousands of kilometres. Data from the VLBA are correlated with the DiFX-2 software correlator (Deller *et al.* 2011), which can generate short integration time (~1 ms) spectrometer data from each antenna during the correlation.

The spectrometer data are fed into a real-time detection pipeline that uses sophisticated statistical and machine-learning techniques (Thompson *et al.* 2011) to evaluate the data quality from each antenna and to excise RFI events. V-FASTR has several advantages over single-dish searches, including the ability to distinguish easily RFI that is local

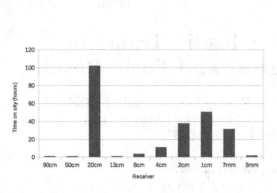
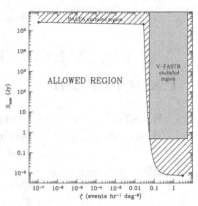

(a) V-FASTR time on sky per VLBA receiver to end of 2011 August.

(b) Event rate limits for V-FASTR compared to PALFA. Figure adapted from Deneva et al. (2009).

Figure 1: Figure adapted from Deneva *et al.* (2009).

to one antenna, and in being able to follow-up candidate events by imaging with VLBI (milli-arcsec) resolution.

2. Results to Date

V-FASTR has been in full-time commensal operation since the deployment of DiFX-2 on the VLBA correlator in mid-2011. Fig. 1 shows the time spent on the sky per VLBA receiver for data processed to the end of 2011 August. For that observing period on the 20-cm receiver, the 5σ detection limit is 0.42 Jy for 10-ms integrations and 64 MHz bandwidth. The upper limit on event rates at that sensitivity in the 20-cm band is 0.06 events hr^{-1} deg^{-2}.

In Fig. 1 we plot the limit on event rates for the 20-cm receiver, assuming isotropically distributed sources, and compare them to the results from the PALFA survey (Deneva *et al.* 2009). V-FASTR is rapidly accumulating hours on the sky. More hours on the sky will push the limit in Fig. 1 to the left, overtaking the PALFA limit in some regions of parameter space. The VLBA is also undergoing an upgrade for broad-band digital backends that will increase the system sensitivity by a factor of a few. That upgrade will push the limit in Fig. 1 down, thereby placing even stricter limits on parameter space.

References

Burke-Spolaor, S., Bailes, M., Ekers, R., Macquart, J.-P., & Crawford, F. 2011, *ApJ*, 727, 18

Deller, A. T., *et al.* 2011, *PASP*, 123, 275

Deneva, J. S., *et al.* 2009, *ApJ*, 703, 2259

Keane, E. F., Kramer, M., Lyne, A. G., Stappers, B. W., & McLaughlin, M. A. 2011, *MNRAS*, 415, 3065

Lorimer, D. R., Bailes, M., McLaughlin, M. A., Narkevic, D. J., & Crawford, F. 2007, *Science*, 318, 777

Macquart, J.-P., *et al.* 2010, *PASA*, 27, 272

Thompson, D. R., *et al.* 2011, *ApJ*, 735, 98

Wayth, R. B., *et al.* 2011, *ApJ*, 735, 97

New Horizons in Time-Domain Astronomy
Proceedings IAU Symposium No. 285, 2011
R.E.M. Griffin, R.J. Hanisch & R. Seaman, eds.

© International Astronomical Union 2012
doi:10.1017/S1743921312001305

Around Gaia Alerts in 20 questions

Łukasz Wyrzykowski[1,2]‡ and Simon Hodgkin[1]

[1]Institute of Astronomy, The Observatories, Cambridge, CB3 0HA, UK
email: wyrzykow@ast.cam.ac.uk, sth@ast.cam.ac.uk

[2] Warsaw University Astronomical Observatory, 00-478 Warszawa, Poland
email: lw@astrouw.edu.pl

Abstract. GAIA is a European Space Agency (ESA) astrometry space mission, and a successor to the ESA Hipparcos mission. GAIA's main goal is to collect high-precision astrometric data (positions, parallaxes, and proper motions) for the 1 billion brightest objects in the sky. Those data, complemented with multi-band, multi-epoch photometric and spectroscopic data observed from the same observing platform, will allow astronomers to reconstruct the formation history, structure, and evolution of the Galaxy.

GAIA will observe the whole sky for 5 years, providing a unique opportunity for the discovery of large numbers of transient and anomalous events such as supernovæ, novæ and microlensing events, GRB afterglows, fallback supernovæ, and other theoretical or unexpected phenomena. The Photometric Science Alerts team has been tasked with the early detection, classification and prompt release of anomalous sources in the GAIA data stream. In this paper we discuss the challenges we face in preparing to use GAIA to search for transient pheonomena at optical wavelengths.

Keywords. space missions: GAIA, supernovae: general, gravitational lensing, novae

---

1. Where, how and when?

GAIAwill be launched in ESA/Kourou (French Guyana) from a Soyuz-Fregat rocket in 2013 June. Deployment will be at the L2 Lagrange Point, with the first community release of alerts expected in mid-2014 (internal verification will begin in early 2014). The mission is scheduled to last until 2018–2019.

2. What telescopes will GAIA have?

GAIA will be equipped with two 1.45 × 0.5m primary mirrors, forming two fields of view separated by 106°.5. The light from the two mirrors will be imaged onto a single focal plane. GAIA's Astrometric Field detectors will reach to $V = 20$.

3. What instruments will GAIA have?

Each object traverses through the focal plane (4.4 sec per CCD); see Fig. 1.
SM: Objects will be detected by Sky Mapper CCDs, and allocated windows for the remaining detectors.
AF: Source positions and G-band magnitudes are to be measured in the Astrometric Field CCDs (plate scale $\sim 0''.04 \times 0''.1$).
BP/RP: Low-dispersion spectrophotometry at 330–680 nm and 640–1000 nm), in 120 samples.
RVS: Intermediate-dispersion spectroscopy (R$\sim$11,500) at 847–874 nm (around the Calcium Infrared triplet) to $V < 17$ mag.

‡ name pronunciation: *Woo-cash Vi-zhi-kov-ski*

Focal Plane

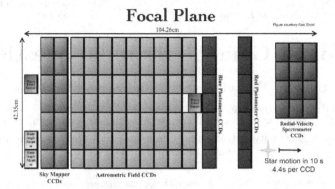

Figure 1. Focal plane of GAIA.

4. What is the data latency?

GAIA will be visible from the Earth for only 8 h a day. All the data from the previous 24 h will be downlinked during a contact. After initial data processing, alerts will be issued during a period from a couple of hours to 48 hours after the observation.

5. What is downloaded?

Most of the sky is empty. GAIA will only transmit small windows around stars that are detected at each transit on the Star Mapper CCDs, plus associated data.

6. How does the scanning law allow for full sky coverage?

GAIA has a pre-defined plan for scanning the sky. The spin axis will be maintained at 45° from the Sun, with a period of 6 h. Details are explained in Fig. 2.

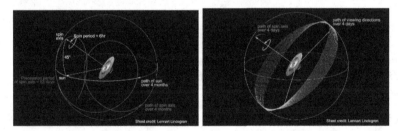

Figure 2. Nominal Scanning Law principles for the GAIA satellite.

7. What is the typical sampling?

On average, each object will be observed 80 times, though at the Ecliptic nodes objects will be scanned > 200 times. Observations will occur in pairs (two FOVs), separated by ∼2 hours. The separation between pairs will be between 6 hours and ∼30 days.

8. What is the precision of the instantaneous photometry and astrometry?

In a single observation (transit), the photometry will reach millimagnitude precision at $G = 14$, and 1% at $G = 19$. The astrometric precision will be in the range 20–80 μas at $G = 8$–15 (Fig. 3 explains the effects of gating), falling to 600 μas at $G = 19$. That level of astrometric precision will only be reached later into the mission.

9. How will anomalies be detected?

Using simple recipes:
1. By comparing the most recent observation with available historic data, and

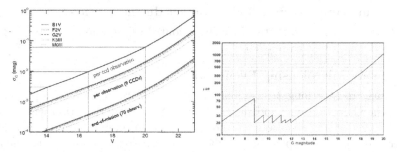

Figure 3. Precision of instantaneous photometry and astrometry of GAIA.

2. inspecting for unexpected changes.
3. No history? New transient!

10. How will the anomalies be classified?

1. From the light-curve,
2. from low-dispersion BP/RP spectroscopy, or
3. by cross-matching with archival data.

11. How will the BP/RP spectra be used?

Self-Organizing Maps (Wyrzykowski & Belokurov 2008) built from low-dispersion spectra can confirm a non-stellar nature, classify supernova types, measure supernova ages and possibly even constrain the redshift.

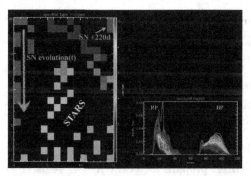

Figure 4. A Self-Organizing Map (left) can distinguish between different spectral types of stars and supernova at different epochs, as built from synthetic GAIA BP/RP spectra (right).

12. How will alerts be disseminated?

Through `skyalert.org`, email, www server, Twitter, iPhone app, etc.

13. What will an alert contain?

The coordinates, a small cut-out image from the SM, the GAIA light curve, a low-resolution spectrum at the trigger, the classification results, and the cross-matching results.

14. What will the main triggers be?

Supernovæ, classical novæ, dwarf novae, microlensing events, Be stars, GRB afterglows, M-dwarf flares, R CrB-type stars, FU Ori-type stars, asteroids—and surprises.

15. How many supernovæ will GAIA detect during 5 years?

6000 SNe are expected down to $G = 19$. About 2000 should be detected before the maximum (Belokurov & Evans 2002).

16. How many Microlensing Events will GAIA detect?

1000+ events (mostly long $t_E > 30\,\mathrm{d}$) are expected to be detected photometrically, mainly in the Galactic bulge and plane. Astrometric centroid motion will be detectable in real time (for larger deviations of about $100\,\mu\mathrm{as}$) in on-going events, and alerts may be triggered to obtain complementary photometry (Belokurov and Evans 2003).

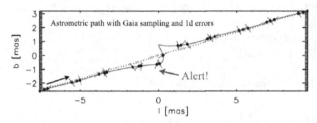

Figure 5. Trajectory of a source due to proper motion and centroid shift during a microlensing event.

17. Will GAIA alert on GRB optical counterparts?

GAIA's sampling and data latency are not appropriate, but detections of 1–2 bright, on-axis afterglows and 5–15 orphan afterglows can be expected (Japelj & Gomboc 2011).

18. How many asteroids is GAIA expected to see?

About 250,000 asteroids (mostly known ones). Alerts on new asteroids and NEO candidates will be based on unsuccessful star matching.

19. What about known anomalous objects?

Such objects can be added to the **Watch List**. Every time GAIA observes them, their data will become available for inspection.

20. How can I get involved now?

Through telescope time: prepare for GAIA Alerts, register at skyalert.org, set-up your alerts for SNe, CVs, blazars, etc. on CRTS stream (Drake *et al.* 2009), follow-up the alerts, and contact us with your data!

Through scientific interests: suggest what would be worth detecting and raising an alert; propose detection algorithms and classification techniques; suggest interesting known targets to be observed.

References

Belokurov, V. A. & Evans, N. W. 2002, *MNRAS* 331, 649
Belokurov, V. A. & Evans, N. W. 2003, *MNRAS* 341, 569
Drake, A. J., *et al.* 2009, *ApJ* 696, 870
Japelj, J. & Gomboc, A. 2011, *PASP* 123, 1034
Wyrzykowski, Ł. & Belokurov, V. 2008, *AIPC* 1082, 201

New Horizons in Time-Domain Astronomy
Proceedings IAU Symposium No. 285, 2011
R.E.M. Griffin, R.J. Hanisch & R. Seaman, eds.

© International Astronomical Union 2012
doi:10.1017/S1743921312001317

Poster Summaries

New Horizons in Time-Domain Astronomy
Proceedings IAU Symposium No. 285, 2011　　　　ⓒ International Astronomical Union 2012
R.E.M. Griffin, R.J. Hanisch & R. Seaman, eds.　　　　doi:10.1017/S1743921312001329

Type II SN Light Curves from the Caltech Core Collapse Project

Iair Arcavi

Department of Particle Physics and Astrophysics, The Weizmann Institute of Science, Rehovot 76100, Israel
email: `iair.arcavi@weizmann.ac.il`

We presented our analysis of a sample of type II supernova (SN) light curves measured by the Caltech Core Collapse Project (CCCP). CCCP is a large observational program which made use of the robotic 60-in and the Hale 200-in telescopes to obtain optical photometry, spectroscopy and IR photometry of 49 nearby core-collapse supernovae (SNe). It provides a fair sample of core-collapse events, with well-defined selection criteria, and uniform, high-quality optical/IR observations. Our goal is to characterize the little-studied properties of core-collapse supernovae as a population. Preliminary data indicate a diverse set of sub-populations including "standard" type IIP supernovæ, declining supernovæ (at different rates) and slowly rising peculiar supernovæ. Work is in progress to map and quantify that diversity better. It is hoped that a single tunable formula will be able to describe most light-curve shapes, thereby helping us attain a better understanding of the physical mechanisms underlying these results.

Stopping and Looking: On Error Bars in Time Series

Guillaume Belanger

ESAC, Villanueva de la Cañada, 28691 Madrid, Spain
email: `guillaume.belanger@esa.int`

In X-ray Astronomy, the error bar or uncertainty associated with the value of a bin in a time series is usually calculated from the square root of the number of events in that bin. This standard practice is erroneous both on theoretical and on practical grounds. On theoretical grounds, it fails to conform to the fact that the uncertainty on a measurement depends upon the accuracy with which the measurement is made, not on the value measured. On practical grounds, it fails to give the correct value of the mean count rate when the error bars are taken into account. Therefore, this practice should be abandoned in favour of an alternative where the error bar is a more appropriate representation of the uncertainty on the measurement, and therefore it does not depend upon the number of events in the given bin.

Photometric Reverberation Mapping of the Broad Emission Line Region in AGN

Doron Chelouche
Department of Physics, University of Haifa, Mount Carmel, Haifa 31905, Israel
email: `doron@ias.edu`

Measuring black hole (BH) masses in galaxies has considerable implications for our understanding of galaxy formation and evolution and accretion physics in active galactic nuclei (AGN). A reliable method for estimating the black-hole mass in AGN uses reverberation mapping, which allows one to measure the size of the broad emission line region in those objects. Presently, the mass of < 50 low-z BHs has been measured in this way. The poster proposed a new method for measuring the size of the broad emission-line region (BLR) using broad-band photometric light curves. Applying the method to a subsample of AGN with previously measured BLR sizes gives consistent results. The poster reported new BH mass measurements that use currently available photometric data, and single-epoch spectral information. This opens up the possibility of measuring BLR sizes and BH masses in numerous objects, and over a broad redshift range, using upcoming photometric surveys.

Identification and Investigation of "Primitive Emitters" that Form Prompt Phases of GRBs

Anton Chernenko
Space Research Institute, Russian Academy of Sciences, 84/32, Moscow 117997, Russia
email: `anton@cgrsmx.iki.rssi.ru`

GRBs are one of the most diverse celestial objects in terms of their time histories and spectral variability during prompt phases. A number of peaks with durations ranging from milliseconds to tens of seconds within an individual GRB randomly overlap, and possibilities to investigate spectral evolution of single unambiguously isolated emission episodes are scarce. Today, no solid theoretical framework for explaining prompt emission of GRBs exists, and no single emission mechanism can explain the spectral evolution of the entire prompt phase of a given GRB. An attempt to fit gamma-rays by a theoretical spectral shape based on particular emission mechanisms imposes rather weak constraints on the model parameters since, owing to the overlapping nature of emission, that fit needs to involve several components. Moreover, the statistics of available gamma-ray data hardly allow one to distinguish between models with the number of variable parameters greater than two. Therefore, a possibility to identify and study emission episodes that are unambiguously produced by single physical emitters is very important from the theoretical perspective. We found that in almost any GRB there is a number of time intervals of considerable duration (up to tens of seconds) during which emission is highly variable in terms of flux, but manifests very little spectral variability—its spectral shape stays constant. On the other hand, spectra of different intervals differ significantly from each other. Such intervals often correspond to the few initial seconds of GRBs and to their tails, but frequently they also span periods with complex multi-peak structures in the middle of the prompt phase. Spectrum-flux correlations are natural in either synchrotron or IC models, and the very existence of periods where such correlations are minimal is important since it implies existence of special "extensive" emission régimes within those models. It is also unlikely that, during such a period, any overlapping takes place. Two or more emitters, even with constant spectra, should be connected too tightly to produce such a coherent picture. It is therefore likely that we are rather frequently able to observe what we call individual "primitive emitters" whose spectra and time histories could be studied independently from other emitters. The poster present the results of this study.

ULTRACAM Observations of Ultra-Compact Binaries

C. Copperwheat[1], T. Marsh[1], V. Dhillon[2], S. Littlefair[2], and D. Steeghs[1]

[1] *Department of Physics, University of Warwick, Coventry CV4 7AL, UK*
email: c.copperwheat@warwick.ac.uk
[2] *Department of Physics and Astronomy, University of Sheffield, Sheffield S3 7RH, UK*

The AM Canum Venaticorum (AM CVn) stars are ultra-compact binaries with the lowest periods of any binary subclass, and consist of a white dwarf accreting material from a donor star that is itself fully or partially degenerate. These objects offer new insight into the formation and evolution of binary systems, and are predicted to be among the strongest gravitational-wave sources in the sky. The poster presented multi-band, high time-resolution light curves of a number of these systems, obtained with the fast photometer ULTRACAM. It discussed the 28-min binary SDSS 0926+3624, the only known eclipsing source of this class. From light-curve models we make the most precise parameter determinations for any AM CVn, and determine the degree of degeneracy of the donor star—a key parameter in differentiating between the proposed formation paths for these objects. The poster also discussed additional phenomena that are apparent in the optical data for this system, such as the superhump, quasi-periodic oscillations, and the outbursting behaviour. In conclusion it outlined our long-term timing programme, through which we are measuring the period evolution of a number of different AM CVn systems. The two most compact systems (HM Cnc and V407 Vul) are known to be decreasing in period, which is the opposite of what was initially expected for accreting AM CVns. For those systems we aim to measure eventually the second time-derivative of the orbital period, which will differentiate between the various formation models. For the 10-min binary ES Cet the poster showed how ten years of timing measurements suggest a double white-dwarf formation path for that system.

Multi-Wavelength Stellar Variability in the Time-Domain Era

James Davenport[1] *et al.*
[1] *Department of Astronomy, University of Washington, Seattle, WA 98195, USA*
email: jrad@astro.washington.edu

We presented results from our studies of (primarily low-mass) main-sequence stars using the SDSS Stripe 82, 2MASS Calibration Fields, and WISE time-domain databases. These photometric catalogues provide measurements from the last decade ranging in wavelength from 0.36 to 22μ. Many forms of variability for these stars are best identified in certain wavelength régimes, and the levels of variability differ across the many photometry surveys. We have studied the data to characterize the rates of flares, periodic sources such as eclipsing binaries, and also the variability structure in the quiescent non-flaring state. Comparisons between wide-field survey data and dedicated ground- and space-based photometric monitoring were also presented.

Contact Binary Population from the RATS Survey

P. Hakala[1], G. Ramsay[2], and T. Barclay[2,3]

[1] FINCA, University of Turku, F-21500 PIIKKIÖ, Finland
email: pahakala@astro.helsinki.fi
[2] Armagh Observatory, College Hill, Armagh BT61 9DG, UK
[3] MSSL, University College London, Holmbury St. Mary, Dorking, RH5 6NT, UK

We presented the first results from our modelling of the Contact Binary population of our variability survey by the 40 deg^2 RATS. We matched the partial light curves against a vast library of pre-computed light-curve models, and derived approximate system parameters for hundreds of sources.

Visibility Stacking in the Quest for Radio Emission from SNIa

Paul Hancock, Bryan Gaensler, and Tara Murphy
Sydney Institute for Astronomy, The University of Sydney, NSW 2006, Australia
email: jancock@physics.usyd.edu.au

The poster described the process of stacking radio interferometry visibilities to form a deep composite image, and its application to the observation of transient phenomena. We applied "visibility stacking" to 46 archival Very Large Array observations of nearby type Ia supernovæ (SNeIa). This new approach provides an upper limit on the SNIa ensemble peak radio luminosity of $1.2 \times 10^{25}\,\mathrm{erg\,s^{-1}\,Hz^{-1}}$ at 5GHz, which is 5–10 times lower than previously measured. That luminosity implies an upper limit on the average companion stellar wind mass-loss rate of $1.3 \times 10^{-7} M_\odot\,\mathrm{yr^{-1}}$. This mass loss rate is consistent with the double degenerate model for SNeIa, and rules out intermediate and high-mass companions in the single degenerate one. In the era of time-domain astronomy, techniques such as visibility stacking will be important in extracting the maximum amount of information from observations of populations of short-lived events.

Stellar Flares in Time Domain Surveys

Eric J. Hilton[1], Suzanne L. Hawley[2], and Adam F. Kowalski[2]
[1] University of Hawaii at Manao, Honolulu, HI 96822, USA
[2] Astronomy Department, University of Washington, Seattle, WA 98915, USA
email: hilton@ifa.hawaii.edu

Stellar flares are powerful magnetic reconnection events that cause transient flux enhancements of up to several magnitudes on low-mass stars. The flares have a characteristic blue colour, and the largest ones may last many hours. Small flares may occur several times per hour. The poster presented simulations of the observational signatures of flares in time-domain surveys using new measurements of flare rates on low-mass stars in a variety of filters. We paid particular attention to the rate of optical transients, where a flare occurs on a previously undetected source. In the poster we compared our model to previous survey data, and predicted flare rates in upcoming surveys. New large surveys will allow us to determine how the flare rates, and hence magnetic fields, change as a function of stellar temperature, age and other physical parameters.

Low-Frequency Searches for Radio Transient Emission using the VLA

T. Jaeger[1], N. Kassim[1], S. Hyman[2], J. Lazio[3], and R. Osten[4]

[1] *US Naval Research Laboratory, Washington, DC 20375, USA*
email: `ted.jaeger.ctr@nrl.navy.mil`

[2] *Dept. of Physics and Engineering, Sweet Briar College, Sweet Briar, VA 24595, USA*

[3] *Jet Propulsion Laboratory, Pasadena, CA 91109, USA*

[4] *Space Telescope Science Institute, Baltimore, MD 21218, USA*

A variety of sources have been predicted to emit at metre wavelengths and would probably appear as transients, including bursts from extrasolar planets and brown dwarfs and prompt emission from gamma-ray bursts. Low-frequency VLA observations are well suited to probe the dynamic radio sky, given the the large field of view (> 150 deg^2) at 74 MHz and sub-mJy sensitivity at 325 MHz. The poster presented preliminary results from multiple low-frequency radio transient searches using data from the VLA archive. In particular, it reported on our searches at 325 MHz using targeted fields, and an all-sky search using fields from the 74-MHz VLA Low-frequency Sky Survey (VLSS), and discussed implications of any transient non-detection for the rate of gamma-ray bursts and/or radio supernovae.

The Century Before Photography

Derek Jones

Institute of Astronomy, Madingley Road, Cambridge, CB3 0HA, UK
email: `dhpj@ast.cam.ac.uk`

The concept of "flux-limited sample" was developed by radio astronomers after World War II, but the same idea can be traced through the work of the optical astronomers: the Herschels (father and son), Lalande, Bessel and Argelander. One motivation was the search for minor planets; variable stars were then regarded as a nuisance. The perennial difficulty was the lack of a consistent definition of stellar magnitude.

Automatic Real-Time Modelling of Anomalous Microlensing Events

N. Kains[1], P. Browne[1], K. Horne[1], and A. Cassan[2]

[1] *School of Physics and Astronomy, University of St Andrews, Fife, KY16 9SS, UK*
email: `nk87@st-and.ac.uk`

[2] *Institut d'Astrophysique de Paris, 98 bis Boulevard Arago, 75014 Paris, France*

Gravitational microlensing is a well-established planet-detection method, with notable discoveries including the first cool, rocky exoplanet. Today, new-generation surveys and robotic telescopes allow us to monitor and follow up thousands of microlensing events per year, of which a significant fraction is "anomalous". Each anomalous microlensing event traditionally requires significant computational and human resources to model, and this is difficult to achieve on the short timescales of the typical microlensing anomalies, particularly those of possible planetary origin. The poster presented the development of automatised methods to model anomalous microlensing events so as to provide feedback to both robotic and human observers about the nature of on-going events.

Wide and Fast: Monitoring the Sky in the Sub-Second Domain

S. Karpov[1] et al.
[1] Special Astrophysical Observatory of RAS, Karachaevo-Cherkesia 369167, Russia
email: karpov@sao.ru

The poster presented our long-term efforts to perform optical wide-field monitoring of the sky with sub-second temporal resolution, in order to look for optical components of GRBs, fast-moving satellites, and meteors. It described the concepts and the results of operating our wide-field cameras, FAVOR and TORTORA, and included observations of the complete development of optical emission from the Naked-Eye Burst, which has provided significant insight into the nature of its central engine. The poster also gave a status report on our new-generation instrument, MiniMegaTORTORA, a multi-objective transforming camera which is able both to detect and to classify fast optical transients and also to perform colour and polarimetric observations of them. The instrument is currently being commissioned.

The Caltech Blazar Monitoring Programme

O. G. King[1] (for the OVRO 40-m Collaboration)
[1] Keck Institute for Space Studies, California Institute of Technology
Pasadena, CA 91125, USA
email: ogk@astro.caltech.edu

The OVRO 40-m Telescope monitoring program currently carries out twice-weekly measurements of the 15-GHz flux density of nearly 1600 blazars and other AGN, including all those associated with northern detections (declination $> -20°$) by the FERMI Large Area Telescope (LAT) and a pre-selected sample ideal for statistical studies. The poster presented some results from the programme, and described the statistical methods we have developed to study the intrinsic variability of blazar populations.

Anomalously Intense Pulses from Pulsars with LOFAR LBAs

V. I. Kondratiev[1] and A. V. Bilous[2] + LOFAR Pulsar Working Group
[1] Netherlands Institute for Radio Astronomy, 7990 AA Dwingeloo, The Netherlands
[2] Department of Astronomy, University of Virginia, Charlottesville, VA 22904, USA
email: kondratiev@astron.nl

The poster reported the results of our observations of anomalously intensive pulses from several pulsars with the LOFAR, in the frequency range 20–70 MHz. Such pulses were first discovered by Ul'yanov et al. in 2006 using the UTR-2 radio telescope at frequencies below 30 MHz; the intensity of that source was shown to be strongly modulated in both time and frequency. LOFAR's much broader frequency range allows one to study such pulses in more detail. The poster discussed the possible origin of the pulses and their relation to the giant-pulse phenomenon observed in other pulsars at higher frequencies.

The "SED" Machine

Nick Konidaris[1], Sagi Ben-Ami[2], Robert Quimby[1], and Shri Kulkarni[1]
[1] *California Institute of Technology, Pasadena, CA 91125, USA*
[2] *The Weizmann Institute of Science, Rehovot 76100, Israel*
email: npk@astro.caltech.exu

The poster described the "SED Machine", a spectrograph designed around the goal of classifying transients efficiently.

Gamma-Ray Variability of Blazars Observed with FERMI LAT

Stefan Larsson[1], Stefano Ciprini, Lise Escande, Benoit Lott, and Greg Madejski (on behalf of the FERMI LAT collaboration)
[1] *Department of Astronomy, Stockholm University, SE-106 91 Stockholm, Sweden*
email: stefan@astro.su.se

Since its launch in 2008 the FERMI Large Area Telescope (LAT) has provided regular monitoring of the gamma-ray emission from blazars on time-scales from hours to years. By combining observations at other wavelengths, it is now possible to study variability and correlation properties in a much more systematic and detailed way than ever before. The poster described FERMI LAT observations of blazars, the properties and limitations of produced light curves, and how those can be handled. Results from timing analyses were also presented; they could be related to variability behaviour at other wavelengths, and to models of those sources.

Ultra-Compact White Dwarf Binaries from the Palomar Transient Factory

David Levitan[1] *et al.* (on behalf of the Palomar Transient Factory Collaboration)
[1] *California Institute of Technology, Pasadena, CA 91125, USA*
email: dlevitan@caltech.edu

The rise of large-area synoptic surveys has enabled the efficient discovery of ultra-compact white-dwarf binaries. Both outbursting systems (including Cataclysmic Variables and AM CVn systems) and eclipsing systems (such as detached white-dwarf binaries) are targets of new surveys such as the Palomar Transient Factory (PTF). The poster described our search for those systems with the PTF. It outlined the motivation for our search and summarized our discoveries to date; the latter included numerous cataclysmic variables and four of the very rare AM CVn systems. It commented on the techniques we are using to conduct our search, and the follow-up efforts involved. It concluded by discussimg the future potential of such surveys, and provided some estimates of the number of ultra-compact systems that the PTF may discover.

Modelling the Light Curve of Tidal Disruption Events

Giuseppe Lodato
Universit degli Studi di Milano, Dipartimento di Fisica, I-20133 Milano, Italy
email: `giuseppe.lodato@unimi.it`

The tidal disruption of stars in the gravitational field of a supermassive black hole is a useful tool to determine the properties of central black holes in quiescent galaxies; it can also probe stellar dynamics in galactic nuclei. For many years the theoretical understanding of such events was largely based on the pioneering models of Rees and Phinney, who predicted that the light curve should scale as $\sim t^{(-5/3)}$. The poster revisited the issue, taking into account two new results: (a) recent numerical simulations have indicated that, especially at early times, the evolution of the infall rate of the debris can show significant departures from the standard profile, depending on the type of star being disrupted, and (b) the monochromatic light curves at specific wavelengths can be quite different from the bolometric one, which should scale with the infall rate but can be affected by the presence of a significant wind emission for strongly super-Eddington events. These issues were also discussed in the light of recently observed candidate TDEs.

ASTROSAT: The Indian Multi-Wavelength Astronomy Satellite

R. K. Manchanda[1] (on behalf of ASTROSAT payload teams)
[1] *Tata Institute of Fundamental Research, Colaba, Mumbai 400 005, India*
email: `ravi@tifr.res.in`

The Indian Astronomy satellite, code named ASTROSAT, is designed to carry out multi-wavelength studies of a variety of Galactic and extragalactic sources from the UV band (1000–3000 Å), the optical band, and the soft and hard X-ray bands covering the range 0.3–80 keV. The planned X-ray studies will span imaging plus spectral and temporal variations of Galactic and extragalactic sources. Five different payloads presently under development will achieve those mission targets. The satellite will be launched in mid-2012.

The Near-Infrared Variability of Quasars

A. A. Miller, J. S. Bloom, and N. R. Butler
Astronomy Department, University of California, Berkeley, CA 94720, USA)
email: `amiller@astro.berkeley.edu`

The poster explored the intrinsic variability of quasi-stellar objects (QSOs) that had been observed as part of the NEWFIRM Medium Band Survey (MBS). We had used data obtained in 6 filters between 1–2.2 μ, and parameterized the structure function for variability in each of the filters for the 40 or so known QSOs that were observed. The poster described how we had compared the variability to that of known variable stars and had examined the possibility of selecting QSOs from near-infrared (NIR) variability alone, regardless of colour. Within the context of upcoming NIR synoptic surveys such as WFIRST and SASIR, we had projected the efficiency with which QSOs can be discovered on the basis of their intrinsic variability.

Fallback Supernovæ: Fast & Faint Domain of Core-Collapse Supernovæ

Takashi Moriya

Institute for the Physics and Mathematics of the Universe, University of Tokyo, Kashiwa, Chiba 277-8583, Japan

email: `takashi.moriya@ipmu.jp`

Supernovæ (SNe) from massive stars (core-collapse SNe) are rich in variety. Many kinds of light curves and spectral types emerge from massive stars depending primarily on their main-sequence mass. Very massive stars are expected not to end up as SNe because all the materials are collapsed into the central black hole. However, if the central black hole is not massive enough, some part of the progenitors can escape from the black hole and can potentially be observed as SNe (fallback SNe). We have performed numerical simulations of fallback SNe, and found that many fallback SNe are predicted to be observable during the rapid brightness rise (which is sometimes less than 10 days) as faint SNe. The poster compared our model with the faint and fast evolved supernova SN 2008ha, and showed that SN 2008ha is consistent with our model of fallback SNe.

Subaru Wide-Field Variability Survey for Active Galactic Nuclei

Tomoki Morokuma[1] *et al.*

[1] *Institute of Astronomy, University of Tokyo, Mitaka, Tokyo 181-0015, Japan*

email: `tmorokuma@ioa.s.u-tokyo.ac.jp`

The poster presented our survey for active galactic nuclei (AGN) found via optical variability with a wide-field optical imager, "Suprime-Cam", on the prime focus of the 8.2-m Subaru telescope. By combining our observations with 50–100-ksec X-ray exposures obtained with XMM-Newton, we confirmed that optical variability is a useful tool for finding AGN. We had also found that it plays a complementary and important role in the study of low-luminosity AGN for which there are X-ray observations. A significant fraction of AGN selected by their optical variability are below the X-ray detection limit; that had also been shown by studies with HST. The poster showed the result of applying our method to the same parent sample at the faint end of the quasar luminosity function at high red-shifts. It concluded with a brief introduction of the next-generation wide-field imager, "Hyper Suprime-Cam" (HSC, which was installed on the Subaru telescope in 2011), a description of our survey plan, and some results expected on the topic of AGN variability.

Real-Time Analysis in the Palomar Transient Factory

Peter Nugent[1,2]
[1] *University of California, Berkeley, Berkeley CA, 94720, USA*
[2] *Physics Division, Lawrence Berkeley National Laboratory, Berkeley, CA 94720, USA*
email: penugent@lbl.gov

The Palomar Transient Factory (PTF) is an experiment designed to explore systematically the optical transient and variable sky. The main goal of this project is to fill the gaps in our present-day knowledge of the optical transient phase space. Besides reasonably well-studied populations (such as classical novæ and supernovæ), there exist many types of either poorly-constrained events (such as luminous red novae and tidal disruption flares) or predicted but not yet discovered phenomena (such as orphan afterglows of gamma-ray bursts). Beginning in May of 2010, we put into action an end-to-end near real-time pipeline for the discovery and automated follow-up of new transient sources. The poster discussed the challenges we faced in enacting such a pipeline, described our successes over the first year since implementation, and outlined directions for future improvements.

The Long-Term Optical Polarization Variability of the BL Lac Object PKS 2155-304

N. Pekeur[1], S. B. Potter[2], P. M. Chadwick[1], and M. Daniel[1]
[1] *Department of Physics, University of Durham, Durham DH1 3LE, UK*
email: n.w.pekeur@durham.ac.uk
[2] *South African Astronomical Observatory, Cape Town, South Africa*

Polarization provides a direct means of disentangling the thermal emission of blazars from the synchrotron emission at optical wavelengths. The poster presented the optical polarization light curve of the high-frequency peaked blazar PKS 2155-304 during 2009 and 2010, when the source experienced both active and quiescent periods. Measurements were obtained with the HIgh Speed Photo-POlarimeter (HIPPO) of the South African Astronomical Observatory as part of an ongoing observing campaign to monitor the source. Complementary observations from the Steward Observatory blazar monitoring program were also described. The optical polarization variability was compared to simultaneous gamma-ray measurements as recorded by the FERMI satellite, enabling us to probe both the low- and high-energy peaks of the source's spectral energy distribution.

Time-Series Data Sets at the NASA Exoplanet Science Institute

P. Plavchan[1] *et al.*
[1] *NASA Exoplanet Science Institute, CalTech, Pasadena, CA 91125, USA*
email: plavchan@ipac.caltech.edu

The NASA Exoplanet Science Institute (NExScI) is home to an exoplanet archive with the aim of providing support for NASA's planet finding and characterization goals. NExScI serves as the official US portal for the public CoRoT data products. In addition, NExScI serves KEPLER public data time-series and KEPLER Object of Interest (KOI) tables, as well as synoptic data from numerous ground-based surveys. NExScI developed a periodogram service to determine periods of variability phenomena and create phased photometric light curves. Through the NExScI periodogram interface, the user may choose three different period detection algorithms to use on any time-series product, or even upload and analyze their own data. For more details, see `http://nexsci.caltech.edu`.

Search for Gravitational Waves Associated with the S5/VSR1 IPN Short Gamma-Ray Bursts

Valeriu Predoi[1] (for LIGO Scientific Collaboration and Virgo Collaboration)
[1] *School of Physics and Astronomy, Cardiff University, Cardiff, CF24 3AA, UK*
email: `Valeriu.Predoi@astro.cf.ac.uk`

One of the preferred progenitor models for short hard gamma-ray bursts (SHB) is the merger of compact binary objects, i.e., either neutron star–neutron star or neutron star–black hole binary systems. Such mergers are also predicted to emit strong gravitational radiation with waveforms that can be described theoretically. The poster described a new search for these known gravitational-wave signatures in temporal and directional coincidence with SHBs detected by the Interplanetary Network (IPN) satellites during LIGO's fifth science run, S5, and Virgo's first science run, VSR1. The IPN uses triangulation of relative times of arrival, so the detected bursts are localized to extended patches of the sky that vary in size and shape. The search is done fully coherently, and a new development needs to be implemented for a search on various error boxes. The poster presented the search targets, the method, and new tools that will be used to perform the search.

The Low-Latency Search for Gravitational Waves from Compact Binary Coalescence

Larry Price[1] (on behalf of the LIGO Scientific Collaboration and Virgo Collaboration)
[1] *LIGO Laboratory, California Institute of Technology, Pasadena, CA 91125, USA*
email: `price_l@ligo.caltech.edu`

During the summer of 2010, the first low-latency search for gravitational waves from compact binary coalescences was performed using the LIGO and Virgo instruments. The aim was to provide triggers for follow-up observations with telescopes operating in all parts of the electromagnetic spectrum. The poster described the low-latency pipeline used to produce the triggers, and the latest results from the pipeline.

The RApid Temporal Survey

Gavin Ramsay and Tom Barclay
Armagh Observatory, College Hill, Armagh BT61 9DG
email: gar@arm.ac.uk

The Rapid Temporal Survey (RATS) explores the faint, variable sky and is sensitive to variability on time-scales ranging from a few minutes to several hours, for sources as faint as $g \sim 24$. Our survey took place between 2003 and 2010 and covered nearly 40 square degrees, with a bias towards the Galactic plane. In the first five years of observations we obtained light curves for 3 million stars and we identified over 100,000 variable stars. Although our prime goal was to detect ultra-compact binary systems, we have discovered pulsating sdB stars, white dwarfs, SX Phe and δ Scuti stars, accreting systems, flare stars and contact binaries. The poster outlined our strategy and main results to date.

Machine-Learning Methods for VAST Transient Detection

Colorado Reed[1,2] *et al.*
[1] *Department of Physics, University of Iowa, Iowa City, IA USA*
[2] *Jet Propulsion Laboratory, CalTech, Pasadena, CA 91109, USA.*
email: colorado-reed@uiowa.edu

The Variables and Slow Transients (VAST) survey is a future radio survey for the Australian Square Kilometer Array Pathfinder (ASKAP). This survey will examine relatively long-lived (slow) astronomical transients—those lasting longer than 5 seconds. In the poster we examined machine-learning methods for the detection of slow transient events in archived light-curve data. Event detection in those light curves is inherently challenging because of uneven sampling, data sparsity, and noise from a multitude of sources. Furthermore, event duration may range from days (supernovæ) to years (novæ and intra-day variables). The poster explored supervised and unsupervised methods for event detection, and evaluated performance both on simulated VAST data and on real data from the VLA.

The Delay-Doppler Spectrum of Interstellar Scintillation from Pulsars

Barney Rickett[1], Bill Coles[1], Dan Stinebring[2], and JJ Gao[1]
[1] *University of California, San Diego, La Jolla, CA 92093, USA*
email: bjrickett@ucsd.edu
[2] *Physics Department, Oberlin College, Oberlin, OH 44074, USA*

The dynamic radio spectrum of the interstellar scintillation of radio pulsars reveals unexpected features of the turbulent ionized interstellar medium, when analyzed in the Fourier domain versus relative delay and Doppler shift. In that domain a primary "parabolic arc" is often seen. For some objects there are also reversed offset parabolæ which imply highly localized concentrations of elongated turbulent plasma. The poster presented the results and methods of analysis from two such pulsars.

How to Simulate a 10-Year LSST Program, and How to Make Sense out of What it Produces

Stephen Ridgway, Srinivasan Chandrasekharan and the LSST Simulator Team
National Optical Astronomical Observatory, Tucson, AX 85719, USA
email: ridgway@noao.edu

A high-throughput survey for transient sources will have to deal with a significant number of transient or variable sources of familiar and less interesting types which may be difficult to distinguish from higher priority and novel sources. The two major astrophysical sources of contaminants are small-solar system bodies and Galactic variable stars. The frequency of solar-system body appearances can be estimated reasonably from existing model populations. Recently, tools and data have become available which can support an estimate of the number of stellar variable sources that will be detected at any survey depth. The essential tool for describing the stellar population is a Galactic model of stellar sources (e.g., the Besançon model). The essential data are empirical statistics on stellar variability from the KEPLER survey and the Sloan Digital Sky Survey. They were applied to the Large Synoptic Survey Telescope (LSST), for which the estimates of the number of variable target detections has ranged over at least an order of magnitude. We estimate that the LSST survey will detect variation in $\sim 10^5$ stars each night, and that this number is only a weak function of the accuracy of photometric calibration.

Classification of Hipparcos Unsolved and Non-Investigated Variables

L. Rimoldini[1] *et al.*
[1] *Observatoire astronomique de l'Université de Genève, 1290 Versoix, Switzerland*
email: Lorenzo.Rimoldini@unige.ch

The Hipparcos catalogue contains thousands of unclassified variable stars whose periodicity could not be determined or confirmed from the literature; to date they remain unsolved, or not investigated. The poster described our development of an automated supervised classification technique, based on Random Forests, that we trained on sources with either periodic or non-periodic light variations, in order to predict variability classes for Hipparcos variables with unreliable or missing classification.

Time Domain Astronomy with JANUS

P. Roming
Southwest Research Institute, San Antonio, TX 78228, USA
email: roming@swri.edu

The Joint Astrophysics Nascent Universe Satellite (JANUS) is a multi-wavelength cosmology mission focused on the cosmic dawn. The primary objectives of JANUS are to discover and observe high-z ($z > 5$) gamma-ray bursts (GRBs) and quasars. An inevitable by-product of searching for these high-z objects is that JANUS will produce X-ray and near-IR surveys that will reveal many other types of time-domain events. Such events include detections of GRB-supernova events, low-z GRBs, stellar super-flares, neutron-star super-bursts, fast X-ray transients, and "tidal disruption events" triggered by the close encounter of a star with a supermassive black hole.

Binary Tidal Disruption by a Supermassive Black Hole: Observational Consequences

Elena M. Rossi
Leiden Observatory, NL-2333 CA Leiden, The Netherlands
email: emr@strw.leidenuniv.nl

Observations of hyper-velocity stars in the Galactic Halo and stellar orbits at the Galactic Centre may be the consequence of tidal disruption of binaries by the supermassive black hole. In this process, one star may be ejected at high velocity, while the companion remains bound to the hole. The poster first presented a new method to calculate the disruption probability of binaries which enter the tidal sphere, and their energy after disruption. It then showed our predictions for the velocity distribution of the ejected stars and for the mass distribution of the stars that remain bound. A tentative comparison with current sparse data was also made.

Viscous Evolution of WD-WD Merger Remnants

Josiah Schwab, Ken Shen, and Eliot Quataert
University of California, Berkeley, CA 94720, USA
email: jwschwab@berkeley.edu

Beginning with the results of 3D smoothed-particle hydrodynamics (SPH) simulations of merging white dwarfs, we performed 2D calculations of the evolution of the post-merger remnant. The poster described the operation of this project.

Newly-Discovered Properties of Cepheids in the Mid-Infrared

Victoria Scowcroft[1] *et al.*
[1] *Carnegie Observatories, Pasadena, CA, 91101, USA*
email: vs@obs.carnegiescience.edu

The Carnegie Hubble Program (CHP) is a Warm Spitzer programme designed to measure the Hubble constant to an accuracy of 2% using Cepheids, supernovæ and the mid-infrared Tully–Fisher relation; Freedman *et al.* (2011; arXiv:1109.3802) present an overview of the CHP. As part of that project we have taken observations of Cepheids in the Milky Way and the Large and Small Magellanic Clouds in a deterministic manner. The stars were targeted individually; each one was observed at 24 (MW and LMC) or 12 (SMC) equally-spaced epochs through a single cycle. We were able to confirm the predicted temperature-dependent effect that the CO bandhead at 4.6 μm has on the [4.5] magnitude (Marengo *et al.* 2010, *ApJ* 709, 120), both through the Cepheid's pulsation cycle and in the period–luminosity (PL) relation.

We find that the 3.6-μm PL relation is a good distance indicator; reddening effects are drastically reduced compared to the optical region, and metallicity differences do not appear to have any influence. In the 4.5-μm PL relation, and additionally in the [3.6]–[4.5] period–colour relation, metallicity does appears to play a role. The amplitude of the effect of the CO bandhead on [4.5] is different between the three galaxies, altering the mean colour of the Cepheids and affecting the amplitude of the Cepheid light curves. We are in the process of using these data to calibrate the photometric metallicity–colour relation and of applying the technique to Cepheids that are not accessible spectroscopically.

The work presented in this poster is being published by Scowcroft et al. (arXiv:1108:4672).

Numerical Simulations of Line-Profile Variation Beyond a Single-Surface Approximation for Oscillations in RoAp Stars

Hiromoto Shibahashi amd Jun Naito
Department of Astronomy, University of Tokyo, Tokyo 113 0033, Japan
email: `shibahashi@astron.s.u-tokyo.ac.jp`

Prior to the last decade, most observations of roAp stars concerned their light variations. Recently some new, striking results of spectroscopic observations with high time-resolution, high spectral dispersion and high signal-to-noise ratios became available. Since the oscillations found in roAp stars are high overtones, the vertical wavelengths of the oscillations are so short that the amplitude and phase of variation of each spectroscopic line are highly dependent on the level of the line profile. Hence the analyses of variation of spectroscopic lines of roAp stars potentially provide us with new information about the vertical structure of the atmosphere of these stars. In order to extract such information, numerical simulation of line-profile variations beyond a single-surface approximation is necessary. We have carried out numerical simulations of line-profile variations by taking into account the finite thickness of the line-forming layer. The poster demonstrated how effective that treatment is, by comparing the simulations with the observed line profiles.

Detecting Variability in Astronomical Time-Series Data: Applications of Clustering Methods in Cloud Computing Environments

Min-Su Shin
Department of Astronomy, University of Michigan, Ann Arbor, MI 48109, USA
email: `msshin@umich.edu`

The poster presented applications of clustering and de-trending methods to detect variability in massive astronomical time-series data. Focusing on variability of bright stars, we used clustering methods to separate possible variables from other time-series sources such as intrinsically non-variable objects and sources with common systematic patterns. We tested how de-trending of systematic patterns can be incorporated into our variability detection approach. The study was conducted in a cloud computing environment provided by KISTI. The poster described our experience using the cloud computing test-bed.

The Collaboration for Astronomy Signal Processing and Electronics Research

Andrew Siemion, Dan Werthimer and The CASPER Collaboration
University of California, Berkeley, Berkeley, CA 94720, USA
email: siemion@berkeley.edu

The poster presented the latest updates on the open source digital signal processing hardware and software being developed by the Collaboration for Astronomy Signal Processing and Electronics Research (CASPER). The CASPER toolkit now includes Xilinx Virtex-6 field programmable gate array (FPGA)-equipped Reconfigurable Open Architecture Computing Boards (ROACH II), as well as infrastructure for heterogenous architectures made up of FPGAs, graphics processing units (GPUs) and commodity CPUs. These tools enable rapid design of high-performance instrumentation by non-experts, enabling the performance of specialized instruments to track closely Moore's law for growth in the electronics industry.

Technique for Low-Latency Detection of Compact Binary Coalescence and its Implications for Multi-Messenger Astronomy

Leo Singer and N. Fotopoulos
California Institute of Technology, Pasadena, CA 01125, USA
email: singer_l@ligo.caltech.edu

The rapid detection of compact binary coalescence with a network of advanced gravitational-wave detectors will offer a unique opportunity for multi-messenger astronomy. Prompt detection alerts to the astronomical community may make it possible to observe the onset of electromagnetic emission from compact binary coalescence. The poster demonstrated a computationally practical analysis strategy which could be used to produce early-warning triggers for astronomical observatories. The work it described was carried out in collaboration with Kipp Cannon, Romain Cariou, Adrian Chapman, Mireia Crispin-Ortuzar, Melissa Frei, Chad Hanna, Erin Kara, Drew Keppel, Laura Liao, Stephen Privitera, Antony Seale and Alan Weinstein. A paper describing the results in full is now being published (arXiv1107.2665).

Hitting the JACPOT: Current and Future Probes of the Accretion Disk—Radio Jet Coupling of X-Ray Binaries

G.R. Sivakoff[1] and the JACPOT XRB Collaboration
[1] *University of Alberta, Edmonton, Alberta, T6G 2G7, Canada*
email: sivakoff@ualberta.ca

The accretion disks and radio jets of X-ray binaries are known to be coupled; however, until recently the prevailing paradigm had not been tested with direct high-resolution imaging of the radio jet over entire outbursts. Moreover, such observations had not previously targeted X-ray binaries where the compact object was either a neutron star or a white dwarf. The Jet Acceleration and Collimation Probe Of Transient X-Ray Binaries (JACPOT XRB) team has recently concluded its first monitoring series, including VLA, VLBA, X-ray, optical and near-IR observations, of entire outbursts of X-ray binaries with a black hole candidate (H1743-322), a neutron star (Aquila X-1) and a white dwarf (SS Cyg). The poster concentrated on the results for H1743-322 and Aql X-1. While we largely confirmed the accretion-disk radio jet paradigm, we also discovered new intriguing behaviours. Neutron stars appear as capable as are black holes at launching jets, but the lack of ejecta in Aql X-1 may suggest a fundamental difference in the jet formation process between neutron star and black hole systems, possibly hinting at an additional role from the ergosphere. The poster discussed how future radio observatories would extend our capacity to study the accretion-disk radio-jet connection.

The Liverpool Telescope

I. A. Steele
Astrophysics Research Institute, LJMU, Birkenhead, CH41 1LD, UK
email: ias@astro.livjm.ac.uk

The Liverpool Telescope is a 2.0-metre fully-robotic telescope sited on La Palma. It specialises in time-domain astrophysics, with strong programmes in SNe, GRBs, exoplanets and QSO monitoring. The telescope is equipped with an optical imaging camera, an integral-field fibre-fed optical spectrograph, a fast-readout wide field camera, a lucky imaging camera and an imaging polarimeter. All instruments are mounted on the telescope at all times, and can be switched over in under 20 seconds. Future instrument plans include a dual beam optical/IR tip-tilt corrected imaging camera, and a multi-band imaging polarimeter.

DASCH variables in the KEPLER field

Sumin Tang[1], Jonathan Grindlay[2], Edward Los[2], and Mathieu Servillat[2]
[1] *Harvard University, Cambridge, MA 02138, USA*
email: stang@cfa.harvard.edu
[2] *Harvard-Smithsonian Center for Astrophysics, Cambridge, MA 02138, USA*

DASCH is a project to digitize and analyze the scientific data contained in the ∼550,000 Harvard College Observatory plates taken between the 1880s and 1990s. The collection is a unique resource for studying temporal variations in the universe over time-scales of 10–100 years. The poster presented a few of the most interesting long-term variables which we found in or near the KEPLER Field, including a group of Be variables showing variations of about 1 mag over years, a group of K giants showing slow variations over 10–100 years, and two other examples.

Towards Higher Precision on Time-Series Differential Photometry through Colour Effects Minimization

Victor Terron[1], Matilde Fernandez[1], and Monika G. Petr-Gotzens[2]

[1] *Institute of Astrophysics of Andalusia, IAA-CSIC, 18008 Granada, Spain*
email: vterron@iaa.es
[2] *European Southern Observatory, D-85748 Garching, Germany*

Common knowledge of differential photometry establishes that in order to improve the photometric precision of light curves it is essential to follow several rules. First and foremost, stars must be placed in the same pixels of the CCD and their light distributed over several of them, while the aperture size must be optimized. Furthermore, algorithms based on a careful selection of the comparison stars, depending on their noise level, go one step further towards achieving the maximum accuracy. Nevertheless, astronomers must still address the fact that, particularly when observing at high air-masses, atmospheric extinction does not affect all the stars equally; blue stars appear dimmer than red ones. Therefore, as the stars in the field of view drift across the sky, going through different air-masses, they suffer from different extinction. As a result a comparison star may look relatively brighter or fainter than it really is. The algorithm that we proposed in the poster tackles this matter by selecting an artificial comparison star for each target so that the colour effects of atmospheric extinction can be minimized. In that manner, and especially in the case of well populated fields, the effects of atmospheric extinction are considerably reduced. Although some issues remain open, we believe that our algorithm would be of decisive help for high-precision photometry with ground-based telescopes.

High-redshift Type II Supernova Survey with Shock Break-out

Nozomu Tominaga[1,2]

[1] *IPMU, University of Tokyo, Chiba 277-8583, Japan*
[2] *Department of Physics, Konan University, Kobe, Hyogo 658-8501, Japan*
email: tominaga@konan-u.ac.jp

Shock break-out is the brightest radiative phenomenon in a supernova (SN), but is difficult to observe owing to its short duration and X-ray/UV-peaked spectra. First observations of shock break-out were reported in 2008 for SN 2008D (Type Ib) in X-ray and optical data from SWIFT, and in SNLS-04D2dc and SNLS-06D1jd (both (Type II SNe) in UV-optical observations made by GALEX and the Supernova Legacy Survey (SNLS). We have constructed a SN II model with a multigroup radiation hydrodynamics code STELLA, which reproduces well the UV-optical multicolour light curves of SNLS-04D2dc. It demonstrates that the peak apparent g-band magnitude of the shock break-out would be $m_g = 26.4$ if an SN identical to SNLS-04D2dc occurred at a red-shift of $z = 1$ (which can be reached by 8-m-class telescopes). We also constructed shock break-out models for SNe II with various main-sequence masses, metallicities and explosion energies, and estimated the observable SN rate and attainable red-shift by convolving an initial mass function with the cosmic star-formation history, intergalactic absorption and host-galaxy extinction; the observable SN rate in the g-band for $m_{g',\mathrm{lim}} = 27.5$ mag is 3.3 SNe deg^{-2} day^{-1}. Half will be found at $z \geqslant 1.2$. That estimate enabled us to propose a realistic survey strategy optimized for shock break-out; multicolour observations in the blue optical bands with $\sim$1 hour intervals are essential for the efficient detection, identification and interpretation of the shock break-out. The poster emphasized that shock break-out is an advantageous tool for probing high-z Type II SNe.

Tidal Resonance in Short Gamma-Ray Burst Precursors and Multi-messenger Constraints on Neutron Star Physics

David Tsang[1], Jocelyn Read[2], Tanja Hinderer[1], and Tony Piro[3]
[1] *California Institute of Technology, Pasadena, CA 91125, USA*
email: dtsang@caltech.edu
[2] *University of Mississippi, Oxford, MS 38677, USA*
[3] *University of California, Berkeley, CA 94720, USA*

The merger of a neutron star with another compact object is a leading model for the progenitor of a short Gamma-Ray Burst (sGRB). The gravitational waves which drive their inspiral are expected to be observed by ground-based gravitational-wave detectors such as the upcoming Advanced LIGO. The poster examined the last seconds of co-alescing binary neutron stars to understand possible sources of X-ray and gamma-ray emission. It described how precursor observations can potentially constrain the structure and equation of state of component neutron stars. Tidal effects are the strongest source of perturbation in this régime. However, the strength of the tidal field is insufficient to break the crust directly until milliseconds before the merger. Resonant modes that are concentrated near the crust–core interface can be excited to large amplitudes, fracturing the crust and leading to a pre-merger signature seconds before the merger itself. The timing of such a precursor flare would measure the resonant frequency of the interfacial mode, probing the structure and equation of state near the base of the neutron-star crust. This is complementary to studying the GW inspiral signal alone; the latter is primarily sensitive to the equation of state in the core, which determines NS mass and radius.

Understanding Pulsation in Luminous M Supergiants with the Aid of Archival Observations

D. G. Turner[1] *et al.*
[1] *Saint Mary's University, Halifax, Nova Scotia, B3H 3C3, Canada*
email: turner@ap.smu.ca

The poster presented a summary of ongoing and archival photometric and radial-velocity observations of the 9th-magnitude F9 Ib supergiant HDE 344787 that were made over the last 120 years, with emphasis on recent trends. HDE 344787 is a double-mode Cepheid variable with a period of 5.4/3.8 days; it has extremely small amplitude (< 0.01 magnitude) with a rapidly-increasing period and sinusoidal light variations of decreasing amplitude that may result in non-variable status within the next half century. The star displays all of the characteristics of a Cepheid that is undergoing a first crossing of the instability strip and is about to depart the cool edge for first crossers. Its existence helps to redefine the observational characteristics normally attributed to Cepheid pulsation.

Detecting Stellar Tidal Disruptions: from Optical to Radio

S. van Velzen, E. Koerding, and H. Falcke
IMAPP, Radboud University, 6500 GL Nijmegen, the Netherlands
email: s.vanvelzen@astro.ru.nl

When a star passes too close to a massive black hole it becomes tidally disrupted. About 10 examples of the flares that occur when the debris of the disruption falls back onto the black hole have been identified in UV and X-ray surveys. The discovery of two tidal-flare candidates in archived SDSS data implies that we can expect tens to hundreds of such events per year from current and near-future optical variability surveys. To explore the potential of detecting tidal disruptions with radio surveys, we used the well-established scaling between accretion power and jet luminosity to construct a robust model of the emission of a tidal disruption jet. The poster illustrated how we reproduced the radio flux of the recent tidal-flare candidate GRB 110328A, and described our finding that future radio surveys will be able to test whether the majority of tidal disruptions are accompanied by a relativistic jet.

IAU "SOFA": Time-scale Transformation Software

Patrick Wallace
STFC Rutherford Appleton Laboratory, Didcot, OX11 0QX, UK
email: Patrick.Wallace@stfc.ac.uk

The software provided through the IAU "Standards of Fundamental Astronomy" (SOFA) initiative now includes a suite of time-scale transformation functions, supporting TAI, UTC, UT1, TT, TCG, TCB and TDB. The SOFA design features two-part Julian dates (to safeguard precision), and individual treatment of specific transformations rather than a single general-purpose routine. Correct handling of UTC leap seconds was the biggest single challenge.

AGILITE: An ATA Survey to Characterize the Population of Galactic Radio Transients and Variables

Peter K. G. Williams and Geoffrey C. Bower
Department of Astronomy, University of California, Berkeley, CA 94720, USA
email: pwilliams@astro.berkeley.edu

Systematic studies of transient and variable radio emission are a relative novelty. Searches for "slow" radio transients—sources that vary over time-scales of days to months—have so far tended to focus on extragalactic fields. Many Galactic sources, however, vary on those time-scales, including X-ray binaries, brown dwarfs, and several objects of unknown nature discovered in previous searches. The poster presented AGILITE, the ATA Galactic Lightcurve and Transient Experiment, an effort to characterize the population of Galactic radio transients and variables more fully. AGILITE has a large overall footprint ($\sim$25 deg^2), a substantial number of epochs (~ 200), and a significant dedication of observatory time (~ 1700 hours over two years), all of which make it sensitive to rare objects as well as to variability on many time-scales. The poster described the AGILITE pipeline and presented preliminary results from 28 observations of a field centred on the Cygnus X-3 high-mass X-ray binary, including epochs during the major flare of 2011 March. It also discussed prospects for the complete survey as well as other applications of the dataset, such as large-scale mapping of extended radio emission toward the Galactic Center.

ThunderKAT: Radio Transients with MeerKAT

Patrick Woudt[1], Rob Fender[2] and the ThunderKAT team
[1] University of Cape Town, Rondebosch 7701, South Africa
email: pwoudt@ast.uct.ac.za
[2] School of Physics & Astronomy, University of Southampton, SO17 1BJ, UK

In 2010 MeerKAT (South Africa's SKA pathfinder telescope) selected ten Large Survey Projects, which collectively were allocated 70% of the observing time in the first 5 years of full operation of MeerKAT. ThunderKAT is the MeerKAT Large Survey Project which studies all aspects of transient radio (synchrotron) emission associated with accretion and explosive events, including relativistic jets, nova explosions, supernovæ and gamma-ray bursts. ThunderKAT's mode of observing will be two-fold; it will be fully commensal with all other Large Survey Projects on MeerKAT, interrogating the data stream at 1-second intervals with the goal of real-time transient detections. In addition a nominal 3000 hours have been reserved in the first five years of operation of MeerKAT for ThunderKAT to study the observed radio transients in detail. Science commissioning observations of transients using the KAT-7 array were planned for the second half of 2011, as part of the pathway to real-time transient detections on MeerKAT.

Exploring Variability of Bright Stars on the Ecliptic Plane with STEREO

K. T. Wraight[1], G. J. White[2], D. Bewsher[3], and A. J. Norton[1]
[1] Dept. of Physical Sciences, The Open University, Milton Keynes, MK7 6AA, UK
email: k.t.wraight@open.ac.uk
[2] STFC Rutherford Appleton Laboratory, Didcot, OX11 0QX, UK
[3] Jeremiah Horrocks Institute, University of Central Lancashire, Preston, PR1 2HE, UK

The STEREO mission's Heliospheric Imagers have considerable potential for use in a wide range of stellar variability studies. With photometry of ∼900,000 stars from 12th to 2nd magnitude over a five-year period within 10° of the Ecliptic Plane, the HI-1 imagers have the capability of finding many new and interesting variables, and of improving considerably the quality of available data for previously known variables in this poorly-studied part of the sky. The moderate cadence of 40 min, maintained for ∼20 days at a time and with repeated visits from each satellite, permits a wide range of potential applications across very different time-scales. Initial analyses of HI-1A data have revealed 122 new eclipsing binaries. Ongoing work includes the determination of rotation rates of magnetic CP stars and the extraction of pulsational modes from bright δ Scuti variables. The initial analyses were published in 2011 by Wraight, K.T., White, Glenn J., Bewsher, D., & Norton, A.J. (*MNRAS*, 416, 2477).

Pair-Instability Supernovæ:A Second Candidate from the PTF Survey

Ofer Yaron, Avishay Gal-Yam, *et al.*
Weizmann Institute of Science, Rehovot 76100, Israel
email: ofer.yaron@weizmann.ac.il

Massive metal-poor stars, attaining core masses in excess of $\sim$50 $M_\odot$sun, may explode via the pair-production instability. The massive oxygen-rich cores of such stars reach high temperatures ($>10^9$K) at relatively low densities, such that pressure-supporting photons are converted to electron-positron pairs. A violent dynamical collapse ensues, triggering a powerful nuclear explosion, which leads to a complete disruption of the star. This notion has been theoretically predicted and understood for almost four decades, but only recently did we begin to obtain more convincing observational proof for the existence of such events. The poster presented additional data and analysis for the single most promising pair-instability supernova (PISN) candidate to date (SN2007bi), and reported the discovery of a second candidate (PTF10nmn), first detected in 2010 May by the Palomar Transient Factory survey. The observed characteristics of PTF10nmn closely resemble those of SN 2007bi in both light curve and spectroscopic signatures. The poster showed the spectral evolution of PTF10nmn, from around peak light till the beginning of a nebular phase. A nebular model for our last spectrum of 2011 July 3, corresponding to an epoch of $>$ 450 days from explosion, revealed a mass of the order of $5M_\odot$ of ejected ^{56}Ni, high ejected masses of C, O, Ca and Si, and a total ejecta mass M_{ej} of $\sim$100 $M_\odot$. The seemingly real double peak (with a rest-frame separation of $\sim$50 days) as apparent from the R-band photometry, is in itself an intriguing finding that requires a suitable physical explanation.

"WISEASS"—A State-of-the-Art Interactive Spectroscopic Database

Ofer Yaron, Avishay Gal-Yam, *et al.*
Weizmann Institute of Science, Rehovot 76100, Israel
email: ofer.yaron@weizmann.ac.il

The poster reported the development of a new spectroscopic database, the Weizmann Institute of Science Experimental Astrophysics Spectroscopy System (WISEASS). The system consists of a MYSQL DB and a web-interface that is implemented mainly in PHP, embedded within a DRUPAL (CMS) framework. The purpose of this system is to serve as an archive of high-quality supernovæ spectra, with respect to both historical ("legacy") data, and data that are accumulated in ongoing modern surveys and programmes (both currently and in the future). Besides serving solely as an archive, with the ability to hold and retrieve spectra and objects alongside all related information (via numerous querying options), we have implemented means by which it is possible to carry out basic online analyses of the spectra. By making use of interactive plots, with zooming and over-plotting capabilities, the database provides graphical interfaces for performing line identifications of the major relevant species, assessing red-shifts and expansion velocities. The Database currently includes $\sim$3000 spectra, of which $\sim$2000 are public, the latter including, for example, published PTF spectra and all of the "SUSPECT" archive. Owing to the increasing interest and use of this system by the SN communities world-wide, it is expected that additional programme and data from various sources and projects will be added and incorporated in the near future.

The Unusual Gamma-Ray Transient Swift J164449.3+573451

B. Ashley Zauderer
Harvard-Smithsonian Center for Astrophysics, Cambridge, MA 02138, USA
email: bzauderer@cfa.harvard.edu

The poster presented data from an observational campaign at radio wavelengths following discovery of the unusual transient source, Swift J164449.3+573451 on 2011 March 25. What made this source unusual was the magnitude and duration of X-ray flaring after the initial burst. The observations described spanned centimetre to sub-mm wavelengths, establishing spatial coincidence with the nucleus of an inactive galaxy at $z = 0.354$. The radio light curves and spectral energy distributions are consistent with the theoretical model of accretion from a tidal disruption event onto a 10^6-$M_\odot$ black hole launching a new relativistic jet. Monitoring this event at radio wavelengths after the initial concerted campaign has continued, and the poster concluded with a summary of the observations to date.

New Horizons in Time-Domain Astronomy
Proceedings IAU Symposium No. 285, 2011
R. E. M. Griffin, R. J. Hanisch & R. Seaman, eds.

© International Astronomical Union 2012
doi:10.1017/S1743921312001330

Afterword

The week's communications and deliberations concentrated on dQ/dt and how we could enhance our detection and interpretation of such occurrences, whether Q were a distance, an emission of energy, a velocity, a brightness, an orbital property or a spectroscopic feature. For us mortals, the derivative of Q is in respect of Time—yet rather little attention was paid during the week to Time itself. Maybe that was just as well. As was said by way of introducing the Astronomer Royal to give the public lecture, *From Microseconds to Æons*, Time is ever-present yet always elusive. It flies and yet it drags. We never seem to have enough of it, yet it can weigh heavily on our hands. It is free but it is priceless. You can't own it but you can spend it. Einstein said bluntly that the only reason for Time is so that everything doesn't happen at once.

Time is the commander of all things. Certainly, Time (like tide) waits for no man (or woman), and it absolutely cannot be driven backwards. Yet its benign features must not be overlooked: Time when cloaked as the seasons governs and assists the growth of plants for food; Time oils and maintains our circadian rhythms, and Time signals the moment to stop work.

Astronomers have struggled to measure Time for a variety of civil and nautical purposes, and some of the first observatories were built in order to provide time-signals for ships and railways. Nowadays we leave the actual measurement of civil Time to a specialist service. But the implications of Time, and its irreversibility, are key features of all discussions and hypotheses about the creation of the world, of the solar system, of the Galaxy, and of the universe itself, however many of it there may be. The physics involved extends from the most rapid we have ever heard of, and then a bit shorter, out to the longest we can possibly imagine, and then some.

As mortals, we are very conscious of Time's command. Generations give way to their successors, and new fashions, new skills, new technology and new ideas, while challenging to the senior populations who grew up not knowing them, are an inseparable part of life to those born into the more endowed systems. The conventions that we tried to overturn in this Symposium were in fact an attempt to unshackle our thought processes from the somewhat entrenched patterns of yesteryear and to open fresh doors that could inspire or capture in other ways. How well we succeeded, only Time will tell, but one of the most blatant banishments of convention was the decision to invite one of the very junior members of the SOC—and a female one at that—to give the "After-Dinner" speech at the banquet evening in Wadham College. That she rose to the occasion superbly with all the required blend of wit, humour and candour belies the fact that she had had only a couple of days to think it up. But what was undoubtedly the climax was her closing message: variability is going to be pursued as Generation Y thinks best. Our very able representative of Generation Y will therefore have the last word.

R.E.M.G.
Victoria, Canada. December, 2011

New Horizons in Time-Domain Astronomy
Proceedings IAU Symposium No. 285, 2011
R.E.M. Griffin, R.J. Hanisch & R. Seaman, eds.

Reflections on a Week in Oxford

Tara Murphy

School of Physics & School of Information Technologies, University of Sydney, Australia
email: `tara@physics.usyd.edu.au`

Invited After-Dinner Speech

When Bob asked me a couple of days ago to give this talk, he said it should be "pithy and wise." When I spoke to Aris the next day he said it should be "informal and funny." Then Elizabeth spoke to me and said it would be great if it was "controversial and challenging." I mentioned to Aris just before dinner that I had put a talk together, and he said "Oh, you've written a talk—I thought it would be something more spontaneous!" Faced with an obviously impossible spec, I did what any reasonable programmer would do and rewrote the spec myself. So what I'm going to talk about this evening is a reflection on my week in Oxford.

Starting in 1985 . . . 1985 was one of those years that, although I didn't realise it at the time, would turn out to be critically important in my career. Two things happened.

First, the primary school I went to decided to upgrade their computer, and so were looking for a home for their old one. Noticing that I was more than a little bit interested in the classroom computer, and obviously under the impression there was no way a primary school could sustain two computers(!), they offered to give it to my parents, hence making me the proud owner of a BBC Micro.

My first programming experience followed shortly after. My mum had a thick book of "programmes", and had obviously figured out that you had to type them in to the computer to get something to happen. Having no prior experience with technology (or maths or science) she picked a programme on "Learning French" and over the next few days we typed in a series of nonsensical lines, which turned out to be, of course, the programming language Basic. The typing was going very well until one day Mum stopped suddenly. "Oh no, Tara", she said. "This is no use. You have to type in the French answers as well as the English. I thought the computer was going to teach us French." I guess this was my first revelation about how programming worked. I suspect Mum has been suspicious of computers ever since.

The second thing that happened that year was that now I was eight years old, my grandparents gave me £5 for my birthday. As befits such an inordinate amount of money, I spent an inordinate amount of time in the bookshop, before selecting "Encyclopedia of Astronomy" from the top shelf. Again, Mum was there, being very supportive, although she did ask rather doubtfully whether I thought the book would be interesting. I can now say: "Yes Mum, it turns out astronomy was pretty interesting."

Little did I know that these two things were the first step along a path that has led me to a joint faculty position in computer science and astrophysics, speaking at a conference that is about some of the most exciting things happening in astronomy, and some of the cutting edge computing required to achieve them. And all happening in Oxford, one of the great academic cities of the world. Mum would be very impressed!

Now you can't visit somewhere like Oxford without reflecting on the past, and a beautiful way of doing that down at the Oxford Museum of Natural History is through the

One Oak project, tracing the history of everything in the life of a single oak tree. One Oak was germinated from a single acorn, just down the road here at Bladon Heath Oxfordshire, in 1788. Some pretty important things happened in 1788. According to the display in the museum:

• The Times newspaper was first published in London
• Europe saw the beginnings of the French revolution
• Mozart composed his last music, and
• Lord Byron, the English poet, was born.

All very interesting, but as an Australian, I kind of felt like something was missing from their list. Aha—the much larger timeline on the wall would presumably be more comprehensive, and indeed it was. The other really important thing that happened in 1788, the year One Oak was born, was

• The Marylebone Cricket Club issued the definitive laws of the game and was confirmed as the sole lawgiver.

At this point I decided to file a bug report in the visitors book, suggesting that the Natural History Museum rectify the obvious omission of the founding of our country!

The age of these trees is impressive, but of course in astronomy we can do much better, with supernovæ and the resulting supernova remnants giving us a history of the last 100s to 1000s of years. This kind of forensic approach was demonstrated beautifully in Armin Rest's talk on light echoes earlier in the week—an incredible way of tracing the history of long dead stars. Like the trees, supernovæ are interesting scientifically, but they also provide a great cultural connection to our astronomical ancestors, who were very aware of the transient and variable universe.

I mentioned 1788 before, an important date in Australian history that marks the founding of New South Wales as a British Colony. For tens of thousands of years before that, Indigenous Australians were looking at the Southern Sky and—like many ancient cultures—they were avid astronomers.

Our ancient ancestors were much more familiar with the sky than most of us. Which reminds me of a bizarre exchange on the way into Britain on Saturday. I was travelling on my Australian passport, so I got quizzed at immigration. The questions from the immigration official started in a fairly standard way:

What are you visiting for?
"Work, a conference"
Where is the conference?
"Oxford"
What is the conference about?
"Astronomy"
But the next question really took me by surprise:
What is the really bright star right next to the moon at the moment?

Wow. British immigration officials are seriously impressive at checking credentials. Wow. In my jetlagged state I attempted an answer. Umm. Could be Venus, could be Jupiter. Ummm. Probably a planet. Is it twinkling? No. OK, yes, its probably a planet then. Ummm. I'm from the Southern Hemisphere, the sky looks different down there. (Even as I said this it sounded weak). Stars twinkle and planets don't, that's how you can tell them apart. Double wow. I'm discussing scintillation with British immigration guy. Major outreach triumph!

Unlike me, our ancestors were very familiar with the night sky and so when they saw a new object, a transient, it took on great mystical and spiritual significance. They of course, immediately released ATels by drawing on the closest cave wall. When I see those inscriptions I feel a deep connection with our ancestors. It's the kind of thing that makes you take a moment and forget about grant applications, and press releases, exam marking and management meetings, VO infrastructure and issue tracking, midnight telecons and unanswered emails, syllabus review meetings and new buildings, and even Nature papers. When you strip all of that away (and for a modern scientist that seems increasingly hard to do), you have the same fundamental thing:

A person. Their awe at experiencing the universe. Their need to communicate it to other people.

Jumping forward in time: our medieval colleagues were also expert astronomers, and when a transient occurred it captured everyone's attention. But oh, the stakes were so much higher. I mean, if you get your transient identification wrong now, what's the worst that can happen? Someone scoops you in Nature. OK, I imagine that is extremely stressful. But nowhere near as stressful as screwing up the King's horoscope and ending up with your head on the chopping block. Literally.

So, taking a final jump forward to the present day, we repeatedly hear that we are entering a new era of astronomy, an era of data-driven science. We also say we need new approaches to IT to deal with this data explosion.

Bob and Elizabeth wanted me to say something controversial. Now I'm not a controversial kind of person, so I asked some friends at the conference what I should talk about. They said "The SKA site decision." OK, so I'm not even going to go near that (except to say, I hope Australia gets it!) They wanted something from my perspective as a person with one foot in IT and one foot in astronomy, and something from my perspective as a person right on the edge of Generation X, the last people who can (just barely) remember a world without the Internet (and what a terrible place it was back then).

Astronomy has often been a driver for computer science. In fact Babbage, trying to raise funds for his famous difference engine, pitched it to the Royal Astronomical Society in 1822, in his "Note on the application of machinery to the computation of astronomical and mathematical tables." However, and this is the controversial bit. I think we have actually got a bit behind. We're extremely good at high-quality data curation, databases and archives. We're great at telescope and data processing software, and we seek out novel algorithms from computer science to solve our data analysis problems. But, when it comes to online communication—which is at the heart of transient science—we've missed a major revolution that's gone on around us.

That revolution is, of course, social media.

Social media. It's either "a verbal sewer that is damaging your children" in the words of a Brisbane school principal earlier this week, or it may have revolutionised the way we think, but for Generation Y and beyond (our current and future students) it is just life. We think in a connected way. We live in a connected way. There isn't "the real world" and "the Internet." The Internet *is* the real world.

Now while some of the current generation have adapted well (I'm sure Bryan Gaensler has tweeted more this week than the rest of the conference put together), we haven't fully adopted this approach when thinking about the way we communicate within science. There's a lot of rubbish out there on the Internet. Nobody could deny that. But there are a lot of brilliant things as well. I don't know about you, but I have no trouble filtering the wheat from the chaff. Britney Spears and lolcats never get in the way of me trying to do science. Finding things we're interested in is something that us Generation X- and Generation Y-ers are really, really good at.

We need to think about automatic dissemination of transient results in the same way. A lot of good comes from codifying standards and building large-scale infrastructure. But large systems are by their nature slow moving. Transients, by their nature, are fast moving. It won't be possible for lengthy standardisation processes to keep up in this rapidly evolving field.

The solution is to develop things organically, staying as agile as possible. This is exactly what cutting edge IT companies are doing. We need to be less conservative. We need to distribute the responsibility for quality control. It sounds scary, but we need to put as much information out there as possible and let the crowd decide what is important. After all, our next generation of scientists are experts at this. They have never lived in any other kind of world.

Author Index

CAMBRIDGE JOURNALS

International Journal of Astrobiology

Volume 9 Issue 3 July 2010   ISSN 1473 5504

International Journal of Astrobiology

Managing Editor
Simon Mitton, University of Cambridge , UK

International Journal of Astrobiology is the peer-reviewed forum for practitioners in this exciting interdisciplinary field. Coverage includes cosmic prebiotic chemistry, planetary evolution, the search for planetary systems and habitable zones, extremophile biology and experimental simulation of extraterrestrial environments, Mars as an abode of life, life detection in our solar system and beyond, the search for extraterrestrial intelligence, the history of the science of astrobiology, as well as societal and educational aspects of astrobiology. Occasionally an issue of the journal is devoted to the keynote plenary research papers from an international meeting. A notable feature of the journal is the global distribution of its authors.

International Journal of Astrobiology
is available online at:
http://journals.cambridge.org/ija

**To subscribe contact
Customer Services**

in Cambridge:
Phone +44 (0)1223 326070
Fax +44 (0)1223 325150
Email journals@cambridge.org

in New York:
Phone +1 (845) 353 7500
Fax +1 (845) 353 4141
Email
subscriptions_newyork@cambridge.org

Price information
is available at: **http://journals.cambridge.org/ija**

Free email alerts
Keep up-to-date with new material – sign up at
http://journals.cambridge.org/ija-alerts

For free online content visit:
http://journals.cambridge.org/ija

CAMBRIDGE
UNIVERSITY PRESS

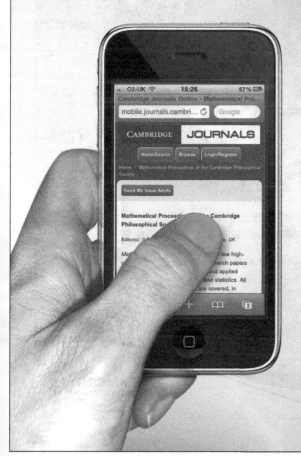

Printed in the United States
by Baker & Taylor Publisher Services